SI System of Measurements

	English	SI
Length	inch (in.)	meter (m)
Mass	pound mass (lb)	kilogram (kg)
Time	minute (min)	second (s)
Electric current	ampere (A)	ampere (A)
Temperature	fahrenheit (F)	celsius (C)
Force	pound-force (lb)	newton (N)
Plane angle	degree (°)	radian (r)
Energy	British thermal unit (BTU)	joule (J)
Horsepower	foot-pounds per minute (ft lb/min)	watt (W)
Velocity	inches per minute (in./min)	meters per second (m/s)
Area	square inch (in.²)	square meter (m²)

Multiple and Submultiple Units

Factor by Which the Unit is Multiplied	Prefix Name	Symbol
10^{12}	tera	T
10^{9}	giga	G
10^{6}	mega	M
10^{3}	kilo	k
10^{2}*	hecto	h
10*	deka	da
10^{-1}*	deci	d
10^{-2}	centi	c
10^{-3}	milli	m
10^{-6}	micro	μ
10^{-9}	nano	n
10^{-12}	pico	p
10^{-15}	femto	f
10^{-18}	atto	a

* Not recommended for engineering practice

MANUFACTURING PROCESSES

EIGHTH EDITION
MANUFACTURING PROCESSES

B. H. AMSTEAD
Professor Emeritus
The University of Texas
at Austin

PHILLIP F. OSTWALD
University of Colorado
Boulder, Colorado

The Late
MYRON L. BEGEMAN
The University of Texas
at Austin

JOHN WILEY & SONS
New York □ Chichester □ Brisbane □ Toronto □ Singapore

Copyright © 1942, 1947, 1952, 1957 by Myron L. Begeman
Copyright © 1969, 1977, 1987 by John Wiley & Sons, Inc.

All rights reserved. Published simultaneously in Canada.

Reproduction or translation of any part of
this work beyond that permitted by Sections
107 and 108 of the 1976 United States Copyright
Act without the permission of the copyright
owner is unlawful. Requests for permission
or further information should be addressed to
the Permissions Department, John Wiley & Sons.

Library of Congress Cataloging in Publication Data:

Amstead, B. H.
  Manufacturing processes.

  Bibliography: p. 688
  Includes index.
  1. Manufacturing processes. I. Ostwald, Phillip F.,
1931-   . II. Begeman, Myron L. (Myron Louis),
1893-1970. III. Title.
TS183.A474   1986   670   86-19094
ISBN 0-471-84236-2

Printed in the United States of America

10 9 8 7 6 5 4 3 2

ABOUT THE AUTHORS

The late **Professor M. L. Begeman** of The University of Texas at Austin was the author of the first editions of this text. His knowledge of the field of manufacturing processes and his ability to communicate has placed him in the history of great American engineers. Few engineers have made the contributions he made to industry, teaching, research, consulting, lecturing, administration, and writing. Professor Begeman had a way of making the complex seem simple and an insight that enabled him to predict which research and development results would be cost effective in production.

Dr. B. H. Amstead has been a practicing engineer, consulting engineer, professor of mechanical engineering, and acting dean of engineering at The University of Texas at Austin. He also served as president of The University of Texas of the Permian Basin. He is now a professor emeritus and president of his own consulting firm. His entire engineering career has focused on manufacturing and industrial processes; the demand for him as a lecturer and consultant attests to his perception as to how things can be made more efficiently and profitably.

Dr. Phillip F. Ostwald is affiliated with the University of Colorado in Boulder, Colorado as a professor of mechanical and industrial engineering. He teaches courses on industrial cost analysis, operations research, engineering statistics, and manufacturing processes. In addition, he has presented numerous lectures and seminars on manufacturing cost analysis to industrial associations, the U.S. Government, and to various organizations throughout the world. The Society of Manufacturing Engineers selected him for their first Sargent Americanism Award on the importance of profit in education. The Institute of Industrial Engineers awarded Dr. Ostwald their Phil E. Carroll Award for an industrial data base for industries to evaluate production operations. His interests include cost estimating and the economic feasibility of design; and his concentration on industrial cost analysis is defining that professional field. He has over 70 publications of various kinds and is the author of a textbook on cost estimating.

TO MYRON L. BEGEMAN

PREFACE

This text introduces the field of manufacturing processes. Engineering and technology students, technicians, practitioners, and management professionals who have a need for a first-course text will find this book useful. Many men and women advance to the profession of manufacturing engineering or supervision from the practical ranks. They often find self-study to be a necessary academic supplement to their intimate grasp of practice. This book has filled a broad educational need since 1942, the date of the first edition. Now printed in several foreign languages, it is studied worldwide. The authors are delighted and proud that this text has served so many.

Instructors will notice that the eighth edition continues to add the most recent technology and has been improved to make the text more easily teachable. Numerous illustrations amplify the discussion. Many questions, problems, and case studies have been added or updated. Because students are now more familiar with calculation, practical formulas have been added with this edition. This is consistent with the advancing sophistication of manufacturing processes.

Instructors will find the text versatile. It may be used with or without a laboratory and rearrangement of chapters is possible to meet specific course requirements. The experience with the previous seven editions indicates that the book can be used for a one-semester or two-quarter course. If this text supplements a laboratory, a full year of material is available.

Prerequisites for the material presented are open to instructor selection. We have maintained that a technical book may be adopted for serious study on the basis of "open opportunity" rather than on narrow restrictions. Because of the renewal of interest in manufacturing processes, classes are more diversified than in the past and prerequisites need to be flexible. Our eighth edition is intended to harmonize with these varied objectives.

The eighth edition includes discussions of many new technologies, processes, and production equipment. For example, computer applications, robots, integrated production planning, and computer-aided manufacturing have been added. Instructors will note that metric and U.S. customary units are used together, and many new problems have been added to encourage understanding. These new technologies and the complete rewriting makes the book useful as a reference.

The authors appreciate the many firms that supplied illustrations, and appropriate credits are noted. The work of the late Professor Myron L. Begeman is evident throughout this book. His contributions are enduring and we are proud to continue that legacy.

Manufacturing is more exciting now than ever before, and this book reflects this new spirit. After all, manufacturing is an employer of millions and a servant to the well-being and standard of living for all nations. It is an important subject. Finally, we are grateful to the many instructors who keep manufacturing processes an exciting and stimulating subject.

B. H. Amstead
Phillip F. Ostwald

CONTENTS

1 MANUFACTURING AND THE ECONOMIC SYSTEM 1
2 NATURE AND PROPERTIES OF MATERIALS 16
3 PRODUCTION OF FERROUS METALS 33
4 PRODUCTION OF NONFERROUS METALS 60
5 CONVENTIONAL FOUNDRY PROCESSES 75
6 CONTEMPORARY CASTING PROCESSES 104
7 HEAT TREATMENT 131
8 WELDING, BRAZING, AND ADHESIVE BONDING 156
9 POWDER METALLURGY 196
10 PLASTIC MATERIALS AND PROCESSES 214
11 METROLOGY AND QUALITY CONTROL 242
12 HOT WORKING OF METAL 286
13 COLD WORKING OF METAL 310
14 PRESS WORK AND TOOLING 338
15 BASIC MACHINE TOOL ELEMENTS 367
16 NUMERICAL CONTROL 390
17 MANUFACTURING SYSTEMS 408
18 METAL CUTTING 449
19 TURNING, DRILLING, BORING, AND MILLING MACHINE TOOLS 479
20 MACHINING OPERATIONS 505
21 SHAPING, PLANING, SAWING, AND BROACHING 546
22 THREADS AND GEARS 571
23 GRINDING AND ABRASIZE MACHINES 599
24 SPECIAL PROCESSES AND ELECTRONIC FABRICATION 625
25 OPERATIONS PLANNING AND COST ESTIMATING 663

BIBLIOGRAPHY 688

PHOTO CREDITS 693

INDEX 697

MANUFACTURING PROCESSES

CHAPTER 1
MANUFACTURING AND THE ECONOMIC SYSTEM

Manufacturing changes the form of materials to create products. If the enterprise makes optimum use of resources to produce products, it earns a profit. A manufacturing system coordinates elements of input, process, and output. An industrial activity requires resources to produce products.

Development of modern manufacturing is dependent on research in materials, and products require a variety of production processes for these materials. Moreover, new and varied materials demand new and improved *manufacturing processes.* Computer-aided design and manufacturing, robotics, new alloys, electronic and aerospace designs, safety, and antipollution laws have stimulated development. In turn, improvements in manufacturing have enhanced the products that consumers use.

The beginning of modern manufacturing processes may be credited to **Eli Whitney** and his cotton gin, or to *interchangeable production,* or to the milling machine during the early 1800s, or to any number of other developments in the world at about the same time. The Civil War and its grim necessities gave impetus to manufacturing processes in the United States. The origin of experimentation and analysis in manufacturing processes is credited in large measure to Fred W. Taylor, whose published papers on the art of cutting metals gave a scientific basis to manufacturing. Others, such as the late Professor Myron L. Begeman, were careful observers, researchers, and reporters of new developments in processing and were prompt to disseminate information to students of manufacturing.

The inventions of power-transforming machines like the waterwheel, steam engine, and electric motor substituted machine energy for human labor. The development of ferrous and nonferrous materials, and plastics and their shaping into useful products moved humanity forward in technology with a giant step. The use of flyball governors, cams, electricity, electronics, and computers enabled people to control machine processes. But these inventions came slowly, and along with them came the necessity for manufacturing. Figure 1.1 is a picture of the first milling machine, which is credited to Eli Whitney. Examples of the most recent technology are illustrated later in this chapter and throughout the book. The student should study these pictures and graphics carefully. Unfortunately, we overlook the history of manufacturing, which is an interesting field of

2 MANUFACTURING AND THE ECONOMIC SYSTEM

Figure 1.1
Eli Whitney's milling machine that is displayed by the New Haven Colony Historical Society. The belt drive was attached to the left spindle and the cutter to the right spindle, 1818.

study and culture. Still, we do want the student to appreciate that manufacturing does have a glorious past and a promising future.

EVOLUTION OF THE ENTERPRISE

Manufacture is a Latin word, *manu factus*, or, literally, "made by hand." But the power of the hand tool is limited. Domesticated animal and water power from the waterwheel were early steps in the evolution. Harnessing water power provided an opportunity for the development of power-driven machine tools and manufacturing. In the United States the small shops were often located near a pond that had a waterfall. In fact, water-driven machine tools were used to build the first practical steam engine. Wilkinson's boring machine was used for machining the cylinders of James Watts' steam engine, which was invented in 1776. This is generally accepted to be the beginning of the Industrial Revolution. This source of power could be located anywhere it was needed for work.

Motive Power

With the availability of power and its transmission by overhead power belts, other machine tool developments followed. Figure 1.2 is a woodcut illustration of a machine shop interior. Overhead line shafts were driven by steam engines. The screw-cutting lathe, invented in 1797 by Henry Maudsley, was one of the first important machines. By the early 1900s the kinematics of the basic machine tools and structures for mechanical cutting methods had been developed. The electrical motor, in the early 1900s, improved the efficiency of power transformation, and power line shafts had disappeared by the 1940s.

The Manufacturing Enterprise

A *manufacturing enterprise* is a factory-based, profit-making organization. Business enterprises, also called companies, firms, and corporations, can be small or large. They have in common a willingness to assume risks. The payoff for these risks is *profit,* which is the amount of money resulting when revenues exceed costs. Profit is an excess that

Figure 1.2
Interior of machine shop, 100 × 300 ft (30 × 90 m) using overhead belt-driven power, circa 1885.

allows the payment of dividends to owners and shareholders, the purchase of new equipment and plants, and the payment of taxes. These taxes, whether they are national, state, or local, are the inescapable trade-off for successful operation—a vital and important contribution to continue the blessings of a democratic society. If dividends, the rent on invested capital and money, are not paid, it would lessen the faith of investors and jeopardize a source of money for growth.

Manufacturing is unable to make everything that everyone wants. In the United States, firms are free to decide which products they will manufacture. People cannot buy all the products they desire, but each buyer is free to select what he or she will buy. Both producers and the buyers have a freedom of choice, and this opportunity is called the *free enterprise system*.

To understand the manufacturing system, we can describe it as having three essential parts: *input, process, and output*. Figure 1.3 shows broad details of a manufacturing system.

In a free enterprise system, *consumer demand* serves as the stimulant to encourage business to provide products. Materials, which are the minerals of nature, such as coal, ores, hydrocarbons, and many more, are converted into these products. It takes financing and money, either by loans from banks or capital investments from stockholders or from plowback of profit into the business, to sustain this activity. *Working capital* is money used to buy materials and pay employees. *Fixed capital* is the money for tools, machines, and factory buildings. A manufacturing enterprise needs money for these and other requirements.

Energy is an important input to manufacturing. Energy exists in many different forms, such as electricity, compressed air, steam, gas, or coal.

Processes are the next step of the enterprise. *Management* provides planning, organization, direction, control, and leadership of the business enterprise to make it productive and profitable. Managers have responsibilities to the owners, employees, customers, general public, and the enterprise itself. The business enterprise must make a profit; otherwise it will fail.

The *design step* consists of creating plans for products so that they are attractive, perform well, and give service at low cost. Manufactured products are designed before they are made. They may be designed by workers in the shop, in the drafting room, or in engineering. However, design is usually handled by trained engineering specialists.

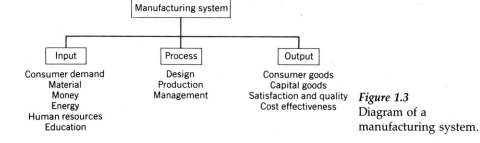

Figure 1.3
Diagram of a manufacturing system.

This book deals with the third step of production. The processes needed to manufacture a product must be designed and engineered in great detail. General plans for the processes are recognized during the design stage. Now the techniques of manufacturing engineering are used. The best combination of machines, processes, and people is selected to satisfy the objectives of the firm, shareholders, employees, and customer.

The output of a manufacturing system is a product. Look around you. Products of manufacture are everywhere. Goods can be classified into consumer and capital kinds. *Consumer goods* are those products that people buy for their personal consumption such as food or cars. *Capital goods* are products purchased by manufacturing firms to make the consumer products. Machine tools, computer-controlled robots, and plant facilities themselves are examples.

Organizing for Manufacturing

Manufacturing depends on ***organization*** to achieve its purpose. Resources such as people, power, materials, machines, and money are necessary. For efficient and economical production, these resources must be organized and coordinated in any successful effort. Many management styles are possible; Figure 1.4 is an example of an organizational chart. Whereas the organizational design depends on the size of the plant and its products, we see that the owners are represented by the shareholders. These in turn establish the policy decisions by a board of directors. Even though companies may differ in how they are organized, there are certain functions that are found in almost any manufacturing organization. These are ownership, general administration, sales and marketing, product engineering and design, manufacturing, human resources, finance and accounting, and purchasing. A smaller plant or a job shop may combine these functions, whereas a large company will indicate a greater diversification of functions and responsibilities.

The manufacturing part of the organization depends on company size, product design, volume, number of employees, sales, and the nature of technology. Figure 1.5, for a medium-sized company that is diversified, shows how the manufacturing engineering

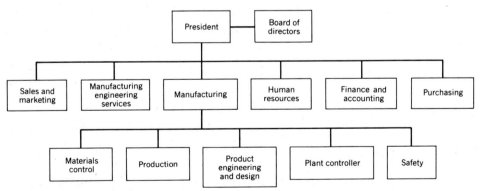

Figure 1.4
Organizational chart for a manufacturing company.

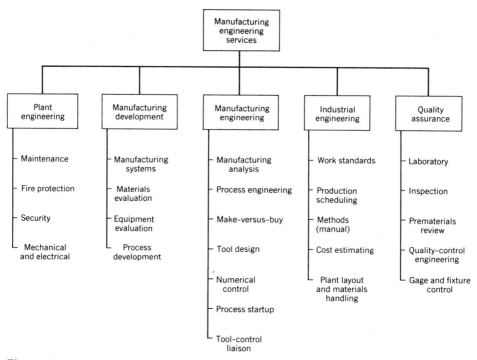

Figure 1.5
Responsibility diagram and organizational chart for manufacturing engineering services department.

services could be organized. Note that subfunctions require education and training, and many of these opportunities are presented in this text.

DESIGN, MATERIALS, AND PRODUCTION

The cost of a product depends on raw materials, production costs for machines and labor, management and sales, warehousing and logistics, and overhead. Machine and labor costs are inexorably related and make up, along with raw materials expenditures, the bulk of production costs. When a material is chosen, the process, including the machine, is frequently specified. Alternatively, if a machine is available, the raw material that can be processed on that machine may be utilized. One could say that the purpose of economical production is to produce a product at a profit. This infers that the cost must be acceptable and competitive; also, a demand for the product must exist or must be created.

Efficiency in Production

Since the first use of machine tools, there has been a gradual trend toward making machines more efficient by combining operations and by transferring more skill to the machine, thus reducing time and labor. To meet these needs, machine tools have become complex both in design and in control. Automatic features have been built into many machines, and some are completely automatic. This technical development has made it possible to attain the high production rate with low labor cost that is essential for any society wishing to enjoy high living standards. Computer-aided design and manufacturing are significant steps of progress. This text devotes a great amount of attention to this development.

Along with the development of production machines, the quality in manufacturing must be maintained. *Quality and accuracy* in manufacturing operations demand that dimensional control be maintained to provide parts that are interchangeable and give the best operating service. For mass production, any one of a quantity of parts must fit in a given assembly. A product made of *interchangeable parts* is quickly assembled, lower in cost, and easily serviced. To maintain this dimensional control, appropriate inspection facilities must be provided.

Three criteria that determine economical production are

1. A functional but simple design that has appropriate aesthetic quality.
2. A material choice that represents the best compromise among physical properties, appearance, cost, and workability or machinability.
3. Selection of the manufacturing processes that will yield a product with no more accuracy or better surface finish than necessary and at the lowest possible unit cost.

Product Engineering and Design

It is important that the product be designed with material, manufacturing, and engineering to be competitive. For any manufactured product it is possible to specify a stronger, a more corrosion-resistant, or a longer life material, for example, but it is the engineer's obligation not to overlook the opportunity of economical production. This leads to *value engineering,* which is the substitution of cheaper materials or elimination of costly materials or of unnecessary operations.

To produce parts of greater accuracy, more expensive machine tools and operations are necessary, more highly skilled labor is required, and rejected parts may be more numerous. Products should not be designed with greater accuracy than the service requirements demand. A good design includes consideration of a finishing or coating operation, because a product is often judged for appearance as well as function and operation. Many products, such as those made from colored plastics or other special materials, are more saleable because of appearance. In most cases the function of the part is the deciding factor. This is particularly true where great strength, wear, corrosion resistance, or weight limitations are encountered.

For mass produced parts the design should be adaptable to mass production-type machines with a minimum of different setups. Whenever a part is loaded, stored, and reloaded into another machine, costs are involved that may not add value to the product.

Engineering Materials

In the design and manufacture of a product, it is essential that the material and the process be understood. Materials differ widely in physical properties, machinability characteristics, methods of forming, and possible service life. The designer should consider these facts in selecting an economical material and a process that is best suited to the product.

Engineering materials are of two basic types: *metallic* or *nonmetallic*. Nonmetallic materials are further classified as *organic* or *inorganic* substances. Since there is an infinite number of nonmetallic materials as well as pure and alloyed metals, considerable study is necessary to choose the appropriate one.

Few commercial materials exist as elements in nature. For example, the natural compounds of metals, such as oxides, sulfides, or carbonates, must undergo a separating or refining operation before they can be further processed. Once separated, they must have an atomic structure that is stable at ordinary temperatures over a prolonged period. In metal working, iron is the most important natural element. Iron has little commercial use in its pure state, but when combined with other elements into various alloys it becomes the leading engineering metal. The nonferrous metals, including copper, tin, zinc, nickel, magnesium, aluminum, lead, and others all play an important part in our economy; each has specific properties and uses.

Manufacturing requires tools and machines that can produce economically and accurately. Economy depends on the proper selection of the machine or process that will give a satisfactory finished product, its optimum operation, and maximum performance of labor and support facilities. The selection is influenced by the quantity of items to be produced. Usually there is one machine best suited for a certain output. In small-lot or job shop manufacturing, *general-purpose machines* such as the lathe, drill press, and milling machine may prove to be best because they are adaptable, have lower initial cost, require less maintenance, and possess the flexibility to meet changing conditions. However, a *special-purpose machine* should be considered for large quantities of a standardized product. A machine built for one type of work or operation, such as the grinding of a piston or the surfacing of a cylinder head, will do the job well, quickly, and at low cost with a semiskilled operator.

Many special-purpose machines or tools differ from the standard type in that they have built into them some of the skill of the operator. A simple bolt may be produced on either a lathe or an automatic screw machine. The lathe operator must know not only how to make the bolt but must also be sufficiently skilled to operate the machine. On the automatic machine the sequence of operations and movements of tools are controlled by cams and stops, and each item produced is identical with the previous one. This "transfer of skill" into the machine, or ***automation,*** allows less skillful operators but does require greater skill in supervision and maintenance. Often it is uneconomical to make a machine completely automatic, because the cost may become prohibitive.

The selection of the best machine or process for a given product requires knowledge of production methods. Factors that must be considered are volume of production, quality of the finished product, and the advantages and limitations of the equipment capable of doing the work. Most parts can be produced by several methods, but usually there is one way that is most economical.

CLASSIFICATION OF BASIC MANUFACTURING

There are a number of ways to classify manufacturing. Broadly speaking, the functions are mass production, moderate production, and job lot production. A part is said to be

CLASSIFICATION OF BASIC MANUFACTURING 9

mass produced if it is produced continuously or intermittently at high volume for a considerable period of time. Some authorities say that over 100,000 parts per year must be produced to qualify as a ***mass-produced part,*** but this is a restrictive definition. In *mass production* industry, sales volume is well established and production rates are independent of individual orders. Machines producing these parts are usually incapable of performing operations on other work. Unit costs must be kept to an absolute minimum. Easily recognized examples of mass-produced items are matches, bottle caps, pencils, automobiles, nuts, bolts, washers, light globes, and wire.

Parts made in *moderate production* operations are produced in relatively large quantitites, but the output may be more variable than for mass-produced parts and more dependent on sales orders. The machines will likely be multipurpose ones, although this is not true in plants producing specialty items with less demand or sales than is the case with mass-produced parts. The number of parts may vary from 2500 to 100,000 per year depending on complexity. Again, examples of this type of industry are more descriptive: printing of books, aircraft compasses, and radio transmitters.

The ***job lot*** industries are more flexible, and their production is usually limited to lots closely attuned to sales orders or expected sales. Production equipment is multipurpose, and employees may be more highly skilled, performing various tasks depending on the part or assembly being made. Lot sizes, customarily varying from 10 to 500 parts per lot, are moved through the various processes from raw material to finished product. The company usually has three or more products and may produce them in any order and quantity depending on demand. In some cases the plant may not have its own product, and if so it then "contracts" work as a subcontractor or vendor. Product changes are rather frequent, and in some the percentage profit per item exceeds other types of manufacture. For example, the following products may be produced in job lot-type industries: airplanes, antique automobile replacement parts, oil field valves, special electrical meters, and artificial hands.

Mass, moderate, and job lot production require different equipment and systems. The specification of the equipment identifies what it is and what it can do. See Figure 1.6, for example. In this introductory chapter, we point out that Figure 1.6 is a vertical turret numerical controlled machine tool. This machine is constructed of close-grain cast iron for its structural members; it has an adjustable carriage, hardened and ground steel wraparound ways for rigidity, electric lead screw drive, and a 20-tool storage capacity. These specifications may seem meaningless to the reader at this point but elaboration is provided in later chapters. The role of this tool is in moderate to job lot production quantity ranges. It is not intended for high volume.

A manufacturing system design may specify a robotic loading station for a mass-produced product. Figure 1.7 shows an isometric drawing of a robotic system inserting a relay into a product presented to it on a pallet. The pallet is synchronized to the robot. Parts are picked up by the gripper from a feeder unit. Special orientation of the part in the gripper and eggshell forces are features of the system. The loading rate is 438 units per hour, which is considered moderate to mass production volume.

Throughout this text pictures, diagrams, and sketches are shown of processes, tools, and machines. The student will want to study these figures for details and overall impressions. Performance statements are listed along with their special purpose.

10 MANUFACTURING AND THE ECONOMIC SYSTEM

Figure 1.6
Vertical turret numerical-controlled machine tool.

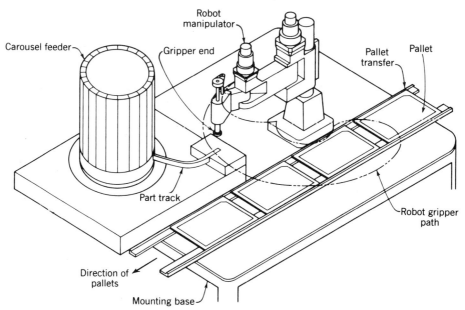

Figure 1.7
Elements of robot loader for a mass-produced electronic product.

ENGLISH METRIC PRACTICE

Le Système International d'Units, officially known worldwide as "SI," is a modernized metric system and incorporates many advanced concepts. Adopting SI is more than converting from inch, pound, gallon, and degree Fahrenheit units to the meter, kilogram, ampere, and degree Celsius units. It is a chance to introduce a new, simplified, coherent, decimalized, and absolute system of measuring units. The practice frequently followed in this book is to provide the English or customary units followed by SI in parentheses.

Base Units of SI

Built on seven base units (meter, kilogram, second, ampere, Kelvin, candela, and mole), *SI* provides a coherent array of units obtained on a direct one-to-one relationship without intermediate factors or duplication of units for any quantity. Thus, *1* newton (N) is the force required to accelerate a mass of *1* kilogram (Kg) at the rate of *1* meter per second squared (M/S^2). *One* joule (J) is the energy involved when a force of *1* newton moves a distance *1* along its line of action. *One* watt (W) is the power that in *1* s gives rise to the energy of *1* J. The SI units for force, energy, and power are the same whether the manufacturing process is mechanical, electrical, or chemical. Some new names receive more usage, including newton, pascal (Pa) joule, and hertz (Hz). Eventually SI will replace such terms and abbreviations as poundals, horsepower, BTU, psi, weight, feet per minute, and thousandths.

Rules for Conversion and Rounding

The front and back endpapers of this book contain conversion factors that give exact or six-figure accuracy and are used for converting **English units** throughout this book. The conversion of quantities should be such that accuracy is neither sacrificed nor exaggerated.

The proper conversion procedure is to multiply the specified quantity by the conversion factor exactly as given in the endpapers, and then *round* to the appropriate number of significant digits. For example, to convert 11.4 ft to meters: $11.4 \times 0.3048 = 3.474$, which rounds to 3.47 m.

The practical aspects of measuring must be considered when using SI equivalents. If a scale divided into sixteenths of an inch is suitable for making the measurement, a metric scale having divisions of 1 millimeter (mm) is suitable for measuring in SI units, and the equivalents should not be closer than the nearest 1 mm. Similarly, a caliper graduated in divisions of 0.02 mm is comparable to one graduated in divisions of 0.001 in. A measurement of 1.1875 in. may be an accurate decimalization of a noncritical $1^1/_{16}$ in., which should have been expressed as 1.19 in. However, the value 3 may mean "about 3" or it may mean "3.0000."

It is necessary to determine the intended precision of a quantity before converting. This estimate of intended precision should never be smaller than the accuracy of measurement, and as a rule of thumb, should usually be smaller than one-tenth the tolerance. After estimating the precision of the dimension, the converted dimension should be rounded to a minimum number of significant digits such that a unit of the last place is equal to or smaller than the converted precision. For example, a piece of stock is 6 in. long. Precision is estimated to be about $1/2$ in. ($\pm 1/4$ in.). This converted precision is 12.7 mm. The converted dimension 152.4 mm should be rounded to the nearest 10 mm as 150 mm.

CAREERS IN MANUFACTURING

Careers in manufacturing industries continue to offer potential for the engineer, technologist, technician, skilled machinist, and machine tender. For example, labor force projections require population estimates. Factors of age, sex, birthrates, migration, new and declining technology, and capital investment and construction are determinants in future and active employment. For example, jobs for numerical computer-aided manufacturing programmers are increasing in numbers, but those for blacksmiths are not.

Productivity, meaning output per worker hour, is a factor suggesting that manufacturing will have dramatic gains in the next decade. Historically, high wages have created the requirement for labor-saving technology. Another determinant is the demand for a product. Some changes occur because of new technologies, and because these technologies are becoming more affordable. Advances in electronic components and computer chips make use of computers in manufacturing more prevalent. Personal computers may be used by the students of this book. As one example, this product growth is leading to career opportunity in adapting electronic controls to production machinery.

During the next decade, the number of persons of prime working age (25–54) in the labor force is expected to grow faster than the total labor force. Young workers will decline in absolute numbers as the rate of growth of the total labor force slows. These growth trends reflect the aging of the baby boom generation and a substantial decline in birthrates. About 130 million are expected to be in the 1995 United States' labor force, which will be the largest number ever.

Growth in exports of consumer and capital goods reflects the expectation that trade in the long run will move toward underdeveloped countries, because these countries tend to require goods with higher technological inputs, such as electronic computers, aircraft parts, telephonic and other electronic parts, and electrical apparatus, to name a few.

In manufacturing, examples of growth industries are customarily thought to be the high-tech industries. Employment in these industrial sectors accounts for most manufacturing gains. Communication equipment, aerospace, and machinery in particular are thought to be aggressive in creating job demands. On the other hand, the steel industry, following a long period of retrenchment, will have job demand in excess of the average.

People are vital resources for a manufacturing enterprise. People perform work or labor. The work may be mental, physical, or a mixture of both. It is evident that educational requirements for human resources have been increasing. The skill requirements of the operator, technician, technologist, engineer, and researcher have become more complex. Special skills and training are necessary as manufacturing moves into computer-aided design and manufacturing. Many schools, colleges, and universities are adding courses in production or manufacturing.

Employment for knowledge-based or skilled manufacturing jobs requires university or college education or specialized postsecondary technical training. With high school education, entry opportunity requires in-plant training.

QUESTIONS AND PROBLEMS

1. Give an explanation of the following terms:
   - Manufacturing processes
   - Eli Whitney
   - Interchangeable production
   - Manufacturing enterprise
   - Profit
   - Organization
   - Engineering materials
   - Special-purpose machines
   - Automation
   - Mass-produced parts
   - Job lot production
   - SI
   - English units
   - Rounding

2. Prepare a list of career opportunities in manufacturing processes from the classified want ads of a newspaper.

3. Indicate the criteria for economical production.

4. Why are materials and design important to manufacturing processes?

5. What purpose does the company's board of directors serve?

6. Contrast the advantages and disadvantages for mass, moderate, and job lot production.

7. Define automation. Is it more suitable for mass, moderate, or job lot production? Discuss.

8. Show the advantages of SI units. Indicate the base units of SI. Why is it important to know English and SI units, conversions, and practices?

Convert the following from customary (English) units to the International System of units, or from SI to English units. Show correct abbreviations. Use conversion factors as found in the book endpapers.

9. List the rules for converting and rounding English units to SI.

10. Area
    a. 15 ft^2, 2.4 ft^2 to square meters (m^2)
    b. 0.15 in.^{2}, 0.035 in.^{2} to square millimeters (mm^2)
    c. 0.51 mm^2, 1.060 mm^2 to in.^{2}

11. Energy
    a. 28,000 British thermal units (BTU) to joules (J)
    b. 8,500,000 kilowatt hours (kWh) to J

12. Force
    a. 19 pound-force (lbf) to newton (N)
    b. 2400 ton-force to meganewton (MN)
    c. 0.6 MN to lbf

13. Length
    a. 1.9 ft, 2.0 ft to meters (m)
    b. 11 in., 18.1 in., 250 in. to m
    c. 10 in., 3.5 in., $^1/_8$ in., 0.500 in., 0.0005 in. to millimeters (mm)
    d. 9 microinches (μin.), 35 μin. to nanometers (nm)
    e. 300 nm, 1250 nm to μ.
    f. 6 mils, 32 mils to mm

14. Mass
    a. 16 ounce-mass (ozm) to kilograms (kg)
    b. 15 grams (g), 1700 kg to ounces
    c. 15 pound-mass (lbm), 2.3 lbm, 1.3 lbm to kg
    d. 35 tons (short) to megagrams (Mg)

15. Density
    a. 14 pound-mass per cubic foot (lbm/ft^3) to kilograms per cubic meter (kg/m^3)
    b. 63 lbm/ft^3 to kg/m^3
    c. 0.2 ounce-mass per cubic inch (ozm/in.^{3}) to kg/m^3
    d. 0.33 kg/m^3 to lbm/ft^3
16. Pressure or stress (force per area)
    a. 2600 pound-force per square foot (lbm/ft^2) to megapascal (MPa; 1 Pa = 1 N/m^2)
    b. 185 pounds per square inch (psi), 1750 psi to Pa (1 Pa =1 N/m^2)
    c. 2500 psi, 1.8 × 10^6 psi to MPa
17. Temperature
    a. −40 degrees Fahrenheit (°F), 180°F, 250°F, 1750°F to degrees Celsius (°C)
    b. 1300°C, 200°C, 1000°C to °F
18. Velocity
    a. 150 feet per minute (fpm), 500 fpm, 750 fpm to meters per second (m/s)
    b. 400 m/s, 150 m/s, 1000 m/s, 2000 mm/s to fpm
19. Volume
    a. 0.35 ft^3, 120 ft^3, 750 ft^3 to m^3
    b. 0.02 in.^{3}, 12 in.^{3}, 150 in.^{3} to mm^3
    c. 155 in.^{3}, 179 in.^{3}, 1900 in.^{3} to m^3
    d. 1 oz, 173 oz, 210 oz to m^3
    e. 25 gallons (gal), 55 gal to m^3
    f. 1000 cubic yards (yd^3) to m^3
    g. 350 mm^3, 80 mm^3, 1500 mm^3 to in.^{3}
20. Volume per time
    a. 90 ft^3/min 64 ft^3/sec to m^3/sec
    b. 24 yd^3/min, 44 yd^3/min to m^3/sec
    c. 11 in.^{3}/min, 67 in.^{3}/min, 0.1 in.^{3}/min to mm^3/s
    d. 1.3 mm^3/s, 0.005 mm^3/s to in.^{3}/min

CASE STUDY
PROFESSOR JAMES SMITH

"Good morning, Professor Smith. I'm Rusty Green, and I'm in your eight o'clock manufacturing processes class."

"Uh huh," replied the professor without looking up from his desk. "What can I do for you, Rusty?"

"Well, it's like this. I'm not sure that I belong in your class." Rusty smiles, and continues as the professor looks up. "Oh, it's not you, Professor, but it's unclear what this course can do for me later on."

"Yes, go on," and Professor Smith, leaning forward in his chair, continues by saying, "What are you wanting?"

"My program is _____, and I am a _____ year student. I came to _____ (school, college, or university) not really sure of what I wanted."

"That's not uncommon, Rusty," the professor reassures.

"I do want a career-oriented program, and is manufacturing processes going to aim in that direction?" Rusty now looks expectantly at the professor.

"That's an important question, Rusty. There are many, many considerations."

Help the professor answer Rusty. Prepare a report that deals with the opportunities in manufacturing for you in your field. Consider the following:

Check the newspaper want ads, trade journals and technical magazines, and your college career office for employment opportunities. Speculate on the training or educational requirements and experience that are necessary for these careers. Consider tangential situations that can develop. List the job titles and their relationship to manufacturing. Do these situations change after 1 year or 5 or 10 years?

Before a person can plan, he and she needs to have a notion of their interests and abilities. The self-esteem, creativity, opportunity, responsibility, and ethical standards for work are also important. Not all people feel these needs with the same intensity. Help Professor Smith answer Rusty's question, except think of yourself as Rusty.

CHAPTER 2
NATURE AND PROPERTIES OF MATERIALS

The great variety of materials available for manufacture has led some to predict that all materials will be manufactured ones like plastics and not iron, aluminum, and other metals that are refined from ores. Although there is some drift toward this position, the commonly used metals will continue to provide the major load-carrying and heat-resistant structures. They are plentiful, relatively cheap, have known properties, and are easily machined or formed.

CLASSIFICATION OF MATERIALS

Materials for manufactured parts or machines have such diversified properties that even when performance and cost are considered, it is often difficult to decide the proper material for a given purpose. One material may have higher strength, another better corrosion properties, and yet another may be more economical. Hence, most choices are a compromise among a number of materials using the best engineering data and judgment available. Copper, for example, may be alloyed in hundreds of ways to produce materials with special properties.

In general, materials may be classified as follows:

A. Metallic
  1. Ferrous
  2. Nonferrous

B. Nonmetallic
  1. Organic
  2. Inorganic

Table 2.1 shows the principal metals used in manufacturing processes and a few of their important physical properties. The values indicate the approximate range of properties that can be expected depending on alloying and heat treatment. Other important commercial metals include tin, silver, platinum, manganese, vanadium, and titanium. Because special properties can be obtained by alloying, few pure metals are used extensively.

The nonferrous metals are generally inferior in strength but superior in corrosion resistance as compared to ferrous materials, and most are more expensive.

Table 2.1 Approximate[a] Properties of Common Metals

Metal	Tensile Strength lb/in.² × 10⁻³	Tensile Strength MPa	Ductility (%)	Melting Point °F	Melting Point °C	Brinell Hardness	Density lb/ft³	Density kg/m³
Ferrous								
Gray Cast Iron	16–30	110–207	0–1	2500	1370	100–150	450	7,209
Malleable Iron	40–50	276–345	1–20	2475	1360	100–145	480	7,689
Steel	40–300	276–2070	15–22	2700	1480	110–500	485	7,769
White Cast Iron	45	310	0–1	2500	1370	450	480	7,689
Wrought Iron	35–47	242–324	30–35	2800	1540	90–110	493	7,897
Nonferrous								
Aluminum	12–45	83–310	10–35	1220	660	30–100	165	2,643
Copper	50–100	345–689	5–50	1977	1080	50–100	556	8,906
Magnesium	12–50	83–345	9–15	1200	650	30–60	109	1,746
Nickel	60–160	414–1103	15–40	2650	1450	90–250	545	8,730
Lead	2–33	18–23	25–40	620	325	3.2–4.5	706	11,309
Titanium	80–150	552–1034	0–12	3270	1800	158–266	282	4,517
Zinc, cast	7–13	48–90	2–10	792	422	80–100	446	7,144

[a]Depending on the alloy.

Nonmetallic materials (Table 2.2) are classified as *organic* if they contain animal or vegetable cells (dead or alive) or carbon compounds. Leather and wood are examples. Materials are classified as *inorganic* if they are other than animal, vegetable, or carbon bearing.

There are fundamental differences between organic and inorganic materials. Organic materials will usually dissolve in organic liquids such as alcohol or carbon tetrachloride, but they will not dissolve in water. Inorganic materials tend to dissolve in water. In general, inorganic materials resist heat more effectively than organic substances.

A number of materials is necessary in manufacturing finished products. Figure 2.1 shows the materials used in making an automobile. The choice between metallic and nonmetallic and between organic and inorganic materials is a compromise between service

Table 2.2 Nonmetallic Materials

Inorganic	Organic
Minerals	Plastics
Cement	Petroleum products
Ceramics	Wood
Glass	Paper
Graphite	Rubber
	Leather

18 NATURE AND PROPERTIES OF MATERIALS

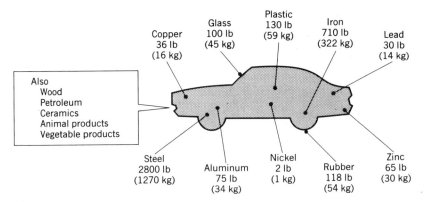

Figure 2.1
Representative materials in a full-sized automobile.

characteristics and cost. In order to increase gasoline mileage, the materials offering the least weight per part have gained favor.

SOURCE FOR MATERIALS

Although some materials used in manufacturing processes are of animal or vegetable origin, all find their source of supply in the crust of the earth. Materials, whether metallic or nonmetallic, organic or inorganic, are seldom found in the state in which they are used. Great difficulty and expense are involved in converting iron ore from a mine into the steel in an automobile body. The ore must be extracted from the extraneous materials, reduced, and often alloyed with other materials and processed to yield the desired properties. *Extraction* and *refining* processes require heat, a chemical reaction, or both. Iron is smelted from its ore in a blast furnace, whereas aluminum is extracted by first converting the bauxite ore to an oxide and then reducing it by an electrolytic process requiring intense heat. The cost of a metal depends on the quality of the ore and on the ease with which the metal can be extracted from the ore. For this reason aluminum is more expensive than iron, although aluminum-bearing ores are 60% more prevalent than iron-bearing ones. Actually about one-twelfth of the earth's crust is aluminum, but the economical recovery of the metal depends on the quality of the ore.

Table 2.3 gives the principal metals, the ores from which each is extracted, and where they are found in commercial quantities. A few metals like copper, gold, and silver can be mined in the pure state, but metals such as aluminum, iron, lead, magnesium, nickel, and tin are seldom found unadulterated.

Processes designed to recover usable metal from an ore yield many by-products that have commercial uses. Examples are the blast furnace slag made in iron production and used in road construction, sulfuric acid produced from the reduction of zinc ore, and iodine recovered when magnesium is extracted from seawater.

Table 2.3 **Metals and Their Ores**

Metal	Principal Ores or Raw Material	Principal Location
Aluminum	Bauxite (a mixture of gibbsite, $Al_2O_3 \cdot 3H_2O$, and diaspore, $Al_2O_3 \cdot H_2O$)	Guianas, Italy, Arkansas
	Cryolite (Na_3AlF_6)[a]	Iceland, Greenland
Iron	Hematite (Fe_2O_3), red ore, 70% iron	Lake Superior District
	Magnetite (Fe_3O_4), black ore, 72.4% iron	New York, Alabama, Sweden
	Siderite $(FeCO_3)$, brown ore, 48.3%	New York, Ohio, Germany, England
	Limonite $[Fe_2O_3X(H_2O)]$, brown ore, 60–65% iron	Eastern United States, Texas, Missouri, Colorado, France
Tin	Cassiterite (SnO_2)	East Indies, Malaya, Bolivia
Magnesium	Magnesium chloride $(MgCl_2)$	Michigan
	Dolomite $(CaCO_3 \cdot MgCO_3)$	United States, Europe
	Seawater	
Zinc	Sphalerite (ZnS)	Missouri, Kansas, Oklahoma, British Columbia
Copper	Chalcocite (Cu_2S)	Arizona
	Bornite (Cu_3FeS_3)	Utah, New Mexico, Michigan, Nevada
Nickel	Miscellaneous sulfides Pentlandite $[(NiS)(FeS)_2]$	Canada
Lead	Galena (PbS)	Colorado, Missouri, Utah, Idaho, Montana, Oklahoma, Mexico
Silver	Argentite (Ag_2S)	Mexico, Utah, Nevada, Colorado, Peru, Bolivia

[a]Necessary to the process.

STRUCTURE OF METALS

Physical State

Although metals may exist as vapor, liquid, or solid, they are generally used in the solid form. Because all substances are composed of atoms, each with its particular characteristic, the same atoms exist whether the metal is in gaseous, liquid, or solid form. The distance between atoms in gases is large, accounting for their low density. In liquids atoms are close together and move about in all directions at random; thus there is no permanent shape to liquid. When the container changes shape, the liquid takes a new form subject to the retaining walls of the container. Some materials, like glass, are called *amorphous.* They are like a liquid that has "solidified," but there is no definite atomic structure, because the atoms exist in a random pattern just as in a liquid.

All solid metals and many other materials are *crystalline* in nature, which means that their atoms align themselves in a geometric pattern upon solidification. This pattern of atoms conforms to the *space lattice* of the materials. The unit cells of several forms of space lattices appear in Figure 2.2. The type of lattice and the distance between the atoms can be determined by X-ray analysis.

Some solid materials such as iron change their type of lattice structure when their temperature reaches a so-called critical temperature. This type of change is called *allotropic*. Any material that may exist in several crystal forms is known as allotropic or *polymorphic*, and such a material will have unique properties depending on the lattice structure. Iron at room temperature has a body-centered cubic lattice structure and is called alpha iron. When iron is heated above approximately 1670°F (910°C), its structure changes to a face-centered lattice structure and it becomes known as gamma iron. This transition may be recognized by changes in the electrical properties, absorption of heat, and dimensions. If examined radiographically the iron will show a different spacing of the atoms. Tin shows a dramatic difference in properties because of its allotropic nature. In its usual form it is silvery white, but when subjected to low temperature it changes gradually to gray.

The **body-centered** *cubic lattice* has atoms at the corners of a cube and one atom at the center, as in Figure 2.2A. Iron (alpha) at room temperature, chromium, molybdenum, vanadium, and tungsten are a few of the more important metals with this lattice structure. The *face-centered cubic lattice* (Figure 2.2B) has atoms at the corners of a cube and an atom in the center of each face. Iron (gamma) at elevated temperature, aluminum, silver, copper, gold, nickel, lead, and platinum are examples. The *hexagonal close-packed lattice* (Figure 2.2C) is characteristic of beryllium, cadmium, magnesium, and titanium.

The properties of a pure metal can be predicted to some degree by the type of lattice structure. The hexagonal close-packed structure generally indicates a material that lacks ductility and becomes even more brittle when it is bent or machined. Face-centered materials are usually the most ductile.

The lattice structure of an alloy is not easily predicted. Any elements added to a pure metal alter the size of the lattice and, depending on the alloy formed, may change the lattice type. The atoms of the added element may take the place of certain atoms in the solvent or pure metals. The resulting alloy is known as a *substitutional solid solution*. **Brass,** an alloy of cooper and zinc, is an example. When the atoms of the added element fit themselves into spaces (interstices) between the solvent atoms, the alloy is called an *interstitial solid solution*. Carbon in iron is an example. *Intermetallic compounds* are

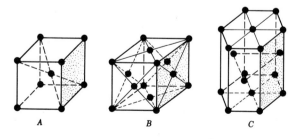

Figure 2.2
Space lattice structures. *A,* Body-centered cubic lattice. *B,* Face-centered cubic lattice. *C,* Hexagonal close-packed lattice.

formed when certain metals are alloyed, and the lattice structure is very complex. Such compounds melt at a fixed temperature and have lower conductivity and ductility but greater strength and hardness than an alloy of face-centered, body-centered, or hexagonal lattice structure. Examples of intermetallic alloys occur in the systems of aluminum–copper, copper–magnesium, and tin–antimony.

GRAIN FORMATION

When a metal solidifies, atoms arrange themselves geometrically. The initial lattice formations in a solidifying liquid form the nuclei for crystals that will grow in orderly form; that is, they maintain their lattice pattern and each successive lattice lines up with the preceding one. Many such nuclei form in a liquid as solidification begins, but the directions in which the initial nuclei are oriented are random. Figure 2.3A schematically illustrates the way the enlarging crystals form. When one crystal comes in contact with another of different orientation, growth of both crystals ceases and the surfaces where they meet, irregular in nature, form part of a *grain* boundary.

Most crystals do not develop uniformly but progress more rapidly in one direction than another. As the crystal growth advances the crystal front branches out in a treelike fashion. Such growth is called *dendritic,* and the crystal formation is called a **dendrite**. The growth is always uneven with branches of the dendrite thickening or new branches forming as solidification progresses. Figure 2.3B shows the completely developed grain boundaries of several crystals. The grains of a metal may be studied with a microscope after the material has been etched to make the boundaries stand out.

The grain size of a metal depends on the rate at which it was cooled and the extent and nature of the hot- or cold-working process. A metal with fine or small grains will have superior strength and toughness as compared to the same metal with large grains. This is because with the atoms closer together there is more "slip interference" in the lattice structure when a deforming force is applied. The larger grained materials are characterized by easier machining, better ability to harden through heat treatment, and superior electrical and thermal conductivity. Although the larger grained metals will harden more uniformly during heat treatment, the fine-grained materials are less apt to crack when quenched. Additives can be made to a molten metal to assure a predetermined grain size; aluminum, for example, may be added to steel to promote fine grains. The desired grain size is usually a compromise depending on the properties sought. For brass,

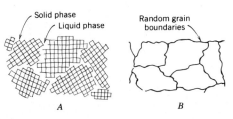

Figure 2.3
Growth of crystals forming grains.
A, Crystal growth. B, Grains.

which is used to make cartridge cases, a large grain enables the case to be formed more easily, but surface finish and strength are improved with a finer grain.

The hardness as well as the grain size is affected by the temperature history of a metal. *Quenching* steel from an elevated temperature will usually harden it, and cooling it slowly will bring out its maximum softness. *Annealing,* the slow cooling of a metal from an elevated temperature, is used to soften, add toughness, remove stresses, and increase ductility of metals.

MICROSCOPIC EXAMINATION

Using a metallurgical microscope to examine a polished specimen reveals the constitutents of some metals as well as surface deformities. The polishing operation creates a mirrorlike surface, but a coating of thin metal known as *smear metal,* which prevents critical analysis, is left on the specimen. If the specimen is **etched** with a suitable chemical solution, the smear metal is removed, the surface becomes slightly dull, grain boundaries are partially dissolved, and certain constituents are revealed because of the selective action of the etching solution. A common etching reagent for steel is a mixture of 3 parts nitric acid and 97 parts alcohol.

Figure 2.4A illustrates the effect of etching on the reflection of light from the specimen. Grains B and D have been attacked by the etching reagent and do not reflect the light in the same way as the other grains. When light is reflected into the lens of the microscope, the surface appears light; when the light is scattered, that area appears dark. The grain boundaries are like small canyons surrounding each grain, and the light is not reflected back to the observer. Thus the grain boundaries are well-defined black lines. Figure 2.4B illustrates the appearance of the specimen in Figure 2.4A when viewed through a microscope. Metal grains are usually examined at a magnification of approximately 100, although the electron microscope enables magnifications of over 1000. Selective etching reagents are used to reveal particular constituents.

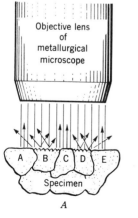
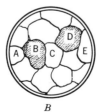

Figure 2.4
Metallurgical examination of an etched specimen. *A,* Etched specimen under examination. *B,* The specimen as viewed through the microscope.

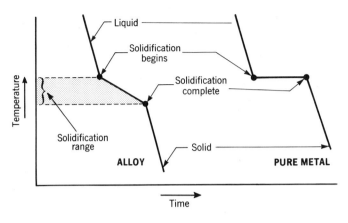

Figure 2.5
Time–temperature curve for an alloy and a pure metal.

SOLIDIFICATION OF METALS AND ALLOYS

Pure metals solidify in a unique manner, as indicated by Figure 2.5. The liquid cools to the point at which the first nucleus forms. From the time solidification begins until it is complete, the temperature of the solid liquid mixture does not change. Once solidification is complete the temperature drops with respect to time. During the freezing of the metal the latent heat of solidification just balances the heat lost by the metal, thus preserving the constant-temperature condition.

Only a few metals are used commercially in their pure state; copper for electrical wiring and zinc for galvanizing are examples. When other elements are added to a pure metal to enhance its properties, the combination is called an **alloy**. *Brass* is an alloy of copper and zinc, *bronze* an alloy of copper and tin, and *steel* an alloy of iron and carbon. Since the number of alloys is infinite, the prediction of their properties and characteristics is impossible.

Although pure metals solidify at a constant temperature, alloys do not, as shown in Figure 2.5. The first nuclei form at a higher temperature than that at which complete solidification occurs. This change in temperature as solidification progresses causes the solid being formed to change in chemical composition, because each element in an alloy has its own peculiarities relative to temperature.

An *equilibrium diagram* shows the way an alloy forms what is called a *solid solution*—that is, a solid that is in effect a solution of two or more materials. Many types of equilibrium diagrams exist depending on the alloys involved, but one of the simplest is the one for an alloy of copper and nickel (Figure 2.6).

Monel is a metal composed of 67% nickel and 33% copper. It is a metal that resists saltwater corrosion and is used in packaging beverages and foods. It has a range of working temperatures from $-100°$ to $400°F$ (-75 to $205°C$). At the dotted line in Figure 2.6 monel begins to solidify when cooled to temperature ℓ_1, but at that point the first material to solidify will have a composition of 23% copper and 77% nickel and as indicated by s_1. The liquid at ℓ_1 will be 67% nickel and 33% copper composition when freezing begins. When the temperature falls to the $\ell_2 s_2$ line, the liquid composition will be that

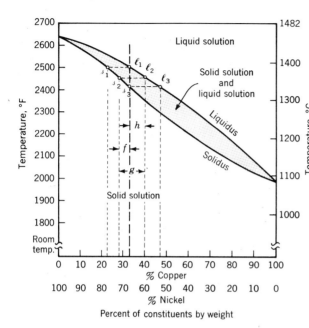

Figure 2.6
Equilibrium diagram for copper–nickel alloys.

indicated by ℓ_2 or 41% copper and 59% nickel, whereas the composition of the solid is indicated by point s_2. At the temperature $\ell_3\,s_3$, the last liquid to solidify is of composition ℓ_3 but the solid solution is monel with the 67% nickel and 33% copper content.

This type of diagram enables the engineer to determine the constituents of an alloy as well as certain other properties of the resulting solid solution.

The percentage of solid (S) at any temperature, s, can be calculated for a given composition. For example, for a 67% nickel and 33% copper alloy corresponding to the s_2–ℓ_2 line, at about 2455°F, S is expressed as

$$S = \frac{\ell_2 - s_3}{\ell_2 - s_2} \times 100$$

where s_3, s_2, and ℓ_2 represent lengths in the horizontal; that is, letting $s_3 - s_2 = f$, $\ell_2 - s_3 = h$, and $\ell_2 - s_2 = g$; then the percentage solid is

$$S = \frac{h}{g} \times 100$$

Similarly, the percentage liquid (L) is calculated

$$L = \frac{s_3 - s_2}{\ell_2 - s_2} \times 100 = \frac{f}{g} \times 100$$

Figure 2.7 shows how physical and mechanical properties of the copper–nickel alloys vary with respect to the two metals. Interestingly enough, the 5¢ coin or "nickel" is 75% copper and 25% nickel.

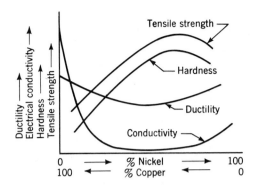

Figure 2.7 Effects of alloy composition on properties.

The equilibrium diagram for the alloy composed of iron and carbon is shown in Chapter 7. It should be noted that one of the differences between wrought iron, steel, and cast iron is the carbon content. The approximate carbon limits for each are as follows:

    Wrought iron   C <0.08%
    Steel   C >0.08% but <2.0%
    Cast iron   C >2.0%

Another characteristic of wrought iron that distinguishes it from other iron–carbon alloys is that it contains 1 to 3% slag. A discussion of the various iron–carbon alloys may be found in Chapter 7.

PROPERTIES OF MATERIALS

The properties of materials include density, vapor pressure, thermal expansion, thermal conductivity, electric and magnetic properties, as well as *engineering properties*. Engineering properties include tensile strength, compressive strength, torsional strength, modulus of elasticity, and hardness. Other properties such as machinability or ease of forming are discussed in subsequent chapters.

Tensile Strength

Tensile strength is determined by pulling on the two ends of a specimen machined like that shown in Figure 2.8. The specimen, about 8 in. (205 mm) in total length, is machined from $3/4$-in. (19 mm) material. The reduced section is $2^1/_4$ in. (57.2 mm) long and $1/_2$ in. (12.7 mm) in diameter and contains center punch or "gage marks" 2 in. (50.8 mm) apart. When the specimen is pulled, the smaller diameter section necks down from an area A to an area A_1 and the gage length increases from L to L_1. For most engineering purposes the area A is used in all calculations because A_1 is difficult to measure. The results are described by determining the changes that take place in length as the force is increased to the breaking point. From the data collected while pulling the specimen a curve can be plotted from two values, the **stress** and **strain,** where

$$\text{Stress} = \frac{\text{Force}}{\text{Area A}}$$

and

$$\text{Strain} = \frac{L_1 - L}{L}$$

Although such curves differ with material and heat treatment, the results generally describe a curve similar to those shown in the insert in Figure 2.9. A family of stress–strain curves, shown in Figure 2.9, is known as *engineering stress–strain curves*.

Referring to Figure 2.9, the point **a** represents the elastic limit of this steel. Any force causing a greater stress or strain would result in a permanent deformation of this material; that is, a tensile specimen would not return to its original length when unloaded. This point **a** also represents the *proportional limit,* and from that point on the stress–strain curve is no longer a straight line. The slope of this straight line is known as the *modulus of elasticity* or *Young's modulus,* which is an indication of the stiffness of the material. At some point **b** the material undergoes more rapid deformation, and it is at this point that its **yield strength** is determined. At point **c** the *ultimate strength* is determined, and this is often referred to as simply the tensile strength of the material. In the case of a tensile specimen this is the stress when necking begins. No single property of a material is more important than its tensile strength. Point **d** notes the point on a stress–strain curve where fracture has occurred after maximum deformation.

The *modulus of elasticity* is an indication of stiffness, and may be determined as the slope of the linear portion of the stress–strain curve.

Shear, Compressive, and Torsional Strength

There are no universally used standard tests for determining shear or torsion characteristics (Figure 2.10), but the relative differences in the properties between materials can be

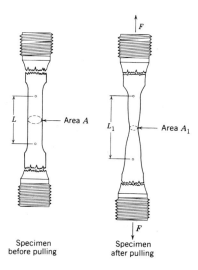

Figure 2.8
Tensile test specimen.

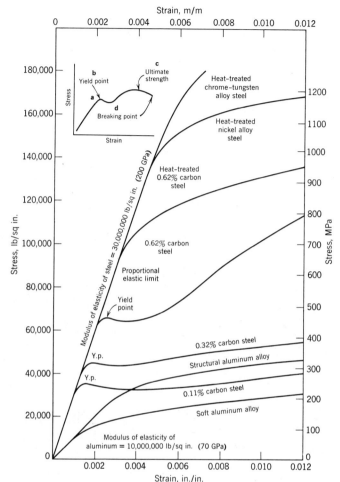

Figure 2.9
Engineering stress–strain curves for various materials.

found in handbooks. The *shear strength* of a material is generally about 50% of its tensile strength and the *torsional strength* about 75%.

Compressive strength (Figure 2.10) is easily determined for brittle materials that will fracture when a sufficient load is applied, but for ductile materials a strength in compression is valid only when the amount of deformation is specified. The compressive strength of cast iron, a relatively brittle material, is three to four times its tensile strength, whereas for a more ductile material like steel predictions cannot be made so easily.

Ductility

Ductility is a property that enables a material to be bent, drawn, stretched, formed, or permanently distorted without rupture. A material that has high ductility will not be brittle

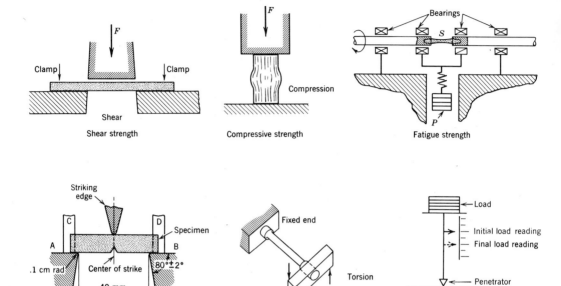

Figure 2.10
Schematic drawings of methods used in determining certain engineering properties of materials.

or very hard. Hard materials, on the other hand, are usually brittle and lack ductility. The tensile test can be used as a measure of ductility by calculating the **percentage elongation** of the specimen upon fracture. Hence,

$$\text{Percentage elongation} = \frac{L_f - L}{L} \times 100$$

where

L = Original gage length
L_f = Separation of gage marks measured on the reassembled bar after fracture

Impact and Endurance Testing

A metal may be very hard and have a high tensile strength yet be totally unacceptable for a use that requires it to withstand impact or sudden load. There are a number of tests that can be used to determine the impact capability of a metal, but the test most generally used is the *Charpy* test. Figure 2.10 shows a notched specimen that is struck by an anvil.

The energy in foot-pounds required to break the specimen is an indication of the impact resistance of the metal.

The yield strength of metals can be used in designing parts that will withstand a static load, but for cyclic or repetitive loading the endurance or *fatigue strength* is useful. An endurance test is made by loading the part and subjecting it to repetitive stress. Figure 2.10 shows one way that the endurance or fatigue strength of a material may be found. Generally, a number of specimens of a metal are tested at various loadings and the number of cycles to failure are noted. A curve of stress in pounds per square inch versus the number of cycles to failure is plotted, and these data can be used for designs involving repetitive loading.

Hardness

Although there are several techniques for determining the **hardness** of a material, most industrial methods measure the resistance to penetration of a small sphere, cone, or pyramid. Figure 2.10 shows a penetration hardness tester.

The first step in obtaining a reading is to force the penetrator and material into contact with the specimen with a predetermined initial load. Then an increased load is applied to the penetrator, and the hardness reading is obtained by noting the difference in penetration caused by the final load as compared to the initial load. Different scales of hardness depend on the shape and type of the penetrator and the loads applied.

The *Rockwell* hardness tester is the most flexible of the many types, because through a variation of different types of penetrators and loads, hardnesses can be measured on a range of materials from thin films to the hardest steel. The "C" scale using a diamond penetrator and a 331-lb load, is used for hard steel, and the "B" scale with a $1/16$-in. diameter ball and 220-lb load is used for softer steels and nonferrous metals.[1]

The *Brinell hardness* is measured by using a 10-mm-diameter ball and a 3000-kg force. To determine the penetration of the ball the diameter of the impression is measured with a microscope fitted with a reticle.

The *Shore Scleroscope* method of measuring hardness utilizes a diamond-tipped hammer weighing approximately 2.3 g that drops on the specimen. The rebound height is a measure of hardness. The *Vickers* hardness test utilizes a diamond pyramid-shaped penetrator that is loaded with from 1 to 120 kg depending on the hardness and thickness of the specimen to be checked. Measurement of the impression leads to Vickers hardness numbers.

Hardness of natural materials can be compared by referring to the **Mohs'** scale, which has a range of 1 to 10 as follows: talc = 1, gypsum = 2, calcite = 3, fluorite = 4, apatite = 5, feldspar = 6, quartz = 7, topaz = 8, titanium nitride = 9, and diamond = 10. The different systems for determining hardness are related, and handbooks provide tables that show the equivalent values of hardness for the various methods and scales.

Measurements of hardness have value in the nondestructive inspection of many in-

---

[1]Customary standards for Rockwell readings are in English units. Metric conversions are not provided. Brinell hardness readings are based on metric units only.

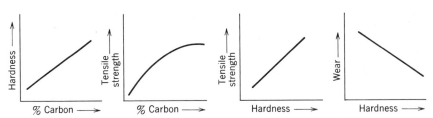

Figure 2.11
General relationships involving tensile strength and hardness for steels.

dustrial parts, because hardness and tensile strength are often proportional to each other and wear is inversely proportional to hardness. Figure 2.11 indicates the general manner in which some properties of ferrous material are related.

Because hardness measurements are relatively easy to make and seldom are destructive to the workpiece, such tests are used to control heat-treating and hot- and cold-working processes. A uniform hardness in parts made of the same ferrous material usually indicates that the engineering properties of the parts are identical.

QUESTIONS AND PROBLEMS

1. Give an explanation of the following terms:

   Amorphous   Solidus line
   Space lattice   Tensile strength
   Body centered   Stress
   Brass   Strain
   Grain   Yield strength
   Dendrite   Percentage elongation
   Etch   Hardness
   Alloy   Moh's hardness
   Liquidus line

2. What are some reasons for not using inorganic materials in outdoor applications?

3. In the automobile shown in Figure 2.1, assume you will replace a steel part that weighs 130 lb with one made of aluminum of the same size. How much weight reduction would result if steel weighs 2.9 times as much as aluminum? Suppose the new part can be made of plastic; steel weighs 4.8 times the weight of plastic; how much weight would be saved?

4. Classify the principal materials used in the manufacture of an automobile. Indicate their principal ores or raw materials.

5. Classify the following materials as organic or inorganic: cardboard, telephone case, rubber cement, polyethylene bottle, copper tubing, eraser, alcohol, peanut butter, salt, scissors, gasoline, cotton cloth, and polyester cloth.

6. Name the desired properties for a metal if the application is an anvil. A dragline bucket?

7. Describe the solidification of the following materials by using a time–temperature curve: cast iron, magnesium, steel, pure iron, and gold.

8. Which characteristics of wrought-iron water pipe enable it to be recognized after 5000 years?

9. From data provided, sketch a map of the United States and show ore-rich areas.

10. Sketch the lattice structures of cadmium, nickel, gold, tungsten, alpha iron, gamma iron, and aluminum.

11. What can be said about a fine-grained metal as compared to large- or coarse-grained metal?

12. Give the purpose of an etching reagent. Why is it applied after polishing?

13. What general statement can be made about the relationship between hardness and tensile strength in steels?

14. Can the lattice structure be used to predict properties? How?

15. How could one build a crude hardness tester using a hand press, a spring, and an industrial diamond?

16. Describe the problems in measuring the hardness of very thin, extremely hard, and very soft metals.

17. For ferrous metals, what alloying element is the most important in determining properties?

18. When a stress–strain test is performed on a steel specimen with 0.32% carbon, at what value of stress and strain does necking of the specimen first occur?

19. The above-mentioned bar is broken and $L = 2.0$ in. and $L_f = 2.3$ in. What is the percentage elongation and what is its significance?

20. A piece of steel is hardened by heat treating. What is the effect on tensile strength and wear resistance?

21. Sketch a golf club driver and identify the materials used in its manufacture. Classify the materials.

22. The harder a steel is the more difficult it is to weld. What is the relationship between weldability and percentage carbon; between weldability and tensile strength?

23. Make a two-dimensional sketch showing the growth of a face-centered lattice structure into grains.

24. What is the difference, if any, between ductility and elasticity?

25. Find the equivalent number of atoms in a single body-centered lattice, taking into account that some of them are shared.

26. Sketch a general time–temperature curve for steel, magnesium, and brass.

27. Using Figure 2.6, determine the composition of the material at ℓ_2, ℓ_3, and s_3.

28. A common alloy used in the past to make nickels had what composition at the liquidus and solidus lines?

29. Give the composition at the liquidus and solidus lines for an alloy of 40% nickel and 60% copper.

CASE STUDY
"THE UNKNOWN MATERIAL"

The Wilsan Machinery Company was asked to repair a foreign-made ski lift that had failed. A 1.13-in. O.D. steel support pin failed by necking down in tension, and subsequent tests showed the tensile strength of the material used in the original lift was about 50,000 psi. A discussion of the failure is held and you are asked to make general recommendations for a replacement part.

The load that can be expected on the pin is 41,000 psi, but it is pragmatic to have a built-in factor of safety of 2.0 (2 × 41,000).

Certain things about the pin are known and recommendations can be made. When steel is requisitioned for the pin, the inspection department checks it by making a standard steel tensile specimen of the material available. It is pulled by a local laboratory, and the results are as follows:

Force		$L_1 - L$
lb	N	in.
32,000	142,336	0.002
62,000	275,776	0.004
91,000	404,768	0.006
105,000	467,040	0.008

From the analysis an estimate of the percentage of carbon in the steel and its suitability for the pin can be made. What conclusions can be made about the carbon content of the failed pin and the material chosen to replace it? What comments can be made about other physical characteristics when the original pin and the proposed pin are compared?

CHAPTER 3
PRODUCTION OF FERROUS METALS

Ferrous metals cost less per pound than any other metal, primarily because of the low cost of ore and the methods of reducing the raw materials. Although the United States is still the leading producer of iron and steel, it no longer dominates the market as it once did. Ferrous metal parts are manufactured from ferrous materials that are classified according to form.

1. *Wrought metal* is rolled or formed to shape after being cast into ingots.
2. *Cast metal* is foundry cast from remelted scrap, pig iron, ingots, or a combination of these.
3. *Powdered metal* is pressed into useable parts by high pressure from finely powdered metal made by a special process.
4. *Extruded metal* is extruded or squeezed at elevated temperature through a die of specific shape.

The worldwide production of ferrous metals has not increased significantly in recent years, primarily because of the effort to hold down the weight of manufactured goods. The production of ferrous castings worldwide and in the United States has declined, but still the tonnage of ferrous materials exceeds by many orders of magnitude the production of all other metals.

PRODUCTION OF PIG IRON

The principal raw material for all ferrous products is *pig iron* or *direct iron,* the product of the *blast furnace*. Pig iron is obtained by smelting iron ore with coke and limestone, the final analysis depending primarily on the kind of ore used.

The principal iron ores used in the production of pig iron are listed in Table 2.3. The metallic contents listed in the table are those of pure ores. Because most ore contains impurities, the actual metallic content is less. Hematite (Fe_2O_3) is the most important iron ore used in the United States. Vast quantities of *iron pyrite* (FeS_2) are available but are not used because of the sulfur content, which must be eliminated by an additional roasting process.

Blast Furnace

The blast furnace is the most widely used method of iron production in the world today. Figure 3.1 is a schematic of the blast furnace. Iron is produced mostly for use in making steel. Iron ore, coke, and limestone are brought to the top of the 25-ft diameter furnace with a skip hoist and then dumped into the *double-bell* hopper. The ore is heated to high temperatures to produce iron. To produce 1000 tons of iron the plant uses 2000 tons of ore, 800 tons of coke, 500 tons of limestone, and 4000 tons of hot air.

The use of hot air instead of cold air enables the coke to burn more effectively, and therefore less coke is used. The hot air is heated in tall cylindrical towers or stoves to about 1000°F. This hot air furthers the production of carbon monoxide, which reacts with ore to produce iron and carbon dioxide. The hot air and ash are cleaned to be used again in the furnace and in-plant fuel. Limestone is added to the furnace to combine with impurities in the ore that rise to the top of the molten ore because they are lighter. The impurities or "*slag*" are withdrawn off the top of the molten mass. The slag is either recycled within the mill or sold as railroad ballast, as landfill, or for road and highway use. More slag is available than is needed, and a majority of it gets stockpiled.

Many different grades of iron can be produced from the blast furnace, depending on what raw materials go into the furnace. It is the manufacturer's choice what grade iron will be produced, depending on economic restrictions and the type of steel that will be produced. Various grades are shown in Table 3.1.

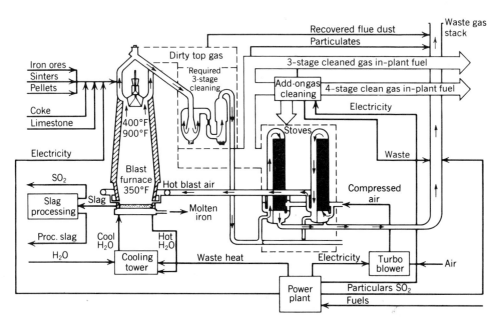

Figure 3.1
Simplified schematic of the blast furnace process.

Table 3.1 Classification of Pig Iron

Grade of Iron	Chemical Content (%)			
	Silicon	Sulfur	Phosphorus	Manganese
No. 1 Foundry	2.5–3.0	<0.035	0.05–1.0	<1.0
No. 2 Foundry	2.0–2.5	<0.045	0.05–1.0	<1.0
No. 3 Foundry	1.5–2.0	<0.055	0.05–1.0	<1.0
Malleable	0.75–1.5	<0.050	<0.2	<1.0
Bessemer	1.0–2.0	<0.050	<0.1	<1.0
Basic	<1.0	<0.050	<1.0	<1.0

Direct Reduction

The direct reduction process employs either a solid or gaseous reducing agent reacting with iron ore to produce a spongelike iron. Although the direct reduction process of making iron was known before the invention of the blast furnace, less than 2% of the world's pig iron is made by this method. The processes consist of crushing the iron ore and reacting it, ususally at an elevated temperature, with the reducing agent. This reducing agent may be coke, natural gas, fuel oil, carbon monoxide, hydrogen, or graphite. The product of the process is a spongy, granular, clinkerlike material that either is used for making metal powders or else is converted in an electric arc furnace to pure iron or steel.

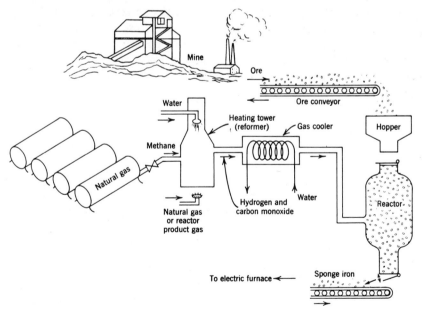

Figure 3.2
Schematic diagram of direct reduction process for producing sponge iron from iron ore.

A Mexican plant uses hydrogen and carbon monoxide, made from the natural gas, methane, to react with magnetite ore, to produce 500 tons (450 Mg) per day. The cost for producing sponge iron is said to be less per pound than for a blast furnace if there is a cheap source of methane. Figure 3.2 is a schematic diagram of the process. The reaction in which 100 tons (90 Mg) of iron ore are reduced to about 63 tons (60 Mg) of sponge iron takes from 10 to 14 h. Approximately 25,000 ft^3 (700 m^3) of methane are required per ton (0.9 Mg) of sponge iron. Because this process is effective in reducing oxygen and sulfur, relatively low-grade ores may be employed.

FURNACES FOR CONVERTING PIG IRON

Pig iron is either cast into permanent iron molds or else is transferred molten in *hot-ladle cars* to a furnace, where it may be refined for manufacturing wrought iron, steel, cast iron, and malleable or ductile iron. The capacity of hot-metal cars may exceed 100 tons (90 Mg), and more than 90% of all pig iron is transported to steel-making furnaces by this method.

The primary difference in ferrous metals is in the amount of carbon that they contain. This difference is shown in Figure 3.3. Although steel may contain as much as 2.0% carbon, practical applications limit the carbon content to about 1.4%. Furnaces used for converting pig iron and remelting ferrous metals are described in Table 3.2.

Three furnaces, basic oxygen, electric arc, and open hearth, account for 60, 25, and 15%, respectively, of steel produced in the United States. These kinds of furnaces are not used interchangeably, because each furnace system requires different materials and energy sources. Steel making by all processes refines the pig iron and aims at precise chemical control.

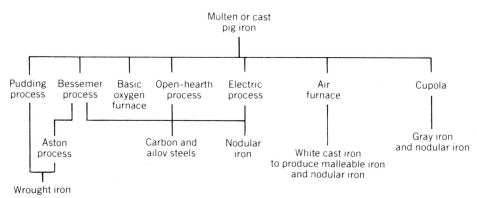

Figure 3.3
Principal processes for remelting or refining pig iron.

Table 3.2 **Furnaces for Ferrous Metals**

Type of Furnace	Primary Fuel	Predominant Metal Charge	Special Atmosphere Available	Product
Air or reverberatory	Pulverized coal, oil	Molten or solid pig iron, scrap		Gray cast iron, white cast iron
Basic oxygen	Oxygen	Molten pig iron and scrap		Steel
Converter	Air	Molten pig iron or molten cupola iron		Raw material for wrought iron and steel
Crucible	Gas, coke, oil	Select scrap		Small quantities of steel and cast irons
Cupola	Coke	Solid pig iron and scrap		Gray cast iron, nodular iron
Electric furnace	Electricity	Scrap	Vacuum or inert gas	Steel, gray iron
Induction	Electricity	Select scrap	Vacuum or inert gas	Steel
Open-hearth furnace	Natural gas, coke oven gas, pulverized coal, oil	Molten pig iron		Steel

Basic Oxygen Furnace

The basic oxygen furnace (***BOF***) process uses as its principal raw material pig iron (65–80%) from a blast furnace. In addition, scrap and lime are added. As the name implies, heat is generated through the use of oxygen. A bottom-blown converter using air instead of oxygen is called a *Bessemer* converter, but it is used only rarely in the United States.

Scrap, about 30% of the total charge, is loaded into a basic refractory lined vessel as shown in Figure 3.4. Hot metal is poured into the mouth of the tilted vessel. A water-cooled ***oxygen-carrying lance*** is lowered 4 to 8 ft (1.2–2.4 m) above the bath in the vertical vessel. With oxygen blowing over the surface, ignition starts immediately and the temperature climbs close to iron's boiling point of about 3000°F (1650°C). Carbon, manganese, and silicon are oxidized. Lime and fluorspar are added to collect impurities such as phosphorus and sulfur in slag form. When the refining process is over the vessel is tilted for tapping. The tap-to-tap time to produce 300 tons (270 Mg) is approximately 45 min. The oxygen needed to produce 1 ton (0.9 Mg) of steel is about 1600 ft^3 (45 m^3).

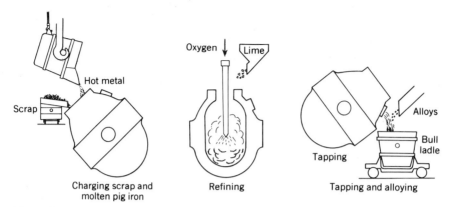

Figure 3.4
The basic oxygen furnace (BOF).

A BOF plant must be designed to have the gases conducted to air treatment facilities. Experienced BOF operators identify the refining progress by a decrease in the flame and change in the sound level and flame color. When a batch of steel is complete, the oxygen is shut off and the lance is retracted through the hood. The furnace is tilted to remove the slag. Following testing for acceptable temperature and carbon, the furnace is tilted to the opposite side for pouring to the ladle car.

Electric Furnace

The *electric furnace* (Figure 3.5) is charged with selected steel scrap rather than molten pig iron, although at one time molten metal was used. By close control of the charge and by adding alloying materials, ingots and castings of stainless steel, heat-resistant steel, tool steel, and many general-purpose alloy steels are poured from the electric furnace. In addition to steel, high-test gray iron is produced to an increasing extent because electric furnaces do not pollute the atmosphere as much as do most other furnaces.

In the past there have been two types of electric furnaces: (1) the **indirect arc** furnace, in which the electrodes are above the metal, the metal being heated by radiation; and (2) the **direct arc** furnace in which current passes from an electrode over the charge to the metal or molten bath and then back to an electrode. The direct arc furnaces are the only ones today that produce steel economically.

The direct arc furnace may have the hearth either **basic** lined or **acidic** lined. The acid-lined furnace, with a hearth of ground gannister and side walls of silica brick, is used in a limited way to produce low-carbon, low-alloy steels provided the scrap is low in phosphorus and sulfur. The basic-lined furnace, with a hearth of magnesite and side walls of magnesite and alumina brick, is used to produce any grade of steel or steel alloy. The basic furnace can control phosphorus, reduce the sulfur, and maintain close control of temperature and composition. The molten metal can be sampled to determine the precise chemical composition of the melt, and before the metal is poured into ingots, the composition may be adjusted.

In an electric furnace recycled scrap iron and steel are loaded through the top after the removable furnace roof is swung to one side. The roof is designed to allow three graphite electrodes to sit above the scrap heap. The three-phase current arcs from one electrode to the charge and then arcs back to another electrode.

Some furnaces average 300 tons (270 Mg) per heat. For a 125-ton (115 Mg) tap-to-tap time, 3 h are required and 50,000 kWh of power. A refining step in the process is to inject high-purity oxygen, which reduces the tap-to-tap time.

Graphite electrodes up to 30 in. (760 mm) in diameter and over 80 ft (24 m) long may be used in a large furnace. All furnaces operate at approximately 40 V and at currents that may exceed 12,000 A.

Open-Hearth Furnace

The *open-hearth furnace,* once the most popular process for making steels, now produces 40 million tons (35 Tg) annually. Figure 3.6 is typical; such furnaces hold from 11 to 600 tons (9.9–540 Mg) of metal in a shallow pool that is heated by action of a gas, tar, or oil flame passing over the charge. The furnace is said to be reverberatory, because the low roof of the furnace reflects the heat onto the long, shallow hearth. It is regenerative, because the chambers on either side of the furnace are capable of being heated by combustion gases, which in turn allow the air and fuel entering the furnace to be increased

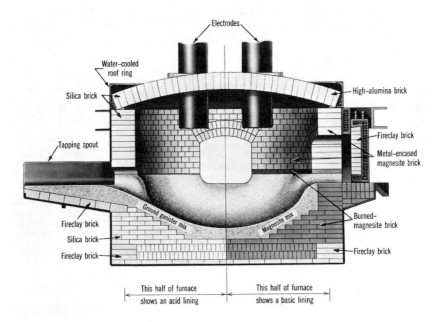

Figure 3.5
Cutaway drawing of electric furnace showing both the acidic- and basic-type linings.

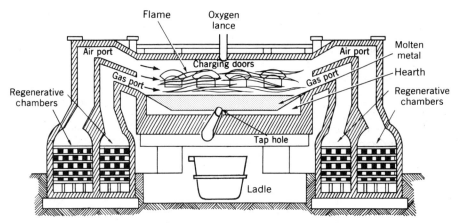

Figure 3.6
Sectional view of open-hearth furnace.

in temperature, thus ensuring an increased combustion efficiency and temperature. The chambers, left and right, are alternately heated so that as one set is being used to raise the temperature of air and fuel, the checkerwork of bricks in the opposite chamber is being heated.

The open-hearth furnace may have a hearth of basic or acidic material, although in practice over 90% are basic open hearths. In the basic unit, which can remove phosphorus, sulfur, silicon, manganese, and carbon, the hearth is lined with magnesite. The acidic open hearth, which can remove only silicon, manganese, and carbon, has a hearth of acid brick or sand whose principal ingredient is silica.

The open-hearth charge can consist of all molten pig iron, all solid steel scrap, or a combination of solid and molten pig iron and steel scrap. Most open hearths use steel scrap, solid pig iron, and molten pig iron from hot-ladle cars. The steel scrap and any solid pig iron are placed in the hearth and melted. Molten iron is added from 2 to 3 h after the scrap has melted, and for the next 6 to 7 h the charge is "boiled" and fluxing agents are added. About 10 h after the initial charge the furnace is ready to be tapped.

The time may be reduced 25% by inserting an oxygen lance through the roof of the furnace into the combustion area after the molten iron is introduced. When using oxygen the fuel cost may be reduced by 30%, but this is offset by the cost of supplying up to 500 ft^3 (14.0 m^3) of oxygen per ton (0.9 Mg) of iron produced. Because of increased chemical activity caused by the addition of oxygen, the charge must be carefully controlled and frequent samples of the molten metal are tested to determine the extent to which additions must be made.

Cupola

Iron castings can be made by remelting scrap along with pig iron in a furnace called a *cupola*. The cupola is simple in construction and economical to operate; it will melt iron continuously with a minimum of maintenance. However, because metal is melted in contact with the fuel, some elements are picked up while others are lost. This affects the final analysis of the metal and necessitates close regulation of the cupola. A modern, water-cooled, unlined cupola is shown in Figure 3.7.

The openings for introducing air to the coke bed are known as *tuyères*. Small windows covered with mica are located at each tuyère, so that conditions in the cupola can be inspected. Surrounding the cupola at the tuyères is an insulted *windbox* or jacket for the air supply. The air blast, furnished by a positive displacement or centrifugal-type blower, enters the side of the wind jacket.

The opening through which the metal flows to the spout is called the *tap hole*. The overflow or slag spout is slightly below the tuyères.

In operation a coke bed is ignited and alternate charges of coke and iron are made in a ratio of 1 part coke to 8 or 10 parts iron, measured by weight. A fluxing material, usually limestone ($CaCO_3$), fluorspar (CaF_2), or soda ash (Na_2CO_3), is added to protect the iron from oxidation and render the slag more fluid. For limestone about 75 lb (35 kg) are added for each ton (0.9 Mg) of iron. The amount of air required to melt a ton

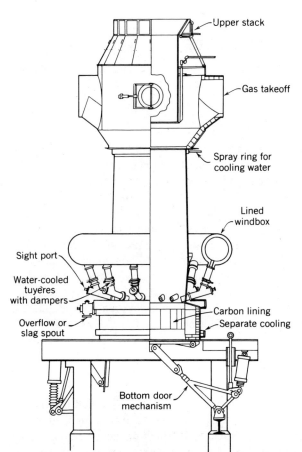

Figure 3.7
Sectional drawing of a cupola.

(0.9 Mg) of iron depends on the quality of coke and the coke–iron ratio. Theoretically, 113 ft^3 (3.19 m^3) of air at 14.7 psi (100 kPa) and 60°F (15.5°C) is required to melt 1 lb (0.5 kg) of carbon.

Combustion may be appreciably improved by preheating the air as in regenerative-type melting furnaces. Such a furnace is called a *hot-blast* cupola. Recently a new method of introducing the charge into the furnace has proved profitable. Pellets of iron ore, coke dust, lime, and silica are cold bonded and fed into the cupola.

Pollution control costs are significant in cupola operation, causing some foundries to use horizontal, oil-fired, *rotary furnaces* (rotary cupolas) for making gray iron. Furnaces of this type use compressed air to atomize the fuel and can melt up to 25 tons (22.7 Mg) per day in a temperature environment of up to 2800°F (1540°C). The units have relatively low initial and maintenance costs and can utilize scrap as small in size as machine borings. A pollution control device utilizing supersonic air flow technology, called the Hydro-Sonic Cyclone and developed by Lone Star Steel Company, is reversing the decreased use of cupolas.

REFINING FURNACES AND VESSELS

Raw steel is sometimes enhanced in furnaces, hearths, ladles, and vessels for certain applications. Small but growing tonnages are involved.

AOD Process

AOD is an acronym for argon oxygen decarburization. This process, patented by the Linde Division of Union Carbide, utilizes a magnesite chrome refractory lining in a converter-shaped vessel. From a proprietary-type vessel referred to as a ***teapot***, metal that has been melted in an electric furnace is transferred to the converter where argon, oxygen, and nitrogen are added by introducing the gases through tuyères at the bottom of the converter. About 80% of the stainless steel manufactured in the western world now employs this process, which has reduced the cost of manufacture by enabling utilization of less expensive feed metals while producing a steel of greater toughness, strength, and ductility as well as a material that can be more easily machined. Instead of using the oxygen lance as in the BOF process, the gases are introduced to decarburize and reduce sulfur and dissolved gas in the steel. By diluting oxygen with argon, less oxidation takes place. The time for sulfur removal is about one-fifteenth as long as in an electric furnace. When the steel treatment is complete, it is transferred to a pouring ladle.

Induction

The electric *induction* furnaces (Figure 3.8) use an induced current to melt the charge. The coreless, induction type is powered by a high-frequency current supplied to the primary water-cooled coil that surrounds the crucible. The high-frequency current, about 1000 cycles per second (cps), is supplied by a motor generator set or a mercury arc

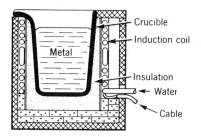

Figure 3.8
Induction furnace.

frequency system. The crucible is charged with a solid piece of metal, scraps, or chips, and a heavy secondary current is induced. The resistance of this induced current in the charge in 50 to 90 min will melt the charge in even large crucibles containing up to 4 tons (3.6 Mg) of steel.

Induction furnaces, available in crucible sizes holding from a few pounds to 4 tons (3.6 Mg), are relatively low in cost, almost noise free, and generate very little heat. The temperature required is no higher than that required for melting the charge, so scrap alloys can be remelted without "burning out" the valuable alloying material. For these reasons they are often found in experimental laboratories or foundries. In electric arc furnaces the high temperature of the arc may refine the metal to the disadvantage of the foundry.

Crucible

The ***crucible*** process is the oldest process for making steel castings, but it is used predominantly in nonferrous foundries. Crucibles (Figure 3.9) are usually made of a mixture of graphite and clay. They are quite fragile when cold and must be handled with care, but they posses considerable strength when heated. Crucibles are heated with coke, oil, or natural gas and are handled with special fitting tongs to prevent damage. Wrought iron, washed metal, steel scrap, charcoal, and ferroalloys constitute the raw materials for steel manufacture by this process. These materials are placed in crucibles having a capacity of about 100 lb (45 kg) and are melted in a regenerative furnace.

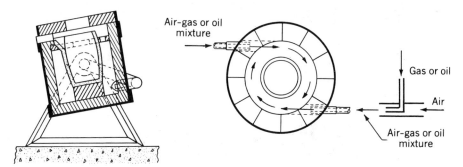

Figure 3.9
Oil- or gas-fired tilting crucible furnace.

Melting in a Vacuum and Special Atmospheres

Molten metals tend to absorb gases because of moisture in the furnace, ladle, and atmosphere or because of entrapped hydrocarbons in the charge. The oxidation rate of metals exposed to the atmosphere increases as the temperature increases. The lowest melting temperature consistent with sufficient fluidity will minimize gas absorption and oxidation, which are both detrimental.

For some metals a slag coating or *dross* is allowed to accumulate over the molten metal to protect it from excessive oxidation. Slag inducements are often added. In aluminum this dross is troublesome during pouring, and care must be exercised to prevent its entrapment in the casting.

If a *vacuum* or *special atmosphere* is to be employed, the furnace and mold must be enclosed in a chamber. The furnace, an induction or electric arc type, is seldom larger than 4 tons (3.6 Mg) in capacity. The steam ejector-type vacuum pumps can reduce the pressure to the equivalent of that supporting 0.000394 in. (0.01 mm) of mercury, as compared to atmospheric conditions of about 29.92 in. (760 mm) of mercury. Vacuum melting and pouring improve the tensile strength and fatigue life of most metals. Some metals such as titanium must be melted in a vacuum.

A high-vacuum, consumable-electrode furnace (Figure 3.10), in which a steel ingot acts as the electrode, is used to remelt and purify steel. The arc between the electrode and the furnace melts the ingot in a progressive manner, and solidification takes place

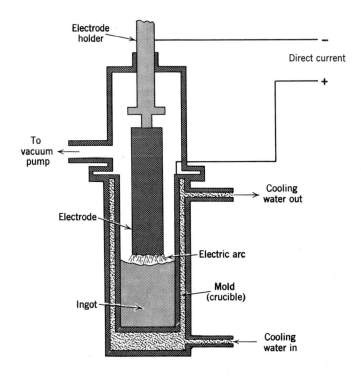

Figure 3.10
Consumable-electrode furnace.

almost as rapidly as the material is melted. Furnaces of this type are capable of making ingots weighing up to 20 tons (18 Mg).

Magnesium is melted and cast under an atmosphere of sulfur dioxide. Castings that must be absolutely free of absorbed gases, pinholes, and entrapped slag are melted and cast under a vacuum.

Some metals, particularly aluminum and magnesium, are often *degassed* after melting by passing an inert gas such as argon through the molten metal. Steel may be degassed with carbon monoxide.

ENERGY REQUIRED FOR MELTING

The energy required for melting a ton of any metal is dependent on the specific heat of the material. **Specific heat** is defined as the ratio of the heat capacity of a material to that of water. Knowing that it takes 1 **BTU** of heat energy to raise 1 lb of water 1°F, aluminum with a specific heat of 0.21 would require roughly only one-fourth as much energy to raise the temperature of 1 lb a single degree. Unfortunately, specific heat of a metal varies with temperature, particularly around its melting point. The specific heat of iron varies from about 0.1228 at 500°F to 0.1666 at 2700°F. Figure 3.11 illustrates the

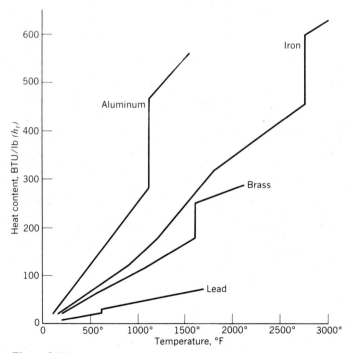

Figure 3.11
Specific heat as a function of temperature.

approximate heat content of several metals as a function of temperature. This graph gives some idea of the melting characteristics of metals and why lead is so easily melted in comparison to brass or steel. However, the metals with higher melting temperatures have much lower efficiencies in employing BTUs; that is, the furnace efficiencies are lower. The energy required to raise the temperature of a metal can be expressed:

$$H_t = f \times h_t \times W$$

where

H_t = energy required to reach a temperature of t (BTU/lb)

f = furnace efficiency (%)

h_t = heat content of material (BTU/lb) at a temperature of t

W = weight of material

Referring to Figure 3.11, it is possible to estimate the energy required to melt 1 lb of metal if you know the furnace efficiency. If one assumes it is 50%, then to melt 1 ton of aluminum requires approximately 560 BTU/lb $\times$ 2000 lb/ton = 11.2 $\times$ 10^5 BTU/ton. If this material is melted in 1 h, the kilowatt hours (kWh) per ton can be calculated.

$$\text{kWh} = (11.2 \times 10^5 \text{ BTU/ton}) (2.928 \times 10^{-4} \text{ kW/BTU})$$
$$= 327.9 \text{ kWh/ton}$$

The 2.928 $\times$ 10^{-4} term is a constant for converting British thermal units to kilowatts. Gaseous and liquified fuels are sold by their BTU content, and electric furnaces are rated in kilowatts.

FERROUS METALS

Following the ore-refining processes, ferrous materials are cast into ingots or other shapes depending on the composition of the metal and the anticipated use. As noted in Chapter 2, the ferrous metals differ in physical properties according to their carbon content.

Wrought Iron

Wrought iron is a ferrous metal containing less than 0.1% carbon with 1% to 3% finely divided slag distributed uniformly throughout the metal. It has been produced for many centuries by a variety of processes, the two in use today being the puddling process and the Aston process. The latter accounts for the greatest tonnage of wrought iron.

In the *puddling process* pig iron and iron scrap are melted in a small, reverberatory, 500-lb (230 kg) capacity *puddling furnace* fired with coal, oil, or gas. Most of the elements are removed by oxidation, because they come in contact with the basic refractory lining of the furnace. This process uses mechanical puddling furnaces of greatly increased capacities and thus eliminates much of the hand labor of stirring and gathering the metal into large balls. Freed from impurities, the product is removed from the furnace as a pasty mass of iron and slag, then mechanically worked both to squeeze out the slag and to form it into some commercial shape.

In the *Aston process* the pig iron is melted in a cupola and refined in a *Bessemer* converter. The blown metal is then poured into a ladle (*shotting*) containing a required amount of slag previously prepared in a small open-hearth furnace. Because the slag is at a lower temperature, the mass solidifies rapidly, liberating the dissolved gases with sufficient force to blow the metal into small pieces. These pieces settle to the bottom of the ladle and weld together as sponge iron. Each ball of iron collected in the ladle weights 3 to 4 tons (2.7–3.6 Mg), and the rate of production is about one ball every 5 min. A press squeezes out the surplus slag and welds the mass of slag-coated particles of iron into a solid bloom that can be rolled into various shapes.

Wrought iron produced by this method usually has a carbon content less than 0.03%, silicon about 0.13%, sulfur less than 0.02%, phosphorus about 0.18%, and manganese less than 0.1%.

A photomicrograph of a piece of polished and etched wrought iron is shown in Figure 3.12. The particles of slag, which constitute about 2.5% by weight, are seen as dark streaks running through the metal. The direction of rolling is clearly visible.

This metal is used principally in the production of pipe and other products subjected to deterioration by rusting such as those used around shipyards, railroads, farms, and oil companies. Advantages other than its resistance to corrosion include ease of welding, high ductility, and ability to hold protective coatings.

Steel

Steel is a crystalline alloy of iron, carbon, and several other elements; it hardens when quenched above its critical temperature. It contains no slag and may be cast, rolled, or forged. Carbon is an important constituent because of its ability to increase the hardness and strength of the steel. More tons of steel are used than all other metals combined. Although steel may be cast into molds to conform to a definite and complex shape and size, it is most often cast into ingots for use in making pipe, bar stock, sheet steel, or structural shapes.

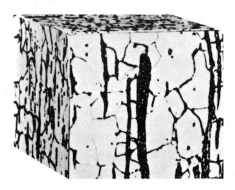

Figure 3.12
Photomicrograph of wrought iron showing slag distribution resulting from rolling. Magnification ×200.

Steel is classified according to the alloying elements it contains. Carbon is the most important element, and for this reason all steels are classified according to carbon content. Plain carbon steels contain primarily iron and carbon, and they are classified as 10XX steels; the first two digits refer to *plain carbon* steel. The third and fourth digits refer to the carbon content in hundredths of a percent. Thus a 1035 steel is a plain carbon steel with 0.35% carbon. There are varying amounts of other materials in carbon steel, but their content is so small that they do not affect physical properties.

As noted in Chapter 2, the maximum hardness of a steel increases with carbon content. If the hardness is known, the tensile strength can be estimated as follows:

$$TS = 500 \times Bhn$$

where

$$TS = \text{tensile strength (psi)}$$
$$Bhn = \text{Brinell hardness number}$$

This formula alone can assist in the choice of a steel an engineer might make for a given product.

Alloy steels have been classified by the Society of Automotive Engineers (SAE) and by the American Iron and Steel Institute (AISI). Some of the designations accepted by them as standard are shown in Table 3.3. Often as many as five or more alloying elements may be present, and it is impossible to describe the alloy correctly by a simple numbering system. Steels may be more broadly classified as follows:

A. Carbon steel
  1. Low carbon—less than 0.30%
  2. Medium carbon—0.30% to 0.70%
  3. High carbon—0.70 to 2.0 (nominally the upper limit is 1.40%)
B. Alloy steel
  1. Low alloys—special alloying elements totaling less than 8.0%
  2. High alloys—special alloying elements totaling over 8.0%

The low-carbon steels are used for wire, structural shapes, and screw machine parts such as screws, nuts, and bolts. Medium-carbon steels are used for rails, axles, gears, and parts requiring high strength and moderate to great hardness. High-carbon steels find use in cutting tools such as knives, drills, taps, and for abrasion-resisting properties.

The alloy steels, which account for only about 15% of the steel produced, are selected for many uses because they have characteristics that are superior to those of plain carbon steel. Although every alloy steel does not contain each characteristic, they include

1. Improvement in ductility without a lowering of tensile strength.
2. Ability to be hardened by quenching in oil or air instead of water, thus lowering the chance of cracking or warping.
3. Ability to retain physical properties at extremes of temperature.
4. Lower susceptibility to corrosion and wear, depending on the alloy.
5. Promotion of desirable metallurgical properties such as fine grain size.

Table 3.3 **Classification of Steels**

Classification	Number	Range of Numbers
A. Carbon Steels		
Carbon steel SAE–AISI	1XXX	
Plain carbon	10XX	1006–1095
Free machining (resulfurized)	11XX	1108–1151
Resulfurized, rephosphorized	12XX	1211–1214
B. Alloy Steels		
Manganese (1.5%–2.0%)	13XX	1320–1340
Molybdenum	4XXX	
C–Mo (0.25% Mo)	40XX	4024–4068
CR–Mo (0.70% Cr; 0.15% Mo)	41XX	4130–4150
Ni–Cr–Mo (1.8% Ni; 0.65% CR)	43XX	4317–4340
Ni–Mo (1.75% Ni)	46XX	4608–4640
Ni–Cr (0.45% Ni, 0.2% Mo)	47XX	
Ni–Mo (3.5% Ni, 0.25% Mo)	48XX	4812–4820
Chromium	5XXX	
0.5% Cr	50XX	
1.0% Cr	51XX	5120–5152
1.5% Cr	52XXX	52095–52101
Corrosion-heat resistant	514XX	(AISI 400 series)
Chromium–vanadium	6XXX	
1% Cr, 0.12% V	61XX	6120–6152
Silicon Manganese		
0.85% Mn, 2% Si	92XX	9255–9262
Triple-alloy steels		
0.55% Ni, 0.50% Cr, 0.20% Mo	86XX	8615–8660
0.55% Ni, 0.50% Cr, 0.25% Mo	87XX	8720–8750
3.25% Ni, 1.20% Cr, 0.12% Mo	93XX	9310–9317
0.45% Ni, 0.40% Cr, 0.12% Mo	94XX	9437–9445
0.45% Ni, 0.15% Cr, 0.20% Mo	97XX	9747–9763
1.00% Ni, 0.80% Cr, 0.25% Mo	98XX	9840–9850
Boron (~0.005% Mn)	XXBXX	

Boron is denoted by addition of B. Boron–Vanadium is denoted by addition of BV. Examples: 14BXX, 50BXX, 80BXX, 43BV14. The letters appearing before the number indicate the following: A, alloy-basic open hearth; B, carbon-acid Bessemer; C, carbon-basic open hearth; D, carbon-acid open hearth; E, electric furnace.

Stainless and heat-resisting steels:

  2XX Chromium–nickel–manganese types

  3XX Chromium–nickel types

  4XX Straight chromium types

  5XX Low chromium types

All stainless steels are produced in the electric furnace.

Stainless Steels

Three types of alloyed steels known as stainless steels are listed here:

1. *Ferritic*. These steels, designated as 405, 430, 430F, and 446, cannot be hardened by heat treatment because their ratio of carbon to chromium is low, but they have good resistance to corrosion.

2. *Martensitic*. These steels, some of which are designated as 410, 414, 416, 431, and 440A, B, and C, are hardenable because their ratio of carbon to chromium is high, and they have high strength and are moderately corrosion resistant. They can be employed as forgings and can be hot worked. They are used for machine parts and cutlery.

3. *Austenitic*. **Austenitic steels,** some of which are 301, 302, 303,, 304, 310, 310S, and 384 (wire), cannot be hardened by heat treatment and are nonmagnetic. They are highly corrosion resistant and can be cold formed. They harden and become stronger when cold worked and can be welded. *Type 303* is the easiest of all stainless steels to machine and is widely used for machine parts and fasteners.

STEEL INGOTS AND CONTINUOUS CASTINGS

To manufacture wire, bar stock, sheet, plate, pipe, or structural shapes, a hot ingot of steel is rolled, pressed, or stretched into a predetermined form. The ingots are poured into molds as shown in Figure 3.13.

The mold may be rectangular, square, or round in cross section, and final casting varying in size from a few hundred pounds to 355 tons (327 Mg). The kind of metal cast and the product are the factors that determine the ingot size. Both rectangular and square-section ingots have rounded corners, and the sides are corrugated. Rounding the corners reduces the tendency for columnar grains to meet and form a plane of weakness. Cooling is accelerated by corrugating the sides, a process that reduces the size of the columnar grains.

The two types of ingot molds in Figure 3.13 are used for top pouring. The big-end-down type shown in Figure 3.13A is easy to strip from the ingot, but the loss in metal is high as a result of the shrinkage cavity or *pipe* that is formed during the cooling operation. This loss is lower when the big-end-up type shown in Figure 3.13B is used. When an ingot is poured the solidification is progressive, starting at the mold surface and progressing toward the center and from the bottom upward. During this period there is considerable shrinkage of the metal. As layer after layer solidifies, the volume of metal

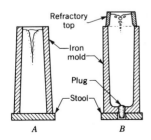

Figure 3.13
Types of Ingot molds. *A,* Big end down. *B,* Big end up.

decreases, resulting in the formation of a pipe when solidification is complete. The rate of cooling is an important factor in the production of a sound ingot.

Ingots made in big-end-up molds have a large volume of hot metal available at the top of the mold during the cooling operation and when solidified show little loss of metal resulting from piping. Losses from pipe formation in ingots can be reduced either by adding metal during cooling or by using refractory risers. Metal in the riser remains molten until the ingot has solidified, and during the solidification period supplies the ingot with needed metal to compensate for shrinkage.

Several types of ingot structures are obtained by controlling or eliminating a gas evolution in the metal during solidification. **Killed steel** has been deoxidized and it evolves no gas during solidification. The top surface of such ingots solidifies immediately as do the walls, and because of the shrinkage of the metal upon solidification, a larger cavity or pipe is formed within the ingot. The process of producing killed steel is complex and depends on starting with a higher carbon steel than finally desired; then, when the carbon is reduced to the exact amount, the steel is deoxidized by furnace or ladle additions of high-silicon pig iron or an alloy high in silicon. All steels having over 0.30% carbon are killed. Such ingots have a minimum of segregation, good structure, and a large cavity in the center.

Another ingot structure is known as *rimmed steel,* which is either not or only lightly deoxidized in the furnace or ladle. This type steel is characterized by a semiboiling action in the ingot after it has been poured resulting from rapid evolution of carbon monoxide gas during the solidification period. This causes the formation of a honeycomb structure that if controlled compensates for most shrinkage loss. These small blow holes do not constitute a defect if they have not had contact with the outside atmosphere and are closed by pressure welding in the hot-working processes. Rimmed-steel ingots have a good surface and there is little or no opportunity for pipes to form. Semikilled and other ingot structures are obtained by controlling the formation of gas during solidification.

Longitudinal and cross sections of a medium-carbon, killed-steel ingot are shown in Figure 3.14. The coarse dendritic crystalline structure, clearly indicated in the cross section, is eliminated by the effect of hot working. The impurities in ingots tend to segregate in the shrink head during the process of solidification. Cutting off the end of the ingot, either before the rolling starts or after, largely eliminates this defect but decreases overall efficiency.

In addition to **teeming** hot metal in ingots, a process that converts molten steel to a slab continuously is shown by Figure 3.15. Solidification begins as the steel cools in passing through the mold. This *continuous-slab* method can convert 300 tons (270 Mg) of molten metal to solid slabs in 45 min instead of 12 h for ingot processing.

Steel not processed from ingots into plates, sheets, bar stock, and wire is often remelted and cast into molds.

The microstructure of a medium-carbon cast steel is shown in Figure 3.16. The light areas are ferrite and the dark areas pearlite. The grain structure of most cast steels is large because of the high casting temperature of the metal combined with relatively slow cooling.

Over 50% of all steel castings are of medium-carbon steel. Such steel castings are used in the transportation industry, the industrial machinery field, and the construction

52 PRODUCTION OF FERROUS METALS

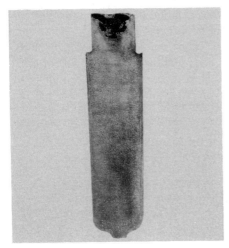 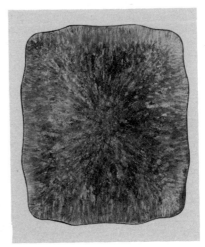

Figure 3.14
Ingot macrographs of longitudinal and cross sections of a medium-carbon killed-steel ingot.

field. They have ductility and good tensile strength in a normalized condition ranging from 60,000 to 100,000 psi (400–690 MPa).

The chemical composition of medium-carbon castings is approximately as follows: carbon, 0.21% to 0.46%; manganese, 0.55% to 0.73%; silicon, 0.28% to 0.45%; phosphorus and sulfur together, less than 0.1%; and ferrite the remainder.

Cast Iron

Cast iron is a general term applied to a wide range of iron–carbon–silicon alloys in combination with smaller percentages of several other elements. It is an iron containing so much carbon or its equivalent that it is not malleable. Cast iron has a wide range of properties, because small percentage variations of its elements may cause considerable change. Cast iron should not be thought of as a metal containing a single element but rather as one with at least six elements in its composition. All cast irons contain iron, carbon, silicon, manganese, phosphorus, and sulfur. Alloy cast iron has still other elements that have important effects on the physical properties.

Gray iron is ordinary commercial iron with a grayish colored fracture. The gray color is caused by flake graphite, the principal form of carbon present. Gray iron is easily machined and has a high compression strength. The tensile strength varies from 20,000 to 60,000 psi (140–415 MPa), but the ductility is usually low. The percentages of the several elements may vary considerably but are usually within the following limits: carbon, 3.00% to 3.50%; silicon, 1.00% to 2.75%; manganese, 0.40% to 1.00%; phosphorus, 0.15% to 1.00%; sulfur, 0.02% to 0.15%; and iron, the remainder.

Figure 3.17A and B are photomicrographs showing the structure of gray cast iron. The dark lines are small flakes of *graphite* that greatly impair the strength of the iron. The

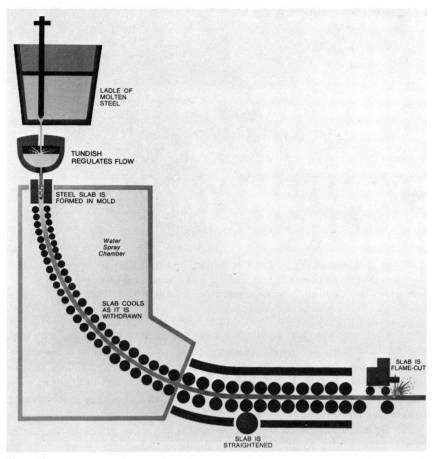

Figure 3.15
Semihorizontal method for continuously casting steel.

Figure 3.16
Structure of medium-carbon cast steel.
Magnification ×200.

54   PRODUCTION OF FERROUS METALS

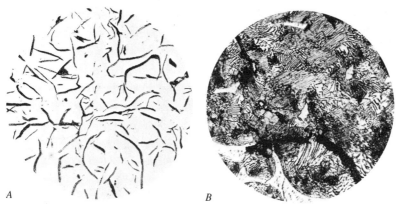

Figure 3.17
Structure of gray cast iron (ASTM Class 40). *A*, Graphite flakes in unetched matrix. Magnification ×125. *B*, Etched in 5% nital acid showing graphite, pearlite, and steadite. Magnification ×562.

strength is greater if these flakes are small and uniformly distributed throughout the metal. The light-colored constituent in the etched specimen is *steadite*, a structural component in cast iron that contains phosphorus; the other constituent is known as *pearlite*. Steadite is identified by its white dendritic formation. It is a eutectic structure of alpha iron and iron phosphide. *Ferrite* or pure iron also appears as a constituent of gray irons having a high silicon content or irons that have been slowly cooled. Pearlite, composed of alternate lamellae of ferrite and iron carbide, is found in most irons and is similar to the pearlite found in carbon steels. This constituent adds to the strength and wear resistance of the iron. The dark graphite flakes may also be seen in Figure 3.17.

White cast iron shows a white fracture because the carbon is in the form of a carbide, Fe_3C. The carbide known as **cementite** is the hardest constituent of iron. White iron with

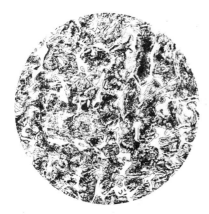

Figure 3.18
Structure of white cast iron as cast. Etched in 5% nital acid showing pearlite and cementite. Magnification ×125.

a high percentage of carbide cannot be machined but can be ground. The principal constituents visible in the micrograph in Figure 3.18 are cementite and pearlite. The dark area is the pearlite and the light area cementite.

White cast iron can be produced by casting against metal *chills* or by regulating the analysis. Chills are used when a hard, wear-resisting surface is wanted for such products as rail wheels, rolls for crushing grain, and jaw crusher plates. The first step in the production of malleable iron is to produce a white iron casting by controlling the analysis of the metal. One specification for the production of these castings is as follows:[2] carbon, 1.75% to 2.30%; silicon, 0.85% to 1.20%; manganese, less than 0.40%; phosphorus, less than 0.20%; sulfur, less than 0.12%; and iron, the remainder.

Mottled cast iron is an intermediary product between gray and white cast iron. The name is derived from the appearance of the fracture. It is obtained in castings in which surfaces subjected to wear have been chilled.

Malleable cast iron or *malleable iron* is made from white cast iron. *White-heart* castings made in the United States are packed in pots and placed in an annealing oven arranged to allow free circulation of heat around each unit. The annealing time lasts 3 to 4 days at temperatures varying from 1500° to 1850°F (815°–1010°C). In this process the hard iron carbides are changed into nodules of *temper* or *graphitic carbon* in a matrix of comparatively pure iron, as shown in the micrograph in Figure 3.19A. Such iron has a tensile strength of around 55,000 psi (380 MPa) and an elongation of 18%. Malleable castings, which have considerable shock resistance and good machinability, are used principally by the railroad, automotive, plumbing, and agricultural implement industries.

[2]ASTM Specifications A47-33, Grade 35018.

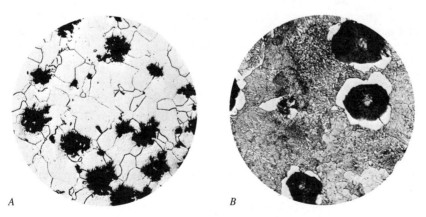

Figure 3.19
A, Structure of malleable iron. Etched in 5% nital acid showing ferrite matrix. Magnification ×125. B, Structure of nodular or ductile cast iron. Magnification ×250.

When the castings are placed in pots with an oxidizing material, the resulting material is called *black-heart* malleable iron, which is sometimes produced in England and Europe. Because of the difficulty in casting white cast iron in large sections, malleable iron's use is predominantly for small castings.

Ductile iron or *nodular iron* is the high-strength, high-ductility iron shown in Figure 3.19B. Ductile iron is classified by three sets of numbers (XX-YY-ZZ), where the first set represents the tensile strength, the second set the yield strength, and the last set the percentage elongation. For example, a 60-50-15 would have 60,000 psi tensile strength, 50,000 psi yield strength, and 15% elongation. It has carbon in the form of graphite nodules and is produced by adding a small amount of a magnesium-containing agent such as magnesium–nickel or magnesium–copper–ferrosilicon alloy to gray iron. The amount of magnesium required to produce graphite depends on the amount of sulfur present. Sulfur is first eliminated by being converted to magnesium sulfide. Additional magnesium present changes the graphite to the nodular form. This type of iron is normally used in the as-cast condition; however, casting followed by a short annealing period is often employed to obtain certain required properties. In this process the time for annealing is much shorter than that used in the manufacture of malleable iron. The improved physical properties of this iron allow it to be used for casting crankshafts and for miscellaneous parts in a wide variety of machines, even though it is more costly than the gray cast iron once used.

EFFECT OF CHEMICAL ELEMENTS ON CAST IRON

Carbon. Although any iron containing over 2.0% carbon is in the cast iron range, gray cast iron has a carbon content of 3% to 4%. The final properties of the iron depend not only on the amount of carbon but also on the form in which it exists. The formation of graphitic carbon depends on slow cooling and on the silicon content. High silicon promotes the formation of graphitic carbon, which reduces the shrinkage and improves machinability. The strength and hardness of iron increases with the percentage of carbon in the combined form. The properties of cast iron may be changed by heat treatment.

Silicon. Silicon up to 3.25% is a softener in iron and is the dominating element in determining the amounts of combined and graphitic carbon. It combines with iron that otherwise would combine with carbon, thus allowing the carbon to change to the graphitic state. After an equilibrium is reached, additional silicon unites with the ferrite to form a hard compound. Silicon above 3.25% acts as a hardener. In melting the average loss of silicon is about 10%. High silicon content is recommended for small castings and low for large castings. When it is used in amounts from 13 to 17%, an alloy having acid and corrosion resistance is formed. Gray irons that are low in silicon respond best to heat treatment.

Manganese. Manganese in small amounts does not have an appreciable effect, but in amounts over 0.5% it combines with sulfur to form a manganese sulfide that has a low specific gravity and is eliminated from the metal with the slag. It acts as a deoxidizer as well as a purifier and increases the fluidity, strength, and hardness of the iron. If the percentage is increased appreciably, it will promote the formation of combined carbon and rapidly increase the hardness of the iron.

Sulfur. Nothing good can be said for sulfur in cast iron. It promotes the formation of combined carbon with accompanying hardness and causes the iron to lose fluidity with resultant blow holes. Each time the iron is remelted there is a slight pickup in sulfur, frequently as much as 0.03%. To counteract this increase, manganese should be added to the charge in the form of *ferromanganese* briquettes or *spiegeleisen*.

Phosphorus. Phosphorus increases the fluidity of the molten metal and lowers the melting temperature. For this reason phosphorus up to 1% is used in small castings and in those having thin sections. There is a slight increase in the phosphorus content during the melting process, about 0.02%. Phosphorus also forms a constituent known as *steadite*, a mixture of iron and phosphide that is hard, brittle, and of rather low melting point. It contains aobut 10% phosphorus, so that an iron with 0.50% phosphorus would have 5% steadite by volume. Steadite appears as a light, structureless area under the microscope but may appear as a network if sufficient phosphorus is present. To control this element, care should be exercised in selecting the grade of scrap.

QUESTIONS AND PROBLEMS

1. Give an explanation of the following terms:

    Bell
    Slag
    BOF
    Oxygen lance
    Indirect arc
    Basic-lined hearth
    Tuyères
    Windbox
    Teapot
    Crucible
    Specific heat
    BTU
    Wrought iron
    Aston process
    Austenitic stainless
    Type 303 stainless
    Killed steel
    Teeming
    Types of cast iron
    Cementite
    Malleable cast iron
    Black heart

2. How is bituminous coal used in the blast furnace?

3. What type of iron ore is predominantly used in the United States?

4. What is an essential ingredient in producing direct reduction pig iron as compared to that in a blast furnace?

5. If the heating value of methane is 1000 BTU/ft^3 and it costs \$3.00 per 1000 ft^3, what is the cost of the methane to produce 1 ton of iron?

6. In problem 5, how many BTUs are required?

7. What are the kinds of cast iron and their distinguishing characteristics?

8. Sketch the direct and indirect types of arc furnaces.

9. How many tons of coke are used to melt 113 tons of iron in a cupola?

10. How many tons of limestone would be required in problem 9?

11. If the heating value of coke is 13,000 BTU/lb and it burns at 50% efficiency, how many pounds of coke would be required in problem 9?

12. Referring to problem 11, how many BTUs would be gainfully employed?
13. How is heat generated in a converter?
14. How can combustion be improved in a blast furnace?
15. Referring to Figure 3.11, what generalizations can be made about the number of BTUs necessary to melt aluminum, iron, brass, and lead?
16. For what reasons do you think such graphs as Figure 3.11 may not tell the whole story? For example, why would melting brass require far more BTUs of fuel than melting aluminum?
17. Describe the process of making steel in the BOF.
18. What is the difference between the process of degassing and controlled-atmosphere melting?
19. What is the general relationship between carbon content and hardenability in ferrous materials?
20. If a steel has a tensile strength of 92,000 psi, what is its Bhn? What does Bhn mean?
21. A steel has a Bhn of 240. What is its approximate tensile strength?
22. If you were trying to make a high-grade steel chain, would a high or low Bhn be sought?
23. Give the range of carbon content that might be used for the following parts: lawnmower blade, bridge railing, padlock, screwdriver, screw for a typewriter, gasoline tank, pliers, and drive shaft.
24. From what type of cast iron is malleable iron produced? What happens to the carbon?
25. Why does an ingot solidify on the bottom first?
26. How is steel "killed"?
27. Iron that is used for remelting has 0.11% sulfur. What is its sulfur content after remelting?
28. What can be said about the cast iron in problem 27 before and after remelting?
29. How can the sulfur content be reduced in the remelting process?
30. Name the metallurgical constituents found in white cast iron, gray cast iron, and malleable iron.
31. How is wrought iron totally unlike the irons described in problem 30?
32. What kinds of steels would be used for the following: knife, car fender, spring, restaurant sink, paper clip.
33. How can a magnet be used to determine if a piece of stainless steel is easy to machine?
34. What would be the classification number of the stainless-steel restaurant sink in problem 32?
35. Why is it important to have phosphorus in a cast iron? What happens to a cast iron that has too much phosphorus?
36. What can you state about the ability of a machinist to machine a cast iron that is almost 100% cementite? What type of cast iron would this be? What are some uses for this type of cast iron?
37. How are malleable iron castings made in Europe as compared to the United States?
38. What is the effect of both the basic and acidic linings in an open-hearth furnace?
39. Zinc-coated wrought iron is used for "tin" roofs. Is there tin in this type of iron? Why will wrought iron stand corrosion so well even if it has no zinc coating?
40. Describe how slag is made and used in the manufacture of wrought iron.
41. Oxygen is an expensive gas. How can its cost be justified in the melting process?

42. What is the purpose of argon gas in the AOD process?

43. Describe the following steels: 1030, 14B30, E4130, 446, 1040.

44. Assign the general classifications of steel, wrought iron, or cast iron, to ferrous materials having the following carbon contents: 0.15%, 0.005%, 0.45%, 2.9%, 1.2%, 0.90%, and 1.17%.

45. Describe the basic differences between the three types of stainless steels.

46. It cost 9¢ per kilowatt-hour of electricity—that is, per volt × amperes × hours = 1 Wh. How many dollars per heat must be spent on electricity to operate the electric furnace described in the text?

47. The use of ductile iron has increased rapidly since the mid-1970s. What are some of the reasons?

48. What are the properties of a nodular iron classified as 80-60-18?

49. What are the steps in producing an iron of the type described in problem 47?

50. Convert the following to metric units: 60,000 psi tensile strength; 30 BTH/lb; 1064 tons; 42 ft; 50 lb/ft^3.

51. What is the approximate electrical cost per pound of electric furnace-produced steel? How does this compare with electrical cost per pound in the production of aluminum? Assume electrical cost is $0.08 per kWh.

CASE STUDY
MELTING COST ESTIMATE FOR A FOUNDRY

You have been employed by Farwell Company to give a ballpark estimate of the fuel costs for a foundry they want to build in connection with their robotics line. These robots hug the floor as they move about and thus need a weight for ballast. The chief engineer wants to make the ballast of gray cast iron because it is relatively easy to machine. The ballast should be 35 in. (89 cm) in diameter and 9.5 in. (24 cm) thick. The production schedule calls for 30 units per working day, and 22 workings days per month.

The electrical energy for a melting furnace will cost $0.11 per kilowatt-hour, and the manufacturer of the furnace expects good operation to yield a furnace efficiency of 49.5%. If electrical cost is one-half the cost of producing a casting, is it practical for Farwell to operate a foundry if they can buy the castings they need for $0.38 per pound? Prepare a report to the chief engineer, including your findings and pointing out what the "break-even" cost of electrical energy would be.

If the chief engineer suggests using malleable iron instead of gray cast iron, what would be your advice? At what price per kilogram can they purchase castings?

CHAPTER 4
PRODUCTION OF NONFERROUS METALS

On the basis of weight, less than 25% of metals used for industrial products are nonferrous. Although nonferrous metals in the pure state possess some useful properties, they are seldom employed for industrial products because they lack structural strength. For this reason they are blended with one or more other elements to form an *alloy* having particular properties. Properties especially characteristic of nonferrous alloys are resistance to corrosion, electrical conductivity, and ease of fabrication.

The choice of a particular alloy is a compromise between adequate strength, ease of fabrication, weight, cost of materials and labor, and the esthetic properties of the product.

Properties

One of the essential differences in metals is density or unit weight, as noted in Table 2.1.

Although few uses of materials are dependent on mass alone, the range of densities of the metals listed in the table does offer certain advantages. Zinc-coated steel roofing known as "corrugated iron" is, for example, 3 times heavier than aluminum roofing. An airplane made of steel would weigh about $3^1/_2$ times as much.

Most nonferrous metals are more resistant to water or moisture and may be used outside without paint or coating. The resistance of each alloy to corrosion must be studied, because each is selective in its qualities. For example, although magnesium is resistant to ordinary atmospheres, it corrodes more rapidly than steel in ocean water. The green oxide on ancient copper artifacts is considered beautiful, but the iron oxide or rust on steel is a sign of a deteriorating part. In general, the nonferrous alloys with the highest densities are the more resistant to corrosion. Aluminum is the exception, however, because it quickly forms an impermeable oxide film on its surface that prevents or slows down attack by most materials other than strong alkalies.

The natural color of aluminum, copper, bronze, brass, tin, and other nonferrous metals allows the engineer to select materials that will esthetically enhance the product. Special coatings on some materials, like the anodized organic coatings on aluminum, add flexibility to color design. The electrical properties of nonferrous materials are usually superior to those made of iron. Copper has 5.3 times and aluminum 3.2 times the electrical conductivity of iron.

As noted in Table 2.1, the melting points of the principal cast nonferrous materials vary from about 621° to 2620°F (327°–1438°C). The pouring temperature is usually about

400° to 600°F (200°–315°C) above the melting point. The nonferrous materials are more difficult to weld than ferrous ones, the lower density ones being the most difficult. The nonferrous materials can be cast, formed, or machined with varying degrees of difficulty. The ease of fabricating the nonferrous alloys varies with the material and the process. Among aluminum alloys alone are many different compositions, some of which can be cold formed. In metal-cutting processes the light nonferrous materials are easier to machine than steel, but a heavier one like nickel is difficult to cut. Handbooks can provide the relative formability and machinability ratings of ferrous and nonferrous materials.

NONFERROUS METALS

Smelting

Unfortunately, the nonferrous metal ores are seldom found in the pure state in commercial quantities. Because they must be separated from the *gangue* before the ore can be reduced, a process known as *ore dressing* is performed. One method of concentrating or "dressing the ore" is familiar to those who have panned for gold. Metals and metal compounds are heavier than the gangue, so they settle to the bottom quicker if such a mixture has been agitated in water. Special methods using this principle have been developed to accelerate the accumulation of metal compounds.

In another method of ore dressing, the ore and gangue are finely powdered and mixed with water. A certain amount of a specific oil is added and violent mixing induced. Frothing occurs and the metallic compounds are suspended in the froth, which is drawn off for processing.

The affinity that most metals have for oxygen increases the difficulty of separation and recovery. The close association of many nonferrous metals with each other leads to difficulties in the smelting process. The ores of copper, lead, and zinc are often found in the same mine, and it has been reported that at least 21 elements can be recovered in useful quantities during their processing, with possibly another 9 that may be discarded and 3 that may be utilized as catalysts in the smelting process. The complexity of a smelting process involving this many elements cannot be overemphasized. The cost associated with recovering certain elements exceeds their value, but they must be removed to assure the purity of the desired product. Although the processes discussed in this chapter are greatly simplified, they serve to illustrate those for ideal ore conditions.

Furnaces for Nonferrous Smelting

The blast furnace used for many years in smelting copper, tin, and other nonferrous metals is of the same proportions but is smaller than the furnace used for manufacturing pig iron. The fuel, usually coke, is mixed with the ore, and a cold-air blast supports combustion. The coke or ore should not be less than $1/2$ in. (12.7 mm) in diameter, or else the updraft will carry it out the flue of the furnace. Selective fluxes are added to the charge in order to have a purer metal and a more fluid slag.

The reverberatory furnace is the one most often used in nonferrous smelting. *Slag inducers* or fluxes are added to reduce oxidation, and all furnaces have fume and dust collectors arranged to trap not only harmful but also valuable by-products.

In addition to furnaces, extensive use is made in nonferrous refining operations of *roasting ovens,* in which sulfide ores are oxidized. In these ovens, oxidizing gases are forced through grates on which the ore lies. Roasting ovens are used for copper and zinc.

PRODUCTION OF ALUMINUM

Although aluminum ores are widely distributed in the earth's crust, only **bauxite** has proved economical as a source of ore from which metal can be smelted. Bauxite is usually mined by the open-pit method, then crushed, sometimes washed to remove clay, and dried. It is then refined into aluminum oxide, or alumina, as it is called in the industry.

The *Bayer process,* named for the German chemist Karl Josef Bayer, is the most widely used method for producing pure alumina. Dried, finely ground bauxite is charged into a digester where it is treated with caustic (NaOH) solution under pressure and at a temperature well above the boiling point. This caustic solution reacts with the bauxite to form sodium aluminate, which is soluble in the liquor.

The pressure is reduced following digestion, and the residue, which contains insoluble oxides or iron, silicon, titanium, and other impurities, is forced out of the digester through filter presses and discarded. The liquor, which contains extracted alumina in the form of sodium aluminate, is pumped to tanks called precipitators.

In the precipitators, fine crystals of aluminum hydroxide from a previous cycle are added to the liquor. These crystals are continuously circulated through the liquor and serve as seed crystals that grow in size as the aluminum hydroxide separates from the solution.

The aluminum hydroxide that settles out from the liquor is filtered, then calcined in kilns at temperatures above 1800°F (980°C). This converts the alumina to a form suitable for smelting.

Metallic aluminum is produced by an electrolytic process that reduces the alumina into oxygen and aluminum. In this process pure alumina is dissolved in a bath of molten cryolite (sodium aluminum fluoride) in large electrolytic furnaces called *reduction cells* or "pots." By means of a carbon anode suspended in the bath, electric current is passed through the bath mixture causing metallic aluminum to be deposited on the carbon cathode at the bottom of the cell. The heat generated by passage of this electric current keeps the bath molten, so that alumina can be added as necessary to make the process continuous. At intervals aluminum is siphoned from the pots, and the molten metal is transferred to holding furnaces for either alloying or impurity removal. It is then cast into ingots of various sizes for further fabrication.

At Aluminum Company of America's Rockdale (Texas) Works, each cell (Figure 4.1) will produce approximately $1/2$ ton (0.5 Mg) of aluminum per day. The production of 1 lb (0.45 kg) of aluminum requires 2 lb (0.9 kg) of alumina [obtained from about 4 lb (3.8 kg) of bauxite], 0.6 lb (0.27 kg) of carbon, small amounts of cryolite and other

Figure 4.1
Electrolytic cells in pot room of aluminum reduction plant.

materials, and approximately 8 kWh of electricity. The cost of a pound of aluminum has increased as the cost of producing electricity has risen.

PRODUCTION OF MAGNESIUM

The largest tonnage of magnesium produced in the United States utilizes seawater as the "ore." Figure 4.2, redrawn from a sketch supplied by the Dow Chemical Company, illustrates the key steps in the production of magnesium ingots. Seawater containing approximately 1300 parts per million (ppm) of magnesium is treated with milk of lime. The lime is made from oyster shells in a kiln that operates at approximately 2400°F (1320°C). When the lime and seawater react, magnesium hydrate settles to the bottom

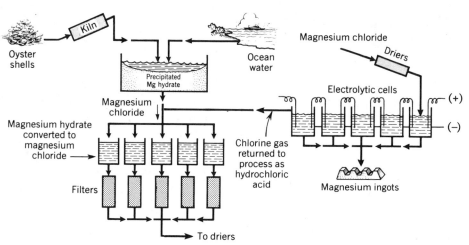

Figure 4.2
Manufacture flow process chart for magnesium from seawater.

of the settling tank and is drawn off as a thin slurry containing about 12% magnesium hydrate. The slurry is filtered and a more concentrated hydrate is obtained; this is converted to magnesium chloride by the addition of hydrochloric acid. The chloride solution is evaporated to remove the water. After subsequent filtration and special drying, the magnesium chloride has a 68% concentration.

The magnesium chloride that has been converted to granular form is then transferred to the electrolytic cells. The cells are approximately 25,000 gal (95 m^3) in capacity and operate at about 1300°F (700°C). The graphite electrodes serve as anodes and the pots are the cathodes. A direct current of 60,000 A causes magnesium chloride to decompose, and the magnesium metal floats to the top. Each pot can produce about 1200 lb (540 kg) of magnesium per day, which is cast into 18-lb (8.1 kg) ingots. Approximately 90% of the magnesium in seawater is recovered. The process generates chlorine gas, which is used to convert the magnesium hydrate to magnesium chloride.

PRODUCTION OF COPPER

The United States produces one-fourth of the world's copper. The ores, known as *chalcopyrite*, contain chiefly Cu_2S and $CuFeS_2$ and usually lie deep underground. The process of producing copper is shown in Figure 4.3. The ore is crushed and mixed with lime and siliceous flux material. To concentrate the copper, flotation tanks or bedding bins are employed. The ores are partially roasted to form a mixture of FeS, FeO, SiO_2, and CuS. This mixture, called calcine, is fused with limestone as a flux in a reverberatory furnace. Most of the iron is removed as slag and the remaining iron and copper—or matte, as it

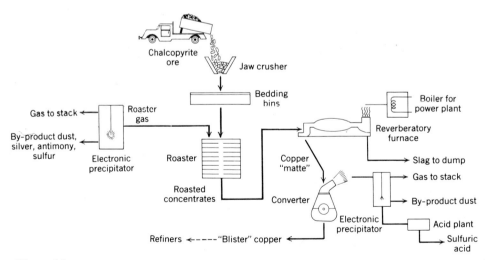

Figure 4.3
Manufacture flow process chart for copper smelting by the reverberatory converter process.

is known—is poured into a converter much like the Bessemer converter used in steel making.

The air supplied to the converter for as long as 4 or 5 h oxidizes the impurities, many of which pass off as volatile oxides; the iron forms a slag that is poured off at intervals. The heat of oxidation keeps the charge molten and the copper sulfide eventually becomes copper oxide or sulfate. When the air is shut off, the cuprous oxide reacts with the cuprous sulfide to form *blister copper* and sulfur dioxide. Blister copper, between 98% and 99% pure, is further electrolytically refined to higher purity.

PRODUCTION OF LEAD

In the production of lead a number of other elements are recovered. Figure 4.4 shows the complexity of the operations involved. The lead concentrate is 65% to 80% lead and must be roasted to remove the sulfides. Limestone, iron ore, sand, and granulated slag are mixed with the lead concentrate before sintering. The sulfur dioxide driven off by the sintering is made into sulfuric acid, and the sintered material is fed into a coke-fired blast furnace. The gases and dust given off contain chloride of cadmium, which can be processed into cadmium. The bullion is drossed. The floating copper dross combines with sulfur to induce separation of copper from the dross, and the liquid lead mixture is oxidized in a furnace known as a softening furnace.

The slag that is skimmed off in the softening furnace contains antimony and arsenic. Zinc is added to the lead mixture in a desilvering kettle, and any gold and silver present become soluble with zinc. The zinc alloy is skimmed off and fed into a retort. The zinc vapor is condensed to produce solid zinc, and any residual liquid is electrolytically separated into gold and silver. The lead from the kettle is cleaned further of zinc before being combined with caustic soda. This is done by shooting a small stream of hot lead into a vacuum chamber causing the zinc to vaporize. The impurities are chemically removed in the final kettle operation, and the lead is cast in 56-lb (25.0 kg) bars or 2000-lb (900 kg) blocks. The process is limited by the blast furnace production, which seldom exceeds 300 tons (270 Mg) of lead bullion per day.

CASTING NONFERROUS MATERIALS

The common elements used in nonferrous castings are copper, aluminum, zinc, tin, and lead. Many alloys, however, have small amounts of other elements such as antimony, phosphorous, manganese, nickel, and silicon.

The foundry method for making nonferrous casting in sand differs little from that used for iron castings. Molds are made in the same way and with the same kind of tools and equipment. The molding sand is usually of finer grain size, because most castings are small, a smooth surface is desired, and the melting temperatures are lower for nonferrous metals. The sand need not be so refractory as for iron and steel castings, because of the lower melting temperatures. Alloying is accomplished by the precise addition of the alloying elements to the base charge. The addition of an alloy to a pure metal always

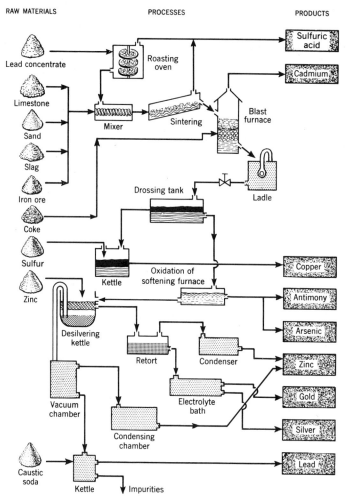

Figure 4.4
Manufacture flow process chart for producing lead.

results in a material with a lower melting point. The pouring temperature is generally about 20% above the melting temperature of the metal, but the type of mold and the section thickness have a bearing on this value.

Crucible furnaces are frequently used for this type of work. They may be either the stationary or the tilting type. Coke is commonly used as the fuel for the stationary pit furnaces, although oil or gas can be used equally well if available. Oil and gas have the advantage of heating more quickly than coke. Electrical resistance, indirect arc, and induction furnaces, having accurate temperature control and low melting losses, can be used under certain conditions. Electric furnaces are widely used for laboratory work as well as for installations requiring mass production.

Two of the most common alloys using copper are brass and bronze. **Brass** is essentially an alloy of copper and zinc. The percentages of each element may vary considerably, but in most cases the zinc content ranges from 10% to 40%. The strength, hardness, and ductility of the alloy are increased as the proportion of zinc is increased to 40%. Zinc contents over 40% are not desirable in that there is a consequent rapid decrease in strength and a tendency to volatilize in melting. Addition of a small percentage of lead (0.5%–5%) increases machinability. Brass is used extensively in industry because of its strength, appearance, resistance to corrosion, and ability to be rolled, cast, or extruded.

Bronze is usually a copper base alloy containing tin with manganese and several other elements. Most elements used as alloys with copper add to the hardness, strength, or corrosion resistance of the metal. Table 4.1 lists properties of some copper base alloys.

Because of light weight and ability to resist many forms of corrosion, aluminum alloys have a wide application in industry today. Many of them respond to heat treatment and are suitable where high strength is needed. Copper has always been one of the principal alloying elements and in amounts up to 8% adds to strength and hardness. Aluminum alloys containing silicon have excellent casting characteristics and increased resistance to corrosion. Magnesium as an alloying element improves machining, makes the castings lighter, and also assists in resisting corrosion. Table 4.2 shows typical properties of aluminum alloys suitable for sand casting.

Magnesium alloys are useful where light weight is essential, because they are about two-thirds the weight of aluminum and one-fourth the weight of cast ferrous metals. They have excellent machinability and respond to certain treatments that improve their physical properties. Aluminum, the principal alloying element, increases hardness and strength. Manganese in small amounts increases the resistance of the metal to salt water. Sand

Table 4.1 **Typical Copper-Based Casting Alloys**

Material	UNS Designation[a]	Melting Temp °F	Melting Temp °C	Tensile Strength (psi)	Machinability Rating[b]
Manganese bronze	C67500	1650	899	65,000–84,000	30
Leaded red brass	C83600	1850	1010	37,000	84
Leaded yellow brass	C85200	1725	941	38,000	80
Manganese bronze	C86200	1725	941	95,000	30
Navy bronze	C92200	1810	988	40,000	42
Aluminum bronze	C95400	1900	1038	85,000–105,000	60
Nickel silver	C97300	1904	1040	35,000	70

[a]The Unified Numbering System for metals and alloys is an extension of the system formerly used by the copper and brass industry and has been accepted by the American Society for Testing and Materials (ASTM).
[b]Based on free-cutting brass, C36000 = 100.

Table 4.2 **Typical Aluminum-Based Casting Alloys**[a]

Alloy	Melting Point °F	Melting Point °C	Tensile Strength (psi)	Type Casting	Uses
201-T4	1135	613	31,000	Sand	Aircraft parts
356-T6	1135	613	33,000	Sand and PM	Manifolds, wheels
A390-5	1140	616	24,000	Sand and PM	Automotive housings

[a]Casting alloys carry Aluminum Association designation, and permanent-mold (PM) alloys use chemical content or nearest Aluminum Association designation.

castings made from magnesium alloys find use in portable tools, aircraft, and other constructions where weight saving is important.

A list of the numerous alloys available for casting purposes and complete information about the analysis and physical properties of both ferrous and nonferrous metals can be found in handbooks.

WROUGHT ALLOYS

A *wrought material* is one that has been or can be formed to a desired shape by hot or cold working. A wrought part may be more or less costly than a casting, but as a rule superior properties can be obtained. Chapters 12 and 13, covering hot and cold working, describe the processes used to fabricate a part by rolling, hammering, or forging. The physical properties of several special wrought alloys are shown in Table 4.3.

Aluminum Alloys. Although aluminum has a strength as shown in Table 2.1, vigorous cold or hot working can double its tensile strength. By judicious use of one or more alloying materials, the application of hot or cold work and heat treating, it is possible to have an alloy strength of over 100,000 psi (690 MPa). Aluminum alloys are available for forgings, extrusions, bending, drawing, spinning, coining, embossing, roll forming, and wire. These alloys are available in commercial forms such as wire, foil, sheets, plates, and bar shapes. All wrought aluminum alloys can be machined, welded, and brazed.

Copper Alloys. Although there are hundreds of copper alloys, most of them can be broadly classified as coppers, brasses, bronzes, and cupronickels and nickel-silvers. The tensile strength of copper alloys varies from approximately 30,000 psi (200 MPa) for almost pure copper to 200,000 psi (1380 MPa) for beryllium copper, and yet the amount of beryllium in such an alloy will be less than 2%. Copper with less than 5% alloying materials is used for low-resistance electrical wiring and conductors, refrigeration, and water tubing. It is relatively soft but becomes harder and brittle when cold worked. The

Table 4.3 Typical Nonferrous Wrought Alloys and Approximate Properties

Material	Designation or Name	Tensile Strength	Elongation (%)	Uses and Remarks
Aluminum[a]	1100-0	13,000	45	Spinning, drawn parts, cooking utensils
	6061-0	18,000	30	Boats
	6061-T6	45,000	17	Rail cars
Copper[b]	C34000	52,000	50	Nuts, rivets, screws, dials
	C36000	55,000	18–53	Gears, screw machine parts; easiest copper alloy to machine
	C65500	58,000–108,000	60	Hydraulic lines, marine hardware
Magnesium[c]	AZ318-B	34,000	10	Excellent machinability, nonmagnetic, often easily corroded
Lead	(0.07% Ca)	4,700	30	Sheet lead; age hardness

[a]Aluminum Association designation.
[b]International Annealed Copper Standard designation.
[c]American Society for Testing and Materials (ASTM) designation.

copper–zinc alloys, commonly known as brasses, are used in heat exchangers and for a variety of parts where corrosion resistance and strength must be balanced against a desired ductility. Nickel, silver, and the bronzes (alloys of copper and tin) are more expensive than brasses. They are used in springs, in bells, and in corrosion atmospheres, particularly where high tensile strength is a factor. Lead can be added to any copper alloy to improve its machining characteristics.

The decision to use copper is based on its unique color, easy forming, machining capabilities, resistance to corrosion, and versatile properties. It can be purchased in all forms, as is the case with aluminum.

Magnesium Alloys. Magnesium is the lightest structural metal. Two-thirds the weight of aluminum, it can be alloyed to produce a metal of good strength, excellent machinability, good weldability, and formability. It is easily extruded if sharp corners are not involved. Both strong and weak acids attack magnesium alloys, and aqueous salt solutions or brines corrode them rapidly. There are few uses for magnesium near the ocean in unprotected environments. Magnesium alloy must be used at temperatures below about 300°F (150°C), because it does not hold its strength at higher temperatures. At cryogenic temperatures magnesium performs well. Magnesium has a very high coefficient of expansion, so care must be taken in delicate assemblies. It is more costly than aluminum or steel and is normally used where savings in weight and machinability result in overall product advantages.

Magnesium alloys are used extensively in aircraft, cameras, binoculars, low-temperature engine parts, portable tools, vacuum cleaners, and for high-speed rotating equipment where it is desirable to minimize inertia. Magnesium can be purchased in bar shapes, plates, and sheets.

DIE-CASTING ALLOYS

A relatively wide range of nonferrous *alloys* can be *die cast*. The principal base metals used in order of commercial importance are zinc, aluminum, magnesium, copper, lead, and tin. The alloys may be further classified as low-temperature alloys and high-temperature alloys; those having a casting temperature below 1000°F (540°C) such as zinc, tin, and lead are in the low-temperature class. The low-temperature alloys have the advantages of lower cost of production and lower die maintenance costs. As the casting temperature increases, ferrous alloy dies in the best treated condition are required to resist the erosion and heat checking of die surfaces. The destructive effect of high temperatures on the dies has been the principal factor in retarding the development of high-temperature die castings.

Other considerations that influence alloy selection are mechanical properties, weight, machinability, resistance to corrosion, surface finish, and, of course, cost. Obviously the least expensive alloy that will give satisfactory service should be selected.

Zinc Base Alloys. Over 75% of die castings produced are the zinc base type. The United States automotive industry uses over 300,000 tons. These alloys cast easily with a good finish at fairly low temperatures, have considerable strength, and are of low cost. The purest grades of commercial zinc, 99.99+% zinc, known as Special High Grade, should be used because such elements as lead, cadmium, and tin are impurities that cause serious casting and aging defects. The usual elements alloyed with zinc are aluminum, copper, and magnesium; all are held within close limits.

Nominal compositions of the two standard zinc die-casting alloys are indicated in Table 4.4. These alloys are much alike in composition except for the copper content, and in most cases they can be used interchangeably. Aluminum in amounts around 4% greatly improves the mechanical properties of the alloys and in addition reduces the tendency of the metal to dissolve iron in the molding process. Copper increases the tensile strength,

Table 4.4 Typical Zinc Die-Casting Alloys

Alloy[a]	Alloy Number	SAE Description	As-Cast Tensile Strength (psi)	As-Cast Elongation (%)
AG40A	3	903	41,000	10
AC41A	5	925	47,600	7
—	7	903	41,000	14

[a]These alloys have multiple designations; Zinc Institute, Society of Automotive Engineers (SAE), and American Society for Testing and Materials (ASTM).

ductibility, and hardness. Magnesium, which is usually held to an optimum of 0.04%, is used because of the beneficial effect it has in making the castings permanently stable.

Zinc alloys are widely used in the automotive industry and for other high-production markets such as washing machines, oil burners, refrigerators, radios, television, business machines, parking meters, and small machine tools.

Aluminum Base Alloys. Many die castings are made of aluminum alloys because of their lightweight and resistance to corrosion, but they are more difficult to die cast.

Because molten alloys of aluminum will attack steel if kept in continuous contact with it, the cold-chamber process is generally used. The melting temperature of aluminum alloys is around 1100°F (540°C).

The principal elements used as alloys with aluminum are silicon, copper, and magnesium. Silicon increases the hardness and corrosion-resisting properties; copper improves the mechanical properties slightly; and magnesium increases the lightness and resistance to impact. In Table 4.2 is shown the nominal composition of the principal aluminum die-casting alloys.

In addition to being light in weight, these alloys have a wide range of helpful properties including resistance to corrosion, high electrical conductivity, ease of applying surface finishes, and good machinability. The two principal aluminum die-casting alloys, as based on the Aluminum Association designation, are 360.0F and 380.0F. They each have a tensile strength of about 47,000 psi.

Copper Base Alloys. Die castings of brass and bronze have presented a greater problem in pressure casting because of their high casting temperatures. These temperatures range from 1600° to 1900°F (870°–1040°C) and make it necessary to use heat-resisting alloy steel for the dies to reduce their rapid deterioration.

Most of the casting alloys listed in Table 4.1 can be die cast, and there are scores of others. Copper base alloys have extensive use in miscellaneous hardware, in electric machinery parts, in small gears, in golf putters, in marine, aircraft, and automotive fittings, in chemical apparatus, and in numerous other small parts. Because the high casting temperatures and pressures cause the die life to be short, the cost of brass die castings is higher than that of other metals. However, these alloys are useful where high strength, resistance to corrosion, or wear resistance is considered important. Since thinner wall sections may be produced, the savings in metal cost coupled with high production rates help offset the short die life disadvantage.

Magnesium Base Alloys. Magnesium is alloyed principally with aluminum but may contain small amounts of silicon, manganese, zinc, copper, and nickel. Its alloys, the lightest in weight of all die cast metals, are about two-thirds the weight of alloys of aluminum. Although the price is slightly higher than for aluminum, the extra cost may be compensated for by light weight and improved machinability.

The corrosion resistance of magnesium alloys is inferior to that of the other die-casting alloys, especially in moist or sea atmospheres, and usually necessitates a chemical treatment as well as the subsequent application of a special priming coat shortly after the casting is produced. These treatments render the casting suitable for a range of applications.

ASTM Specification B94, alloy AZ91B, is the principal die-casting alloy. It has good casting characteristics and fairly high mechanical properties. This alloy contains 9% aluminum, 0.5% zinc, 0.13% manganese, 0.5% maximum silicon, 0.3% copper, 0.03% nickel, and the remainder magnesium. It is desirable that copper and nickel content be kept very low to minimize corrosion.

Magnesium alloys are cast in much the same manner as aluminum alloys and require a casting temperature between 1200° and 1300°F (650°–700°C). Best results are obtained in so-called cold-chamber machines, and it is necessary to ladle the alloy from a crucible that is hooded, keeping the metal covered by a nonoxidizing atmosphere. The lightness of these alloys combined with good mechanical properties and excellent machinability merits their use for aircraft, motor and instrument parts, portable tools, textile machinery, household appliances, and many other similar applications.

Lead Base Alloys. Pure lead, which melts at 621.3°F (327°C), will melt at around 470°F (240°C) when alloyed with about 16% antimony. This element is the principal one used with lead and its percentage ranges from 9.25 to 16%. Antimony hardens lead and reduces its shrinkage value. Lead alloys have poor mechanical properties but are inexpensive and easily cast. They are used principally for light-duty bearings, weights, battery parts, X-ray shields, and applications requiring a noncorrosive metal.

Tin Base Alloys. Die-casting alloys based on tin are in about the same category as the lead alloys as far as mechanical properties are concerned, but are high in price. They are high in corrosion resistance, and some of them are well suited for use in contact with foods and beverages. Also, tin alloys have excellent bearing properties and can be cast within remarkably close dimensional tolerances. This fact together with high corrosion resistance accounts for their use in small parts such as number wheels, especially where contact with corrosive inks may be involved. Tin alloys can also be used for low-cost jewelry, and certain grades are classed as *pewter*.

QUESTIONS AND PROBLEMS

1. Give an explanation of the following terms:

   Gangue
   Slag inducer
   Roasting oven
   Bauxite
   Reduction cells
   Chalcopyrite
   Brass
   Wrought material
   Alloy
   Die casting
   Pewter

2. What generalizations can be made concerning nonferrous materials with regard to cost, strength, electrical properties, and resistance to corrosion? Explain.

3. Explain how the corrosion resistance of magnesium would compare to that of wrought iron.

4. As tensile strength of an alloy increases, what can be generally stated about its ductility?

5. For what might a large lime plant use its surplus in the production of a nonferrous metal?

6. Compare the weight of an 8-in. sphere made of steel, aluminum, and magnesium. The volume V of a sphere is $V = \frac{4}{3}\pi r^3$, where r = radius in feet and V is expressed in cubic feet. Use Table 2.1.

7. How and why does the cost of aluminum depend to some extent on the price of electricity?

8. What is the general rule for the metal temperature if it is to be used in a sand casting? To what temperature would aluminum be raised before pouring?

9. For what purposes are the electrostatic precipitators used in the manufacture of blister copper?

10. What are precipitators used for in the manufacture of aluminum? How do they differ from the electrostatic type?

11. How much weight reduction could be made for a 14.6-lb aluminum auto part if it were made of magnesium? What about the use of magnesium in a high-temperature application?

12. Examine your local newspaper or *The Wall Street Journal* and tabulate the prices for ferrous and nonferrous materials.

13. Why would you think brass costs less per pound than bronze?

14. Explain the difference in properties between an aluminum alloy that is used for wrought products instead of cast ones.

15. List 10 products made from wrought aluminum alloys.

16. List 10 products made from wrought copper-based alloys.

17. Describe the decorative color characteristics of copper and aluminum and give three uses for each that would increase the salability of a product.

18. What are the principal reasons for the high cost of brass die castings?

19. What is the principal advantage of using tin-coated materials?

20. Describe the principal process for producing aluminum and indicate every step that requires energy.

21. What is the essential difference in a blast furnace used for making pig iron as compared to one used in smelting tin? Describe the method of heating in a blast furnace.

CASE STUDY

PROFIT ANALYSIS

As president of Forward Die Castings, Inc., you have been asked to bid on a new olympic symbol, the *unity cross* (Figure 4.5). Suppliers will want the solid cross for their packaging and promotion in most nonferrous metals. Of course, they want a relatively smooth finish but no machining, so the units must be die cast to their correct shape and the sprues cut off. The sprue is the area through which metal is fed to the casting. You are authorized to charge 5.13 times the cost of the metal in an individual cross. Cost analysis reveals the following: aluminum die cast

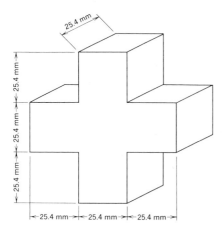

Figure 4.5
Case-study.

alloy, $1.10/lb; brass die cast alloy, $1.44/lb; zinc die cast alloy, $0.59/lb; magnesium die cast alloy, $1.43/lb; and lead, $0.29/lb. Figure your cost for each metal; then figure your production and overhead cost as

Overhead and production costs per unit
= 2.3 (metal cost per unit + $0.24/lb)

The $0.24/lb is for special handling costs. The bid quantities are 50,000, 100,000, and 300,000 units. Densities are aluminum, 165 lb/ft^3; brass 535 lb/ft^3; zinc 446 lb/ft^3; magnesium 109 lb/ft^3; and lead 706 lb/ft^3.

Tabulate the essential figures including metal cost, handling cost, production and overhead, selling price, and profit for each quantity and each material. Let your last column be the profit per unit. If you have a computer available, the problem can be programmed and the printout submitted.

Prepare a short discussion on the advantage and disadvantage of the use of each material for the supplier's edification.

CHAPTER 5
CONVENTIONAL FOUNDRY PROCESSES

Foundry processes consist of making molds, preparing and melting the metal, pouring the metal into the molds, cleaning the castings, and reclaiming the sand for reuse. Machining a block of metal to an intricate shape can be very expensive, which encourages a foundry worker to say, "Why whittle when you can cast." High rates of production, good surface finish, small dimensional tolerances, and improved properties of materials have enabled both large and small intricate parts to be cast of almost all metals and their alloys.

Molds may be made of metal, plaster, ceramics, or other refractory substances, but this chapter is concerned primarily with the preparation of sand molds. Although it is true that sand castings are not produced in as large a quantity as they once were, the principles used in sand casting are important to understand if more contemporary processes are to be employed.

TYPES OF SAND CASTINGS

There are two different methods by which sand castings can be produced. Classified according to the type of pattern used, they are (1) removable pattern and (2) disposable pattern.

In the method employing a removable pattern, sand is packed round the pattern. Later the pattern is removed and the cavity produced is filled with molten metal. Disposable patterns are made from polystyrene and, instead of being removed from the sand, are vaporized when the molten metal is poured into the mold.

To understand the foundry process, it is necessary to know how a mold is made and what factors are important to produce a good casting. The principal factors are molding procedure, patterns, sand, cores, mechanical equipment, the metal, and pouring and cleaning the casting.

MOLDING PROCEDURE

Molds are classified according to the materials used.

1. *Green-sand molds.* This most common method, consisting of forming the mold from damp molding sand, is used in both of the processes previously described. The term

green sand does not refer to the color of the sand, which is a dark brown or black, but rather to the fact that the sand is uncured. Figure 5.1 illustrates the use of green sand.

2. *Skin-dried molds.* Two general methods are used in preparing the skin-dried molds. In one the sand around the pattern to a depth of about $1/2$ in. (12.7 mm) is mixed with a binder, so that when it is dried it will leave a hard surface on the mold. The remainder of the mold is ordinary green sand. The other method makes the mold of green sand and then coats its surface with a spray or wash that hardens when heat is applied. Sprays used for this purpose include linseed oil, molasses water, gelatinized starch, and similar liquid solutions. In both methods the mold must be dried either by air or by a torch to harden the surface and drive out excess moisture.

3. *Dry-sand molds.* These molds are made entirely from fairly coarse molding sand mixed with a binding material similar to those already mentioned. Since they must be oven-baked before being used, the flasks are of metal. A dry-sand mold holds its shape when poured and is free from gas problems caused by moisture. Both the skin-dried and dry-sand molds are widely used in steel foundries.

4. *Loam molds.* Loam molds are used for large castings. The mold is first built up with bricks or large iron parts. These parts are plastered over with a thick loam mortar, the shape of the mold being obtained with sweeps or skeleton patterns. The mold is then allowed to dry thoroughly, so that it can resist the heavy rush of molten metal. Such molds take a long time to make and are not used extensively.

5. *Furan molds.* This process is good for making molds using disposable patterns and cores. Dry sharp sand is thoroughly mulled with phosphoric acid, which acts as an accelerator. Furan resin is added and mulling is continued only long enough to distribute the resin. The sand material begins to air-harden almost immediately, but the time delay is sufficient to allow molding. In use with disposable patterns, furan resin sand can be employed as a wall or shell around the pattern supported by green or sharp sand, or it can be used as the complete molding material.

6. *CO_2 molds.* In this process clean sand is mixed with sodium silicate and the mixture is rammed about a pattern. When CO_2 gas is pressure-fed to the mold, the sand mixture

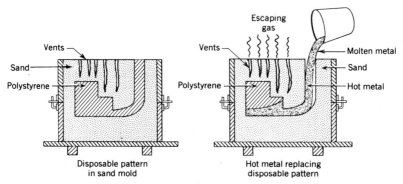

Figure 5.1
Disposable-pattern mold.

hardens. Very smooth and intricate castings are obtained by this method, although the process was originally developed for making cores.

7. *Metal molds.* Metal molds are used mainly in die-casting low-melting-temperature alloys. Castings are accurately shaped with a smooth finish, thus eliminating much machine work.

8. *Special molds.* Plastics, cement, plaster, paper, wood, and rubber are all mold materials used to fit particular applications. These are discussed in more detail in Chapter 6, "Contemporary Casting Processes."

Molding processes in the conventional foundry may be classified as

1. *Bench molding.* Bench-type molding is for small work done on a bench at a height convenient to the molder.

2. *Floor molding.* When castings increase in size with resultant difficulty in handling, the work is done on the foundry floor. This type of molding is used for practically all medium-sized and large castings.

3. *Pit molding.* Extremely large castings are frequently molded in a pit instead of a flask. The pit acts as the drag part of the flask, and a separate cope is used above it. The sides of the pit are brick-lined, and on the bottom there is a thick layer of cinders with connecting vent pipes to the floor level.

4. *Machine molding.* Machines have been developed to do a number of the operations that the molder ordinarily does by hand. Ramming the sand, rolling the mold over, forming the gate, and drawing the pattern can be done by these machines much better and more efficiently than by hand.

Removable Patterns

A simple procedure for molding a cast-iron gear blank is illustrated in Figure 5.2. The mold for this blank is made in a flask that has two parts. The top part is called the *cope* and the lower part the *drag*. If the flask is made in three parts the center is called a *cheek*. The parts of the flask are held in a definite relation to one another by pins on either side of the drag that fit into openings in angle clips fastened to the sides of the cope.

The first step in making a mold is to place the pattern on a *molding board* that fits the flask being used. Next, the drag is placed on the board with the pins down (Figure 5.2A). Molding sand is then riddled in to cover the pattern. The sand should be pressed around the pattern with the fingers and the drag completely filled. Small molds are packed by a hand rammer, and mechanical ramming is used for large molds and in high-production molding. If the mold is not sufficiently rammed, it will fall apart when handled or when the molten iron strikes it, and if it is rammed too hard, steam and gas cannot escape when the molten metal enters.

After ramming has been completed, the excess sand is leveled off with a straight bar called a *strike rod*. To ensure the escape of gases when the casting is poured, small vent holes are made through the sand to within a fraction of an inch of the pattern.

The lower half of the mold is then turned over, so that the cope may be placed in

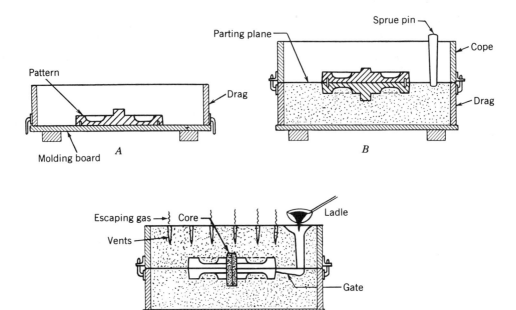

Figure 5.2
Procedure for making green-sand molds. *A*, Pattern on molding board ready to ram up drag. *B*, Drag rolled over and pattern assembled ready to ram cope. *C*, Mold complete with dry-sand core in place.

position and the mold finished. Before turning, a little sand is sprinkled over the mold and a *bottom board* placed on top. The drag is then rolled over and the molding board removed, exposing the pattern. The surface of the sand is smoothed over with a trowel and covered with a fine coating of dry **parting sand.** Parting sand is a fine-grained, dry silica sand without strength. It prevents bonding of sand in the cope with sand in the drag.

Next, the cope is placed on the drag (Figure 5.2*B*). To provide a place for the iron to enter the mold, a tapered pin known as a *sprue pin* is placed approximately 1 in. (25 mm) to one side of the pattern. The operations of filling, ramming, and venting the cope proceed in the same manner as in the drag.

At this point the mold is complete except for removal of the pattern and sprue pin. The sprue pin is first withdrawn and a funnel-shaped opening is scooped out at the top, so that there will be a fairly large opening into which to pour the metal. The cope half of the flask is then carefully lifted off and set to one side. Before the pattern is withdrawn, the sand around the edge of the pattern is usually moistened with a *swab,* so that the edges of the mold hold firmly together when the pattern is removed. To loosen the pattern a *draw spike* is driven into it and rapped lightly in all directions. The pattern can then be withdrawn by lifting the draw spike.

Before the mold is closed, a small passage known as a *gate* must be cut between the cavity made by the pattern and the sprue opening. This passage is shallowest at the mold, so that after the metal has been poured the metal in the gate may be broken off close to the casting. To allow for metal shrinkage, a hollow is sometimes cut into the cope, which provides a supply of hot metal as the casting cools; this opening is called a *riser*.

The mold surfaces may be sprayed, swabbed, or dusted with a prepared coating material. Such coatings often contain silica flour and graphite, but their composition varies considerably depending on the material being cast. A mold coating improves the surface finish of the casting and reduces possible surface defects. The completed mold is shown in Figure 5.2C. Before the mold is poured, a weight should be put on top to prevent the hydrostatic pressure of the liquid metal from floating the cope and allowing metal to run out of the mold at the parting line.

The net force tending to float the cope is

$$F_N = A(D \times h) - (W_c + W_{sc})$$

where

F_N = Net force or the weight that must be added, lb (kg)

A = Area of metal surface at the parting line, ft² (m²)

D = Density of the metal, lb/ft³ (kg/m³)

h = Height of cope, ft (m)

W_c = Weight of cope, lb (kg)

W_{sc} = Weight of sand in cope, lb (kg)

If $W_c + W_{sc}$ is greater than $A(D \times h)$, no weight is necessary.

Disposable Patterns

In making a disposable-pattern mold the pattern, usually one piece including the gate, is placed on a follow board and the drag is molded in the conventional way. Vent holes are added and the drag is turned over for molding the cope. Although green sand is the most used material, other special-purpose sands can be used, particularly as facing immediately around the pattern. No parting sand is applied, for the cope and drag will not be separated until the casting is removed. Instead, the cope is filled with sand and rammed. Either the sprue is cut into the gating system or else, as usually happens, it is a part of the disposable pattern. Vent holes are added and a hold-down weight is placed upon the cope. The polystyrene pattern, including the gating and pouring system, is left in the mold. Molten metal (Figure 5.1) is poured rapidly into the sprue, and the polystyrene vaporizes. The mold is poured sufficiently fast to prevent combustion of the polystyrene with the resulting carbonaceous residue. Instead, the gases caused by vaporization of the material are driven out through the permeable sand and vent holes. A refractory coating is usually applied to the pattern to ensure a better surface finish for the casting and added pattern strength. Considerable hold-down weight and at times even side binding are necessary to accommodate the relatively high pressures within the mold.

The advantages of this process include the following.

1. For one of a kind or non-machine-molded castings the process requires less time.
2. Allowances are unnecessary in removing the pattern from the sand.
3. Finish is uniform and reasonably smooth.
4. A complex wooden pattern with loose pieces is unnecessary.
5. Cores are seldom required.
6. Molding is greatly simplified.

The disadvantages include the following.

1. The pattern is destroyed in the process.
2. Patterns are more delicate to handle.
3. There is no opportunity to inspect the finish cavity.

Estimating Casting Weight

The weight of a complex casting is not easily determined but can be estimated by weighing the pattern. The following relationship is used.

$$W_p/D_p = W_c/D_c$$

where

W_p and W_c = Weight of pattern and casting, respectively, lb (kg)

D_p and D_c = Density of pattern and casting, respectively, lb/ft^3 (kg/m^3)

The density of polystyrene foam is about 21 lb/ft^3 (336 kg/m^3) and that of soft pattern pine is 25 lb/ft^3 (400 kg/m^3).

GATES, RISERS, AND SOLIDIFICATION CHARACTERISTICS

The passageway for bringing the molten metal to the mold cavity, which is known as the *gating system*, is usually made up of a *pouring basin*, a downgate or vertical passage known as a *sprue*, and a *gate* through which the metal flows from the sprue base to the mold cavity. In large castings a *runner* may be used that takes the metal from the sprue base and distributes it to several gate passageways around the cavity. The design of the gating system is important and involves a number of factors.

1. Metal should enter the cavity with as little turbulence as possible at or near the bottom of the mold cavity when pouring small castings.

2. Erosion of the passageway or cavity surfaces should be avoided by properly regulating the flow of metal or by the use of dry-sand cores. Formed gates and runners resist erosion better than those that are cut.

3. Metal should enter the cavity so that directional solidification is provided if possible. The solidification should progress from the mold surfaces to the hottest metal, so that there is always molten metal available to compensate for shrinkage.

GATES, RISERS, AND SOLIDIFICATION CHARACTERISTICS 81

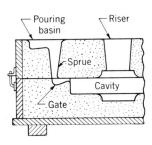
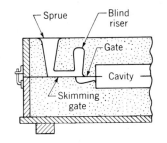

Figure 5.3
Methods used in introducing metal to mold cavity.

4. Slag or other foreign particles should be prevented from entering the mold cavity. A *pouring basin* next to the top of the sprue hole is often provided on large molds to simplify pouring and to keep slag from entering the mold. *Skimming gates* such as the one shown in Figure 5.3 trap slag or other light particles into the second sprue hole. The gate to the mold is restricted somewhat to allow time for floating particles to rise into the skimmer. A *strainer* made of baked dry sand or ceramic material can also be used at the pouring basin to control the metal flow and to allow only clean metal to enter.

Risers are often provided in molds to feed molten metal into the main casting cavity to compensate for the shrinkage. They should be large in section so as to remain molten as long as possible, and should be located near heavy sections subject to large shrinkage. If they are placed at the top of the section (Figure 5.3), gravity will assist in feeding the metal into the casting proper.

Blind risers are domelike risers found in the cope half of the flask that are not the complete height of the cope. They are normally placed directly over the gate where the metal feeds into the mold cavity and thus supply the hottest metal when pouring is completed.

Volumetric shrinkage usually occurs when metal solidifies, and a shrinkage cavity results if the solidification is not directed, so that any voids caused by shrinkage take place in the gate, risers, or sprue. The shrinkage occurs in the area where the metal stays molten the longest. Figure 5.4 illustrates temperature gradient or isotherm lines in a casting and the directions of heat flow from the solidifying metal to the sand. In each instance shrinkage-caused voids will occur in the areas of highest temperature, and mold design must be modified so as to remove this tendency if such a void is detrimental to a casting.

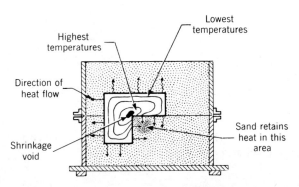

Figure 5.4
Isotherms indicating the area of shrinkage void.

PATTERNS

Types of Removable Patterns

Figure 5.5 shows seven types of pattern construction. The simplest form is the solid or single-piece pattern shown in Figure 5.5A. Many patterns cannot be made in a single piece because they cannot be removed from the sand. To eliminate this difficulty some patterns are made in two parts (Figure 5.5B), so that half of the pattern will rest in the lower part of the mold and half in the upper part. The split in the pattern occurs at the parting line of the mold. Figure 5.5C shows a pattern with two loose pieces that are necessary to facilitate withdrawing it from the mold.

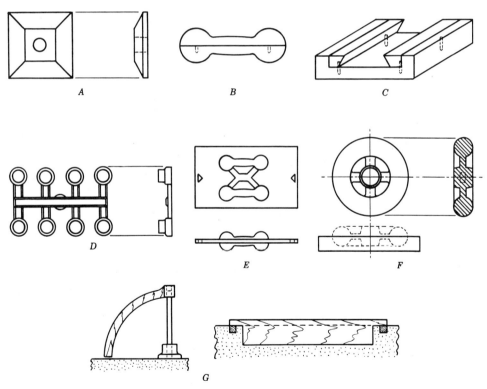

Figure 5.5
Types of patterns. A, Solid pattern. B, Split pattern. C, Loose-piece pattern. D, Gated pattern. E, Match plate. F, Follow board for wheel pattern. G, Sweep patterns: curved sweep for shaping large green-sand core, and straight sweep.

In production work where many castings are required, *gated* patterns as shown in Figure 5.5D may be used. Such patterns are made of metal to give them strength and to eliminate any warping tendency. *Match plates* provide a substantial mounting for patterns and are widely used with machine molding. Figure 5.5E shows such a plate, upon which are mounted the patterns for two small dumbbells. The *follow board,* shown in Figure 5.5E, may be used with either single- or multiple-gated patterns. Patterns requiring follow boards are usually somewhat difficult to make as a split pattern. The board is routed out, so that the pattern rests in it up to the parting line; this board then acts as a molding board for the first molding operation. Many molds of regular shape may be constructed by the use of *sweep* patterns as illustrated in Figure 5.5G. The principal advantage of this pattern is that it eliminates expensive pattern construction. Practically all high-production work on molding machines uses the match plate pattern.

Pattern Allowances

In pattern work the question asked is why a finished gear blank or any other object could not be used for making molds without the trouble and expense of making a pattern. In some cases they might be used, but in general this procedure is not practical because certain allowances must be made in the pattern. These allowances are *shrinkage, draft, finish, distortion,* and *shake.*

Shrinkage. When any pure metal and most alloy metals cool, they shrink. To compensate for shrinkage a **shrink rule** must be used in laying out the measurements for the pattern. A shrink rule for cast iron is $1/8$ in. longer per foot $(1.04\%)^1$ than a standard rule, the average shrinkage for cast iron. If a gear blank was planned to have an outside diameter of 6 in. (152 mm) when finished, the shrink rule would actually make it $6^{1}/_{16}$ in. (154 mm) in diameter, thus compensating for the shrinkage. The shrinkage for brass varies with its composition but is usually close to $3/_{16}$ in./ft (1.56%), steel $1/_{4}$ in./ft (2.08%), and aluminum and magnesium $5/_{32}$ in./ft (1.30%). When metal patterns are to be cast from original patterns, double shrinkage must be allowed.

Draft. When a removable pattern is drawn from a mold, the tendency to tear away the edges of the mold in contact with the pattern is greatly decreased if the surfaces of the pattern, parallel to the direction it is being withdrawn, are slightly tapered. This tapering of the sides of the pattern, known as *draft,* is done to provide a slight clearance for the pattern as it is lifted up. Draft is added to the exterior dimensions of a pattern and is usually $1/_{8}$ to $1/_{4}$ in./ft (1.04%–2.08%). Interior holes may need draft as large as $3/_{4}$ in./ft (6.25%).

Finish. On a drawing of the details of a part to be cast, each surface that will ultimately be machined is indicated by a finish mark. This mark shows the pattern maker where additional metal must be provided so that there will be some metal to machine, hence a

---

[1]Indicates percentage of shrinkage for the SI metric.

finish allowance. The amount that is to be added to the pattern depends on the size and shape of the casting, but in general the allowance for small and average-sized castings is $1/8$ in. (3.2 mm). When patterns are large this allowance must be increased.

Distortion. Distortion allowance applies only to those castings of irregular shape that are distorted in the process of cooling because of metal shrinkage. A horseshoe-shaped piece would be an example.

Shake. When a removable pattern is rapped in the mold before it is withdrawn, the cavity in the mold increases slightly. In an average-sized casting this increase in size can be ignored. In large castings or in ones that must fit together without machining, a *shake* allowance should be considered by making the pattern slightly smaller.

Materials for Removable Patterns

The first step in making a casting is to prepare a model known as a *pattern* that differs in a number of respects from the resulting casting. These differences, known as *pattern allowances,* compensate for metal shrinkage, provide sufficient metal for machined surfaces, and facilitate molding.

Most patterns are made of wood, which is inexpensive and can be easily worked. Because only a small percentage of patterns go into quantity production work, the majority do not need to be made of material that will stand hard usage in the foundry.

Many of the patterns used in production work are made of *metal* because it withstands hard use. Metal patterns do not change their shape when subjected to moist conditions, and they require minimum maintenance to keep them in operating condition. Metals used for patterns include brass, white metal, cast iron, and aluminum. Aluminum is probably the best all-around metal because it is easy to work, light in weight, and resistant to corrosion. Metal patterns are usually cast from a master pattern constructed of wood. The *plastics* are especially well adapted for pattern materials because they do not absorb moisture, are strong and dimensionally stable, and have a smooth surface.

CONSTRUCTION OF A REMOVABLE PATTERN

The details for laying out a cast-iron V block are shown in Figure 5.6, where the end view is drawn first using a shrink rule. Because the detail calls for "finish" all over, more metal must be provided; this is indicated by the second outline of the V block on the layout. In providing for the draft, the method of molding the pattern must be considered. The final outline on the layout board represents the actual size and shape that are used for constructing the pattern. Sharp interior corners are filleted to eliminate the development of metal shrinkage cracks.

A *fillet* is a concave connecting surface or the rounding out of a corner at two intersecting planes. Rounded corners and fillets assist materially in molding, in that the sand is not likely to break out when the pattern is drawn.

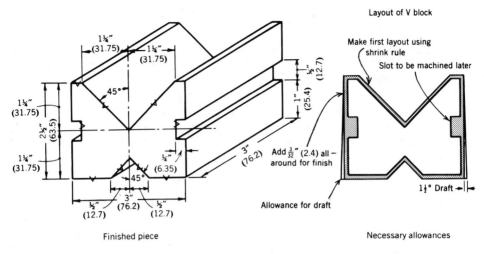

Figure 5.6
Method of making a pattern layout for a cast-iron block.

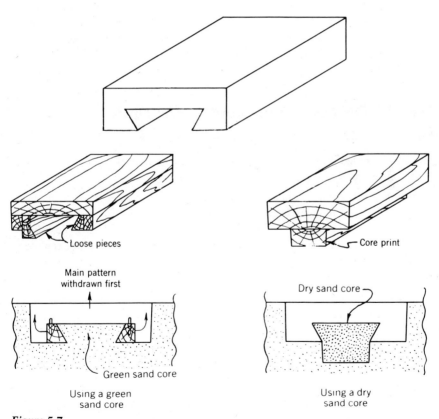

Figure 5.7
Two methods of molding a gib block.

Wood patterns are usually given at least three coats of shellac or a synthetic varnish that will not redissolve upon contact with moisture. This finish fills the pores of the wood, creates a moisture seal, and gives the pattern a smooth surface.

Loose-piece patterns must be made when projections or overhanging parts occur making it impossible to remove them from the sand even though they are parted. In such patterns the projections have to be fastened loosely to the main pattern by wooden or wire dowel pins. When the mold is being made such loose pieces remain in the mold until after the pattern is withdrawn. They are then withdrawn separately through the cavity formed by the main pattern. The use of loose pieces is illustrated in the pattern for a **gib block** casting that fits over a dovetailed slide; a detail is shown in Figure 5.7.

How the pattern will be molded must be decided. Two methods are possible, as illustrated in the figure. In the first method two loose pieces facilitate the withdrawal of the pattern from the sand.

The loose pieces may be eliminated by a dry-sand core. If this construction is used the pattern would then be made using a core print as shown. In addition a core box would be necessary. This latter method is less economical because of the expense in making the core box and core.

To improve the surface finish of the completed castings the mold is often brushed, swabbed, or sprayed with a zirconite wash. The facing sand placed next to the pattern can be green sand, **sodium silicate**-bonded sand, furan sand, and sometimes unbonded sand or even dirt.

CONSTRUCTION OF A DISPOSABLE PATTERN

Disposable patterns can be made in one of two ways.

1. They may be glued and hot-wire or hot-knife carved. Sprues, runners, and gates may be glued on or attached with wire or nails.

2. They may be made from expandable or foam polystyrene in the form of small beads that contain penthane as a blowing agent. The polystyrene thus transformed into foam is injected under heat and pressure into a metal die. The die is then cooled. Complex shapes can be formed at high production rates. Because no allowance is necessary for removing the pattern from the mold, the only allowances necessary are for shrinkage, finish, and distortion.

TYPES OF SAND

Silica sand (SiO_2), found in many natural deposits, is well suited for molding purposes because it can withstand a high temperature without decomposition. This sand is low in cost, has long life, and is available in a wide range of grain sizes and shapes. Disadvantages are that it has a high expansion rate when subjected to heat and has some tendency to fuse with the metal. If it contains a high percentage of fine dust, it may constitute a health hazard.

Pure silica sand is not suitable for molding because it lacks binding qualities. The binding qualities may be obtained by adding 8% to 15% clay. The three types of clay commonly used are kaolinite, illite, and bentonite. The latter, used most often, is weathered volcanic ash.

Some natural molding sands are bonded with clay when quarried, and only water must be added to have an adequate molding sand for nonferrous castings. The large amount of organic material found in natural sands prevents them from being sufficiently refractory for high-temperature applications such as in the molding of higher melting point metals and alloys.

Synthetic molding sands are composed of washed, sharp-grained silica (*sharp sand*) to which 3% to 5% clay is added. Less gas is generated with synthetic sands because less than 5% moisture is necessary to develop adequate strength.

The size of the sand grains will depend on the type of work to be molded. For small and intricate castings a fine sand is desirable to allow all the details of the mold to be brought out sharply. As the casting size increases the sand particles should be coarser to permit the gases that are generated in the mold to escape. Sharp, irregular-shaped grains are usually preferred because they interlock and add strength to the mold.

SAND TESTING

Periodic tests are necessary to determine the essential qualities of foundry sand. The properties change by contamination from foreign materials, by washing action in pouring, by the gradual change and distribution of grain size, and by continual subjection to high temperatures. Tests may be either chemical or mechanical, but, aside from determining undesirable elements in the sand, chemical tests are little used. Most mechanical tests are simple and do not require elaborate equipment. Various tests are designed to determine the following properties of a molding sand.

1. *Permeability*. Porosity of the sand enables the escape of gas and steam formed in the mold.

2. *Strength*. Sand must be cohesive to the extent that it has sufficient bond; both water and clay content affect the cohesive properties.

3. *Refractoriness*. Sand must resist high temperatures without fusing.

4. *Grain size and shape*. Sand must have a grain size commensurate with the surface to be produced, and grains must be irregular to the extent that they will have sufficient bonding strength.

Mold and Core Hardness Test

The mold hardness tester shown in Figure 5.8 operates on the principle that the depth a steel ball penetrates into sand is a measure of the hardness or firmness of the sand. A spring-loaded (2.3 N) steel ball 0.2 in. (5.08 mm) in diameter is pressed into the surface of the mold, and the depth of penetration is indicated on the dial in thousandths of an inch. Medium-rammed molds give a value around 75.

Figure 5.8
Mold hardness tester for measuring the surface hardness of green-sand molds.

Fineness Test

This test to determine the percentage distribution of grain sizes in sand is performed on a dried-sand sample from which all clay substance has been removed. A set of standard testing sieves is used having U.S. National Bureau of Standards meshes 6, 12, 20, 30, 40, 50, 70, 100, 140, 200, and 270. These sieves are stacked and placed in one of the several types of motor-driven shakers. The sand is placed on the coarsest sieve at the top, and after 15 min of vibration the weight of the sand retained on each sieve is obtained and converted to a percentage basis.

To obtain the American Foundry Association (AFA) *fineness number* (Table 5.1), each percentage is multiplied by a factor as given in an example. The fineness number is obtained by adding all the resulting products and dividing the total by the percentage of sand retained.

This number is a useful means of comparing different sands in the foundry.

Table 5.1 Example of AFA Fineness Calculation

Mesh	Percentage retained	Multiplier	Product
6	0	3	0
12	0	5	0
20	0	10	0
30	2.0	20	40.0
40	2.5	30	75.0
50	3.0	40	120.0
70	6.0	50	300.0
100	20.0	70	1400.0
140	32.0	100	3200.0
200	12.0	140	1680.0
270	9.0	200	1800.0
Pan	4.0	300	1200.0
Totals	90.5		9815.0

$$\text{Grain fineness number} = \frac{9815}{90.5} = 108$$

Test for Moisture Content

The moisture content of foundry sands varies according to the type of molds being made and the kind of metal being poured. For a given condition there is a close range within which the moisture percentage should be held to produce satisfactory results.

The *moisture teller* contains electric heating units and a blower for forcing warm air through the filter pan containing the sand sample. By weighing the sand after it is dried and noting the difference in the initial and final readings, the percentage of moisture can be determined. The moisture content should vary from 2% to 8%, depending on the type of molding being done.

Clay Content Test

The equipment necessary for determining the percentage of clay in molding sands consists of a drying oven, a balance and weights, and a sand washer. A quantity of sand is dried and a water-based caustic soda solution is added. Following a timed mixing the caustic solution, which has absorbed the clay, is siphoned off. This process is repeated three times. The sample is dried, weighted, and compared to the original sample weight to determine the loss in clay.

Permeability Test

One of the essential qualities of molding sand is sufficient porosity to permit the escape of gases generated by the hot metal. This depends on several factors including the shape of sand grains, fineness, degree of packing, moisture content, and amount of binder present. *Permeability* is measured by the quantity of air that passes through a given sample of sand in a prescribed time and under standard conditions. Coarse-grained sands are naturally more permeable, but when coarse grains of sand are added to a fine-grained sand, the permeability initially decreases and then increases. Permeability increases with moisture content up to approximately 5% moisture.

Figure 5.9 illustrates a schematic laboratory arrangement for determining the permeability of a specimen. Equipment is available in which permeability may be read directly from a scale on the instrument.

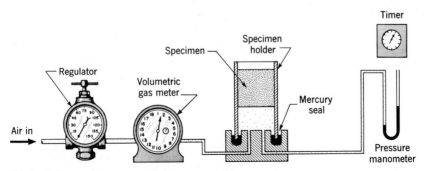

Figure 5.9
Equipment used to measure sand permeability.

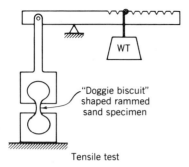 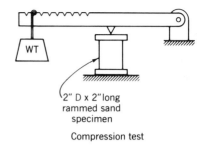

Figure 5.10
Principle of sand tester.

Sand Strength Test

Several strength tests have been devised to test the holding power of various bonding materials in green and dry sand. Compression tests are the most common, although tension, shear, and transverse tests are sometimes used. Procedure varies according to the type of equipment, but, in general, the tests are similar to those used for other materials. The fragile nature of sand requires special consideration in handling and loading test specimens. The principle of the sand strength machine is shown in Figure 5.10.

SAND-CONDITIONING EQUIPMENT

Properly conditioned sand is an important factor in obtaining good castings. New sand as well as used sand properly prepared accomplishes the following results.

1. Binder is distributed more uniformly around the sand grains.
2. Moisture content is controlled and particle surfaces are moistened.
3. Foreign particles are eliminated from the sand.
4. Sand is aerated so that it is not packed and is in proper condition for molding.
5. Sand is cooled to about room temperature.

Because conditioning by hand is difficult, most foundries have appropriate equipment for this operation. The mixer for preparing the sand (Figure 5.11) has two circular pans in which are mounted a combination of plows and *mullers* driven by a vertical shaft. The two mullers are arranged to process the sand continuously. The mullers provide an intensive kneading and rubbing action.

A representative sand reclamation and conditioning installation is illustrated in Figure 5.12. After the metal in a mold is solidified and is shaken out at the ends of the roller conveyor line, the sand falls through a grate on a belt conveyor as in the figure.

This conveyor carries the used sand to a smaller belt conveyor equipped with a magnetic

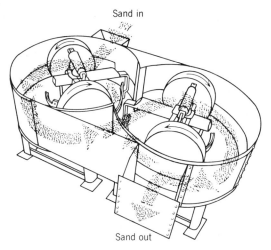

Figure 5.11
Continuous Simpson multimull.

separator. The sand is then discharged onto a bucket elevator from which it passes through an enclosed revolving screen into the storage bin. It is delivered from this bin to one or more mullers and conditioned for reuse. From here it is discharged to an overhead belt conveyor through an aerator that separates the sand grains and improves their flowability for molding. The cycle is completed when the sand is discharged into the several hoppers serving the molding stations where the molds are made and poured.

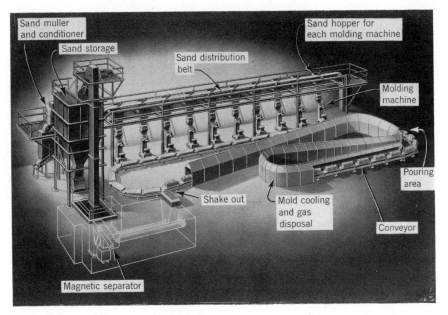

Figure 5.12
Progressive foundry unit.

CORES

When a casting is to have a cavity or recess in it such as a hole for a bolt, some form of *core* must be introduced into the mold. A core is sometimes defined as "any projection of the sand into the mold." This projection may be formed by the pattern itself or made elsewhere and introduced into the mold after the pattern is withdrawn. Either internal or external surfaces of a casting can be formed by a core.

Types of Cores

Cores may be classified as *green-sand cores* and *dry-sand cores*. Figure 5.13 shows various types of cores. Green-sand cores (Figure 5.13A) are those formed by the pattern and made from the same sand as the rest of the mold. This drawing shows how a flanged casting can be molded with the hole through the center "cored out" with green sand.

Dry-sand cores are those formed separately to be inserted after the pattern is withdrawn but before the mold is closed. They are usually made of clean river sand that is mixed with a binder and then baked to give the desired strength. The box in which they are formed to proper shape is called a *core box*.

Several types of dry-sand cores are also illustrated. The usual arrangement for supporting a core when molding a cylindrical bushing is shown in Figure 5.13B. The projections on each end of the cylindrical pattern are known as *core prints* and form the seats that support and hold the core in place. A *vertical core* is shown in Figure 5.13C, the upper end of which requires considerable taper so as not to tear the sand in the cope when the flask is assembled. Cores that have to be supported only at one end must have

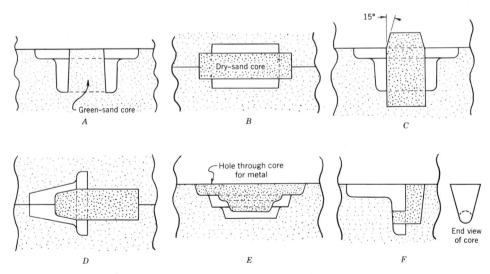

Figure 5.13
Typical cores. A, Solid pattern with green-sand core. B, Dry-sand core supported on both ends. C, Vertical dry-sand core. D, Balanced dry-sand core. E, Hanging dry-sand core. F, Drop core.

the core print of sufficient length to prevent the core from falling into the mold. Such a core (Figure 5.13D) is known as *balanced core*. A core supported above and hanging into the mold is shown in Figure 5.13E. This type usually requires a hole through the upper part to permit the metal to reach the mold. A *drop core* (Figure 5.13F) is required when a hole is not in line with the parting surface and must be formed at a lower level.

In general, green-sand cores should be used where possible to keep the pattern and casting cost to a minimum. Separate cores naturally increase the production cost. *Core boxes* must be made, and the cores must be formed separately, baked, and properly placed in the molds. More accurate holes can be made with dry-sand cores, for they give a better surface and are less likely to be washed away by the molten metal.

In setting dry-sand cores into molds, adequate supports must be provided. Ordinarily these supports are formed into the mold by the pattern, but for large or intricate cores additional supports in the form of **chaplets** (small metal shapes made of low-melting-point alloy) are placed in the mold to give additional support to the core until the molten metal enters the mold and fuses the chaplets into the casting. The use of chaplets should be limited as much as possible because of the difficulty in fusing the chaplet with the metal.

Essential Qualities

A core must have sufficient *strength* to support itself. *Porosity* or permeability is also an important consideration in making cores. As the hot metal pours over the cores, gases are generated by the heat being in contact with the binding material. Provision must be made to carry away these gases.

The core must have a *smooth surface* to ensure a smooth casting. Cores require *refractoriness* to resist the action of the heat until the hot metal has stabilized.

Core Making

The core is formed by being rammed into a core box or by the use of sweeps. Fragile and medium-sized cores should be reinforced with wires for added strength to withstand deflection and the hydrostatic action of the metal. In large cores, perforated pipes or arbors are used. In addition to giving the core strength, they also serve as a large vent. Cores having round sections are often made in halves and glued together after baking.

Binders and Core Mixtures

Among several types of binders used in making cores are those classified as oil binders. one of these, linseed oil, is frequently used in small cores. The oil forms a film around the sand grain that hardens when oxidized by the action of the heat. Such cores should be baked for 2 h at a temperature between 350° and 425°F (180°–220°C). A common mixture uses 40 parts river sand and 1 part linseed oil. An advantage of this proportion is its strength and its resistance to water absorption.

Another group of water-soluble binders includes wheat flour, dextrin, gelatinized starch, and many commercial preparations. The ratio of binder to sand in these mixtures is rather high (i.e., 1 : 8 or more parts of sand). Frequently a small percentage of old sand is used in place of new sand. In addition, pulverized pitch or rosin may be used.

Several types of thermosetting plastics including urea and phenol formaldehyde are used as core binders. They are made in both liquid and powder form and are mixed with silica flour, cereal binder, water, kerosene, and a parting liquid. Urea resin binders are baked at 325° to 375°F (165°–190°C), and the phenolic binders at 400° to 450°F (200°–230°C). Both respond to dielectric heating and are combustible under the heat of the metal. Their success as core binders is based on their high adhesive strength, moisture resistance, burnout characteristic, and ability to provide a smooth surface to the core.

The use of furfuryl alcohol resin binders with sand is replacing many of the core binders that require baking. These *resin binders* either are air-dried or are blown or tamped into a hot core box maintained at about 425°F (220°C). The *hot box*-made cores can be removed from the mold in 10 to 20 s. If hot boxes are not used, the furfuryl alcohol resin is mixed with formaldehyde or urea formaldehyde resins from which air-dried cores can be made. They are known as furan or no-bake[2] cores.

Many cores are made from a mixture of sand and sodium silicate. After being rammed into a core box they are hardened by the application of carbon dioxide gas. This is the CO_2 process. Because cores made this way do not need to be baked, they can be produced rapidly at low cost in an air-conditioned environment.

CORE-MAKING MACHINES

Cores can be made not only in hand-filled boxed but also on a variety of molding machines including many of the conventional types such as the jolt, squeeze, rollover, jolt-squeeze, and sandslinger machines.

Pneumatic *core-blowing machines* also offer a rapid means of producing small and medium-sized cores in quantity production work. In this method sand is blown under pressure and at high velocity from a sand magazine into the core box.

Figure 5.14 illustrates the process. Suitable vents are built into the core box or blow plate to permit the air to escape. These vents must be small enough to prevent sand seepage. Their purpose is to allow air venting. This equipment is especially adapted to production work in which the expense of metal core boxes is justified.

Stock cores of uniform cross section may be produced continuously by an extrusion process. The machine consists of a hopper in which the sand is mixed. Below it, in a horizontal position, is a spiral screw conveyor that forces the prepared sand through a die tube at uniform speed and pressure.

MECHANICAL MOLDING EQUIPMENT

Machines can eliminate much of the labor in molding and at the same time produce better molds. Molding machines, varying considerably in design and method of operation, are named according to the way the ramming operation is performed. Figure 5.15 illustrates the principles used in packing a mold. The shading indicates the density or uniformity

---

[2]Quaker Oats Company product.

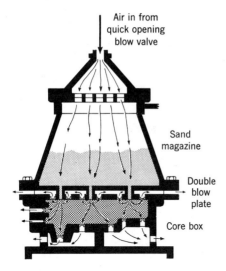

Figure 5.14
Schematic diagram of core-blowing operation.

of sand packing for each process. Machines utilizing these principles will now be described.

Jolt Machine

The plain jolt-molding machine is equipped with adjustable flask-lifting pins to permit the use of flasks of various sizes within the capacity of the machine. Molds weighing up to 13,000 lb (5850 kg) can be made on the larger machines. In the operation of this machine the table is raised a short distance by means of air pressure and then dropped. This action causes the sand to be packed evenly about the pattern. The density of the sand is greatest around the pattern and at the parting line and varies according to the height of the drop and depth of the sand in the flask. The uniform ramming about the pattern gives added strength to the mold and reduces the possibility of swells, scabs, or runouts. Castings produced under such conditions vary little in size or weight. The lifting pins on the machine engage the flask and raise it from the match plate after the mold is complete. Jolt machines quite obviously take care of only one part of a flask at a time and are especially adapted to production runs.

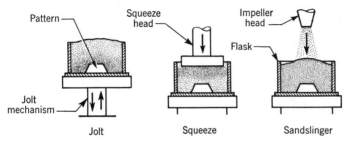

Figure 5.15
Machine molding principles.

Squeezer Machine

Squeezer machines press the sand in the flask between the machine table and an overhead platen. Greatest mold density is obtained at the side of the mold from which the pressure is applied. Because it is impossible to obtain uniform mold density by this method, squeezing machines are limited to molds only a few inches in thickness.

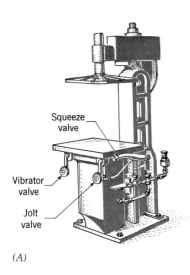

(A)

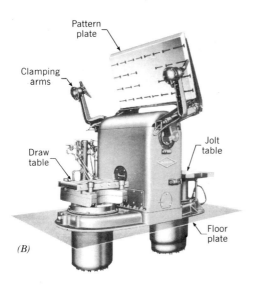

(B)

Figure 5.16
Types of mechanical molding machines. *A*, Jolt-squeeze molding machine. *B*, Joint rollover pattern–draw molding machine. *C*, Contour diaphragm molding machine. *D*, Motive-type sandslinger.

Jolt-Squeeze Machine

Many machines, Figure 5.16 use both jolt and squeeze. To produce a mold in this machine the flask is assembled with the match plate between the cope and drag, and the assembly is placed upside down on the machine table. Sand is shoveled into the drag and leveled off and a bottom board is placed on top. The jolting action then rams the sand in the drag. The assembly is turned over and the cope filled with sand and leveled off. A pressure board is placed on top of the flask and the top platen of the machine is brought into position. By application of pressure the flask is squeezed between the platen and table, packing the sand in the cope to the proper density. After the pressure is released the platen is swung out of the way. The cope is then lifted from the match plate while the plate is vibrated, after which the plate is removed from the drag. This machine eliminates six separate hand operations: ramming, smoothing the parting surface, applying parting sand, swabbing around the pattern, rapping the pattern, and cutting the gate.

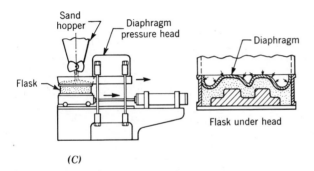

(C)

Figure 5.16 (continued)

Jolt-Squeeze Rollover Machine

This machine, similar to the conventional jolt-squeeze molding machine, has two arms that engage the flask after jolting and lift it a sufficient height so that it can be rolled over. The cope is then filled with sand and rammed by squeezing action. Next it is clamped by two air clamps on the upper platen and drawn from the match plate. This lifting device handles the cope while the match plate is manually removed from the drag. When the mold is ready to be closed, the cope is swung back into position and the drag raised until the mold halves are together. This machine is designed to handle larger flasks than can be conveniently handled on the usual jolt-squeeze machine.

Diaphragm Molding Machines

A recent development in molding machines utilizes a pure gum rubber diaphragm for packing the sand over the pattern contour as illustrated in Figure 5.16C. The process uses a uniform air pressure to force the rubber diaphragm over the surface of the mold regardless of the pattern contour.

In Figure 5.16C the flask is shown at the filling position. The flask and sand chute are then moved to the right under the diaphragm pressure head. Air is admitted to the pressure head and the diaphragm is forced against the molding sand in the flask. The flask is then returned to its original position, striking off any sand above the flask. A pin lift removes the match plate from the flask. The entire process is very rapid, and close tolerances can be maintained because of the uniform packing of the sand.

Jolt Rollover Pattern Draw Machine

For large molds up to 12,000 lb (5400 kg) that are difficult to handle, machines such as shown in Figure 5.16B have been developed. The sand is first packed by jolting. After the sand is leveled off, a bottom board is placed on the mold and clamped in position; then the assembly is rolled over and the pattern drawn from the mold hydraulically. This machine is used for separate cope or drag molding; normally the cavity is in the drag only.

Sandslinger

Uniform packing of sand in molds is an important operation in producing castings. For large molds a mechanical device known as the sandslinger has been developed. Figure 5.16D shows a motive-type sandslinger that is a self-propelled unit operating on a narrow-gage track. The supply of sand is carried in a large tank with a capacity of about 300 ft^3 (8.5 m^3), which may be refilled at intervals by overhead-handling equipment. A delivery belt feeding out of a hopper on the frame at the fixed end conveys the sand to the rotating impeller head. The impeller head, which is enclosed, contains a single rotating, cup-shaped part that slings the sand into the mold. This part, rotating at high speed, slings more than 1000 small buckets of sand a minute. The ramming capacity of this machine is 7 to 10 ft^3 (0.2–0.28 m^3), or 1000 lb (450 kg) sand/min. The density of the packing can be controlled by the speed of the impeller head. For high production, machines of this type are available having a capacity of 4000 lb sand/min (30 kg/s).

POURING AND CLEANING CASTINGS

In jobbing and small-production foundries, the molds are lined up on the floor as they are made and the metal is taken to them in small hand ladles. When more metal is required or if heavier metal is poured, ladle tongs designed for two workers are used. In large foundries engaged in mass-producing castings the problem of handling molds and molten metal is solved by placing the molds on conveyors and passing them slowly by a pouring station. The pouring station may be located permanently next to the furnace, or metal may be brought to certain points by overhead-handling equipment. The conveyor serves as a storage place for the molds while they are being transported to the cleaning room.

After a casting has solidified and cooled to a suitable temperature for handling, it is shaken from the mold. Very often this is done at a ventilated mold shakeout. The dust is collected by a cyclone dust collector while the sand is collected underneath and transported to the conditioning station. All castings are retained on grate bars of the shakeout.

Nonferrous castings do not present much of a cleaning problem, because they are poured at lower temperatures than iron or steel and the sand has little tendency to adhere to the surface. Gates and sprues are cut off either in a sprue press or with a metal band saw. Hand or rotary machine brushing is usually sufficient to prepare the casting for machining operations.

Iron and steel castings are covered with a layer of sand and scale that is somewhat difficult to remove. The gates and risers on iron castings may be broken off, but to remove them from steel castings a cutting torch or a high-speed cutting-off wheel is necessary. The process of removing gates, runners, risers, flash, and cleaning the casting is termed *fettling*. The fettling operation accounts for 15% to 25% of the labor cost of manufacturing a casting.

To clean castings several methods may be used depending on the size, kind, and shape of the castings. The most common piece of equipment used is the rotating cylindrical *tumbling mill*. It cleans by the tumbling action of the castings upon one another as the mill rotates. A similar piece of equipment is shown in Figure 5.17. It cleans 65 to 100 lb (30–45 kg) of gray iron or malleable castings in 5 to 8 min. Larger machines of this type have capacities of over 1 ton (0.9 Mg) per charge. The machine consists of cleaning barrel formed by an endless apron conveyor. The work is tumbled beneath a blasting unit located just above the load, and a metallic shot is blasted onto the castings. After striking the load the shot falls through holes in the conveyor and is carried overhead to a separator and storage hopper. From there it is fed by gravity to the blasting unit. The unit is unloaded by reversing the apron conveyor. A dust collector is installed with the machine to eliminate dust hazards.

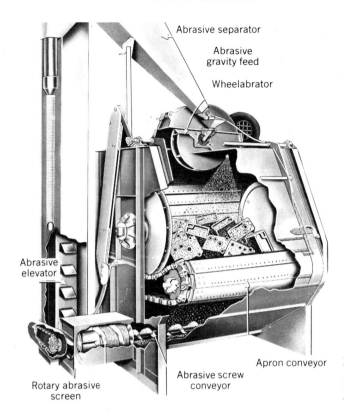

Figure 5.17 Tumbler and blasting machine for cleaning castings.

Sandblasting units may be used separately for cleaning castings. Sharp sand is blown against castings inside a blasting cabinet, removing all foreign matter completely and giving the casting a clean surface appearance. Castings that are to be plated or galvanized are frequently pickled in a weak acid solution and then rinsed in hot water. Large castings that are difficult to handle are often cleaned by hydraulic means. The casting is placed on a rotating table, and streams of water under considerable pressure wash away the sand.

In addition to these cleaning processes, many castings require a certain amount of chipping or grinding to remove surface and edge defects. Stand, portable, and swing frame grinders are used for this work. Fast, free-cutting abrasive wheels, operating at a cutting speed of around 9500 ft/min (48 m/s), are also recommended.

QUESTIONS AND PROBLEMS

1. Give an explanation of the following terms:

   Green sand
   Parting sand
   Blind riser
   Sweep
   Shrink rule
   Draft
   Gib block
   Sodium silicate

   Sharp sand
   Fineness number
   Permeability
   Muller
   Chaplet
   Resin binder
   Core blower

2. If approximately 53% of metal poured as castings is actually used for the casting(s) desired, what makes up the remainder of the metal poured?

3. Describe the step-by-step process to mold the V block shown in Figure 5.6 (a) by the disposable-pattern method and (b) by the solid-pattern method.

4. Sketch the isothermal lines for a dumbbell-shaped casting.

5. How would the gear blank shown in Figure 5.2 be molded (a) by the CO_2 technique and (b) by use of a styrofoam pattern?

6. How could a sweep-type pattern cast a solid vase-shaped casting?

7. Explain the problems in slowly pouring a mold with a polystyrene pattern.

8. What happens to conventional molds if they are poured too slowly?

9. Why is a blind riser more apt to "force-feed" a possible shrinkage area?

10. Explain the use of risers in problem 4, and redraw the isothermal lines with risers employed.

11. Where and how could exothermic materials be used in casting a dumbbell? What would be the isothermal lines?

12. What are the disadvantages and advantages of a skimming gate?

13. Sketch a pattern requiring a cheek.

14. If an aluminum casting were to be used for a steel casting, what allowances would be necessary in the wood pattern used to make the aluminum casting?

15. Outline the steps necessary to mold the split pattern shown in Figure 5.5B.

16. Should an allowance be made for draft in a disposable mold casting? Why?

17. For a steel casting, state the differences in allowances between a wood and a polystyrene pattern.

18. To what does the "full mold" process refer?

19. What surfaces on the gib block patterns (Figure 5.7) should have draft?

20. Describe a method of producing 5000 polystyrene patterns for an electric motor base.

21. How are gates, risers, and sprues accommodated in problem 20?

22. Sketch a part that would require a chaplet to effect the mold.

23. Give the advantage for a match plate in problem 20.

24. Would one normally use a match plate with polystyrene patterns?

25. Discuss the factors relating to permeability.

26. How does grain size of a molding sand affect permeability?

27. How does grain size distribution of a molding sand affect strength and permeability?

28. How is permeability artificially introduced into the cope of a mold?

29. How does clay content affect molding sand strength, permeability, and surface finish?

30. What are some ways to improve the surface finish of a mold?

31. Why are weights employed on top of the cope before pouring? Why are weights more necessary when pouring cast iron as compared to aluminum?

32. Sand weights 100 lb/ft³ and a cope contains 1.4 ft³ of sand and 8 lb of wood. The cross section at the parting line of a casting is 14 in.² If the cope is 10 in. high, how much weight is required to keep the cope from "floating" if aluminum is being poured? In SI units?

33. In problem 32 assume steel is being cast; how much weight on the cope is required?

34. Discuss the effects of moisture on molding sand.

35. How does moisture in a molding sand affect a wood pattern, a core, polystyrene, permeability, strength?

36. For 1 ton of sharp clean sand, how many pounds of bentonite need to be added for a nonferrous molding sand?

37. Why are cores seldom necessary in disposable-pattern molding?

38. In problems 36, how many pounds of water? How many gallons?

39. What choices can be made in selecting a core binder?

40. State the principal use for a furan binder.

41. What is the purpose and advantage of dielectric heating ability for a core binder?

42. Why is it best to use green sand in making a core if the mold permits?

43. Explain the differences in permeability that come from jolting, squeezing, or sandslinging a mold.

44. In problem 43, what are the differences in casting surface that can be expected?

45. Explain the steps in fettling.

46. What are the comparative advantages and disadvantages of sandblasting?

47. Why is coarser sand better for steel castings than a fine-grained one? Why is it that as castings increase in size it is often better to use increasingly coarser sand?

48. What is a core print and what is its purpose?

49. Explain the process of making a core by the CO_2 process. Why is the process employed so little in making molds?

50. Explain the purpose of parting sand and the mechanism by which it works.

51. If 40% of a casting is sprues, gates, and risers, how much aluminum must be melted per day to produce 350 finished castings weighing 1.9 kg each, in SI and English units?

52. What is the actual length measurement if a cast iron shrink rule is used for laying off the following lengths: 14 in. (355 mm) and 3 in. (76 mm)? In English and SI units?

53. Make a graph indicating how the impeller speed on a sandslinger is affected by the impeller diameter.

54. Calculate the grain fineness number of a sand that when passed through a set of standard testing sieves gives the following percentages retained on each sieve: No. 6, 0%; No. 12, 0%; No. 20, 4.0%; No. 30, 4.0%; No. 40, 2.0%; No. 50, 3.5%; No. 70, 5.0%; No. 100, 18%; No. 140, 25.0%; No. 200, 17.0%; No. 270, 9.0%; and pan, 5.0%.

55. If an abrasive grinder rotates at 1260 rpm, what would be the approximate diameter of the wheel if it operates at the optimum cutting speed? Also in SI units.

56. If a 10-in.-diameter sandslinger is to deliver sand at 60 ft/s velocity to a mold, how fast must the impeller turn?

CASE STUDY

FOUNDRY ECONOMICS

The Try Castings Company has been asked by Bitterhand Company to bid on the manufacture of a casting in the form of a soft drink can except for a 1.5-in. (38 mm) hole down through the center and along the longitudinal axis. The can is 2.5 in. (63.5 mm) in diameter and 4.75 in. (120.7 mm) long. The quantity required for store displays is 5000 units in steel and 100 units in aluminum. Do an analysis of the foundry process you recommend from pattern to delivery, including fettling for each item. Make a step-by-step analysis and give your reason for each step.

Organize your presentation to Bitterhand for maximum clarity, brevity, and understandability. Because the aluminum cans are for display, they need to be machined. Sketch the pattern for this unit and the method of molding to help Bitterhand purchasing agents understand the added costs per unit of the aluminum units. It will help your company explain how much the aluminum units will cost as compared to the steel units if steel costs $0.26/lb and aluminum costs $1.45/lb. Explain to Bitterhand why the same molds and molding sand cannot be used for both castings.

CHAPTER 6
CONTEMPORARY CASTING PROCESSES

The process used for making a casting depends on the quantity to be produced, the metal to be cast, and the intricacy of the part. Most commercial metals can be cast in sand molds, and sizes vary from small to large. However, sand molds are used once and are destroyed after metal solidification. *Permanent* molds offer considerable saving in cost for large production quantities when the size of the casting is not large.

A summary of the various special casting methods that will be discussed in this chapter is as follows.

A. Casting in metallic molds
  1. Die
  2. Low pressure
  3. Gravity or permanent mold
  4. Slush
  5. Pressed or corthias

B. Centrifugal casting
  1. True centrifugal
  2. Semicentrifugal
  3. Centrifuging

C. Precision or investment casting
  1. "Lost wax" method
  2. Ceramic shell process
  3. Plaster molds
  4. Shell molding
  5. CO_2 mold hardening process
  6. Molds of wood, paper, and rubber

D. Continuous casting
  1. Reciprocating molds
  2. Draw casting
  3. Stationary molds
  4. Direct sheet casting

METHODS OF CASTING IN METALLIC MOLDS

Permanent molds must be made of metals capable of withstanding high temperatures. Because of their high cost they are recommended only when many castings are to be produced. Although permanent molds are impractical for large castings and alloys of high melting temperatures, they can be used advantageously for small and medium-size castings.

Die Casting

In die casting, molten metal is forced by pressure into a metal mold known as a die. Because the metal solidifies under a pressure from 80 to 40,000 psi (0.6–275 MPa), the casting conforms to the die cavity in shape and surface finish. The usual pressure is from 1500 to 2000 psi (10.3–14 MPa).

Die casting is the most widely used of the permanent-mold processes. Two methods are employed: (1) hot chamber and (2) cold chamber.

The principal distinction between the two methods is determined by the location of the melting pot. In the **hot-chamber method** a melting pot is included with the machine and the injection cylinder is immersed in the molten metal at all times. The injection cylinder is actuated by either air or hydraulic pressure, which forces the metal into the dies to complete the casting. Machines using the cold-chamber process have a separate melting furnace, and metal is introduced into the injection cylinder by hand or mechanical means. Hydraulic pressure then forces the metal into the die.

The process is rapid because both the dies and cores are permanent. A smooth surface not only improves the appearance but also minimizes the work required to prepare the castings for plating or other finishing operations. The wall thickness can be more uniform than in sand castings and, consequently, less metal is required. The optimum production quantity ranges from 1000 to 200,000 pieces. The maximum weight of a brass die casting is about 5 lb (2.3 kg), but aluminum die castings of over 100 lb (45 kg) are common. Small to medium-size castings can be made at a cycle rate of 100 to 800 die fillings/h. The size is so accurately controlled that little or no machining is necessary. The scrap loss is low because the sprue, runners, and gates can be remelted. The process largely eliminates secondary operations such as drilling and certain types of threading. Die-casting tolerances vary according to the size of the casting and the kind of metal. For small castings the tolerance ranges from ±0.001 to 0.010 in. (±0.03–0.25 mm). The closest tolerances are obtained when zinc alloys are die cast.

One of the limitations of die casting is the high cost of the equipment and dies. This is not an important factor in mass production, but it does limit its use in short-run jobs. In some cases there is an undesirable chilling effect on the metal unless high temperatures are maintained. Metals having a high coefficient of contraction must be removed from the mold as soon as possible because of the inability of the mold to contract with the casting.

Die castings were once limited to low-melting alloys, but with a gradual improvement of heat-resisting metals for dies, this process can now be used for numerous alloys. Whereas gray cast iron and low-carbon and alloy steels have been produced in dies made of unalloyed sintered molybdenum, the process is commercially limited to nonferrous alloys.

Dies

Dies for the hot- and cold-chamber machines are similar in construction because there is little difference in the method of holding and operating them. They are made in two sections to allow casting removal; they are usually equipped with heavy dowel pins to maintain the halves in proper alignment. Metal enters the stationary side when the die is locked in the closed position. As the die opens, the ejector plate in the movable half of the die is advanced, so that pins project through the die half and force the casting from the cavity and fixed cores. The dies are provided with a separate mechanism for moving the ejector plate or movable cores. The life of these molds depends on the metal cast and may range from 10,000 fillings, if brass castings are made, to several million if zinc is used.

It is always desirable to provide vents and small overflow wells on one side of a die to facilitate the escape of air and to catch surplus metal that has passed through the die cavity. There is a certain amount of *flash* metal found at the sectional mating surface that must be trimmed off in the finishing operation.

For large or complex castings a single-cavity mold is used. The casting and gate from such a mold are shown in Figure 6.1. The part shown is cast with a steel insert from magnesium in a 600-ton (540-Mg) machine. If the quantity of castings to be produced is large and they are relatively small in size, a multiple-cavity die can be used. Figure 6.2 shows a number of castings with the flash, gates, and sprue from such a die. The parts are produced from aluminum in a 400-ton (360-Mg) machine. A combination die is one that has two or more cavities, each of which is different. They are frequently made of insert blocks that can be removed so that other die blocks can be substituted. Most dies are provided with channels for water cooling to keep the die at the correct temperature for rapid production.

Hot-Chamber Die Casting

Low-melting alloys of zinc, tin, and lead are the most widely used materials cast in hot-chamber machines. Chapter 4 includes a discussion of these alloys. Most other materials either have too high a melting point, an affinity for iron, or cause other problems that will reduce the life of the machine. Hot-chamber castings vary in size from an ounce to 90 lb (0.03–40 kg), although very small castings are usually cast in multiple-mold dies.

In this method metal is forced into the mold and pressure is maintained during solidification either by a *plunger* or by *compressed air*. The plunger-type machine, illustrated in Figure 6.3, is hydraulically operated for both the metal plunger and the mechanism for opening and closing the die. In this machine the plunger operates in one end of a gooseneck casting that is submerged in the molten metal. With the plunger in the upper position, metal flows by gravity into this casting through several holes just below the plunger. On

Figure 6.1
Gate and casting of a chain saw crankcase from a single-cavity die.

Figure 6.2
The part is an end cap for an aluminum connecting rod used in air compressors.

the downstroke these holes are closed by the plunger and pressure is applied on the entrapped metal, causing it to be forced into the die cavity. Pressures over 5000 psi (35 MPa) are used in some machines of this type, resulting in castings of dense structure. As soon as the casting is solidified, the pressure is relieved, the dies are forced open, and the casting is ejected by means of knockout pins. The sprue is removed with the runner and castings.

The air-operated machines have a gooseneck operated by a lifting mechanism. In the starting position the casting is submerged in the molten metal and is filled by gravity. It is then raised so that the nozzle is in contact with the die opening and locked in position. Compressed air at pressures ranging from 80 to 600 psi (0.5–4.0 MPa) is applied directly on the metal, thus forcing it into the die. When solidification is nearly complete, the air pressure is turned off and the gooseneck lowered into position to receive more metal. The operation of opening the dies, withdrawing cores, and ejecting the castings is the same as for the plunger-type machine.

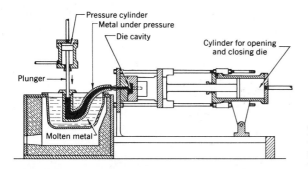

Figure 6.3
Plunger-activated actuated hot-chamber die-casting machine.

Cold-Chamber Die Casting

Die casting brass, aluminum, and magnesium requires higher pressures and melting temperatures and necessitates a change in the melting procedure from that previously described. Chapter 4 has a discussion of the alloys used in cold-chamber casting. These metals are not melted in a self-contained pot, because the life of the pot would be very short. The usual procedure is to heat the metal in an auxiliary furnace and ladle it to the plunger cavity next to the dies. It is then forced into the dies under hydraulic pressure. Machines operating by this method are built to be very strong and rigid to withstand the heavy pressures exerted on the metal as it is forced into the dies. Of the two machines in general use, one has the plunger in a vertical position, the other in a horizontal position.

A diagrammatic sketch illustrating the operation of horizontal-plunger cold-chamber machines is shown in Figure 6.4. In Figure 6.4A the dies are shown closed with cores in position and the molten metal ready to be poured. As soon as the ladle is emptied the plunger moves to the left and forces the metal into the mold (Figure 6.4B). After the metal solidifies, the cores are withdrawn and the dies are opened. In Figure 6.4C the dies are open and the casting is ejected from the stationary half. To complete the process of opening, an *ejector rod* comes into operation and ejects the casting from the movable half of the die (Figure 6.4D). This operating cycle is used in a variety of machines that operate at pressures ranging from 5600 to 22,000 psi (39–150 MPa). These machines

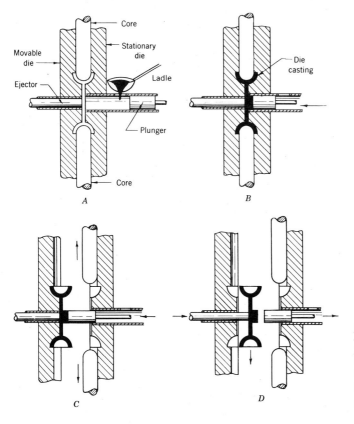

Figure 6.4
The die casting of brass, aluminum, or magnesium in horizontal-plunger cold-chamber machine.

are fully hydraulic and semiautomatic. After the metal is poured, the rest of the operations are automatic.

The 2500-ton (22-MN) hydraulically operated, cold-chamber die-casting machine shown in Figure 6.5 is for making die castings up to 84 lb (38 kg) of aluminum, brass, or magnesium.

The manufacture of brass die castings is an important achievement. The difficulties of the high temperatures and the rapid oxidation of the steel dies have been largely overcome by improvements in die metals and casting at as low a temperature as possible. One machine is designed to use metal in a semiliquid or plastic state to permit operation at lower temperatures than those used for liquid metal. To protect the dies further from overheating, water is circulated through plates adjacent to the dies. Metal is maintained under close temperature control and is ladled by hand to the compression chamber. The pressure used in this machine is 9800 psi (68 MPa); 100 to 200 shots/h can be made depending on the size of the machine.

Two variations of this process, each with the injection plunger in a vertical position, are diagrammatically illustrated in Figure 6.6. In Figure 6.6*B* the compression chamber into which the metal is ladled is separate from the dies. The metal is poured into this cavity onto a spring-backed plunger. As the ram descends this plunger is forced down until the gate opening is exposed, permitting the metal to be forced into the die cavity. As the ram returns to its upper position the ejector plunger also moves upward, carrying with it any surplus metal. As the die opens the casting is ejected.

A variation of this machine with the compression chamber a part of the die is illustrated in Figure 6.6*A*. Metal is poured into this chamber at the upper part of the die and forced by pressure into the die cavity as the ram descends. As soon as the ram moves up the dies open and the casting is ejected by means of the ejector pins. The sprue and excess metal are later trimmed off in the finishing operation.

Low-Pressure Permanent-Mold Casting

In the low-pressure permanent-mold process a mold made of metal is mounted over an induction furnace as shown in Figure 6.7. The furnace is sealed and an inert gas under

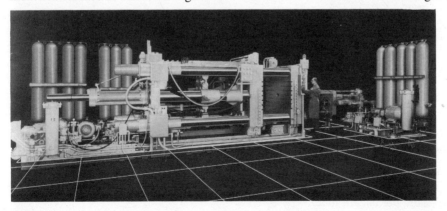

Figure 6.5
A 2500-ton (22-MN) cold-chamber die-casting machine.

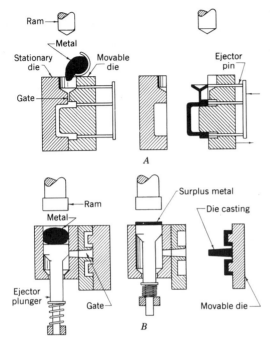

Figure 6.6
Two methods of die construction for pressing brass castings. *A,* Compression chamber in dies. *B,* Compression chamber separate from dies.

pressure forces the molten metal in the furnace up through a heated refractory *"stalk"* into the cavity. Vacuum pumps are sometimes utilized to remove entrapped air from the mold and to ensure a more dense structure and faster filling. Small castings may be left to cool in the mold for a minute or less, but castings up to 65 lb (29 kg) in weight are reported to have a cycle time of only 3 min. The process is most economical if the production rate is at least 5000 to 50,000 parts per year. Castings produced by this method

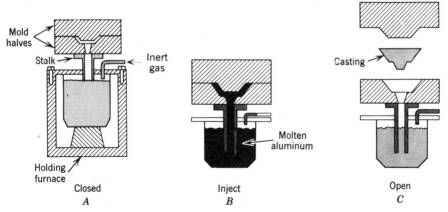

Figure 6.7
Low-pressure permanent mold.

are dense, free from inclusions, have good dimensional accuracy, and the scrap loss is usually less than 10% and can be as low as 2%.

Gravity or Permanent-Mold Casting

This method utilizes a permanent mold made of metal or graphite. The molds are usually coated with a refractory wash and then lampblack, which reduces the chilling effect on the metal and facilitates the removal of the casting. No pressure is used except that obtained from the head of metal in the mold. The process is used successfully for both ferrous and nonferrous castings, although the latter type does not present so many problems as ferrous castings because of the lower pouring temperatures. The simplest type of permanent mold hinges at one end of the mold with provision for clamping the halves together at the other. Some production machines, as illustrated in Figure 6.8, are circular in arrangement and have molds placed at a number of stations. The cycle of events consists of pouring, cooling, ejecting the casting, blowing out the hot molds, coating them, and in some cases setting the cores. Both metal and dry-sand cores can be used in molds of this type. If metal cores are used they are withdrawn as soon as the metal starts to solidify.

Permanent molds produce castings free from embedded sand and with good finish and surface detail. They are especially adapted to the quantity production of small and medium-size castings and are capable of maintaining tolerances ranging from 0.0025 to 0.010 in. (0.064–0.25 mm). The high initial cost of equipment and the cost of mold maintenance might be listed as disadvantages of this process, which turns out such products as aluminum pistons, cooking utensils, refrigerator parts, electric irons, and small gear blanks.

Cast iron is being increasingly cast by the gravity method. Often the dies are lined with sand about $1/4$ in. (6 mm) thick by utilizing a core blower. In this process only a small amount of sand is used, compared to sand casting, the cleaning costs are lower, and the accuracy better.

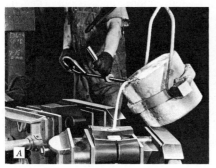

Figure 6.8
Multistation machine for permanent mold casting. *A*, Pouring into metal mold. *B*, Mold opening for removal of casting.

Slush Casting

Slush casting is a method of producing hollow castings in metal molds without the use of cores. Molten metal is poured into the mold, which is turned over immediately so that the metal remaining as liquid can run out. A thin-walled casting results, the thickness depending on the chilling effect from the mold and the time of the operation. The casting is removed by opening the halves of the mold. This method of casting is used only for ornamental objects, statuettes, toys, and other novelties. The metals used for these objects are lead, zinc, and various low-melting alloys. Parts cast in this fashion are either painted or finished to represent bronze, silver, or other more expensive metals.

Pressed or Corthias Casting

This method of casting resembles both the gravity and the slush processes but differs somewhat in procedure. A definite amount of metal is poured into an open-ended mold, and a close-fitting core is forced into the cavity causing the metal to be forced into the mold cavities with some pressure. The core is removed as soon as the metal sets, leaving a hollow, thin-walled casting. This process, developed in France by *Corthias,* is limited in use mainly to ornamental casting of open design. Nonmetallic molds are available that can be used with both high- and low-temperature alloys.

ELECTROSLAG CASTING

The electroslag casting process is unusual in that it does not employ a furnace. Instead, consumable electrodes melting or striking beneath a slag layer furnish molten metal to fill a water-cooled permanent mold. The molten metal continually drips or runs into the mold. It does not come in contact with the atmosphere because of the slag layer. No gates or risers are necessary, and usually the electrodes are caused to withdraw from the mold in concert with its filling from bottom to top. Studies indicate metal cast in this way may be superior to forgings. One interesting application comes about when the electrode material is changed in carbon content to effect a varying property in the casting.

CENTRIFUGAL CASTING

Centrifugal casting is the process of rotating a mold while the metal solidifies, so as to utilize centrifugal force to position the metal in the mold. Greater detail on the surface of the casting is obtained, and the dense metal structure has superior physical properties. Castings of symmetrical shape lend themselves particularly to this method, although many other types of castings can be produced.

Centrifugal casting is often more economical than other methods. Cores in cylindrical shapes and risers or feedheads are both eliminated. The castings have a dense metal structure, with all impurities forced back to the center where frequently they can be machined out. Because of the pressure exerted on the metal, thinner sections can be cast than would be possible in static casting.

Although there are limitations on the size and shape of centrifugally cast parts, piston rings weighing a few ounces and paper mill rolls weighing over 42 tons (38 Mg) have been cast in this manner. Aluminum engine blocks utilize centrifugally cast iron liners. If a metal can be melted, it can be cast centrifugally, but for a few alloys the heavier elements tend to be separated from the base metal. This separation is known as *gravity segregation*.

The methods of centrifugal casting may be classified as follows: (1) true centrifugal casting, (2) semicentrifugal casting, and (3) centrifuging

True Centrifugal Casting

True centrifugal casting is used for pipe, liners, and symmetrical objects that are cast by rotating the mold about its horizontal or vertical axis. The metal is held against the wall of the mold by centrifugal force, and no core is required to form a cylindrical cavity on the inside. There are two types of horizontal-axis molds used for producing cast-iron pipe. Massive, thick metal molds with a thin refractory coating allow the molten metal to begin solidification faster and the solidification to proceed from the wall of the mold toward the inside of the cast pipe. Such a mold encourages a preferred solidification that assures a more solid casting with any impurities on the inside wall. Figure 6.9 illustrates such a casting machine. The mold is spinning rapidly at the time the molten metal is introduced, and the spinning action is not stopped until solidification is complete. The wall thickness of the pipe produced is controlled by the amount of metal poured into the mold.

Another type of horizontal centrifugal casting uses a thick, highly insulating sand interface between the mold and the casting. Such a sand lining is spun into the mold. When metal is introduced the insulating nature of the sand prevents directional solidification, and hence the metal solidifies from the wall and from the inside pipe face at the same time. This can cause a spongy, less dense midsection that has entrapped inclusions.

Another example of true centrifugal casting is shown in Figure 6.10, which illustrates two methods that may be used for casting radial engine cylinder barrels. The horizontal

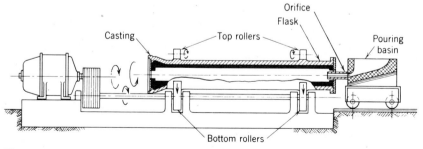

Figure 6.9
Centrifugal casting machine for casting steel or cast-iron pipe.

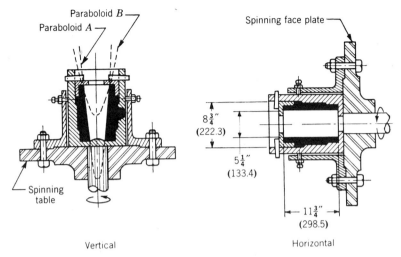

Figure 6.10
True centrifugal method of casting radial engine cylinder barrels.

method of casting is similar to the process followed in casting pipe lengths, and the inside diameter is a true cylinder requiring a minimum amount of machining. In vertical castings the inside cavity takes the form of a paraboloid as illustrated by the figure. The slope of the sides of the paraboloid depends on the speed of rotation, the dotted lines at A representing a higher rotational speed than shown by the paraboloid B. In order to reduce the inside-diameter differences between the top and bottom of the cylinder, spinning speeds are higher for vertical than for horizontal casting.

If the centrifugal force is too low the metal will slip, slide, or rain. If the force is too great the surface will develop abnormalities that are detrimental. Most horizontal castings are spun so that a force of about 65 g's or 65 × (force of gravity) will be developed. Vertically cast parts are usually spun at 90 to 100 g's. Vertical castings are much smaller in size and weight because of the instability of a spinning vertical cylinder, the higher g force necessary to overcome the paraboloidal shape and the increased pressures on the mold. The revolutions per minute necessary to produce a given number of g's are independent of the density of the metal or the total weight being cast. The centrifugal force CF is expressed as

$$CF = \frac{mv^2}{r}$$

where

CF = Centrifugal force, lb

$$m = \text{Mass} = \frac{W}{g} = \frac{\text{Weight, lb}}{\text{Acceleration of gravity (ft/s)}^2}$$

$$= \frac{W}{32.2}$$

v = Velocity, ft/s = $r \times w$

r = Radius, ft = $\frac{1}{2} D$

w = Angular velocity, rad/s

  = $2\pi/60 \times$ rpm

D = Inside diameter, ft

The number of g's is

$$g\text{'s} = CF/W$$

Hence

$$g\text{'s} = \frac{1}{W} \times \left[\frac{W}{32.2 \times r}\left(\frac{r \times 2\pi}{60}\right)^2\right]$$

$$= r \times 3.41 \times 10^{-4} \text{ rpm}^2$$

$$= 1.7 \times 10^{-4} \times D \times (\text{rpm})^2$$

The spinning speed for horizontal-axis molds may be found in English units from the equation

$$N = \sqrt{(\text{Number of } g\text{'s}) \times \frac{70{,}500}{D}}$$

where

N = rpm

D = inside diameter of mold, ft

Semicentrifugal Casting

In semicentrifugal casting the mold is completely full of metal as it is spun about its vertical axis, and risers and cores may be employed. The center of the casting is usually solid, but because the pressure is less there, the structure is not so dense and inclusions and entrapped air are often present. This method is normally used for parts in which the center of the casting will be removed by machining. The **stack mold** shown in Figure 6.11 can produce five semicentrifugally cast track wheels. The number of castings made in a mold depends on the size of the casting and the convenience in handling and assembling the molds. Rotational speeds for this form of centrifugal casting are not so great as for the true centrifugal process. The process produces a dense structure at the outer circumference, whereas the center metal is usually removed.

Centrifuging

In the centrifuge method several casting cavities are located around the outer portion of a mold, and metal is fed to these cavities by radial gates from the center. Either single or stack molds can be used. The mold cavities are filled under pressure from the centrifugal

116 CONTEMPORARY CASTING PROCESSES

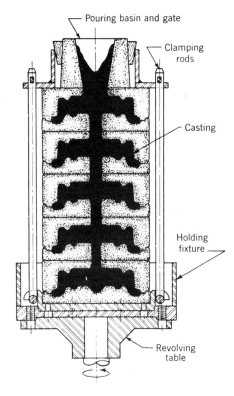

Figure 6.11
Semicentrifugal stack molding of track wheels.

force of the metal as the mold is rotated. In Figure 6.12 are shown five castings made in one mold by this process. The internal cavities of these castings are irregular in shape and are formed by dry-sand cores. The centrifuge method, not limited to symmetrical objects, can produce castings of irregular shape such as bearing caps or small brackets. The dental profession uses this process for casting gold inlays.

PRECISION OR INVESTMENT CASTING

Precision or *investment casting* employs techniques that enable very smooth, highly accurate castings to be made from both ferrous and nonferrous alloys. Figure 6.13 shows

Figure 6.12
Centrifuged castings with internal cavities of irregular shape.

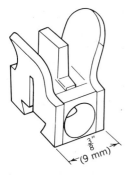

Figure 6.13
Investment casting of a front rifle sight.

a small investment casting made from a chrome–molybdenum–steel alloy. No other casting method other than die casting can assure production of so intricate a part. The process is useful in casting unmachinable alloys and radioactive metals. There are a number of processes employed, but all incorporate a sand, ceramic, plaster, or plastic shell made from an accurate pattern into which metal is poured. Although most castings are small, the investment process has been used to produce castings weighing over 100 lb (45 kg).

Advantages of precision or investment techniques are

1. Intricate forms with undercuts can be cast.
2. A very smooth surface is obtained with no parting line.
3. Dimensional accuracy is good.
4. Certain unmachinable parts can be cast to preplanned shape.
5. It may be used to replace die casting where short runs are involved.

The investment process is expensive, is usually limited to small castings, and presents some difficulties where cores are involved. Holes cannot be smaller than $1/16$ in. (1.6 mm) and should be no deeper than about 1.5 times the diameter.

"Lost Wax" Precision Casting Process

The *lost wax* process derives its name from the fact that the wax pattern used in the process is subsequently melted from the mold, leaving a cavity having all the details of the original pattern. The process, as originally practiced by artisans in the sixteenth century, consisted of forming the object in wax by hand. The wax object or pattern was then covered by a plaster investment. When this plaster became hard the mold was heated in an oven, melting the wax and at the same time further drying and hardening the mold. The remaining cavity, having all the intricate details of the original wax form, was then filled with metal. Upon cooling the plaster investment was broken away, leaving the casting. In large casting such as statuary,[1] plaster cores were used to provide relatively thin walls in the casting.

---

[1] Benvenuto Cellini's famous chapter on the casting of his bronze statue, *Perseue*, contains much interesting information on methods of molding used during the Renaissance. Cellini used a form of "lost wax" process.

118 CONTEMPORARY CASTING PROCESSES

Present practice requires that a replica of the part to be cast be made from steel or brass. From this replica a bismuth or lead alloy split mold is made. After wax is poured into the mold and solidification takes place, the mold is opened and the wax pattern removed. In the forming operation the mold is held in a water-cooled vise and the heated wax is injected into it under considerable pressure. Thermoplastic polystyrene resin is sometimes used in place of wax.

Several patterns are usually assembled with necessary gates and risers by heating the contact surfaces (wax welding) with a hot wire. This cluster is then supported in a metal flask. A finely ground refractory material, thinned by a mixing agent such as alcohol or water, is poured into the flask after first spraying the pattern with a fine silica–flour mixture. After the plaster sets, the mold is placed upside down and heated in an oven for several hours to melt out the wax and to dry the mold. The casting can be produced by gravity, vacuum, pressure, or centrifugal casting. Pressures ranging from 3 to 30 psi (0.02–0.2 MPa) are generally used in the casting operation. When the mold has cooled the plaster is broken away. After gates and feeders are cut off the castings are cleaned by grinding, sandblasting, or other finishing operations.

Ceramic Shell Process

This process, similar to the "lost wax" one, also involves the removal of a heat-disposable pattern from a refractory investment. The pattern is made from wax or a low-melting-point plastic; often a number of them are joined by "wax welding" into a cluster, as shown in Figure 6.14. The cost of producing plastic patterns is less than that for wax. A handle is usually mounted in one end of the wax sprue that forms the pouring basin. The pattern cluster is repeatedly dipped into a ceramic slurry and dusted with refractory material. This process, called stuccoing, is repeated until the shell is $3/16$ to $1/2$ in. (4.8–12.7

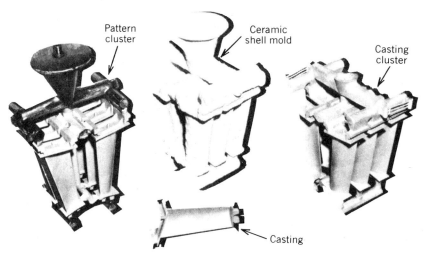

Figure 6.14
Ceramic shell process.

mm) thick. The pattern is then melted out of the mold, which is first dried and then fired at 1800° to 2000°F (980°–1095°C) to remove all moisture and organic material. The mold, free of any parting lines, is usually poured immediately after it is removed from the furnace. The shell breaks away from the casting as cooling takes place. Very good accuracy and surface finish are obtained with both ferrous and nonferrous metals. Tolerances of ±0.005 in. (±0.13 mm) are common, and as-cast tolerances can be improved by coining or sizing, but the cost is increased.

Frozen mercury is sometimes used in the place of wax or plastic patterns. A metal mold or die is made of the part to be cast with the necessary gates and sprue hole. When assembled and ready for pouring it is partially immersed in a cold bath and filled with acetone, which acts as a lubricant. As the mercury is poured into the mold the acetone is displaced. Freezing takes place in a liquid bath held at around −76°F (−60°C) and is complete in about 10 min.

The patterns are then removed from the mold and invested in a cold ceramic slurry by repeated dippings until a shell about $1/_8$ in. (3.1 mm) thick is built up. Mercury is melted and removed from the shell at room temperature and, after a short drying period, is fired at a high temperature resulting in a hard permeable form. The shell is then placed in a flask, surrounded by sand, preheated, and filled with metal. Casting is usually done by the centrifugal method. Although the process yields castings of high accuracy, it is limited because of high production costs and the hazards associated with handling mercury.

Plaster Mold Casting

The gypsum-based plasters used as a casting investment dry quickly with good porosity but are not permanent, being destroyed when the casting is removed.

Patterns are made of a free-machining brass and are held to a close tolerance. They are assembled on bottom plates of standard-size flasks as shown in Figure 6.15A. Before receiving the plaster they are sprayed with a parting compound. The plaster, which is

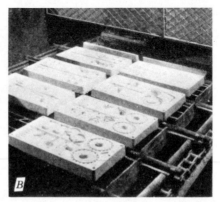

Figure 6.15
A, Assembling metal patterns in flask for plaster mold. B, Finished plaster molds emerging from drying oven.

made of gypsum with added strengtheners and setting agents, is dry-mixed and water added. It is then poured over the patterns and the mold is vibrated slightly to ensure that the plaster fills all small cavities. The plaster sets in a few minutes and is removed from the flask by a vacuum head. All moisture is driven from the molds by baking them in an oven conveyor at a temperature around 1500°F (815°C). These molds are shown in Figure 6.15B as they are emerging from the drying oven. After pouring, the castings are removed by breaking up the mold. Any surplus plaster is removed by a washing operation.

Mold porosity for the removal of any gases developed in the mold is controlled by the water content of the plaster. When the mold is dried, the water driven out leaves numerous fine passageways that act as vents. In addition to having adequate porosity, plaster molds have the necessary structural strength for casting plus enough elasticity to allow some contraction of the metal during cooling.

Plaster molds are suitable only for nonferrous alloys. The wide variety of small castings made by this process includes miscellaneous airplane parts, small gears, cams, handles, pump parts, small housings, and numerous other intricate castings.

One of the principal advantages of plaster mold casting is the resulting high degree of dimensional accuracy. This, coupled with the smooth surface obtained, enables the process to compete favorably with sand castings requiring machining to obtain a smooth surface. Because of the low thermal conductivity of plaster, the metal does not chill rapidly and very thin sections may be cast. There is little tendency toward internal porosity in plaster mold castings, and there is no difficulty with sand or other inclusions. In general, the process competes more successfully with die casting using high-temperature alloys such as brass rather than metals such as zinc and aluminum. At high temperatures metal molds have a relatively short life; with plaster molds, which are used only once, the temperature is no problem.

A tolerance of ±0.005 in. (±0.13 mm) can be maintained for simple castings; slightly more is required if the dimension crosses the parting line. The process can be used for both small and quantity production runs.

The complex aluminum impellers shown in Figure 6.16 may be cast by vacuum pouring into a cast-iron mold containing a plaster core. Vacuum melting and pouring techniques along with a moistureless mold and a low pouring temperature yield high strength and good surface finish.

The two principal variations of this process are (1) foam plaster molds and (2) the

Figure 6.16
Shapes of vacuum-cast, plaster-cored aluminum impellers.

Antioch process. In the foam process a foaming agent is added to the water, then the plaster is added and the materials mixed until the desired slurry and density is obtained. After oven-drying the more permeable molds are poured. Material is saved in this process and permeability is increased.

In the Antioch process a typical plaster mold is steam-autoclaved for about 10 h to cause the plaster to granulate and become more permeable. Then the mold is dipped in water and allowed to granulate further at room temperature, then oven-dried. Even though the mold granulates internally, the surface is only moderately roughened.

Shell Molding Process

The mold in this process is made up of a mixture of dried silica sand and phenolic resin formed into thin, half-mold shells that are clamped together for pouring as illustrated by the series of sketches in Figure 6.17. The sand, free from clay, is first mixed with either urea or phenol formaldehyde resin; the mix is then put into a ***dump box*** or blowing machine. A metal pattern must be used because it is preheated to a temperature around 450°F (230°C) and sprayed with a silicon release agent before being placed on top of the dump box. The box is then inverted, causing the sand mix to drop on the pattern, and is held there for 15 to 30 s before it is returned to its original position. The pattern, with a thin shell of sand $1/8$ to $3/16$ in. (3.1–4.7 mm) thick adhering to it, is then placed in an oven and the shell cured $1/2$ to 1 min until it is rigid. The shell is finally removed from the pattern by ejector pins and the mold halves assembled with clamps, resin adhesives,

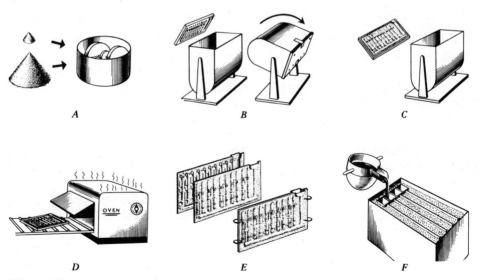

Figure 6.17
Diagram of shell molding process. A, Mulling sand and resin. B, Resin sand mixture applied to pattern. C, Excess resin sand mixture falls back into dump box. D, Curing shells on pattern. E, Mold halves are aligned and joined. F, Molds supported and poured.

or other devices. They may then be put in a flask supported against one another or by some backing material such as shot or gravel. Some are poured while they are resting flat on the floor with a weight on top.

The advantages of this process include fine tolerances of 0.002 to 0.005 in./in. (0.2%–0.5%), low cleaning costs, and smooth surfaces. Little molding skill is necessary and the sand requirements are low. Shell molding also can be readily adapted to automation. Disadvantages are that the process requires metal patterns and fairly expensive equipment for making and heating the molds.

CO_2 Mold Hardening Process

The process of hardening molds and cores using CO_2 and a *sodium silicate* liquid base binder is widely used. Because of inherent advantages and rapid sand hardening, it is used in many foundries. Briefly the process consists of thoroughly mixing clean, dry silica or other dry conventional sand (fineness number around 75) with 3.5% to 5% sodium silicate liquid base binder in a muller. It is then ready for use and may be packed in flasks and core boxes by standard molding machines, by core blowers, or by hand. The sand should be free from moisture and clay, but other ingredients such as coal dust, pitch, graphite, or wood flour may be added to improve certain properties like collapsibility.

When the packing is complete, CO_2 is forced into the mold or core at a pressure of around 20 psi (0.1 MPa). The reaction is complex but is usually represented by the following chemical equation.

$$Na_2SiO_3 + CO_2 = SiO_2(aq) + Na_2CO_3$$

The silica gel that is formed hardens and acts as a cement to bond the sand grains together. The method of introducing the gas, important to the success of the process, must be simple, rapid, and uniform throughout the sand body and not be cumbersome to apply. The time to harden a small or medium-size body of sand is 15 to 30 s. For small cores a gasketed, funnel-shaped head may be placed over the core box. Larger molds may be hardened by placing a hood over the mold, by running small tubes into the mold as illustrated in Figure 6.18, or by introducing the gas into a hollow, vented pattern. The figure shows diagrammatically the steps followed in preparing a mold. In some cases the sprue is at the ends of the cope and drag, and a number of molds may be book-staked between end braces for pouring.

A CO_2 mold or core can be made quickly; no baking is required. Semiskilled help can be used. The surface finish of the casting is good, and the same sand can be used for both cores and molds.

Some sodium silicate bonding agents permit self-setting action without CO_2 and permit shear strengths of 90 psi (0.6 MPa) and compression strengths of 300 psi (2 MPa). Seacoal, coal in the amount of 2%, is the most used additive to improve shakeout, sand removal from the casting, and low pollution. Special shakeout sands having carbonaceous materials are also used. Most silicate-bonded sands can be shaken out relatively easily if organic esters are used in the mix. Sugar up to 12% is added as breakdown agent so the core will more fully deteriorate after solidification of the mold. The sand can be

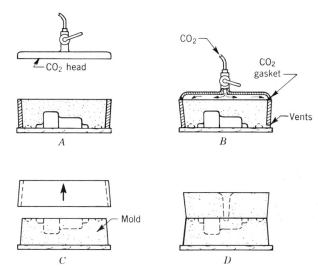

Figure 6.18
Diagram of CO_2 mold hardening process. *A*, Rammed mold. *B*, Hardening mold with CO_2. *C*, Mold jacket removed. *D*, Assembled mold.

reused if about 30% new sand is added, although the economics of reclamation are questioned.

The molds have relatively short storage life, and sometimes poor collapsibility gives trouble. The process is used in both the ferrous and nonferrous casting industry.

Molds of Other Materials

Various materials such as rubber, paper, and wood can be used for molds for low-melting-temperature metals. Costume jewelry and similar small items are successfully cast in *rubber molds*. An alloy of 98% tin, 1% copper, and 1% antimony is frequently used in this work.

Figure 6.19 illustrates a mold made of a Dow Corning silicone rubber product known as Silastic. This material can be used to cast wax patterns, plastics, or low-melting-point alloys. The molds will stand 500°F (260°C) and will reproduce as fine a detail as found on a high-fidelity record. The material is so flexible that it can be removed from intricate shapes without difficulty.

The *Shaw* process uses a mixture of sand, hydrolyzed ethyl silicate, and other ingredients that permit the investment mold to be "peeled" from the pattern. The pattern need not be wax or mercury, because in the "as-poured" state the mold material is rubberlike. Once removed from the pattern the mold is ignited and later baked to provide a rigid, permeable, high-quality surface finish mold. The Shaw process, adaptable to complex shapes and reusable patterns, can be adapted to automatic operation but is relatively time-consuming and costly except for certain castings.

Full-page newspaper type is cast in a mold called a "mat," upon which the type and illustration impressions have been made on damp paper. The type metal is poured into the mold after the paper is dry. End-grain wood may also be used as a mold material for low-melting alloys where only a limited number of simple castings are required.

Figure 6.19
Silastic RTV silicone rubber mold.

CHARACTERISTICS OF VARIOUS MOLDING PROCESSES

In the last chapter and this one various methods of casting have been discussed. Characteristics of each have been described but they have not been compared to each other in very definitive ways. In order to compare some of the different processes Table 6.1 has been prepared.

CONTINUOUS CASTING

Research and experimental work have proved that there are many opportunities for cost economies in the continuous casting of metals. In addition, metals starting as continuous castings have a degree of soundness and uniformity not possessed by other methods of producing bars and billets. Briefly, the process consists of continuously pouring molten metal into a mold that has the facilities for rapidly chilling the metal to the point of solidification, and then withdrawing it from the mold. The following processes are typical.

Reciprocating Mold Process

A reciprocating, water-cooled copper mold is used, the downstroke being synchronized with the discharge rate of the slab.[2] Molten metal is poured into the holding furnace and

---

[2]Process used by the Scovill Manufacturing Company and a development of the Junghans-Rossi process.

Table 6.1 Characteristics of Casting Processes[1]

Type Process	Surface Finish, (μin.)	Dimensional Accuracy for Small Casting[a]	Dimensional Accuracy for Large Casting[b]	Intricacy[c]	Tooling Cost[d]	Cost per Unit[d]
High-pressure die casting	30	1	1	3	1	8
Low-pressure die casting	50	4	5	5	2	7
Plaster casting	32	3	2	2	7	1–5
Investment casting	60	2	4	1	6	4
Gravity casting	70	5	7	7	3	6
Shell	125	6	3	4	5	3
Sand	500	7	6	6	4	2

[a]Order, 1 being best.
[b]Order, 1 being able to produce most accurate.
[c]Order, 1 the most intricacy.
[d]Order, 1 being lowest.

is discharged to the mold after being metered through a 7/8-in. (22-mm) orifice at the needle valve. The downspout tube is 1 1/8 in. (28 mm) in diameter and delivers metal at the rate of 30,000 lb/h (225 kg/s).

The molten metal is distributed across the mold from a submerged horizontal crosspiece. The level of the metal is held constant at all times. The pouring rate of the molten metal is controlled by a needle valve through the top of the holding furnace. As the metal becomes chilled in the lower part of the mold, it is discharged at a constant rate and enters the withdrawing rolls. These are synchronized with the downward movement of the mold and are mounted just above a circular saw that cuts the slab to required lengths. Brass slabs thus produced are further processed by cold rolling into sheets and strips. Large quantities of 7- to 10-in. (178- to 254-mm) round billets for hot-extrusion processes are also produced in this manner.

Asarco Process

The process (Figure 6.20) differs from other continuous processes in that the forming die or mold is integral with the furnace, and there is no problem of controlling the flow of metal. The metal is fed by gravity into the mold from the furnace as it is continuously

[1]*Source*: G. Richards, Precision Casting, *The British Foundryman*, July 1979.

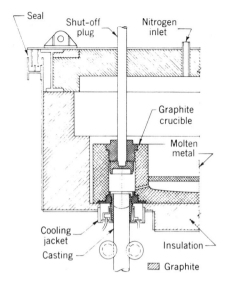

Figure 6.20
Process for continuous-cast shapes.

solidified and withdrawn by the rolls below. An important feature of this process is the water-cooled graphite forming die, which is self-lubricating, is resistant to thermal shock and is not attacked by copper base alloys. The upper end in molten metal acts as a riser and compensates for any shrinkage that might take place during solidification, while simultaneously acting as an effective path for the dissipation of evolved gases. These dies are easily machined to shape, and products may be produced ranging from $7/16$ to 9 in. (11–229 mm) in diameter. Multiple production from a single die permits casting small–cross section rods.

In starting the process a rod of the same shape as that to be cast is placed between the drawing rolls and inserted into the die. This rod is tipped with a short length of the alloy to be cast. As the molten metal enters the die it melts the end surface of the rod, forming a perfect joint. The casting cycle is then started by the drawing rolls, and the molten metal is continuously solidified as it is chilled and withdrawn from the die. When the casting leaves the furnace it ultimately reaches the sawing floor where it is cut to desired length while still in motion. A tilting receiver takes the work and drops it to a horizontal position, and from there it goes to inspecting and straightening operations.

The process has proved successful for phosphorized copper and many of the standard bronzes. The alloy composition may be produced with satisfactory commercial finish as rounds, tubes, squares, or special shapes. Physical properties are superior to those obtained from permanent-mold and sand castings.

Brass Mold Continuous Process

A continuous-casting process for carbon and alloy steels uses brass or copper molds of a thickness that permits a heat flow rate that is sufficient to prevent the mold from being damaged by the metal being cast. The brass or copper mold has high heat conductivity and is not easily wetted by molten steel. The mold cross sections used to date vary from 12 to 90 in.^{2} (7–57 mm^2). Metal is supplied to the mold by a nozzle located in a tundish or pouring box. The tundish in turn is supplied from a conventional ladle.

Rapid mold cooling is essential for the success of this process and results in improved mold life, less segregation, smaller grain structure, and a better surface. Actually, the metal next to the mold wall solidifies only a few inches below the top surface and shrinks slightly from the mold sides. As the cast section leaves the cooled mold it passes through a section that controls the rate of cooling where roller guides prevent expansion of the casting. Below this area are the withdrawal, speed-controlling rolls through which the casting passes to the cutting area. The castings are cut to length by an oxyacetylene torch. The castings are rolled, forged, or extruded into blooms, billets, or slabs.

Direct-Chill Process

In this process aluminum and aluminum alloy ingots are continuously cast by forming a shell in a vertical, stationary, water-cooled mold. Solidification is completed by direct water application beneath the mold. The mold is closed at the start by a block on an elevator or by a dummy ingot. Molten metal is fed from a furnace through troughs and spouts with flow regulated manually or automatically by float control to coincide with the casting rate controlled by the elevator or driven rolls. The process is shown diagrammatically in Figure 6.21. Cross sections up to 1500 in.^{2} (0.97 m^2) are produced. Lengths 100 to 150 in. (2.54–3.81 m) long and limited by the stroke are cast where an elevator is employed, or ingots are sawed to finish length where rolls are used. Surface quality is adequate as cast for certain alloys and products, or it may require scalping for other alloys or more critical applications.

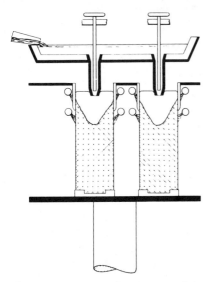

Figure 6.21
Casting aluminum ingots by the direct-chill process.

QUESTIONS AND PROBLEMS

1. Give an explanation of the following terms:

Hot chamber	Stack mold
Flash	Investment casting
Ejector rod	Lost wax
"Stalk"	Dump box
Corthias	Sodium silicate

2. What effect does the melting temperature have on die life?

3. How is pressure applied to a cold-chamber die-cast part?

4. What special casting process might be used to produce a cast-iron dumbbell?

5. Sketch a die that could be used for casting a small dumbbell out of aluminum and label the principal parts.

6. Using Chapters 4 and 6, write a paragraph on the characteristics of zinc.

7. Give the principal reasons that account for 75% of all die castings being made from zinc.

8. Compare the differences in characteristics of a small zinc casting made by the processes listed in Table 6.1.

9. Prepare a list of metals and alloys normally cast in a hot-chamber machine.

10. Why must an inert gas be used in low-pressure, permanent-mold castings?

11. Prepare a list of metals and alloys normally cast in a cold-chamber machine.

12. What can be done with gates and risers that are trimmed from die castings?

13. What is the difference between a multiple-cavity die and a combination die?

14. Why are ejector pins or rods required in a die casting? Do they mar the surface in any way? Why?

15. Describe three products that could be made by the slush casting process.

16. Sketch a design for a small vase that is to be cast of lead by the slush casting process.

17. Why are brass die castings so difficult and expensive to produce?

18. How is air removed from a die in order that metal may completely fill it?

19. What is the function of the "stalk" in a low-pressure permanent mold? What can be done with sprues and gates in a low-pressure permanent molding?

20. Explain the difficulties of making large castings with the vertical true centrifugal casting method.

21. A small aircraft cylinder liner 9 in. (228.6 mm) OD and 7 in. (177.8 mm) ID is to be spun in a vertical, true centrifugal mold. How many revolutions per minute should be used?

22. The same casting described in problem 21 is to be spun horizontally. How many revolutions per minute should be used?

23. If too low a speed were used in problem 22, what difficulties could be encountered?

24. Describe the mold and coatings that would be used in problems 21 and 22.

25. Compare the advantages between gravity-type permanent molds and die casting. How are molds protected in making a gravity-type permanent-mold casting?

26. Would alloys or pure metals be most adaptable to slush casting? Why? Refer to Chapter 2.

27. Describe the steps in order and in necessary detail to make a bronze bust from a clay carving.

28. Describe the methods available for manufacturing 24-in. (609.6-mm) diameter, 20-ft (6096-mm) lengths of cast-iron pipe.

29. How many pounds of cast iron are required in problem 28?

30. Why are ferrous metals unsuitable for molds into which plaster would be poured?

31. Referring to Table 6.1, how would you rank the horizontal true centrifugal casting process?

32. Referring to Table 6.1, how would you rank the vertical true centrifugal casting process?

33. Discuss the manufacture of a hollow brass decorative head for a walking cane using the lost wax process.

34. How are small wax patterns attached to the gates in a multiple-cavity investment pattern?

35. What is meant by "vacuum pouring?" Would the use of an inert gas instead of a vacuum do the same thing?

36. Using the principles of shrinkage examined in Chapter 5, what can be said about continuous casting as influenced by shrinkage?

37. Comparing the shell mold and CO_2 processes, which would be the more adaptable to mass production and why?

38. What special expenses are incurred in a shell molding operation? Give the reasons it cannot be employed in job lot or small production runs.

39. Estimate the time to make a shell mold half. Write down times for each step and multiply your total time by 2.5 to allow for breakdowns, bad molds, and coffee breaks.

40. What process should be used to make the following: small zinc castings, statuettes, aluminum pistons, aluminum ingot, small brass gears, carburetors, automobile door handles, automobile grills, and aluminum alloy ingots?

41. Why are ferrous metals more difficult to cast continuously? What are the advantages of continuous-casting processes?

42. Describe the method of making a Silastic mold from a decorative brass duck.

43. For what types of products is continuous casting most successful?

44. Discuss the truism that "all metals begin with a casting."

45. If 35% of a die casting is trimmed away and remelted with a 5% loss during melting, how many metric tonnes of metal would be required per day in a plant making and shipping 3500 nut crackers per day that weigh 200 g each?

46. Estimate total monthly costs involved in producing the 3500 nutcrackers in problem 45.

47. Discuss the advantages and disadvantages of the various ways of casting the nutcracker in problem 45. Refer to Table 6.1.

48. If a 48-in. diameter metal pipe 10 ft long is centrifugally cast with a force of 60 g's, what rotational speed in revolutions per minute should be used?

49. In problem 48 what is the effect of length of pipe on the rotational speed? Why?

50. Plot a curve of revolutions per minute versus inside diameter for a vertical true centrifugal process. Use diameters of 1 in. to 20 in.

CASE STUDY

WATER PIPE

The LaGrande Company in South America has been casting 15 ft (4.6 m) long, 8 in. (203 mm) diameter gray cast-iron pipe in sand molds for 75 years using conventional techniques employing a long core supported by chaplets. As a consultant you have been asked to compare the cost with a proposed centrifugal casting process. The Bolivar is the monetary unit and it is worth U.S. $0.133. The iron costs 1.2 Bolivars/kg, and labor is paid, on the average, 10.7 Bolivars/h. Tabulate your cost estimates. Advise LaGrande the type of centrifugal process to use and the revolutions per minute it must maintain during pouring and solidification. Can you make an estimate of the savings that would benefit LaGrande if they wish to produce 1000 units/month? What about the space requirements for the two types of operations as well as any logistic problems?

CHAPTER 7
HEAT TREATMENT

The understanding of heat treatment is embraced by the broader study of metallurgy. Metallurgy is the physics, chemistry, and engineering related to metals from ore extraction to the final product. Heat treatment is the operation of heating and cooling a metal in its solid state to change its physical properties. According to the procedure used, steel can be hardened to resist cutting action and abrasion, or it can be softened to permit machining. With the proper heat treatment internal stresses may be removed, grain size reduced, toughness increased, or a hard surface produced on a ductile interior. The analysis of the steel must be known because small percentages of certain elements, notably carbon, greatly affect the physical properties.

Alloy steels owe their properties to the presence of one or more elements other than carbon, namely nickel, chromium, manganese, molybdenum, tungsten, silicon, vanadium, and copper. Because of their improved physical properties they are used commercially in many ways not possible with carbon steels.

The following discussion applies principally to the heat treatment of ordinary commercial steel known as plain-carbon steels. With this process the rate of cooling is the controlling factor; rapid cooling from above the critical range results in hard structure, whereas very slow cooling produces the opposite effect.

IRON–IRON CARBIDE DIAGRAM

Under conditions of equilibrium the knowledge of steel and its structure is best summarized in the partial *iron–iron carbide diagram* shown in Figure 7.1. If a piece of 0.20% carbon steel is slowly and uniformly heated and its temperature recorded at definite intervals of time, a curve as shown in Figure 7.2 may be obtained. Such a curve is called an *inverse rate curve*. The abscissa is the heating rate or the time required to heat or cool the steel 10° F or C. The curve is a vertical line except at those points where the heating or cooling rates show marked change. It is evident that at three temperatures there is a definite change in the heating rate. In a similar fashion these same three points again show upon cooling but occur at slightly lower temperatures. Where structural changes occur these points are known as *critical points* and are designated by the symbols Ac_1, Ac_2, and Ac_3. The letter "c" is the initial letter of the French word **chauffage,** meaning "to heat." The points on the cooling curve are designated by Ar_1, Ar_2, and Ar_3. The "r" is taken from the word **refroidissement,** meaning "to cool." The Ac_2 and Ar_2 values are believed to have little effect on the understanding of the metallurgy of iron and steel.

132 HEAT TREATMENT

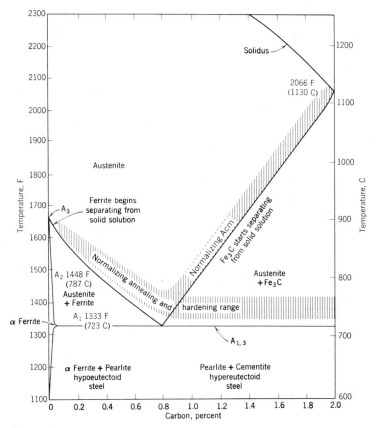

Figure 7.1
Partial iron–iron carbide phase diagram.

Certain changes that take place at these critical points are called *allotropic changes*. Although the chemical content of the steel remains the same, its properties are changed. Principal among these are changes in electrical resistance, atomic structure, and loss of magnetism. An allotropic change is a reversible change in the atomic structure of the metal with a corresponding change in the properties of the steel. These critical points should be known, because most heat-treating processes require heating the steel to a temperature above this range. Steel cannot be hardened unless it is heated to a temperature above the lower critical range and in certain instances above the upper critical range.

If a series of time–temperature heating curves are made for steels of different carbon contents and the corresponding critical points plotted on a temperature–percentage carbon curve, a diagram similar to Figure 7.1 is obtained. This diagram, which applies only under slow cooling conditions, is known as a partial iron–iron carbide diagram. The proper quenching temperatures for any carbon steel may be observed in this diagram.

Consider again the piece of 0.20% carbon steel that has been heated to a temperature

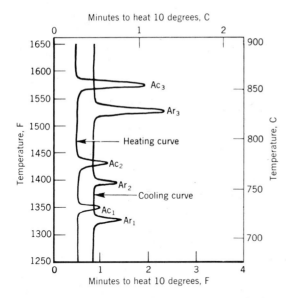

Figure 7.2
Inverse rate curve for SAE 1020 steel.

around 1600°F (870°C). Above the Ar_3 point this steel is a solid solution of carbon in gamma iron and is called *austenite*. The iron atoms lie in a face-centered cubic lattice and are nonmagnetic. Upon cooling this steel the iron atoms start to form a body-centered cubic lattice below the Ar_3 point. This new structure being formed is called *ferrite* or alpha iron and is a solid solution of carbon in alpha iron. The solubility of carbon in alpha iron is very much less than in gamma iron. At the Ar_2 point the steel becomes magnetic and, as the steel is cooled to the Ar_1 line, additional ferrite is formed. At the Ar_1 line the austenite that remains is transformed to a new structure called *pearlite*. This constituent is lamellar in appearance under high magnification. The lamellae are alternately ferrite and iron carbide. Called pearlite because of its "mother-of-pearl" appearance, it is shown under high magnification in Figure 7.3.

As the carbon content of the steel reaches about 0.20%, the temperature at which the ferrite is first rejected from the austenite drops until at about 0.80% carbon no free ferrite is rejected from the austenite. This steel is called *eutectoid* steel and is 100% pearlite in structure composition. The eutectoid point in any metal is the lowest temperature at which changes occur in a solid solution. It marks the lowest temperature for equilibrium de-

Figure 7.3
Structure of SAE 1095 steel, furnace cooled from 1550°F. Etched in 5% picral. Showing lamellae of cementite and ferrite in pearlite. Magnification ×1200.

composition of austenite to ferrite and cementite. If the carbon content of the steel is greater than the eutectoid, a new line is observed in the iron–iron carbide diagram labeled *Acm*. The line denotes the temperature at which iron carbide is first rejected from the austenite instead of ferrite. The iron carbide (Fe_3C) is known as *cementite* and is extremely hard and brittle. Steels containing less carbon than the eutectoid are known as **hypoeutectoid steels,** and those with more carbon content are called **hypereutectoid steels.**

The structures of these steels are shown in a series of photomicrographs in Figure 7.4. Figure 7.4A shows pure iron or ferrite. As the carbon content increases up to 0.80% carbon (Figure 7.4E), the dark areas of the pearlite form and increase in quantity while the white background area of ferrite decreases and the sample is nearly all pearlite. In the sample containing 1.41% carbon (Figure 7.4F) the pearlitic area is smaller and the white background area is now cementite and ferrite. The maximum amount of cementite at 1.41% carbon would be about 11%. All these iron–carbon alloys have been cooled slowly to produce the constituents just described.

So far as steel treating is concerned, the partial diagram shown in Figure 7.1 is sufficient because 2.0% is the limit of carbon content in steel. If the diagram is extended to include the cast irons with carbon contents up to 6.67%, it will appear as in Figure 7.5. It is not shown beyond this point because 6.67% carbon is the carbon content of cementite. Actually, most commercial cast irons have a carbon content from 2.25% to 4.50%.

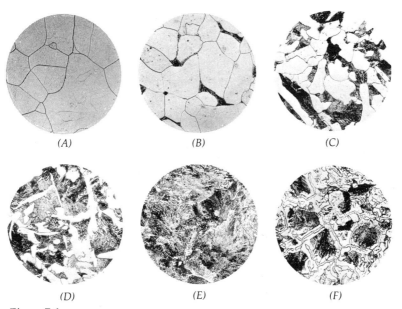

Figure 7.4
Photomicrographs of iron–carbon alloys showing the effect of increasing amounts of carbon on structure of the metal. A, High-purity iron; B, 0.12% carbon; C, 0.40% carbon; D, 0.62% carbon; E, 0.79% carbon; F, 1.41% carbon.

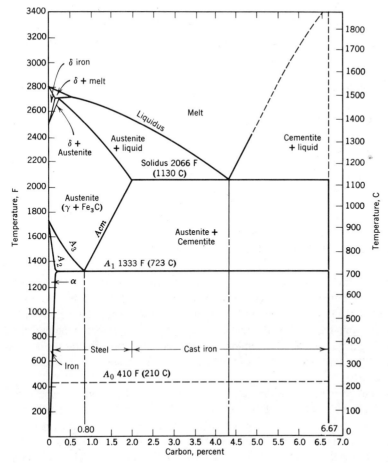

Figure 7.5
Iron–iron carbide diagram.

The amount of ferrite, pearlite, and cementite can be calculated as follows. For a hypoeutectoid steel, the percentage pearlite is

$$\%Pe = \frac{\%C}{0.8}$$

where %C is the percentage carbon in the specimen. The remainder is ferrite.

For a hypereuctectoid steel the percentage cementite can be expressed as

$$\%Ce = 100\left[1 - \frac{(6.67 - \%C)}{5.87}\right]$$

The remainder is pearlite.

GRAIN SIZE

Upon cooling, molten steel starts solidifying at many small centers of nuclei. The atoms in each group tend to be positioned similarly. The irregular grain boundaries, seen under the microscope after polishing and etching, are the outlines of each group of atomic cells that have the same general orientation. The size of these grains depends on a number of factors, the principal one being the furnace treatment it has received.

Course-grained steels are less tough and have a greater tendency for distortion than those having a fine grain; however, they have better machinability and greater depth-hardening power. The fine-grained steels, in addition to being tougher, are more ductile and tend less to distort or crack during heat treatment. The control of grain size is possible through regulation of composition in the initial manufacturing procedure, but after the steel is made the control is through proper heat treatment. Aluminum, when used as a deoxidizer, is the most important controlling factor during the manufacturing period, because it raises the temperature at which rapid grain growth occurs.

When a piece of low-carbon steel is heated, there is no change in the grain size up to the Ac_1 point. As the temperature increases through the critical range, the ferrite and pearlite are gradually transformed to austenite, and at the upper critical point Ac_3, the average grain size is at a minimum. Further heating of the steel causes an increase in the size of the austenitic grains, which in turn governs the final size of the grains when cooled. Quenching from the Ac_3 point would result in a fine structure, whereas slow cooling or quenching from a higher temperature would yield a coarser structure. The final grain size depends to a large extent on the prior austenitic grain size in the steel at the time of quenching.

Not all steels start growing large crystals immediately upon being heated above the upper critical range; some steels can be heated to a higher temperature with little change in their structure. A temperature known as a *coarsening temperature* is eventually reached, and grain size increase becomes rapid. This is characteristic of medium-carbon steels, many alloy steels, and steels that have been deoxidized with aluminum. The coarsening temperature is not a fixed temperature and may be changed by prior hot or cold working and heat treatment. Hot work on steel is started at temperatures well above the critical range with the steel in a plastic state. It refines the grain structure and eliminates any coarsening effect caused by the high temperature. Hot forging or rolling should not continue below the critical temperature.

The principal method of determining grain size is by microscopic examination, although it may be roughly estimated by examination of a fracture. Low-carbon steels have ferrite precipitated from the austenite upon slow cooling, and the outlines of these grains can be clearly brought out by polishing and etching. Because a very slow cooling rate may produce too much primary ferrite to permit evaluation of prior austenitic grain size, a cooling rate must be employed such that the proeutectoid constituent is restricted merely to outlining the pearlitic regions. Likewise, for medium-carbon steels the former austenitic grain size would be represented roughly by the pearlitic area plus one-half the surrounding ferrite. Hypereutectoid steels will have the grain boundaries outlined by the cementite that is precipitated.

An example of a large-grained steel is shown in the photomicrograph in Figure 7.6.

Figure 7.6
Crystalline separation and excessive grain size. Magnification ×300.

This specimen has been heated to an excessively high temperature resulting in large-grain growth and some crystalline separation. Steel that has been "burnt" shows this separation as a result of oxidation at the grain boundaries, and this cannot be remedied by heat treatment. The steel can be rendered fit for commercial use only by remelting.

ISOTHERMAL TRANSFORMATION DIAGRAMS

The iron–iron carbide phase diagram in Figure 7.1 is useful in selecting temperatures for parts to be heated for various treating operations, and it also shows the type of structure to expect in slowly cooled steels. Though very useful in all heat-treating operations, it does not give much information concerning effects of cooling rate, time, grain structure, or structures obtainable when the quench is interrupted at certain elevated temperatures. *Isothermal transformation diagrams*, also known as *time–temperature transformation diagrams* or *S curves*, have been developed that indicate this information (Figure 7.7). This diagram shows the way austenitized steel changes if held at some constant temperature. Knowing this temperature, the times at which the transformation starts and ends may be determined; the resulting structure is indicated on the diagram. To obtain a martensitic structure the steel must be quenched with sufficient rapidity so that the cooling curve does not intersect the nose of the transformation curve. This is indicated in Figure 7.7, which shows the cooling curve passing through the M_s and M_f lines (start and finish of austenite transforming to martensite).

The general shape of a time–temperature transformation curve differs for each steel depending on the carbon content, alloys present, and austenitic grain size. Most alloying elements in steel shift the curves to the right, thus allowing more time to harden the steel fully without hitting the bend in the curve. This increase in hardenability of the steel permits the hardening of thicker sections than would otherwise be possible. In carbon steels lowering the carbon content moves the curve to the left and also raises the M_s and M_f temperature lines. This makes it very difficult to produce martensite by quenching a

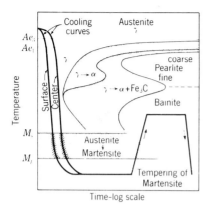

Figure 7.7
Transformation diagram illustrating the formation of tempered martensite.

hypoeutectoid steel. Carbon steel having an eutectoid composition responds well to hardening treatments. Steels having a fine austenitic grain size will displace the curve to the left, thus making it more difficult to harden a fine-grained steel than one having coarse grains. However, coarse-grained steels are more apt to crack or distort during quenching, so their increased hardenability is of little advantage.

HARDENING

Hardening is the process of heating a piece of steel to a temperature within or above its critical range and then cooling it rapidly. If the carbon content of the steel is known, the proper temperature to which the steel should be heated may be obtained by reference to Figure 7.1, the iron–iron carbide phase diagram. However, if the composition of the steel is unknown, a little preliminary experimentation may be necessary to determine the range. A good procedure to follow is to heat-quench a number of small specimens of the steel at various temperatures and observe the results, either by hardness testing or by microscopic examination. When the correct temperature is obtained, there will be a marked change in hardness and other properties.

In any heat-treating operation the rate of heating is important. Heat flows from the exterior to the interior of steel at a definite rate. If the steel is heated too fast, the outside becomes hotter than the interior and uniform structure cannot be obtained. If a piece is irregular in shape, a slow rate is all the more essential to eliminate warping and cracking. The heavier the section, the longer must be the heating time to achieve uniform results. Even after the correct temperature has been reached, the piece should be held at that temperature for a sufficient period of time to permit its thickest section to attain a uniform temperature.

The hardness obtained from a given treatment depends on the quenching rate, the carbon content, and the work size. In alloy steels the kind and amount of alloying element influences only the hardenability (the ability of the workpiece to be hardened to depths) of the steel and does not affect the hardness except in unhardened or partially hardened steels.

Reference to the isothermal transformation diagram in Figure 7.7 indicates that a very rapid quench is necessary to avoid intersecting the nose of the curve and to obtain a martensitic structure. For low and medium plain-carbon steels quenching in a water bath is a method of rapid cooling that is common practice. For high-carbon and alloy steel, oil is generally used as the quenching medium because its action is not so severe as water. Various commercial oils such as mineral oil have different cooling speeds and, consequently, impart different hardness to steel on quenching. For extreme cooling, brine or water spray is most effective. Certain alloys can be hardened by air-cooling, but for ordinary steels such a cooling rate is too slow to give an appreciable hardening effect. Large parts are usually quenched in an oil bath, which has the advantage of cooling the part down to room temperatures rapidly and yet not being too severe. The temperature of the quenching medium must be kept uniform to achieve uniform results. Any quenching bath used in production work should be provided with means for cooling.

Steel with low carbon content will not respond appreciably to hardening treatments. As the carbon content in steel increases up to around 0.60%, the possible hardness obtainable also increases. Above this point the hardness can be increased only slightly, because steels above the eutectoid point are made up entirely of pearlite and cementite in the annealed state. Pearlite responds best to heat-treating operations; any steel composed mostly of pearlite can be transformed into a hard steel.

As the size of parts to be hardened increases, the surface hardness decreases somewhat even though all other conditions have remained the same. There is a limit to the rate of heat flow through steel. No matter how cool the quenching medium may be, if the heat inside a large piece cannot escape faster than a certain critical rate, there is a definite limit to the inside hardness. However, brine or water quenching is capable of rapidly bringing the surface of the quenched part to its own temperature and maintaining it at or close to this temperature. Under these circumstances there would always be some finite depth of surface hardening regardless of size. This is not true in oil quenching, when the surface temperature may be high during the critical stages of quenching.

Hardenability of Steel

Hardenability refers to the response of a metal to quenching and may be measured by the Jominy end-quench test as illustrated in Figure 7.8. A normalized specimen of the steel to be tested is machined to a diameter of 1 in. (25 mm) and a length of 4 in. (100 mm) and then heated to its austenitizing temperature. It is quickly placed in the quenching fixture where it is held $1/2$-in. (12.7 mm) above a $1/2$-in. (12.7-mm) diameter orifice. Water is directed against the bottom surface until the entire specimen is cool.

Upon removal from the fixture two flats, each 0.015 in. (0.38 mm) deep, are ground on opposite sides of the specimen. Rockwell hardness readings are then taken at $1/16$-in. (1.6-mm) intervals from the bottom of the specimen and plotted as shown on the figure. The reading next to the bottom will show the greatest hardness because that portion of the specimen has been most severely quenched. Readings away from the quenched end show progressively lower hardnesses as the heat must pass through the specimen by conduction during cooling. A steel with high hardenability shows high hardness readings

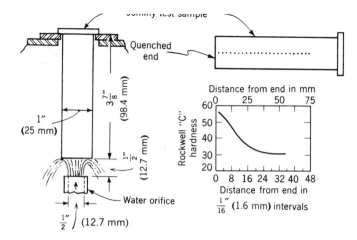

Figure 7.8
The Jominy end-quench test for measuring hardenability.

for some distance from the quenched end, while the hardness readings for low-hardenability steel show a sharp drop only a short distance from the end.

This test provides means of comparing the depth-hardening ability of different steels. Alloys increase the hardenability of steel and make it possible to harden small pieces uniformly from the outside to the inside. Because the effect of an alloy on the isothermal transformation diagram is to move the curve to the right, it is easier to quench such a steel without intersecting the curve. Because of this characteristic it is possible to harden alloy steels at a slower rate than plain-carbon steels. This permits alloy steels to be effectively hardened by quenching in oil instead of water.

Constituents of Hardened Steel

As has been previously stated, austenite is a solid solution of carbon in gamma iron. All carbon steels are composed entirely of this constituent above the upper critical point. The appearance of austenite under the microscope is shown in Figure 7.9 at a magnification of $\times 125$. Extreme quenching of a steel from a high temperature will preserve some of the austenite at ordinary temperatures. This constituent is about half as hard as martensite and is nonmagnetic.

Figure 7.9
Structure of 18-8 stainless steel, water-quenched to show austenite. Lines caused by hot rolling. Magnification $\times 125$.

Figure 7.10
Structure of SAE 1095 steel water quenched. Etched with Villella's reagent to show martensite. Magnification ×562.

If a hypoeutectoid steel is cooled slowly, the austenite is transformed into ferrite and pearlite. Steel having these constituents is soft and ductile. Faster cooling will result in a different constituent and the steel will be harder and less ductile. A rapid cooling such as a water quench will result in a martensitic structure, which is the hardest structure that can be obtained. Cementite, though somewhat harder, is not present in its free state except in hypereutectoid steels and then only in such small quantities that its influence on the hardness of the steel can be ignored.

The essential ingredient of any hardened steel is *martensite*. Martensite is obtained by the rapid quenching of carbon steels and is the transitional substance formed by the rapid decomposition of austenite. It is a supersaturated solution of carbon in alpha iron. Under the microscope it appears as a needlelike constituent, as seen in Figure 7.10. The hardness of martensite depends on the amount of carbon present and varies from Rockwell C45 to C67.[1] It cannot be machined, is quite brittle, and is strongly magnetic.

If steel is quenched at slightly less than the critical rate, a dark constituent with somewhat rounded outlines will be obtained. The name of this constituent is *fine pearlite*. Under the microscope at usual magnifications it appears as a dark unresolved mass, but at very high magnification a fine lamellar structure is seen. Fine pearlite is less hard than martensite. It has a Rockwell C hardness varying from 34 to 45, but is quite tough and capable of resisting considerable impact. As the quenching rate is still further reduced, the pearlite becomes coarser and is definitely lamellar under high magnification at slow rates of cooling.

Maximum Hardness of Steel

The maximum hardness obtainable in a given piece of steel depends on the carbon content. Although various alloys such as chromium and vanadium increase the rate and depth-hardening ability of alloy steels, their maximum hardness will not exceed that of a carbon steel having the same carbon content. This fact is illustrated in the curve shown in Figure 7.11, where Rockwell C hardness is plotted against percentage of carbon. This curve shows the maximum hardness that is possible for a given carbon percentage. To obtain maximum hardness the carbon must be completely in solution in the austenite when

---

[1]Values stated in U.S. customary units are regarded as standard. Metric conversions are not provided. ASTM E18(67).

142 HEAT TREATMENT

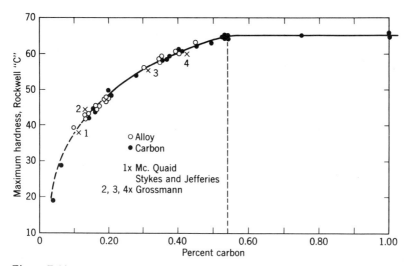

Figure 7.11
Maximum hardness versus carbon content.

quenched. The *critical quenching rate,* which is the slowest rate of cooling that will result in 100% martensite, should be used. Finally, austenite must not be retained in any appreciable percentages because it will soften the structure.

The curve in Figure 7.11 is made up of test points from both alloy and carbon steels, and little variation in the results may be seen. However, the same quenching rate cannot be used for both alloy and carbon steels of the same carbon content. The maximum hardness obtained in any steel represents the hardness of martensite and is approximately 66 to 67 Rockwell C. Carbon content equal to or in excess of 0.60% is necessary to achieve this level.

TEMPERING

Steel that has been hardened by rapid quenching is brittle and not suitable for most uses. By *tempering* or **drawing,** the hardness and brittleness may be reduced to the desired point for service conditions. As these properties are reduced there is also a decrease in tensile strength and an increase in the ductility and toughness of the steel. The operation, as depicted in Figure 7.12, consists of reheating quench-hardened steel to some temperature below the critical range followed by any rate of cooling. Although this process softens steel, it differs considerably from annealing in that the process lends itself to close control of the physical properties and in most cases does not soften the steel to the extent that annealing would. The final structure obtained from tempering a fully hardened steel is called tempered martensite.

Tempering is possible because of the instability of the martensite, the principal constituent of hardened steel. Low-temperature draws, from 300° to 400°F (150°–205°C), do not cause much decrease in hardness and are used principally to relieve internal strains. As the tempering temperatures are increased, the breakdown of the martensite takes place

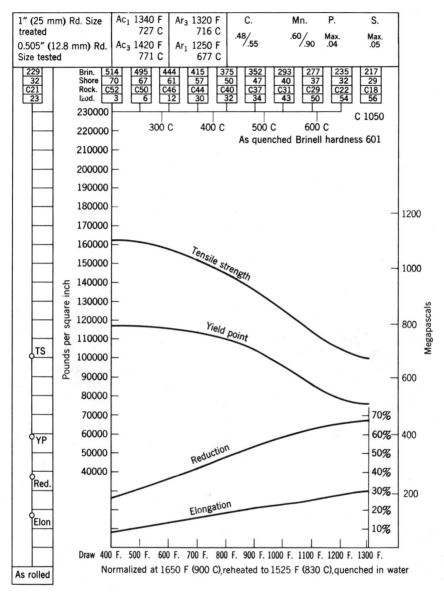

Figure 7.12
Physical properties chart (average values) of AISI steel, fine grain, water quenched.

at a faster rate, and at about 600°F (315°C) the change to a structure called tempered martensite is very rapid. The tempering operation may be described as one of precipitation and agglomeration or coalescence of cementite. A substantial precipitation of cementite begins at 600°F (315°C), which produces a decrease in hardness. Increasing the temperature causes coalescence of the carbides with continued decrease in hardness. Figure 7.12 shows a typical set of property curves for AISI 1050 steel, giving the tensile strength, hardness, percentage elongation, and percentage reduction in area for various tempering or draw temperatures. The influence of tempering on the physical properties of the steel is clearly shown by these curves. Alloying elements have a profound influence on tempering, the general effect being to retard the softening rate so that alloy steels will require a higher tempering temperature to produce a given hardness.

In the process of tempering, some consideration should be given to time as well as to temperature. Although most of the softening action occurs in the first few minutes after the temperature is reached, there is some additional reduction in hardness if the temperature is maintained for a prolonged time. Usual practice is to heat the steel to the desired temperature and hold it there only long enough to have it uniformly heated.

Two special processes using interrupted quenching are a form of tempering. In both, the hardened steel is quenched in a salt bath held at a selected lower temperature before being allowed to cool. These processes, known as austempering and martempering, result in products having certain desirable physical properties.

Austempering

The interrupted quenching process (Figure 7.13A) is known as austempering. It is an isothermal transformation that converts austenite to a hard structure called *bainite*. Parts to be treated must be rapidly quenched to the correct holding temperature so that the cooling curve is not permitted to intersect the nose in the transformation diagram. The steel is held at a temperature above the M_s line but below 800°F (430°C). When held at the constant temperature for sufficient time to complete the transformation, a structure called bainite is obtained. Under the microscope this structure resembles martensite. Although this microstructure exhibits approximately the same hardness, it is relatively tougher than in the quenched and tempered condition. The process is limited to small parts with good hardenability such that pearlite is not formed during the initial quench.

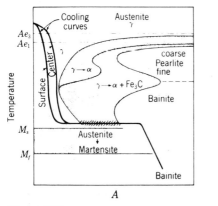 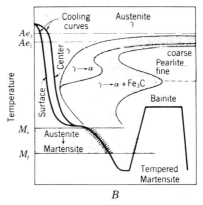

Figure 7.13
Transformation diagrams illustrating the interrupted quenching processes for (A) austempering and (B) martempering.

Martempering

In the process known as martempering the steel is rapidly quenched from the austenite region to a temperature just above the M_s line (see Figure 7.13B). Here the steel is held long enough to enable the surface and the center of the piece being treated to arrive at the same temperature. When this occurs the piece is usually cooled in air to room temperature, thus forming martensite. The steel is reheated to a temperature varying with the carbon and alloy content, although for plain-carbon steels containing around 0.40% carbon, the temperature is 700°F (370°C). The main purpose of martempering is to minimize distortion, cracking, and internal stresses that result from quenching in oil or water. Although the resulting product is similar to tempered martensite, a subsequent tempering operation is generally performed.

ANNEALING

The primary purpose of ***annealing*** is to soften hard steel so that it may be machined or cold worked. This is usually accomplished by heating the steel to slightly above the Ac_3 critical temperature, holding it there until the temperature of the piece is uniform throughout, and then cooling at a slowly controlled rate so that the temperature of the surface and that of the center of the piece are approximately the same. This process, illustrated in Figure 7.14A is known as *full annealing* because it wipes out all trace of previous structure, refines the crystalline structure, and softens the metal. Annealing also relieves internal stresses previously set up in the metal.

When hardened steel is reheated to above the critical range the constituents are changed back into austenite, and slow cooling then provides ample time for complete transformation of the austenite into the softer constituents. For the hypoeutectoid steels these constituents are pearlite and ferrite. It may be noted by referring to the equilibrium diagram that the

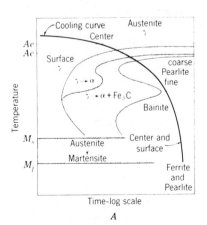
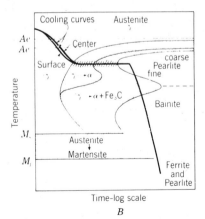

Figure 7.14
Transformation curves for (A) full and (B) isothermal annealing.

annealing temperature for hypertectoid steels is lower, being slightly above the A_1 line. There is no reason to heat above the *Acm* line as it is at this point that the precipitation of the hard constituent cementite is started. All martensite is changed into pearlite by heating above the lower critical range and slowly cooling. Any free cementite in the steel is unaffected by the treatment.

The temperature to which a given steel should be heated in annealing depends on its composition; for carbon steels it can be obtained readily from the partial iron–iron carbide equilibrium diagram in Figure 7.1. The heating rate should be consistent with the size and uniformity of sections, so that the entire part is brought up to temperature as uniformly as possible. When the annealing temperature has been reached, the steel should be held there until it is uniform throughout. This usually takes about 45 min for each inch (25 mm) of thickness of the largest section. For maximum softness and ductility the cooling rate should be very slow, such as allowing the parts to cool down with the furnace. The higher the carbon content, the slower this rate must be.

Isothermal Annealing

Isothermal annealing, as illustrated in Figure 7.14B, provides a short annealing cycle. Steel is rapidly quenched to the temperature at which austenite transforms to a relatively soft ferrite carbide aggregate in the shortest possible time. It is then held for the time necessary to transform the austenite completely to pearlite. After the transformation is complete the part may be cooled in any manner. Isothermal annealing results in giving pearlite a more uniform structure than that obtained by other annealing processes. Fineness depends on the transformation temperature used.

Process Annealing

Process annealing practiced in the sheet and wire industry between cold-working operations consists of heating the steel to a temperature a little below the critical range and then cooling it slowly. This process is more rapid than spheroidizing and results in the usual pearlitic structure. It is similar to the tempering process but will not give so much softness and ductility as a full anneal. Also, at the lower heating temperature there is less tendency for the steel to scale or decarburize.

NORMALIZING AND SPHEROIDIZING

The process of *normalizing* consists of heating the steel about 50° to 100°F (10°–40°C) above the upper critical range and cooling in still air to room temperature. This process is principally used with low- and medium-carbon steels as well as alloy steels to make the grain structure more uniform, to relieve internal stresses, or to achieve desired results in physical properties. Most commercial steels are normalized after being rolled or cast.

Spheroidizing is the process of producing a structure in which the cementite is in a spheroidal distribution, Figure 7.15. If a steel is heated slowly to a temperature just below the critical range and held there for a prolonged period of time, this structure will be

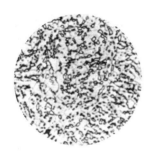

Figure 7.15
SAE 1095 steel quenched from 1550°F and tempered at 1250°F for 8 h. Structure is spheroidized cementite in a ferritic matrix. Magnification ×900.

obtained. It may also be accomplished by alternately heating and cooling between temperatures that are just above and below the Ac_1 range. The globular structure obtained gives improved machinability to the steel. This treatment is particularly useful for hypereutectoid steels that must be machined.

SURFACE HARDENING

Carburizing

The oldest known method of producing a hard surface on steel is case hardening or carburizing. Steel is heated above Ac_1 while in contact with some carbonaceous material, which may be solid, liquid, or gas. Iron at temperatures close to and above its critical temperature has an affinity for carbon. The carbon is absorbed into the metal to form a solid solution with iron and converts the outer surface into a high-carbon steel. The carbon is gradually diffused to the interior of the part. The depth of the case depends on the time and temperature of the treatment. *Pack carburizing* consists of placing the parts to be treated in a closed container with some carbonaceous material such as charcoal or coke. It is a long process and used to produce fairly thick cases of from 0.030 to 0.160 in. (0.76–4.06 mm) in depth.

For shallower case depths of 0.005 to 0.030 in. (0.13–0.76 mm), *gas carburizing* is frequently used, employing such hydrocarbon fuels as natural gas or propane. Gas carburizing is adapted to the case hardening of small parts that may be surface hardened by direct quenching from the furnace at the end of the heating cycle.

In *liquid carburizing* the steel is heated above the Ac_1 in a cyanide salt bath, causing the carbon and some nitrogen to diffuse into the case. It is similar to cyaniding, except that the case is higher in carbon and lower in nitrogen. Liquid carburizing can be used for case thickness up to 0.250 in. (6.35 mm), although the thickness seldom exceeds 0.025 in. (0.64 mm). This method is best suited for case hardening small and medium-sized parts.

Steel for carburizing is usually a low-carbon steel of about 0.15% carbon that would not in itself respond appreciably to heat treatment. In the course of the process the outer layer is converted into a high-carbon steel with a content ranging from 0.9% to 1.2% carbon.

A steel with varying carbon content and, consequently, different critical temperatures requires a special heat treatment. Because there is some grain growth in the steel during the prolonged carburizing treatment, the work should be heated to the critical temperature of the core and then cooled, thus refining the core structure. The steel should then be reheated to a point above the transformation range of the case (Ac_1) and quenched to produce a hard, fine structure. The lower heat-treating temperature of the case results from the fact that hypereutectoid steels are normally austenitized for hardening just above the lower critical point. A third tempering treatment may be used to reduce strains.

Carbonitriding

Carbonitriding, sometimes known as dry cyaniding or nicarbing, is a case-hardening process in which the steel is held at a temperature above the critical range in a gaseous atmosphere from which it absorbs carbon and nitrogen. Any carbon-rich gas with ammonia can be used. The wear-resistant case produced ranges from 0.003 to 0.030 in. (0.08–0.76 mm) in thickness. An advantage of carbonitriding is that the hardenability of the case is significantly increased when nitrogen is added, permitting the use of low-cost steels.

Cyaniding

Cyaniding, or liquid carbonitriding as it is sometimes called, is also a process that combines the absorption of carbon and nitrogen to obtain surface hardness in low-carbon steels that do not respond to ordinary heat treatment. The part to be case hardened is immersed in a bath of fused sodium cyanide salts at a temperature slightly above the Ac_1 range, the duration of soaking depending on the depth of the case. The part is then quenched in water or oil to obtain a hard surface. Case depths of 0.005 to 0.015 in. (0.13–0.38 mm) may be readily obtained by this process. Cyaniding is used principally for the treatment of small parts.

Nitriding

Nitriding is somewhat similar to ordinary case hardening, but it uses a different material and treatment to create the hard surface constitutents. In this process the metal is heated to a temperature of around 950°F (510°C) and held there for a period of time in contact with ammonia gas. Nitrogen from the gas is introduced into the steel, forming very hard nitrides that are finely dispersed through the surface metal.

Nitrogen has greater hardening ability with certain elements than with others; hence, special nitriding alloy steels have been developed. Aluminum in the range of 1% to 1.5% has proved to be especially suitable in steel, in that it combines with the gas to form a very stable and hard constituent. The temperature of heating ranges from 925° to 1050° (495°–565°C).

Liquid nitriding utilizes molten cyanide salts and, as in gas nitriding, the temperature is held below the transformation range. Liquid nitriding adds more nitrogen and less carbon than either cyaniding or carburizing in cyanide baths. Case thicknesses of 0.001 to 0.012 in. (0.03–0.30 mm) are obtained, whereas for gas nitriding the case may be as thick as 0.025 in. (0.64 mm). In general the uses of the two nitriding processes are similar.

Nitriding develops extreme hardness in the surface of steel. This hardness ranges from 900 to 1100 Brinell, which is considerably higher than that obtained by ordinary case hardening. Nitriding steels, by virtue of their alloying content, are stronger than ordinary steels and respond readily to heat treatment. It is recommended that these steels be machined and heat-treated before nitriding, because there is no scale or further work necessary after this process. Fortunately, the interior structure and properties are not affected appreciably by the nitriding treatment and, because no quenching is necessary, there is little tendency to warp, develop cracks, or change condition in any way. The surface effectively resists corrosive action of water, saltwater spray, alkalies, crude oil, and natural gas.

INDUCTION HARDENING

In recent years induced electric current has been widely accepted by industry. Principal applications for this method of heating include melting of metals, hardening and other heat-treatment operations, preheating metals for hot work, and heating for sintering, brazing, and similar operations. High-frequency alternating current obtained from motor generator sets, mercury arc converters, or spark gap oscillators is used for this type of heating. Although there are different powers and frequency limitations for each type of equipment, most do not exceed 500,000 cps (Hz). For thin cases high frequencies are used, whereas for intermediate and thick cases low frequencies give better results.

Induction heating has proved satisfactory for many surface-hardening operations as required on crankshafts and similar wearing surfaces. It differs from ordinary case-hardening practice in that the analysis of the surface steel is not changed, because the hardening is accomplished by an extremely rapid heating and quenching of the wearing surface, which has not effect on the interior core. The hardness obtained in induction hardening is the same as that obtained in conventional treatment and depends on carbon content.

An inductor block acting as a primary coil of a transformer is placed around but not touching the journal to be hardened. A high-frequency current is passed through this block inducing a current in the surface of the bearing. The heating effect is due to induced eddy current and hysteresis losses in the surface material. As the steel is heated to the upper critical range, the heating effect of these losses is gradually decreased, thereby eliminating any possibility of overheating the steel. The inductor block surrounding the heated surface has water connections and numerous small holes on its inside surface; as soon as the steel has been brought up to the proper temperatures, it is automatically spray-quenched under pressure.

An important feature of this method of hardening is its quickness because it requires only a few seconds to heat steel to a depth of 1/8 in. (3.2 mm). The actual time will depend primarily on the frequency, power input, and depth of hardening required. Al-

150 HEAT TREATMENT

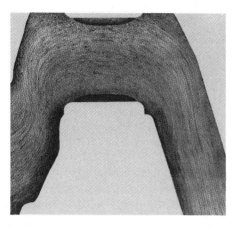

Figure 7.16
Section of an induction-hardened crankpin bearing.

though the equipment cost is high, it is offset by the advantages of the process which include fast operation, freedom from scaling, clean operation, little distortion, no manual handling of hot parts, and low treating costs. Medium-carbon steel has proved satisfactory for parts, and the nature of the process has practically eliminated the necessity for using costly alloy steels. Figure 7.16 illustrates the local heating obtained in a hardened crankpin bearing that has been induction hardened.

Flame Hardening

Flame hardening, like the induction-hardening process, is based on rapid heating and quenching of the wearing surface. The depth of hardness of the case depends entirely on the hardenability of the material being treated, as no other elements are added or absorbed during the process. The heating is done by an oxyacetylene flame, which is applied for enough time to heat the surface above the critical temperature range of the steel. Integral

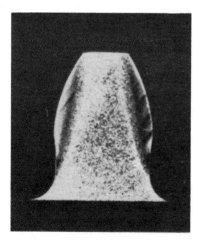

Figure 7.17
Section through a gear tooth showing the structure obtained by flame hardening.

with the flame head are water connections that cool the surface by spraying as soon as the desired temperature is reached. With proper control the interior surface is not affected by the treatment. The depth of the case is a function of the heating time and flame temperature. Figure 7.17 shows an etched cross section of a gear tooth and the hardened area. With this process hard surfaces with a ductile backing are obtained, large pieces are treated without heating the entire part, the case depth is easily controlled, the surface is free of scale, and the equipment is portable.

There are various torch combinations that allow coating powders to be applied after the flame, and surfaces can be laid on that vary in hardness, strength, and permeability. Usually when materials are added the part must be machined or ground after application. The advantage of this technique is that parts like shafts can be built up at the same time they are treated.

HARDENING NONFERROUS MATERIALS

Precipitation hardening, often called age hardening, can be achieved only with those alloys in which there is a decreasing solubility of one material in another as the temperature is reduced. Such a situation is illustrated in Figure 7.18, where an enlarged portion of an aluminum–copper constitutional diagram is shown. If the alloy is cooled slowly from a temperature at A, the Al_2Cu material is precipitated out of the solid solution because the solubility of copper is greatly decreased at low temperature. An alloy in this condition will not respond to a hardening treatment.

In order to harden the alloy shown in Figure 7.18, two treatments are necessary. First, the alloy must be heated to A above the solubility line CD to allow the constituents to enter into a solid solution. The alloy must be held there for sufficient time until a homogeneous alloy is obtained. It is then cooled rapidly to room temperature leaving the

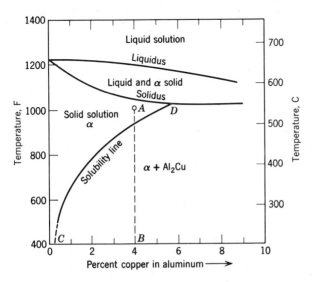

Figure 7.18
Partial constitutional diagram for aluminum-cooper alloy system.

alloy in a supersaturated, unstable state. In this state the alloy is soft but, being unstable, the excess constituents will precipitate out of solution with the lapse of time. This process, if accomplished at room temperature, is known as *natural aging*. Particles that are precipitated from the solid solution form at grain boundaries and slip planes, producing a "keying" action that reduces slippage between the crystals. The hardness resulting from the phenomenon is influenced by the size of the precipitated particles. It is increased as the particle size is increased, but after a critical size is reached further growth results in brittleness and loss of strength. The process of precipitation hardening can be delayed by refrigeration. An example of this is holding aluminum rivets in a refrigerator until they are ready for use. In this way they remain soft for the riveting operation and then gradually harden with time.

Artificial aging differs slightly in that an additional heat treatment is used. In this case the alloy is heated to some elevated temperature that accelerates the precipitation of some constituent from the supersaturated solution. As the temperature is increased, the more rapid is the precipitation and increase in hardness. An equilibrium condition is finally reached that decreases the strength as a result of the large size of the precipitate crystals. This condition, which should be avoided, is known as *overaging*.

Numerous alloys are of a composition that can be age hardened, the only requirement being that solubility must decrease with decreasing temperature, permitting the formation of a supersaturated solid solution. In this group the nonferrous metals, which are most important, include many of the aluminum, copper, nickel, and magnesium alloys. Some stainless steels also respond to this treatment.

FURNACES FOR HEAT TREATING

The *car bottom furnace* (Figure 7.19A) is used for large parts that are to be heat treated and particularly for annealing and normalizing. Such furnaces get their name from their design, which employs a flatbed car on railroadlike wheels that is pushed into either a

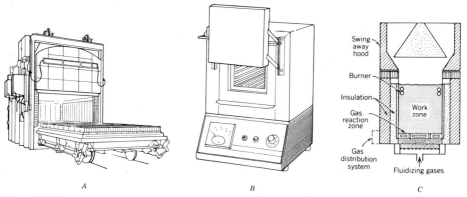

Figure 7.19
Typical heat-treating furnaces. *A,* Car bottom. *B,* Small box. *C,* Fluidized bed.

gas-fired or electrically heated, well-insulated, roomlike furnace. They vary in size from about 320 to 8000 ft^2 (30–740 m^2).

The *bell top* furnace is made so that once parts are loaded on a platform the furnace can be lifted up and placed over the parts. The *box furnace,* in which steel pellets are loaded into a furnace by a forklift truck, is a variation of the car bottom furnace. Small versions of the *box furnace* are called a *muffle furnace* (Figure 7.19B) and can sit on a bench.

Continuous furnaces employ a set of rollers, skids, or a walking beam on which the castings are systematically moved through the furnace. One type utilizes a carousel arrangement. Continuous furnaces are seldom used for normalizing. Because of the size of the part regulates the time spent in the furnace, such furnaces are often single-purpose units.

Salt bath furnaces are resistant heated using salt as the resistance, and the magnetic field around the electrodes causes the bath to circulate. Because of the environmental problems associated with the vapors and the disposal of the chlorides of barium, cadmimum, and sodium salts, these furnaces are declining in favor.

For small parts the *fluidized bed* (Figure 7.19C) is increasingly used. It consists of a container filled with pellets of alumina or sand, which are heated by a high-velocity, air–gas flame that not only heats the bed but "fluidizes" the particles. The particles become almost airborne, and parts immersed in the pellets are surrounded by the pellets much as if they were in a liquid. If a stoichiometric mixture of air and gas is used, an almost inert atmosphere is present in the container. Some furnaces are electrically heated and an inert gas is forced upward to fluidize the pellets.

When parts are *vacuum heat treated,* a furnace is heated by electrical means or hot gases are circulated around it. The parts are placed inside and a vacuum is pulled. These furnaces offer operator safety, economy, uniform heating, and low capital cost, and they present no environmental problems. This type of operation is expensive but is often used for heat-treating die-casting dies and similar highly machined parts.

QUESTIONS AND PROBLEMS

1. Give an explanation of the following terms:

   Chauffage Anneal
   Hypoeutectoid steel Normalize
   Hardenability Spheroidizing
   Fine pearlite Nitriding
   Draw Car bottom furnace

2. How much pearlite is in pure iron? In a 1080 steel?

3. Sketch the microstructure of an etched specimen of 1040 steel and a 1080 steel.

4. What is the percentage of pearlite in a 1030 steel?

5. To what temperature must a 1050 steel be elevated if it is to be hardened? Sketch the constituents of an annealed 1050 steel.

6. To what temperature must the following steels be elevated for hardening: 1040, 1060, 1080, 10110?

7. To what temperature must the following steels be elevated for normalizing: 1040, 1060, 1080, 10110?

8. What is martensite? How does it appear under the microscope?

9. Describe the lattice structure of a steel with 0.4% carbon as it goes from a molten to a cool condition. At what temperature do the transformations take place?

10. What determines the maximum hardness that can be obtained in a piece of steel?

11. What happens if more than 6.67% carbon is present in a batch of molten steel?

12. What is the difference between the eutectic and euectoid points?

13. Why is oil instead of water used as a quenching medium for many steels? What other materials or processes can be used as quenching media?

14. What is the relationship between the amount of pearlite in a piece of steel and its ability to be hardened?

15. Explain what happens to surface hardness for a given quench if the part is greatly increased in size. Why?

16. What is the maximum attainable hardness to which the following steels can be heat treated: 1020, 1040, 1060, 1080?

17. Suppose in problem 16 the steels were alloy steels but with the same carbon content. What is the maximum attainable hardness?

18. How can a 1050 steel be heat treated to yield a tensile strength of 160,000 psi? 120,000 psi?

19. In problem 18 what is the approximate Rockwell hardness of the two strengths of steel?

20. In problem 18 what other physical characteristics can you predict for the two pieces of steel as heat treated to their respective tensile strengths?

21. What are the critical temperatures of a 1050 steel?

22. On what size specimen is Figure 7.12 based? What would be the results for a much larger specimen? A much smaller specimen?

23. What is the value of an isothermal transformation diagram?

24. Describe the process of carburizing. For what type of work is cyaniding used?

25. How is a time–temperature transformation diagram changed by varying the carbon content or by adding alloys?

26. Why does tempering cause a decrease in tensile strength?

27. What effect does carbon content have on hardenability?

28. What can be said about the physical characteristics of martensite?

29. Discuss the instability of martensite with temperature.

30. Calculate the percentages of pearlite and cementite in the following annealed steels: 1020, 1040, 1080, 10110.

31. After steels are rolled or cast, what heat treatment is usually applied?

32. How does liquid nitriding differ from cyaniding?

33. What factors affect hardness as obtained by induction hardening?

34. Explain why Duralumin (4% copper) can or cannot be age hardened.

35. A box of aluminum rivets located in the storeroom was found to be age hardened. Explain whether or not the rivets can be salvaged.

36. Plot a curve showing percentage pearlite as a function of carbon content.

37. Sketch a design for an electrically heated, fluidized-bed furnace.

38. Describe the advantages of a fluidized-bed furnace.

CASE STUDY

HEAT TREATING

As chief engineer of Real Temperature Company you have been asked to bid on heat-treating 500,000 parts per year for an automotive parts company. The parts weigh about 1.9 lb (0.86 kg) each and are shaped like a 4-in. (101.6-mm) OD saucer. The material is AISI 1050 steel and must be heat treated to have a tensile strength of 145,000 psi. It is your job to reply to the bid request by stating the method you will use to heat treat parts and to report on the physical properties of the finished product. In your report should be included the type of furnace you have selected.

CHAPTER 8
WELDING, BRAZING, AND ADHESIVE BONDING

Welding is a metal-joining process in which coalescence is obtained by heat and pressure. It may also be defined as a metallurgical bond accomplished by the attracting forces between atoms. Before these atoms can be bonded together, absorbed vapors and oxides on contacting surfaces must be overcome. The number one enemy to welding is oxidation, and, consequently, many welding processes are performed in a controlled environment or shielded by an inert atmosphere. If force is applied between two smooth metal surfaces to be joined, some crystals will crush through the surfaces and be in contact. As more and more pressure is applied, these areas spread out and other contacts are made. The brittle oxide layer is broken and fragmented as the metal is deformed plastically. Coalescence is obtained when the boundaries between the two surfaces are mainly crystalline planes. This process, known as *cold welding*, will be discussed further in this chapter. The breaking through or elimination of surface oxide layers happens when a weld is made.

If temperature is added to pressure, the welding of two surfaces will be facilitated, and coalescence is obtained in the same manner as cold-pressure welding. As temperature is increased the ductility of the base metal is increased and atomic diffusion progresses more rapidly. Nonmetallic materials on interfacial surfaces are softened, permitting them to be removed or broken up by plastic flow of the base materials. Hot-pressure welds are more efficient but not necessarily stronger if the atom-to-atom bond is the same.

Many welding processes have been developed. They differ widely in the manner that heat is applied and in the equipment used. The principal processes are listed here.

WELDING PROCESSES

  I. Braze welding
    A. Torch
    B. Furnace
    C. Induction
    D. Resistance
    E. Dip
    F. Infrared
 II. Forge welding
    A. Manual
    B. Machine

    1. Rolling
    2. Hammer
    3. Die
III. Gas welding
   A. Oxyacetylene
   B. Oxyhydrogen
   C. Air acetylene
   D. Pressure
IV. Resistance welding
   A. Spot
   B. Projection
   C. Seam
   D. Butt
   E. Flash
   F. Percussion
   G. High frequency
V. Induction welding, high frequency
VI. Arc welding
   A. Carbon electrode
     1. Shielded
     2. Unshielded
   B. Metal electrode
     1. Shielded
       a. Shielded arc
       b. Atomic hydrogen
       c. Inert gas
       d. Arc spot
       e. Submerged arc
       f. Stud
       g. Electroslag
     2. Unshielded
       a. Bare metal
       b. Stud
VII. Special welding processes
   A. Electron beam
   B. Laser welding
   C. Friction welding
   D. Thermit welding
     a. Pressure
     b. Nonpressure
   E. Flow welding
   F. Cold welding
     a. Pressure
     b. Ultrasonic
   G. Explosive welding
   H. Diffusion welding

Some of these processes require hammering, rolling, or pressing to effect the weld; others bring the metal to a fluid state and require no pressure. Processes that use pressure generally require bringing the surfaces of the metal to a temperature sufficient for cohesion to take place. This is usually a subfusion temperature but if the fusion temperature is reached, the molten metal must be confined by surrounding solid metal. Most welds are made at fusion temperature and require the addition of weld metal in some form. Welds are also made by casting when metal is heated to a high temperature and poured into a cavity between the two pieces to be joined.

In any welding, coalescence is improved by cleanliness of the surfaces to be welded. Surface oxides should be removed, because they tend to become entrapped in the solidifying metal. *Fluxes* are often employed to remove oxides in fusible slags that float on the molten metal and protect it from atmospheric contamination. In electric-arc welding, flux is coated on the electrode, whereas in gas or forge welding a powder form is used. In other processes a nonoxidizing atmosphere is created at the point where the welding is done. Since oxidation takes place rapidly at high temperature, speed in welding is important.

SOLDERING AND BRAZING

Soldering and brazing are processes that unite metals with a third joining metal introduced into the joint in a liquid state and allowed to solidify. These processes have wide commercial use in uniting small assemblies and electrical parts.

Soldering

Soldering is the uniting of two pieces of metal with a metal that is applied between the two in a molten state at a temperature not exceeding 840°F (450°C). In this process, a little alloying with the base metal takes place and additional strength is obtained by mechanical bonding. Lead and tin alloys having a melting range of 350° to 700°F (180°–370°C) are principally used. The strength of the joint is largely determined by the adhesive quality of the alloy, which never reaches the strength of the materials being joined. Because solder has little strength, the solder should be as thin as possible. The strongest joints are made by wetting or tinning the surfaces with solder and then bringing them together with heat. The wetting is effective because the solder reacts with the base metal to form a compound. A flux helps clean the base metal to allow compounding to take place.

The characteristics of most solders are shown in the lead–tin phase diagram in Figure 8.1. A 50-50 composition, common in most soldering, flows at 430°F (220°C) with complete solidification at the eutectic. To obtain higher physical properties or to reduce price, some solders contain other elements such as cadmium, silver, copper, or zinc. Heating can be accomplished by dipping, furnace, torch, electrical resistance, induction, or ultrasonic heating, or by a soldering iron. Materials that are easy to solder include copper, tin, lead, silver, and gold. Iron and nickel are more difficult to solder.

Brazing

In brazing a nonferrous alloy is introduced in a liquid state between the pieces of metal to be joined and allowed to solidify. The filler metal, having a melting temperature of over 840°F (450°C) but lower than the melting temperature of the parent metal, is dis-

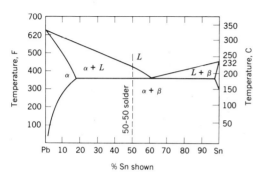

Figure 8.1
Lead–tin phase diagram.

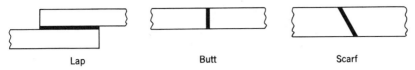

Figure 8.2
Common welding and brazing joints.

tributed between the surfaces by capillary attraction. *Braze welding* is similar to ordinary brazing except that the filler metal is not distributed by capillary attraction. In this process filler metal is melted and deposited at the point where the weld is to be made. In both, special fluxes are required to remove surface oxide and to give the filler metal the fluidity to wet the joint surfaces completely. Alloys of copper, silver, and aluminum are the most common brazing filler metals.

The basic joint types found in brazing are the lap, butt, and scarf designs, Figure 8.2. Many modifications of these joints are used in production brazing, depending on the shape of the parts to be joined. Of the three joints shown, the lap joint is the strongest because it has greater contact area. Joint clearance is important, because sufficient space must be allowed for capillary attraction to provide filler metal distribution. The processes that have wide commercial use in uniting small assemblies and electrical parts are listed in Table 8.1.

In any brazing operation the joint must first be cleaned of all dirt, oil, or oxides. Then the pieces are properly fitted together with appropriate clearance for the filler metal. Mechanical or chemical cleaning may be necessary in the joint preparation in addition to the flux used during the process. Borax either alone or in combination with other salts is commonly used as a flux.

To facilitate speed in brazing, the filler metal is frequently prepared in the form of rings, washers, rods, or other special shapes to fit the joint being brazed. This ensures having the proper amount of filler metal available for the joint as well as having it placed in the correct position.

Advantages of the brazing process are the effecting of joints in materials difficult to weld, in dissimilar metals, and in exceedingly thin sections of metal. In addition, the process is rapid and results in a neat-appearing joint requiring a minimum of finishing. Brazing is used for the assembly of pipes to fittings, carbide tips to tools, radiators, heat exchangers, electrical parts, and the repair of castings.

Table 8.1 **Brazing and Braze Welding**

Dipping	Furnace	Torch	Electric	Welding
1. Metal	1. Gas	1. Oxyacetylene	1. Resistance	1. Torch
2. Chemical	2. Electric	2. Oxyhydrogen	2. Induction	2. Arc
			3. Infrared	

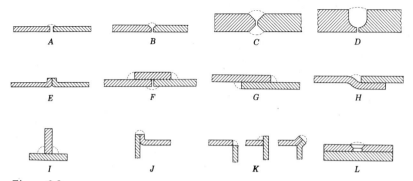

Figure 8.3
Types of welded joints. *A*, Butt weld. *B*, Single vee. *C*, Double vee (heavy plates). *D*, U-shaped (heavy casting). *E*, Flange weld (thin metal). *F*, Single-strap butt joint. *G*, Lap joint (single- or double-fillet weld). *H*, Joggled lap joint (single or double weld). *I*, Tee joint (fillet welds). *J*, Edge weld (used on thin plates). *K*, Corner welds (thin metal). *L*, Plug or rivet butt joint.

WELDED JOINTS

To be efficient a welded joint must be properly designed to the service for which it is intended. A few of the common types may be seen in Figure 8.3. Some such as butt welds may be further subdivided as they vary in form according to the thickness of the material. Joints for forge welding differ in their manner of preparation and do not resemble those shown in the figure. Lap and butt joints are ordinarily used in resistance welding. Resistance-welded joints must be prepared more accurately and be cleaner than those for other processes. Both gas and arc welding use the same kind of joints. The minimum welding thickness for aluminum is about 0.030 in. (0.76 mm) and 0.015 in. (0.38 mm) for steel.

FORGE WELDING

Forge welding was the first form of welding and for many centuries the only one in general use. Briefly, the process consists of heating the metal in a forge to a plastic condition and then uniting it by pressure. The heating is usually done in a coal- or coke-fired forge, although modern installations frequently employ oil or gas furnaces. The manual process is naturally limited to light work, because all forming and welding is accomplished with a hand sledge. Before the weld is made the pieces are formed to correct shape, so that when they are welded they will unite at the center first. As they are hammered together from the center to the outside edges, any oxide or foreign particles will be forced out. The process of preparing the metal is known as *scarfing*.

Forge welding is rather slow, and there is considerable danger of an oxide scale forming on the surface. Oxidation can be counteracted somewhat by using a thick fuel bed and by covering the surfaces with a fluxing material that dissolves the oxides. Borax in combination with sal ammoniac is commonly used. Heating must be slow when there are unequal section thicknesses. When heated to a desired temperature, the parts are removed to the anvil and pounded together.

For this type of welding, low-carbon steel and wrought iron are recommended, because they have a broad welding temperature range. Weldability decreases as the carbon content increases.

GAS WELDING

Gas welding includes all the processes in which gases are used in combination to obtain a hot flame. Those commonly used are acetylene, natural gas, and hydrogen in combination with oxygen. *Oxyhydrogen welding* was the first gas process to be commercially developed. The maximum temperature developed by this process is 3600°F (1980°C). *Hydrogen* is produced by either the electrolysis of water or passing steam over coke. The most used combination is the oxyacetylene process, which has a flame temperature of 6300°F (3500°C).

Oxyacetylene Welding

An oxyacetylene weld is produced by heating with a flame obtained from the combustion of oxygen and acetylene with or without the use of a filler metal. Most often the joint is heated to a state of fusion and as a rule no pressure is used.

Oxygen is produced by both electrolysis and liquification of air. *Electrolysis* separates water into hydrogen and oxygen by passing an electric current through it. Most commercial oxygen is made by liquefying air and separating the oxygen from the nitrogen. It is stored in steel cylinders at a pressure of 2000 psi (14 MPa), as noted in Figure 8.4.

Acetylene gas (C_2H_2) is obtained by dropping lumps of calcium carbide in water. The

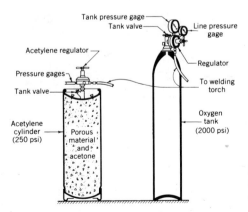

Figure 8.4
Cylinders and regulators for oxyacetylene welding.

gas bubbles through the water, and any precipitate is slaked lime. The reaction that takes place in an acetylene generator is

$$\underset{\text{calcium carbide}}{CaC_2} + \underset{\text{water}}{2H_2O} = \underset{\text{slaked lime}}{Ca(OH)_2} + \underset{\text{acetylene gas}}{C_2H_2}$$

Acetylene gas can be either obtained from acetylene generators that generate the gas by mixing the carbide with the water or purchased in cylinders ready for use. Because this gas may not be safely stored at pressure much over 15 psi (0.1 MPa), it is stored with acetone. Acetylene cylinders are filled with a porous filler saturated with acetone in which the acetylene gas can be compressed. These cylinders hold 300 ft^3 (9 m^3) of gas at pressures up to 250 psi (1.7 MPa).

A schematic sketch of a welding torch and its gas supply is shown in Figure 8.5. Gas pressures are controlled by regulating valves, and final adjustment is done manually at the torch. Regulation of the proportion of the two gases is extremely important because the characteristics of the flame may vary.

The perfect gas laws describe the amount of gas available at a regulated pressure from a pressurized cylinder. That is

$$\frac{P_1V_1}{T_1} = \frac{P_2V_2}{T_2}$$

where P_1, V_1, and T_1 refer to the pressure, volume, and absolute temperature in the cylinder; P_2, V_2, T_2 refer to the regulated pressure, volume, and temperature. The absolute temperature is $T = 460 + t°F$. If the cylinder and regulated temperatures are the same or nearly so, then

$$P_1V_1 = P_2V_2$$

Thus if an acetylene cylinder contains 300 ft^3 of gas at 250 psi, it can deliver a much larger volume of gas at a regulated pressure of 7 psi above the atmospheric pressure of 14.7 psi. Thus

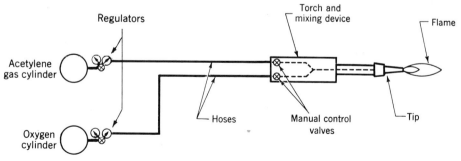

Figure 8.5
Schematic sketch of oxyacetylene welding torch and gas supply.

$$V_2 = \frac{P_1 V_1}{P_2} = \frac{250 \times 300}{7 + 14.7}$$

$$= 3456 \text{ ft}^3$$

The chemical reaction that occurs when acetylene and oxygen are burned in a neutral flame occurs in two stages, one at the inner cone and one at the outer envelope (Figure 8.6). The heat generated in the inner cone is approximately

$$H_{ic} = 150 \text{ BTU} \times V$$

where

H_{ic} = Heat generated in the inner cone

V = Volume of regulated gas, ft^3

and

$$H_{oc} = 960 \text{ BTU} \times V$$

where H_{oc} = heat generated in the outer envelope.

These equations are not a paradox because the inner cone is much smaller, thus displaying the higher temperature.

When there is an excess of acetylene used there is a decided change in the appearance of the flame. Three zones will be found instead of the two just described. Between the luminous cone and the outer envelope there is an intermediate, white-colored cone that has its length determined by the amount of excess acetylene. This flame, known as a reducing or carburizing flame, is used in the welding of Monel metal, nickel, certain alloy steels, and many of the nonferrous, hard surfacing materials.

If the torch is adjusted to give excess oxygen, a flame similar in appearance to the neutral flame is obtained except that the inner luminous cone is much shorter and the outer envelope appears to have more color. This oxidizing flame may be used in fusion welding brass and bronze but is undesirable in other applications.

The advantages and uses of oxyacetylene welding are numerous. The equipment is comparatively inexpensive and requires little maintenance. It is portable and can be used with equal facility in the field and in the factory. With proper technique practically all metals can be welded and the equipment used for cutting as well as for welding.

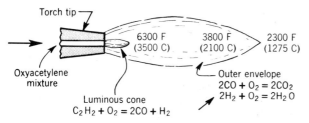

Figure 8.6 Sketch of neutral flame showing temperatures attained.

Oxyhydrogen Welding

Because oxyhydrogen burns at 3600°F (2000°C), a much lower temperature than oxygen and acetylene, it is used primarily for welding thin sheets and low-melting alloys and in some brazing work. While the same equipment can be used for both processes, flame adjustments are more difficult in hydrogen welding because there is no distinguishing color to judge the gas proportions. A reducing atmosphere is recommended and the process is characterized by the absence of oxides formed on the surface of the weld. The quality of these welds is equal to that obtained by other processes.

Air Acetylene Welding

The torch used in this process is similar in construction to a Bunsen burner, in which air is drawn into the torch as required for proper combustion. Because the temperature is lower than that attained by other gas processes, this type of welding has a limited use (e.g., for lead welding and low-temperature brazing or soldering operations).

Pressure Gas Welding

In pressure gas welding the abutting areas of parts to be joined are heated with oxyacetylene flames to a welding temperature of about 2200°F (1200°C) and pressure is applied. Two methods are in common use. In the first, known as the closed-joint method, the surfaces to be joined are held together under pressure during the heating period. Multiflame, water-cooled torches designed to surround the joint are used. During the heating operation the torches are oscillated slightly to eliminate excessive local heating. As the heating progresses the slightly beveled ends close. When the correct temperature is reached an additional upsetting pressure is applied. For low-carbon steel the initial pressure is below 1500 psi (10 MPa) and the upsetting pressure around 4000 psi (28 MPa).

The second or open-joint method (Figure 8.7) employs a flat multiflame torch placed between the two surfaces to be joined. This torch uniformly heats these surfaces until there is a film of molten metal over each of them. The torch is quickly withdrawn, and the surfaces are forced together and held at about 4000 psi (28 MPa) until solidification takes place. No filler metal is used. The quality of the weld is determined by the properties of the metal being joined.

Oxyacetylene Torch Cutting

Cutting steel with a torch is an important production process. A simple hand torch for flame cutting differs from a welding torch. It has several small holes for preheating flames surrounding a central hole through which pure oxygen passes. The preheating flames are exactly like the welding flames and are intended to preheat the steel before cutting. The principle of flame cutting is that oxygen has an affinity for iron and steel. At ordinary temperatures this action is slow, but eventually an oxide in the form of rust materializes. As the temperature of the steel is increased this action becomes much more rapid. If the steel is heated to a red color, about 1600°F (870°C), and a jet of pure oxygen is blown

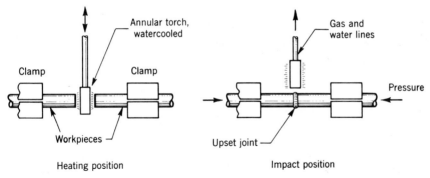

Figure 8.7
Schematic sketch illustrating pressure-gas butt welding.

on the surface, the action is almost instantaneous and the steel is actually burned into an iron oxide.

The amount of oxygen at regulated pressure necessary to burn or remove a cubic inch of iron can be expressed approximately as

$$V_o = 1.3 \times V_s$$

where

V_o = Oxygen required, ft³ or mm³

V_s = Amount of steel removed, ft³ or mm³

or

$$V_o = 1.3 \times d \times \ell$$

where

d = Thickness of steel, ft or mm

ℓ = length of cut, ft or mm

Metal up to 30 in. (762 mm) in thickness can be cut by this process.

Underwater cutting torches are provided with connections for three hoses: one for preheating gas, one for oxygen, and one for compressed air. The latter provides an air bubble around the tip of the torch to stabilize the flame and displace the water from the tip area. Hydrogen gas is generally used for the preheating flame, as acetylene gas is not safe to operate under the high pressures necessary to neutralize the pressure created by the depth of water.

Many cutting machines have been developed that automatically control the movement of the torch to cut any desired shape. Such a machine cutting several parts simultaneously is shown diagrammatically in Figure 8.8. In all such machines some control or sensing device is provided to guide the torches along a predetermined path. This control in its simplest form may be a hand-guided pointer following a drawing or held against a template. Usually these machines are electrically driven and are provided with a knurled,

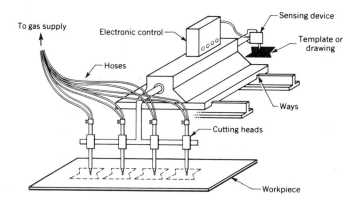

Figure 8.8 Schematic sketch of oxyacetylene torch cutting machine.

magnetized drive spindle that follows a steel template and controls the movement of the machine at the proper cutting speed. There is also an electronically controlled sensing device that is provided with an electric eye capable of following a line drawing, thus eliminating the need for template construction.

Numerical tape control can be adapted to cutting machines to provide greater accuracy and productive output in flame-cutting operations. All functions of the machine such as speeds, control of preheating, sequence of cutting, piercing, regulation of torch height, and travel from one piece to the next can be programmed. These operations are translated into control language from simple sketches, and tapes are punched on a special typewriter. Certain modifications of the machine permit plasma arc cutting.

Many parts that previously required shaping by forging or casting are now cut to shape by this process. Flame-cutting machines, which replace many machining operations where accuracy is not paramount, are widely used in the shipbuilding industry, structural fabrication, maintenance work, and the production of numerous items made from steel sheets and plates. Cast iron, nonferrous alloys, and high-manganese alloys are not readily cut by this process.

RESISTANCE WELDING

In this process a strong electric current is passed through the metals causing local heating at the joint; then when pressure is applied the weld is completed. A transformer in the welding machine reduces the alternating-current (a-c) voltage from either 120 or 240 V to around 4 to 12 V and raises the amperage sufficiently to produce a good heating current. When the current passes through the metal, most of the heating takes place at the point of greatest resistance in the electrical path which is at the interface of the two sheets. It is here that the weld forms. The amount of current necessary is 30 to 40 kVA/in.^{2} (47–62 MVA/m^2) of the area to be united based on a time of about 10 s. The necessary pressure to effect the weld will vary from 4000 to 8000 psi (28–55 MPa).

Resistance welding is essentially a production process adapted to the joining of light-gage metals, which can be lapped. Usually the equipment is suitable for only one type of weld, and the work must be moved to the machine. It is the only process that uses a pressure action at the weld during an accurately regulated heat application. The operation is rapid. Practically all metals can be welded by resistance welding, although a few such as tin, zinc, and lead can be welded only with great difficulty.

In all resistance welding the three factors that must be given consideration are expressed in the formula

$$H = I^2Rt$$

where

H = heat, BTU's

I = welding current, A

R = resistance, ohms

t = time, s

but the power is expressed as

$$P = I^2R$$

Thus

$$H = P \times t$$

The amperage of the secondary or welding current is determined by the transformer. To provide possible variation of the secondary current, the transformer is equipped with a regulator on the primary side to vary the number of turns on the primary coil. This may be seen in Figure 8.9. For good welds these three variables (current, resistance, and time) must be carefully considered and determined by such factors as material thickness, kind of material, and type and size of electrode.

The timing of the welding current is important. There should be an adjustable delay after the pressure has been applied until the weld is started. The current is turned on by the timer and held a sufficient time for the weld. It is then stopped but the pressure remains until the weld cools, thus the electrodes do not arc and the weld is protected from discoloration. The pressure on the weld may be obtained manually, by mechanical means, air pressure, springs, or hydraulic means. Its application must be controlled and coordinated with the welding current.

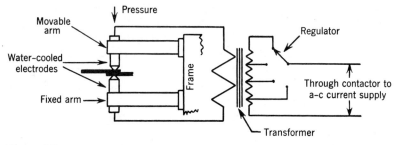

Figure 8.9
Diagram of spot welder.

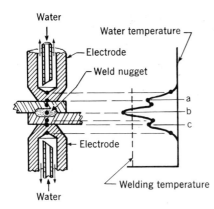

Figure 8.10
Temperature distribution in spot welding.

Referring to Figure 8.10, the greatest heat is generated at (b) because the resistance is much greater at that point than at (a) and (c); thus the far greater power is consumed in making the weld. The resistance of the parent metal between (a) and (c) is partially utilized in effecting the weld. Power is dissipated at (a) and (c) by the water-cooled electrodes.

Resistance-welding processes include (1) spot welding, (2) projection welding, (3) seam welding, (4) butt welding, (5) flash welding, and (6) percussion welding.

Spot Welding

In this form of resistance welding two or more sheets of metal are held between metal electrodes (Figure 8.9). The welding cycle is started with the electrodes contacting the metal under pressure before the current is applied for a period known as the *squeeze time*. A low-voltage current of sufficient amperage is then passed between the electrodes causing the metal in contact to be rapidly raised to a welding temperature. As soon as the temperature is reached, the pressure between the electrodes squeezes the metal together and completes the weld. This period, usually 3 to 30 cycles of 60-Hz current, is known as the *weld time*. Next, while the pressure is still on, the current is shut off for a period called the **hold time** during which the metal regains some strength by cooling. The pressure is then released and the work is either removed from the machine or moved so that another portion can be welded. This is the ***off time***.

Spot welding is the simplest form of resistance welding. In spot welding there are five zones of heat generation: one at the interface between the two sheets, two at the contact surfaces of the sheets with the electrodes, and the other two in each piece of metal (Figure 8.10). The contact resistance at the interface of the two sheets is the point of highest resistance where the weld formation starts. Contact resistance at this point as well as between the electrodes and the outsides of the sheets depends on surface conditions, magnitude of electrode force, and the size of the electrodes. If the sheets are the same in both thickness and analysis, the heat balance will be such that the weld nugget will be at the center. Uneven sheet thicknesses or welding sheets of different thermal conductivities may necessitate using electrodes of different size or conductivity to obtain a proper heat balance.

Machines for spot welding are made in three general types: *stationary single spot, portable single spot,* and *multiple spot.* Stationary machines may be further classified as *rocker arm* and *direct-pressure* types. The rocker arm type is generally limited to machines of small capacity. It is so designated because the motion for applying pressure and raising the upper electrode is made by rocking the upper arm. The larger machines usually employ direct straight-line motion of the upper electrode. This arrangement permits them also to be used for projection welding.

Portable spot welders connected to the transformer by long cables and capable of being moved to any position are then used. Where all welds cannot be made by a single machine setup, welding jig assemblies are served best by portable welders. The principal differences among them are the manner of applying the pressure and the shape of the tips. Pressure is applied manually, pneumatically, or hydraulically depending on the size and type of gun.

For production work, multiple spot welding machines have been developed that are capable of producing two or more spots simultaneously. A system known as *indirect welding* is used, where two electrodes are in series and the current passes through a heavy plate underneath the sheets and between the electrodes.

Projection Welding

Projection welding is illustrated in the line diagram shown in Figure 8.11. Projection welds are produced at localized points in workpieces held under pressure between suitable electrodes. Sheet metal is first put through a punch press that makes small projections or buttons in the metal. These projections are made with a diameter on the face equal to the thickness of the stock and extend from the stock about 60% of its thickness. Such projection spots or ridges are made at all points where a weld is desired. This process is also used for crosswire welding and for parts where the ridges are produced by machining. With this form of welding a number of welds can be made simultaneously. The only limit to the number is the ability of the press to furnish and distribute equally the correct current and pressure. Results are generally uniform and weld appearance is often better

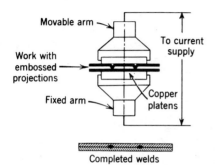

Figure 8.11
Projection welding.

than in spot welding. Electrode life is long, because only flat surfaces are used and little electrode maintenance is required.

Seam Welding

Seam welding consists of a continuous weld on two overlapping pieces of sheet metal. Coalescence is produced by heat obtained from the resistance of current. The current passes through the overlapping sheets, which are held together under pressure between two circular electrodes. This method is in effect a continuous spot-welding process, because the current is not usually on continuously but is regulated by the timer of the machine. In high-speed seam welding using continuous current, the frequency of the current acts as an interrupter.

The three types of seam welds used in industry are illustrated in Figure 8 12. The most common is the simple lap seam weld shown in the upper part of the figure. This weld consists of a series of overlapping spot welds with sufficient overlap of the weld nuggets to provide a pressure-tight joint. If the pressure-tight quality of the lap seam weld is not required, the individual nuggets may be spaced to give a stitch effect. This process is known as **roll spot welding.** A *mash seam weld* is also shown, which is produced by reducing the amount of sheet lapping to a small value. Broad-faced, flat electrodes are used that forge the sheets together while welding. This forging action by the electrodes is known as "mashdown" and occurs simultaneously with the fusing on the sheets. Because the joint is covered above and below by the electrodes and on either side by the sheets, any extrusion or spitting from the weld is prevented. Micrographs showing longitudinal transverse sections of two pieces of 0.050-in. (1.27-mm) steel are shown in Figure 8.13. Alterations can be made in electrode face contour and the amount of lap increased so that the mashdown takes place on only one side, which can be finished later leaving no

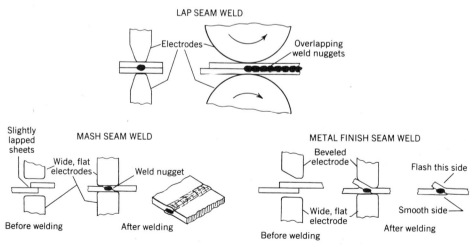

Figure 8.12
Types of seam welds.

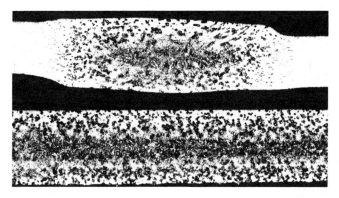

Figure 8.13
Longitudinal and cross section of mash seam welds made with 0.050-in. (1.27-mm) steel at 80 in./min (33.8 mm/s). Current 19,000 A, electrode force 1500 lb (6600 N), and initial overlap 150%. Magnification ×12.5.

trace of the joint. This type of mash seam welding, shown in the lower part of the figure, is often referred to as *finish seam* welding. It finds application where the product is normally viewed from one side only.

The heat at the electrode contact surfaces is kept to a minimum by the use of copper alloy electrodes and is dissipated by flooding the electrodes and weld area with water. Heat generated at the interface by contact resistance may be increased by decreasing the electrode force. Another variable that influences the magnitude of the heat is the weld time, which in seam welding is controlled by the speed of rotation of the electrodes. The amount of heat generated is decreased with an increase in welding speed.

Seam welding is used in manufacturing metal containers, automobile mufflers and fenders, refrigerator cabinets, and gasoline tanks. Advantages of this type of fabrication include improved design, material saving, tight joints, and low-cost construction.

Butt Welding

This form of welding, illustrated in Figure 8.14, is accomplished by gripping two pieces of metal that have the same cross section and pressing them together while heat is being

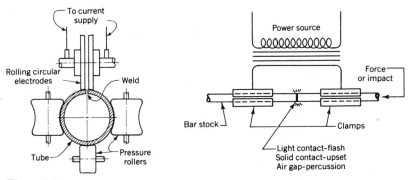

Figure 8.14
Butt-welding methods. *A*, Continuous-resistance butt welding of steel pipe. *B*, Sketch illustrating forms of butt-welding bar stock.

generated in the contact surface by electrical resistance. Although pressure is maintained while the heating takes place, at no time is the temperature sufficient to melt the metal. The joint is upset somewhat by the process, but this defect can be eliminated by subsequent rolling or grinding. Both parts to be welded should be of the same resistance to have uniform heating at the joint. If two dissimilar metals are to be welded, the metal projecting from the die holders must be in proportion to the specific resistance of the materials to be welded. The same treatment must be used where materials of different cross sections are butt welded.

In actual operation, the work is first clamped in the machine and pressure is applied on the joints. The welding current is then started and heating takes place; the rate depends on the pressure, the material, and the condition of surfaces. Because the contact resistance varies inversely with the pressure, the pressure is less at the start and is then increased to whatever is necessary to effect the weld. The pressure is usually about 2500 to 8000 psi (17–55 MPa) when the welding temperature is reached. This type of welding is especially adapted to rods, pipes, small structural shapes, and many other parts of uniform section. Areas up to 70 in.^{2} (0.05 m^2) have been successfully welded, but generally the process is limited to small areas because of current limitations.

Figure 8.14 illustrates a special type of butt seam welding used in pipe manufacturing. Two rolling electrodes bring a high-amperage current across the joint that generates the heat in the contact surfaces by electrical resistance. For thin-walled tubes high-frequency current from an induction coil can be used to generate the heat in place of electrodes.

Flash Welding

Butt and flash butt welding, though similar in application, differ somewhat in the manner of heating the metal. For flash welding the parts must be brought together in very light contact. A high voltage starts a flashing action between the two surfaces and continues as the parts advance slowly and the forging temperature is reached. The heat generated for welding results from the arcing between surfaces. The weld is completed by the application of a sufficient forging pressure of 5000 to 25,000 psi (35–170 MPa) to effect a weld. A small fin or projection left around the joint can be easily removed.

Welding of small areas is usually done by the butt-welding method, large areas are done by the flash butt method; however, there is no clear demarcation between the two. The shape of the piece and the nature of the alloy are frequently the determining factors. Areas ranging from 0.002 to 50 in.^{2} (1–32,000 mm^2) have been successfully welded by flash welding. In this process less current is required than in ordinary butt welding; there is less metal to remove around the joints; the metal that forms the weld is protected from atmospheric contamination; the operation consumes little time; and end-to-end welding of sheets is possible. Because of these advantages flash welding is more widely used than the ordinary butt or upset process. While many nonferrous metals can be flash welded satisfactorily, alloys containing high percentages of lead, zinc, tin, and copper are not recommended for this process.

Percussion Welding

Like the flash weld process, *percussion welding* relies on arc effect for heating rather than on the resistance in the metal. Pieces to be welded are held apart, one in a stationary holder and the other in a clamp mounted in a slide and backed up against heavy spring pressure as in Figure 8.14. When the movable clamp is released it moves rapidly, carrying with it the piece to be welded. When the pieces are about $1/16$ in. (1.6 mm) apart there is a sudden discharge of electric energy, causing intense arcing over the surfaces and bringing them to a high temperature. The arc is extinguished by the percussion blow of the two parts coming together with sufficient force to effect the weld.

The electric energy for the discharge is built up in one of two ways. In the electrostatic method, energy is stored in a capacitor and the parts to be welded are heated by the sudden discharge of a heavy current from the capacitor. The electromagnetic welder uses the energy discharge caused by the collapsing of the magnetic field linking the primary and secondary windings of a transformer or other inductive device.

The action of this process is so rapid (about 0.1 s) that there is little heating effect in the material adjacent to the weld. Heat-treated parts may be welded without being annealed. Parts differing in thermal conductivity and mass can be successfully joined because the heat is concentrated only at the two surfaces. Examples are welding Stellite tips to tools, copper to aluminum or stainless steel, silver contact tips to copper, cast iron to steel, lead-in wires on electric lamps, and zinc to steel. These welds are made without any upset or flash at the joint. The principal limitation of the process is that only small areas up to $1/2$ in.^{2} (650 mm^2) of nearly regular sections can be welded.

High-Frequency Resistance Welding

This process, used primarily to manufacture structural steel shapes, utilizes an electric current at about 400,000 Hz. Such shapes as I-beams can be welded at high speeds with relatively low heat input, thus reducing grain distortion. Such a beam would consist of two flanges welded to the web section by means of applying high-frequency current and passing the flanges between pressure rolls and effecting the weld to the separating web section. No special cleaning or fluxing is required. Even dissimilar materials may be joined at rates up to 170 ft/min (51 m/min). The carbon content of steels welded by this process must be under 0.40%, but structural shapes are typically low in carbon. So recent is this process that patents are still controlled by Thermatool Corporation.

INDUCTION WELDING

Coalescence in induction welding is produced by the heat obtained from the resistance of the weldment to the flow of an induced electrical current. Pressure is frequently used to complete the weld. The inductor coil is not in contact with the weldment; the current is induced into the conductive material. Resistance of the material to this current flow results in the rapid generation of heat.

In operation a high current is induced into both edges of the work close to where the weld is to be made. Heating to welding temperature is extremely rapid and the joint is completed by pressure rolls or contacts.

A form of induction welding known as *high-frequency welding* is similar except that the current is supplied to the conductor being welded by direct contact. Frequencies ranging from 200,000 to 500,000 cps (Hz) are used in high-frequency work, whereas frequencies of 400 to 450 cps (Hz) are satisfactory for induction welding of most metals. High-frequency current flows near the surface of the metal and, because heating is almost instantaneous, there is no chance for harmful oxides to form. Vacuum tube oscillators are the source of power for most high-frequency welding.

Induction welding can be used for most metals and has been successfully employed for some dissimilar metals. Applications include the butt and seam welding of pipe, sealing containers, welding expanded metal, and fabricating various structural shapes from flat stock.

ARC WELDING

Arc welding is a process in which coalescence is obtained by heat produced from an electric arc between the work and an electrode. The electrode or filler metal is heated to a liquid state and deposited into the joint to make the weld. Contact is first made between the electrode and the work to create an electric circuit. Then an arc is formed by separating the conductors. The electric energy is converted into intense heat in the arc, which attains a temperature around 10,000°F (5500°C).

Either direct or alternating current can be used for arc welding, direct current (d-c) being preferred for most purposes. A d-c welder is simply a motor generator set of constant energy type (constant potential may also be used) having the necessary characteristics to produce a stable arc. There should not be too great a current surge when the short circuit is made, and the machine should compensate to some extent for varying lengths of the arc. Direct-current machines are built in capacities up to 1000 A, having an open-circuit voltage of 40 to 95 V. A 200-A machine has a rated current range of 40 to 250 A according to the standard of the National Electrical Manufacturers Association (NEMA). While welding is going on, the arc voltage is 18 to 40 V. In *straight polarity* the electrode is connected to the negative terminal, whereas in *reverse polarity* the electrode is positive.

Carbon Electrode Welding

The first methods of arc welding, which employed only carbon electrodes, are still in use to some extent for both manual and machine operation. The carbon arc is used only as a source of heat and the flame is handled in a fashion similar to that used in gas welding. Filler rods supply weld metal if additional metal is necessary. The twin-carbon arc method was one of the first used. The arc was between the two electrodes and not with the work. In operation, the arc is held $1/4$ to $3/8$ in. (6.4–9.5 mm) above the work and best results are obtained with the work in a flat position. The use of this method is limited to brazing and soldering.

A second process utilizing a single carbon electrode is considerably simpler. The arc is created between the carbon electrode and the work, and any weld metal needed is supplied by a separate rod. Such an arc is easy to start as the electrode does not stick to the metal. Straight polarity must always be used, because a carbon arc is unstable under reverse polarity.

Metal Electrode Welding

Shortly after the development of carbon electrode welding, it was discovered that a metal electrode with the proper current characteristics could be melted to supply the necessary weld metal. A patent for this process, still in general use, was issued to Charles Coffin in 1889. An arc is started by striking the work with an electrode and quickly raising it a short distance. New powdered-metal electrode designs involve a drag technique where the coated electrode rides lightly on the work. In both the electrode end is melted by the intense heat. Most of the heat is transferred across the arc in the form of small globules to a molten pool. A small amount of metal is vaporized and lost. Some globules are deposited beside the weld as spatter. The arc is maintained by uniformly moving the electrode along the work at a rate that compensates for metal that has been melted and transferred to the solid weld. At the same time the electrode is gradually moved along the joint.

For ordinary welding there is little difference in the weld quality made by a-c and d-c equipment, but polarity causes great variation in weld quality. For a-c machines consist principally of static transformers, which are simple equipment without moving parts. Their efficiency is high, their loss at no load is negligible, and their maintenance and initial costs are low. Welders of this type are built in six sizes as specified by NEMA and are rated at 150, 200, 300, 500, 750, and 1000 A. For hand welding requiring 200 A or higher, a-c equipment is preferred. The fact that there is less magnetic flare of the arc or "arc blow" with a-c than d-c equipment is important in welding heavy plates or fillet welding. Most of the nonferrous metals and many of the alloys cannot be welded with a-c equipment because electrodes have not been developed for this purpose.

Welding speed and the ease of welding with a-c and d-c welders are similar. However, on heavy plate using large-diameter rods, a-c welding is faster. Direct-current machines may be used with all types of carbon and metal electrodes, because the polarity can be changed to suit the electrode. With the a-c welder the alternating current is constantly reversing with every cycle, and electrodes have to be selected that will operate on both polarities. Alternating-current welders operate at slightly higher voltages, hence the danger of shock to the operator is increased.

Electrodes. The three types of metal electrodes (or "rods") are *bare, fluxed,* and *heavy coated*. Bare electrodes have limited use for welding wrought iron and low or medium carbon steel. Straight polarity is generally recommended. Improved welds may be made by applying a light coating of flux on the rods with a dusting or washing process. The flux assists in eliminating undesirable oxides and in preventing their formation. However, the heavily coated arc electrodes are the most important ones used in commercial welding. Over 95% of the total manual welding done today is with coated electrodes.

Figure 8.15 is a diagrammatic sketch showing the action of an arc using a heavily coated electrode. In the ordinary arc with bare wire the deposited metal is affected by the oxygen and nitrogen in the air. This causes undesirable oxides and nitrides to be formed in the weld metal. The effect of heavy coatings on electrodes provides a gas shield around the arc to eliminate such conditions and covers the weld metal with a protective slag coating that prevents oxidation of the surface metal during cooling. Welds made from rods of this type have superior physical characteristics.

Electrode Coatings. Electrodes coated with slagging or fluxing materials are necessary in welding alloys and nonferrous metals. Some of the elements in these alloys are unstable and are lost if there is no protection against oxidation. Heavy coatings permit the use of larger welding rods, stronger current, and higher welding speeds. In summary, the coatings perform the following functions.

1. Provide a protecting atmosphere.
2. Provide slag of suitable characteristics to protect the molten metal.
3. Facilitate overhead and position welding.
4. Stabilize the arc.
5. Add alloying elements to the weld metal.
6. Perform metallurgical refining operations.
7. Reduce spatter of weld metal.
8. Increase deposition efficiency.
9. Remove oxides and impurities.
10. Influence the depth of arc penetration.
11. Influence the shape of the bead.
12. Slow down the cooling rate of the weld.

These functions are not common to all *coated electrodes* because the coating put on an electrode is largely determined by the kind of welding. It is interesting that the coating composition is also a determining factor in electrode polarity. By varying the coating, rods may be used with either pole.

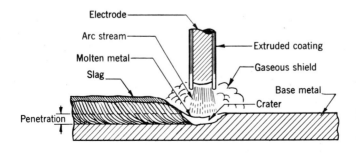

Figure 8.15
Diagrammatic sketch of arc flame.

Coating compositions may be classified as organic and inorganic, although sometimes both types may be used. Inorganic coatings can be further subdivided into flux compounds and slag-forming compounds. Some of the principal constituents used are

1. *Slag-forming constituents.* SiO_2, MnO_2, and FeO. Al_2O_3 is sometimes used but it makes the arc less stable.
2. *Constituents to improve arc characteristics.* Na_2O, CaO, MgO, and TiO_2.
3. *Deoxidizing constituents.* Graphite, aluminum, and wood flour.
4. *Binding material.* Sodium silicate, potassium silicate, and asbestos.
5. *Alloying constituents to improve strength of weld.* Vanadium, cesium, cobalt, molybdenum, aluminum, zirconium, chromium, nickel, manganese, and tungsten.

The term *contact electrode* has been given to electrodes that have a thick coating with a high metal powder content and that are suitable for welding with a drag or contact technique. Most electrodes have an automatic striking or self-igniting characteristic owing to the iron in the coating. By adding metal powder to the coating and increasing its thickness, the deposition rate is increased.

Atomic Hydrogen Arc Welding. In this process, a single-phase a-c arc is maintained between two tungsten electrodes and hydrogen is introduced into the arc. As the hydrogen enters the arc the molecules are broken into atoms, which recombine into molecules of hydrogen outside the arc. This reaction is accompanied by intense heat of about 11,000°F (6100°C). Weld metal may be added to the joint by a welding rod. The operation is similar to the oxyacetylene process. The atomic hydrogen process differs from other arc-welding processes because the arc is formed between two electrodes (Figure 8.16) rather than between one electrode and the work. This makes the electrode holder a mobile tool because it can be moved without the arc being extinguished.

The advantage of this process is its ability to provide high heat concentrations. The hydrogen also acts as a shield and protects the electrodes and molten metal from oxidation. Filler metal of the same analysis can be used with manual and automatic equipment. Many alloys difficult to weld by other processes can be successfully treated. Most applications of atomic hydrogen arc welding are accomplished by the inert-gas–shielded-arc process.

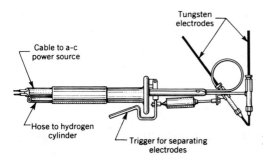

Figure 8.16
Torch for atomic hydrogen welding.

Inert-Gas–Shielded-Arc Welding. In this process, coalescence is produced by heat from an arc between a metal electrode and the work shielded by an atmosphere of argon, helium, CO_2, or a mixture of gases. Although CO_2 is not an inert gas, it ionizes at welding temperatures and acts like one. Two methods are employed: one uses a tungsten electrode with filler metal as in gas welding (***TIG welding,*** tungsten inert gas), and the other uses a consumable metal wire as the electrode (***MIG welding,*** metal inert gas). Both methods are adaptable to manual or automatic machine welding, with no flux or wire coating required for protection of the weld.

The first method using a single tungsten electrode (nonconsumable), also known as *gas tungsten arc welding,* is illustrated in Figure 8.17. The weld zone is protected by inert gas fed through the water-cooled electrode holder. Argon is frequently used, although helium or a mixture of the two may be employed. Alternating current or direct current can be used, the selection being determined by the kind of metal to be welded. Direct current with straight polarity is required for steel, cast iron, copper alloys, and stainless steel, while reverse polarity is not widely used. Alternating current is used for aluminum, magnesium, cast iron, and a number of other metals. Because of the cost this process is best used in welding light-gage work. It is not competitive with processes selected for welding heavier materials. Most inert-gas nonconsumable welding is done manually.

Another method of inert-gas–shielded-arc welding using consumable electrodes is accomplished by employing a shielded arc between the consumable bare wire electrode and the workpiece. Because filler material is transferred through the protected arc, greater efficiency is obtained resulting in more rapid welding. The metal is deposited in an atmosphere that prevents contamination.

In this process (Figure 8.18), a wire is fed continuously through a gun to a contact surface that imparts a current to the wire. Direct-current reverse polarity provides a stable arc and offers the greatest heat input at the workpiece. It is generally recommended for aluminum, magnesium, copper, and steel. Straight polarity with argon has a high burnoff rate but the arc is unstable with high spatter. Alternating current is also inherently unstable and seldom used in this process.

Carbon dioxide gas is widely used in welding plain-carbon and low-alloy steels. Having excellent penetration, it produces sound welds at high speed. Because the gas decomposes into carbon monoxide and oxygen at high temperature, some gas-evolving flux is generally provided in or on the wire. Gas metal arc welding may be manual or automatic, but the process is especially adapted to automatic operation.

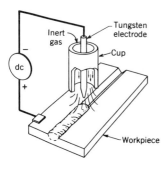

Figure 8.17 Schematic diagram of the gas tungsten arc welding process.

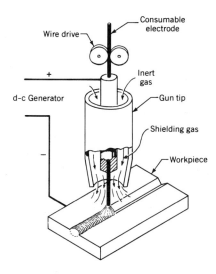

Figure 8.18
Schematic diagram of the gas metal arc welding process.

Automatic Arc-Welding Machines. Much metal electrode welding is done by machine or automatic units. This welding is essentially the same as manual welding, except that machine units are supplied with controls that feed the wire and move the welding head or workpiece along at a proper welding rate. Some control systems vary the feed rate of the electrode as the voltage across the arc varies. If the voltage increases indicating a longer arc than desired, the rate of feed increases. A lower voltage indicates a shorter arc and the feed is reduced, the goal being to hold a uniform arc length. Other systems feed the electrode at a constant rate of speed. This also gives a uniform arc length when used with a power source having constant voltage output. The electrode and molten base metal are protected from oxidation by shields of argon, helium, or carbon dioxide gas.

Prerequisites for economical machine welding are a sufficient volume of production to justify expensive equipment and a uniform product. Assemblies must fit readily and pieces must be of the same size and contour. Often jigs and handling devices are necessary to position the work. Under these conditions machine welding may be economical. The use of automatic welding machines results in increased welding speed and uniform quality of the weld.

Arc Spot Welding. An interesting application of inert-gas arc welding is making spot, plug, or tack welds by an argon-shielded electric arc using consumable electrodes. To effect a weld a small welding gun with pistol grip is held tightly against the work. As the trigger is released the argon valve is opened; the current is allowed to pass through the electrode for a preset interval (2–5 s); then both are shut off. Gas tungsten arc welding guns using direct current are also employed in making spot welds. An advantage of this equipment is that spot welds can be made on thin sheets from one side of the work. Being a low-cost process, it is particularly useful in welding large or irregularly shaped assemblies that are difficult to spot weld with resistance equipment.

Submerged-Arc Welding. This process is so named because the metal arc is shielded by a blanket of granular, fusible flux during the welding. Aside from this feature its operation is similar to other automatic arc-welding methods. A bare electrode is fed through the welding head into the granular material (Figure 8.19). This material is laid along the seam to be welded and the entire welding action takes place beneath it. The arc is started by striking beneath the flux on the work or by initially placing some conductive medium like steel wool beneath the electrode. The intense heat of the arc immediately produces a pool of molten metal in the joint and melts a portion of the granular flux. This material floats on top, forming a blanket that eliminates spatter losses and protects the welded joint from oxidation. Upon cooling the fused slag solidifies and is easily removed; granular material not fused is recycled and used again.

This process is limited to flat welding, although welds can be made on a slight slope or on circumferential joints. It is advisable to use a backing strip of steel, copper, or some refractory material on the joint to avoid losing some of the molten metal. The process uses strong current (300–4000 A), which permits a high rate of metal transfer and welding speeds. Deep penetration is obtained and most commercial-thickness metal plate can be welded with one pass. As a result thin plates can be welded without preparation, whereas a small vee is required on others. Most submerged-arc welding is done on low-carbon and alloy steels, but it may be used on many of the nonferrous metals.

Stud Arc Welding. Stud welding is a d-c arc-welding process developed to end-weld metal studs to flat surfaces. It is accomplished with a pistol-shaped welding gun that holds the stud or fastener to be welded. When the trigger of the gun is pressed, the stud is lifted to create an arc and then forced against the molten pool by a backing spring. The entire operation is controlled by a timer, preset according to the size of the stud being welded. The arc is shielded by surrounding it with a ceramic ferrule, which also confines the metal to the weld area and protects the operator from the arc.

A percussion-type gun has a small projection on the end of the stud. As the stud moves forward, the projection touches the work and is vaporized causing an arc to form. The weld is completed as the stud is driven onto the work surface.

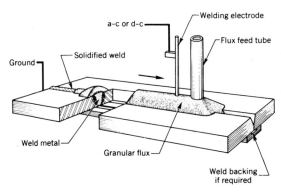

Figure 8.19
Schematic sketch of submerged-arc welding.

Transferred-Arc Cutting. In a plasma torch gas is heated by a tungsten arc to such a high temperature that it becomes ionized and acts as a conductor of electricity. In this state the arc gas is known as *plasma.* The torch is generally designed so that the gas is closely confined to the arc column through a small orifice. This increases the temperature of the plasma and concentrates its energy on a small area of the workpiece, which rapidly melts the metal. As the gas jet stream leaves the nozzle it expands rapidly, removing the molten metal continuously as the cut progresses. Because the heat obtained does not depend on a chemical reaction, this torch can be used to cut any metal. Temperatures approach 60,000°F (33,000°C), roughly 10 times that possible by the reaction of oxygen and acetylene.

Plasma-generating torches are of two general designs, one known as a *transferred plasma torch* and the other a *nontransferred plasma torch*. In the nontransferred torch the arc circuit is completed within the torch and the plasma is projected from the nozzle. Such torches utilized for metal spraying are described in Chapter 24. Transferred-arc plasma torches used for cutting are diagrammed in Figure 8.20. the workpiece becomes the anode and the arc continues to the workpiece in the jet of gas. With the workpiece as one of the electrodes, the intensity of heat transfer and efficiency are increased making it more suitable for metal cutting than the nontransferred plasma torch.

The Heliarc cutting torch illustrated in Figure 8.20A does not constrict the arc and may be used for either welding or cutting most common metals. When used as a cutting torch, the current density is increased over good welding conditions and an argon–hydrogen gas mixture is used. Good kerf quality is obtained on one side of the cut, but the cutting thickness is limited to $1/2$ in. (12.7 mm). For high-speed accurate cutting the constricted-arc torch as illustrated in Figure 8.20B is more satisfactory. The arc is constricted in a narrow opening at the end of the torch, creating a high-velocity arc that readily metals a narrow kerf through both ferrous and nonferrous metals. Higher tem-

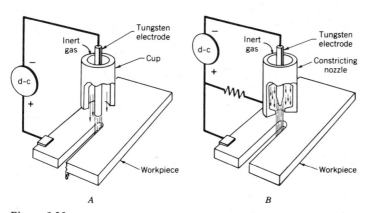

Figure 8.20
Schematic comparison of the two basic gas tungsten electrode arc cutting processes. A, Heliarc cutting—nonconstricted transferred-arc cutting. B, Plasma arc cutting—constricted transferred-arc cutting.

peratures are obtained than in the nonconstricted torch, and thicknesses up to 4 in. (100 mm) may be cut in any metal.

While gases used in plasma torch cutting include argon, hydrogen, and nitrogen, a combination of argon and nitrogen gives the best results. For manual operation the arc is started in an atmosphere of argon followed by the right proportion of hydrogen blended into the gas stream. A mixture of 80% argon and 20% hydrogen is used in cutting operations up to 400 A. For higher current proportions of 65% and 35% are recommended. Nitrogen is recommended only for the mechanized cutting of stainless steels. Because of its toxic fumes an exhaust system must be provided.

Plasma arc is used for either manual or mechanized operations. The process has little effect on the metallurgical characteristics or physical properties of adjacent metal because of the rapidity of its action. It is particularly useful in cutting materials such as aluminum, stainless steel, copper, and magnesium.

Electroslag Welding. *Electroslag welding,* another metal arc welding process, is preferred in welding heavy plates, the joint to be welded being in a vertical position. Heat is obtained from the resistance of current in an electrically conductive molten flux. One or more electrodes are continuously fed into a pool of molten slag, which maintains a temperature in excess of 3200°F (1800°C).

A schematic diagram of the process is shown in Figure 8.21. In starting the weld, an arc is created between the electrode and the bottom plate and continues until a sufficiently thick layer of molten slag is formed. The current then flows through the slag, which maintains sufficient temperature to melt the wire electrodes and the surfaces of the workpiece. On either side of the joint are water-cooled copper slides that confine the molten metal and slag. As the metal solidifies, the copper plates automatically move upward. The rate is determined by the speed at which the electrodes and base metal are melted. The lower part of the metal bath is solidified by cooling from the plates; thus the welded joint is formed. Slag may be added continuously from an overhead hopper or by using a flux-cored wire. The latter method is preferred, because the entering flux is always carried to the hottest part of the pool.

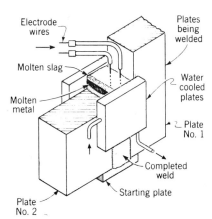

Figure 8.21
Diagram of the electroslag welding process.

Advantages of this process include ability to weld metals of great thickness in a single pass, minimum joint preparation, high welding speed, and good stress distribution across the weld with little distortion. The weld metal is at all times protected from contamination.

SPECIAL WELDING PROCESSES
Electron Beam Welding

In electron beam welding coalescence is produced by bombarding the workpiece with a dense beam of high-velocity electrons. The metal is joined by melting the edges of the workpiece or by penetrating the material. Usually no filler metal is added. This process may be used not only to join common metals but also refractory metals, highly oxidizable metals, and various super alloys that have previously been impossible to weld.

Figure 8.22 is a diagrammatic sketch of an electron gun. This gun is placed within a vacuum chamber and so arranged that it may be raised, lowered, or moved in a horizontal plane. The gun can be positioned while the chamber is evacuated prior to the welding operation. After the chamber is evacuated to a pressure of around 10^{-4} mm Hg, the beam circuit is energized and directed to the desired spot on the weldment. The beam generally remains stationary, and the work is moved at a desired speed past the electron beam. The temperature range of this electron gun is sufficient to vaporize tungsten or any known material.

In this process heat energy is imparted to the weldments at a much higher rate than

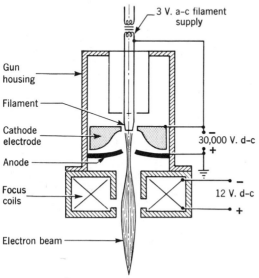

Figure 8.22
Diagrammatic sketch of electron gun used for electron beam welding.

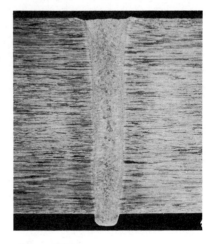

Figure 8.23
Photomicrograph of a 0.5-in. (12.7-mm)-thick joint in 2024-T4 aluminum accomplished by electron beam welding. Magnification ×4.

it can be conducted away from the electron beam zone, resulting in a high depth:width ratio (up to 20:1). Figure 8.23 is a photomicrograph of a 0.5-in (12-mm) joint in 2024-T4 aluminum that was welded at 32 in./min (2 m/s). Three-inch 5083 aluminum has been welded at 30 kV, 500 mA, and 15 in./min (0.9 m/s). It is interesting to note that the adjacent parent metal has been almost unaffected by the weld because of the high welding speeds. The process is adapted to numerical control for many electronic components such as the "cans" for condensers. Several units are placed in a special holding fixture and, after evacuation, the beam is directed around each unit at high speed to effect the weld.

Electron beam welding may be performed in air or under a blanket of inert gas with some limitations. The electron beam is formed in a chamber arrangement similar to the vacuum machine, then passes through a special orifice, and finally through argon or helium to the workpiece. The maximum effectiveness of the beam is 1 in. (25 mm), with a workpiece limitation of $1/2$ in. (12.7 mm). Although welding speed is increased, welds are not free from contamination and the weldment size is much smaller than those obtained by the vacuum method. The nonvacuum unit supplements the conventional vacuum-welding process and increases the range of welding that can be done with electron beam welding.

Laser Welding

A brief description of an optical laser and its range of capabilities is given in Chapter 11. The laser is finely focused as a high-collimated beam of photons that is referred to as a coherent beam. This monochromatic beam is capable of delivering up to 30,000 W/in.^{2}. The theory of the various types of lasers is beyond the scope of this book. The most used is the CO_2 type, which can weld $1/32$-in. (0.8-mm)-thick stainless steel. The newer *gas dynamic* laser can weld up to $3/4$-in. (19-mm)-thick stainless. Because the energy bond is so small, deep penetration welds can be made while the base metal itself is relatively unaffected because broad heating does not take place. Thus very thin materials can be welded, as well as thicker ones. Joint designs for laser welds are similar to those discussed earlier except that the tolerances must be less and the surface finish better, because the laser is a highly accurate beam. This type of welding makes excellent, precise welds, usually no more than 0.001 in. (0.025 mm) wide, on all metals (even dissimilar ones), and the base metal temperature rises almost imperceptibly. The principal disadvantages are cost and size of equipment and the safety hazards associated with high-energy lasers.

The Apollo laser shown in Figure 8.24 is an 80-W unit with a wavelength of 10.6 λ. The lasing material is a mixture of carbon dioxide, helium, and nitrogen. This unit will accomplish soldering, welding, and cutting as well as other operations such as drilling, slitting, and perforating.

Friction Welding

In *friction welding* coalescence is produced from the heat generated by rotating one piece against another under controlled axial pressure. The two surfaces are heated to a melting temperature; the adjacent material becomes plastic. The relative motion between the two

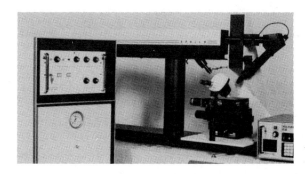

Figure 8.24
Apollo industrial laser.

is then stopped and a forging pressure applied that upsets the joint slightly. This pressure may be equal or in excess of the pressure during the heating, its value depending on the material being welded. Flash developed during the process tends to carry out surface oxides and impurities from the joint.

In this process no special preparation of the weld surfaces is necessary except to have them clean and reasonably smooth. Equipment must be capable of holding one piece securely while the other rotates, then stopping rotation quickly when the welding temperature is reached. Figure 8.25 shows diagrammatically the setup for welding two round shafts. A similar arrangement would be used in welding tubes. In its present state of development no configuration can be welded except a rod or tube abutted to a flat surface. Rotational speeds and contact pressure depend on the work size and type of material. For example, recommendations for a 1-in. (25-mm) carbon steel bar call for a relative rotational speed of 1500 rpm and an axial pressure of 1500 psi (10 MPa), while the same size stainless-steel rod requires 3000 rpm and 12,000 psi (85 MPa).

The chief limitation of the process is that few configurations can be welded. Advantages claimed for the process are simple equipment, rapid production of sound welds, little preparation required for joints, and low total energy requirements. Also, many dissimilar metals can be welded and the proper welding cycle easily programmed into the machine. Friction welding finds considerable use in welding plastics.

Thermit Welding

Thermit welding is the only welding process employing an exothermal chemical reaction for the purpose of developing a high temperature. It is based on the fact that aluminum

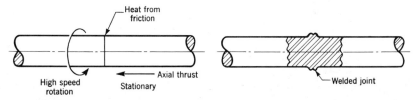

Figure 8.25
Illustrating a process that uses heat generated by friction to produce a weld.

has a great affinity for oxygen and can be used as a reducing agent for many oxides. The usual Thermit mixture or compound consists of finely divided aluminum and iron oxide mixed at a ratio of about 1:3 by weight. The iron oxide is usually roll mill scale. This mixture is not explosive and can be ignited only at a temperature of about 2800°F (1500°C). A special ignition powder is used to start the reaction. The chemical reaction requires about 30 s and attains a temperature around 4500°F (2500°C). The mixture reacts according to the chemical equation

$$8Aln + 3Fe_3O_4 = 9Fe + 4Al_2O_3$$

The resultant products are a highly purified iron (actually steel) and an aluminum oxide slag that floats on top and is not used. Other reactions also take place, as most *Thermit metal* is alloyed with manganese, nickel, or other elements.

Figure 8.26 illustrates the method of preparing the material for such a weld. Around the break where the weld is to be made a wax pattern of the weld is built up. Refractory sand is packed around the joint and necessary provision is made for risers and gates. A preheating flame is used to melt and burn out the wax, to dry the mold, and to bring the joints to a red heat. The reaction is then started in the crucible and, when it is complete, the metal is tapped and allowed to flow into the mold. Because the weld metal temperature is approximately twice the melting temperature of steel, it readily fuses in the joint. Such welds are sound because the metal solidifies from the inside toward the outside, and all air is excluded from around the mold.

There is no limit to the size of welds that can be made by Thermit welding. It is used primarily for repairing large parts that would be difficult to weld by other processes.

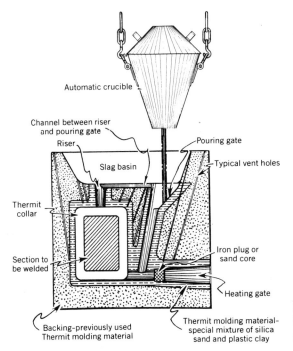

Figure 8.26
Line drawing of mold and crucible for a Thermit weld.

Flow Welding

Flow welding is defined as a welding process in which coalescence is produced by heating with molten filler metal poured over the surfaces until the welding temperature is attained and the required filler metal has been added. Flow welding is used in joining thick sections of nonferrous metals using a filler of the same composition as the base metal.

In operation the weld area is first properly prepared and preheated. Molten filler metal is then poured between the ends of the material until melting starts. At this point the flow is stopped and the joint, filled with metal, is allowed to cool slowly. To ensure the fusion of the top edges and provide a good weld, the level of the molten filler metal must be kept higher than the surfaces being welded.

Cold Welding

Cold welding is a method of joining metals at room temperature by the application of pressure alone. The pressure applied causes the surface metals to flow, producing the weld. It is a solid-state bonding process in which no heat is supplied from an external source. The type of bond obtained in a lap weld is shown in Figure 8.27. Butt welds of wire and rods can also be made by clamping the ends in special dies and bringing them together under a load sufficient to produce plastic flow at the joint. Before a weld is made the surfaces or parts to be joined must be wire-brushed thoroughly at a surface speed around 3000 ft/min (15 m/s) to remove oxide films on the surface. Other methods of

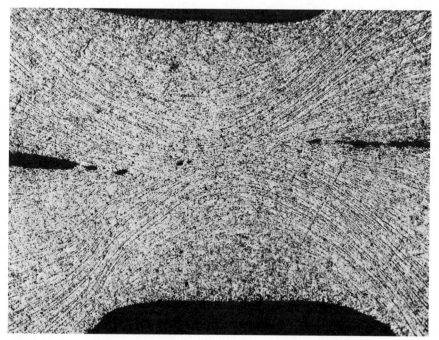

Figure 8.27
A macrograph of an aluminum cold weld showing lines of flow.

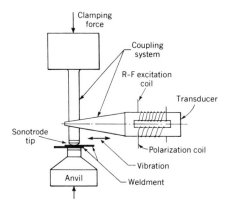

Figure 8.28
Ultrasonic welding.

cleaning seem to be unsatisfactory. In making a weld, the pressure is applied over a narrow area so that the metal can flow away from the weld on both sides. It may be applied either by impact or with a slow squeezing action, both methods being equally effective. Pressure required for aluminum is 25,000 to 35,000 psi (170–240 MPa). Spot welds are rectangular in shape and are approximately $t \times 5t$ in area, where t represents the metal thickness. In addition, both ring welds and continuous-seam welds can be made. This method of welding is used with aluminum and copper; however, lead, nickel, zinc, and monel can also be joined by cold pressure.

Ultrasonic Welding. *Ultrasonic welding* is a solid-state bonding process for joining similar or dissimilar metals, generally with an overlap-type joint. High-frequency vibratory energy is introduced into the weld area in a plane parallel to the surface of the weldment. The forces set up oscillating shear stresses at the weld interface that break up and expel surface oxides. This results in metal-to-metal contact permitting the intermingling of the metal and the forming of a sound weld nugget. No external heat is applied, although the weld metal does undergo a modest temperature rise.

A diagrammatic sketch of one spot-type welding system is shown in Figure 8.28. The machine is preset for clamping force, time, and power and the overlapping pieces are placed upon the anvil or support member. As the welding cycle is started the *sonotrode*[1] is lowered to the weldment and the clamping force builds to the desired amounts. Ultrasonic power of sufficient intensity is then introduced through the sonotrode for the preset time. The power is cut off automatically and the weldment is released.

Continuous-seam welding is accomplished with a rotating disk tip and with either a counter-rotating roller anvil or table-type anvil. The entire transducer coupling tip assembly rotates, so that the peripheral speed of the tip matches the traversing speed of the workpiece. Continuous seams are produced with essentially no slippage between the tip and the workpiece.

---

[1]Sonotrode is a name given to the vibrating element to distinguish it from an electrode used in resistance welding.

The only machine settings for welding are power, clamping force, and weld time or welding rate for continuous-seam welding. These settings vary according to the type and thickness of the material to be joined. Thin foils or fine wires may require only a few watts of power and a few ounces of clamping force, while heavier and harder materials may require several thousand watts and several hundred pounds. The weld time for spot-type welds is very short, usually less than 1 s.

This process can be used for materials up to about $1/8$ in. (3 mm) in thickness, although equipment is being developed to extend this range. There is no minimum limit to the sheet thickness that can be welded. Ultrasonic welding is excellent for joining thin sheets to thicker sheets because the thickness limitation applies only to the thinner piece.

The welds obtained are characterized by local plastic deformation at the interface and mechanical intermingling of the contacting surfaces. They are metallurgically sound and often exhibit strengths superior to those obtained by other joining processes. Ring-type and continuous-seam welds can be used for hermetic sealing. The process has a variety of applications in the electrical and electronic industries, in sealing and packaging, in foil splicing, in aircraft, in missiles, and in the fabrication of nuclear reactor components.

Explosive Welding

Explosive welding or cladding, as it is often called, brings together two metal surfaces with sufficient impact and pressure to bond. Pressure is developed by a high-explosive shot placed in contact with or in close proximity to the metals as illustrated in Figure 8.29. In some instances a protective material such as rubber is placed over the upper panel to prevent damage to the surface. The entire assembly is placed upon a buffer plate or anvil to absorb energy generated during the joining operation. Of the two arrangements showing cladding or laminating of metals, the left one is preferred.

To obtain a metallurgical bond, atoms from both surfaces must come into intimate contact. The oxides and films always present on the surface of metals are broken or dispersed by the high pressure or dissolved in the molten region. The explosive force brings the clean surfaces together and produces a sound bond.

Figure 8.30 shows the development of a high-velocity jet emanating from the collision point, a phenomenon that occurs in most explosive welds. The jet is formed by the surfaces of both plates flowing ahead of the collision point into the space between them. High pressure is required. Such extreme plastic deformation clears away nonmetallic films and produces metal bonding.

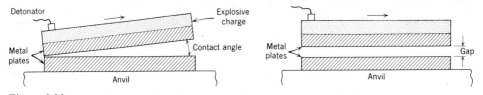

Figure 8.29
Common arrangements used to produce explosive clads. Buffer materials such as rubber are often used between the explosive and metal.

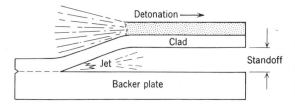

Figure 8.30
Explosion-bonding process showing high-velocity jet emanating from the collision point because of upstream pressure.

The general appearance of most metal bonds is shown in Figure 8.31. The wavy profile developed as a result of the surface jetting effect is a combination of direct metal-to-metal bonding and periodic melted areas. The surface jetting is a result of a compressive stress wave progressing across the surface of the plates as they collide. This action flushes the impurities from the surfaces and enhances the weld. In general the wavy-type bond is preferred because it localizes any solidification defects. A rather straight layer bond occurs with some metals as a result of a previously continuous molten surface zone that has solidified rapidly.

The principal use of explosive cladding is the uniting of large-area sheets. Areas up to 7 ft × 20 ft (2m × 6m) have been bonded. In addition to area welds, seam, spot, lap, and edge welds are possible variations, as well as internal cladding of tubes and pressure vessels. Advantages for this process include simplicity, rapidity, close thickness tolerance, and the ability to unite dissimilar metals. Metals with low melting points and low impact resistance cannot be bonded effectively.

Diffusion Welding

In *diffusion welding*, clean flat parts with fine surface finish are brought together in a vacuum or in an inert gas atmosphere with pressure. Usually the temperature is lower than the melting temperature of the base metal. The process temperature is about

$$T_p = 0.7 T_m$$

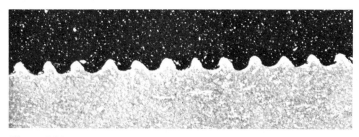

Figure 8.31
Direct metal-to-metal bond between explosively welded Monel 400 (top) to ASM-A 516-70 steel showing sawtooth bond. Magnification ×10.

where

T_p = Process temperature

T_m = Melting point of the lowest temperature of base metal being welded.

Actually the equation is accurate only if absolute temperatures are employed; however, for practical purposes conventional temperatures may be used. Pressure required varies from 6 to 50 psi (41 to 345 MPa), but pressures just under the yield stress at the operating temperature are preferred.

In liquid diffusion welding the temperature is slightly above the melting temperature. The process creates a metallic bond as a result of (1) the pressure causing microdeformation of the surfaces into each other, and (2) diffusion of atoms between the two materials being joined. The process is slow and costly because (1) surfaces must be flat and prepared to a roughness of about 8 μm (Figure 11.15), which is costly; (2) extreme cleanliness must be maintained, which requires a "clean room"; (3) use of an inert atmosphere requires a containment vessel; and (4) time required varies from seconds to 4 to 20 h.

Variations of this process use a thin film of a third metal to effect a weld between dissimilar materials. This is not always necessary because almost any metal and many plastics can be joined by this method.

The principal applications are in high technology, particularly in the atomic, aerospace, and electronic industries. One application is "coating" cutting tools by diffusion welding as opposed to brazing.

ADHESIVE BONDING

Adhesive bonding is rapidly replacing other joining operations because (1) the operation can be more economical, (2) machining operations such as threading can be eliminated, (3) there is no change in material properties or surface roughness, (4) dissimilar materials can be joined, (5) lighter gage materials can be employed, and (6) the elastomeric nature of some adhesives gives shock and vibration protection. The principal disadvantages are that (1) the surfaces to be joined must usually be chemically or mechanically scrubbed, (2) some adhesives are toxic and flammable, (3) the joints assembled are seldom effective at more than 400°F (240°C) and may start losing strength at much lower temperatures, (4) the use of adhesives has been difficult to incorporate in high-production lines, primarily because of curing time, and (5) some of the adhesives are unstable and have a short pot or shelf life.

The surface finish of parts to be bonded depends to a large extent on the ability of the plastic to wet the surface without asperities or air voids; cleanliness is usually more important than surface finish. Often the best surfaces are obtained by sandblasting or bead blasting followed by ion etching, which is accomplished by bombarding the surface with inert-gas ions.

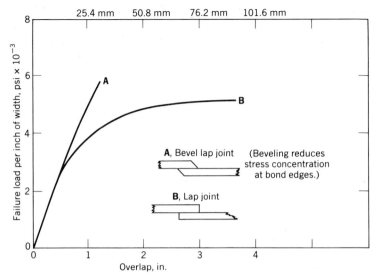

Figure 8.32
Strength and effect of adhesive joints.

Most of the plastics used for bonding are discussed in more detail in Chapter 10. Thermosetting plastics are more often used than thermoplastics because of the temperature softening characteristics of the latter. Design for adhesive bonds should be such that the joint is in shear rather than tension. The ideal joint is shown in Figure 8.32 as a bevel lap joint. There is less stress concentration and less cleavage stress at the bond's edges. A joint with less adhesive area may be stronger than one with a larger area because of this phenomenon, but in general the more area involved the greater the strength. It is preferable to increase the width of the joint rather than the overlap for maximum strength per unit of bonded area. Most of the adhesives have a shear strength from 3500 to 7000 psi (25–48 MPa), although some heat-cured ones have a shear strength of over 10,000 psi (68.9 MPa). The materials most used are

1. Epoxies (one and two part).
2. Anaerobics (one part).
3. Acrylic (one part).
4. Cyanoacrylates (one part).
5. Urethanes (one part thermoplastic and two part thermosetting).

The one-part types usually require temperatures on the order of 300°F (149°C) for curing or an absence of oxygen. Many of the newer adhesives are made in films or pressure-sensitive tapes.

QUESTIONS AND PROBLEMS

1. Give an explanation of the following terms:

   Flux
   Filler metal
   Scarfing
   Electrolysis
   Acetylene gas
   Hold time
   Roll spot welding
   Straight polarity
   Coated electrode
   TIG welding
   Plasma
   Electroslag welding
   Thermit metal
   Sonotrode
   Diffusion welding

2. Define the term coalescence. How is it obtained in electric-arc welding and percussion welding?

3. Which welding joint is the strongest? Why?

4. What effect does carbon content of steel have on weldability?

5. Describe the process of furnace brazing two parts.

6. List the distinctive steps required for a spot weld.

7. Describe the differences between oxygen and acetylene cylinders.

8. Of the types of welded joints shown in Figure 8.3, which is suitable for spot welding?

9. What type of oxyactylene flame is used for welding (a) Monel; (b) nickel; (c) steel; (d) bronze?

10. Compare the processes of laser, electron beam, and ultrasonic welding.

11. What welding temperatures are attained in oxyhydrogen welding, oxyacetylene welding, plasma arc cutting, electric-arc welding, and Thermit welding?

12. Calculate the heat generated in a spot weld if the current is 300 A, the resistance is 2400 ohms, and the time is 20 cycles of 60-cycle current.

13. How is acetylene made?

14. Plot a curve for the heat generated as a function of current for problem 12 if the current is 100, 150, 200, 250, 300, and 350 A.

15. Describe the chemical process that enables the cutting of ferrous materials with oxyacetylene.

16. How does the oxyacetylene cutting torch operate?

17. What are the advantages and disadvantages of an a-c welding machine?

18. What are the advantages and disadvantages of a d-c welding machine?

19. Describe the applicability of oxyacetylene cutting to gray cast iron.

20. What are the purposes of a coating on an arc-welding electrode?

21. How is underwater oxyacetylene cutting accomplished?

22. How does the exothermic chemical reaction apply to Thermit welding?

23. An electron beam welding machine costs about $300,000. What applications would justify its use?

24. What are the hazards associated with oxyacetylene welding, electron beam welding, and laser welding?

25. List, sketch, and discuss the five zones of heat generation in a spot weld.

26. What process do you recommend for each of the following: butt weld band saw blades; lap weld foil material; welding two 12-in. (305-mm)-diameter ferrous alloys; welding a crack in an aluminum automobile crankcase?

27. What are the materials for the spot-welding electrodes?

28. How and why are spot-welding electrodes cooled?

29. Describe the steps in oxyacetylene cutting a $^1/_2$-in. (12.7-mm) steel plate.

30. If the kerf in problem 29 is $^1/_4$ in. (6.4 mm) wide, how many pounds of steel are wasted in cutting 1000 parts of 24-in. (610-mm) circles?

31. Describe how two cylindrical cylinders can be joined together, one inside the other, by explosive welding.

32. Why do welds made in steel tend to harden otherwise soft materials?

33. What safety precautions are necessary when cutting stainless steel within a nitrogen atmosphere?

34. What causes weldments to crack? Explain the reasons and suggest remedies.

35. What happens to weldability of steel as the carbon content is increased?

36. How can a lathe be equipped to weld short pieces of $^1/_4$-in. (6.3-mm) round bar stock end to end frictionally?

37. What advantages do coated electrodes give to electrode arc welding? How does manual skill prevent problems in arc welding?

38. What is a contact electrode? List its advantages.

39. Calculate the amount of oxygen to cut 1000 pieces of 30 in. (76mm) by $1^1/_4$-in. (318- mm) steel plate 23 in. (584 mm) long. Assume that the cut is $^3/_8$ in. (9.5 mm) wide.

40. Specify the projection dimensions on stock 0.125 in. (3.18 mm) thick that is to be projection welded.

41. How many watts (volts × amperes) are necessary to weld $^1/_4$ in. (6.4 mm) diameter with a spot welder?

42. What is the range in seconds to spot weld?

43. What are the principal disadvantages of diffusion welding?

44. How does the strength of an adhesive bond compare with diffusion welding?

45. What is the principal reason that prevents adhesive joining of ceramic-to-titanium heat shields for spacecraft?

46. If a 1.0-in. (25.4-mm) solid ceramic shaft needed to attached to a 1.0-in. (25.4-mm) solid aluminum shaft in end-to-end fashion, what adhesive joint design would you recommend? Which would probably be more expensive, diffusion welding or adhesives?

47. What are the commercial and economical applications for electron beam welding?

48. If on each pass of a $^1/_8$-in. (3.18-mm)-diameter, 12-in. (305-mm)-long steel welding rod, 1 in. (25.4 mm) of rod is used for every $^1/_2$ in. (12.7 mm) of weld, how many pounds of rod are used in making four passes on 600 lineal feet of welding? The last 2 in. of each rod are unusable.

CASE STUDY

WELDING SPACECRAFT HEAT SHIELDS

In your capacity as a designer in the Genius Aerospace Company, a new heat shield for a spacecraft is being designed that requires a thin section of titanium and a thick aluminum plate to be joined together by a welding process.

The thickness of titanium is 0.031 in. (0.79 mm) and the aluminum is 0.75 in. (19.1 mm) thick. The unit makes a 12 × 12 in. (305 × 305 mm) plate about 1.06 in. (26.9 mm) thick. Six thousand units are required. Some units are curved, and buckling and blisters must be avoided.

Describe the method for producing these heat shields. Sketch the production line and indicate the equipment.

CHAPTER 9
POWDER METALLURGY

Powder metallurgy is the art of producing products from metallic powders by pressure. Heat, which is almost always used in the process, must be kept at a temperature below the melting point of the powder. The application of heat during the process or subsequently is known as *sintering* and the results in bonding the fine particles together, thus improving the strength and other properties of the finished product. A *green* compact is unsintered and has little strength or resistance to abrasion. Products made by powder metallurgy are frequently mixed with different metal powders or contain nonmetallic constitutents to improve the bonding qualities of the particles and properties of the final product. Cobalt or some other metal is necessary in bonding tungsten carbide particles, whereas graphite is added with bearing metal powders to improve the lubricating qualities of the finished bearing.

Metal in powder form is higher in cost than in solid form, and the process most adaptable to mass production requires expensive dies and machines. This higher cost is often justified by the unusual properties obtained. Some products cannot be made by any other process; others may compete favorably with their counterparts made by other methods because the close tolerances obtained eliminate the necessity of any further processing.

IMPORTANT CHARACTERISTICS OF METAL POWDERS

The two important characteristics of a metal powder part are strength and machinability. Even though powder metallurgy is a process that has its chief economical advantage in minimizing machining, parts generally need some machining operation such as broaching or a tapping.

The particle size, shape, and size distribution of metal powders affect the characteristics and physical properties of the compacted product. Powders are produced according to specifications such as shape, fineness, particle size distribution, flowability, chemical properties, compressibility, apparent specific gravity, and sintering properties.

The *shape* of a powder particle depends largely on how it is produced. It may be spherical, ragged, dendritic, flat, or angular. A powder with good green strength can only be produced from irregularly shaped particles that will interlock on compacting.

Fineness refers to the particle size and is determined by passing the powder through standard sieves or by microscopic measurement. Standard sieves ranging from 100 to

325 mesh (45–150 μm) are used for checking size and also for determining particle size distribution within a certain range. Because most powders are of irregular shape and not spheres, the size of individual units cannot be specifically stated. Perhaps the most used size is about 100 μm. Finer powders sinter better and are particularly important for slip casting.

Particle size distribution has reference to the amount of each standard particle size in the powder. It influences flowability, apparent density, compressibility, final porosity, and mechanical properties such as strength and elasticity. It cannot be varied appreciably without affecting the size of the compact.

Flowability is that characteristic of a powder which permits it to flow readily and conform to the mold cavity. It can be described as the rate of flow through a fixed orifice. Smaller and more regularly shaped, nearly spherical powders flow best, so if the design requires increased flowability they would be used. Very fine particles on the order of 1 to 10 μm have characteristics similar to a liquid when they are pressed and tend to "flow" into complex molds. The addition of stearate to the mix also increases flowability.

A specification of *chemical properties* has to do with the *purity* of the powder, amount of oxides permitted, and the percentages of other elements allowed. Alloys can be made in the crucible prior to manufacture of the powder, and various additions can be made to change the characteristic of the final product.

Compressibility is the ratio of the volume of initial powder to the volume of the compressed piece. It varies considerably and is affected by particle size distribution and shape. The green strength of a compact is dependent on compressibility. There is no formula for compressibility, but a general "rule of thumb" is that most fine powders compress to about two-thirds of their original filling depth. Every mold, depending on the powder, powder shape and size, lubrication, and compacting pressure, must be specifically designed.

The *apparent density* of a powder is expressed in pounds per cubic inch (kilograms per cubic meter). It should be kept constant so that the same amount of powder can be fed into the die each time.

Sintering is the bonding of particles by the application of heat. A broad temperature range is preferred. The final strength and hardness of a part is largely determined by heat treatment, that is, quenching and tempering. This separate step follows sintering. Sintering is so rapid that diffusion is slight; otherwise, additional strength could be developed by this means. When a part requires a specific porosity, say 20% porous, the particle size distribution can be adjusted to accomplish the desired result. Mechanical properties are controlled by particle size, compaction pressure, sintering, and heat treatment.

METHODS OF PRODUCING POWDERS

Although all metals can be produced in the powder form, only a few are widely applied in manufacturing pressed-metal parts. Some lack the desired characteristics or properties described above, which are necessary for economical production. The two principal kinds are iron and copper base powders. Both lend themselves to powder metallurgy. Whereas bronze is used in porous bearings, brass and iron are more often found with small machine parts. Other powders of nickel, silver, tungsten, and aluminum have a limited but important application in the field of powder metallurgy.

All metal powders, because of their individual physical and chemical characteristics, cannot be manufactured in the same way. The procedures vary widely as do the sizes and structures of the particles obtained from the various processes. *Machining* results in coarse particles and is used principally for producing magnesium powders. *Milling and grinding* processes, utilizing various types of crushers, rotary mills, stamping mills, and grinders, break down the metals by crushing and impact. Brittle materials may be reduced to irregular shapes of almost any fineness by this method. The process is also used in pigment manufacturing of ductile materials where flake particles are obtained. An oil is used in the process to keep them from sticking together. *Atomization* or the operation of metal spraying is an excellent means of producing powders from many of the low-temperature metals such as lead, aluminum, zinc, and tin. Iron powders with high carbon content have been produced by this process, but the green strength is lower than for other methods. Iron powders are also produced by the following steps: (1) melt in a crucible; (2) atomize using air or water to break up the particles; (3) grind to about 100 mesh, anneal, grind to 100 mesh a second time. The particles are irregular in shape and are produced in many sizes. A few metals can be converted into small particles by rapidly stirring the metal while it is cooling. This process, known as *granulation*, depends on the formation of oxides on the individual particles during the stirring operation.

Electrolytic deposition is a common means for processing silver, tantalum, and several other metals. The *reduction method* reduces metal oxides to powder form by contact with a gas at temperatures below the melting point. For making iron powder, millscale (a form of iron oxide) is fed into a rotating kiln along with crushed coke. Near the discharge end the mixture is heated to around 1900°F (1050°C), causing the carbon to unite with the oxygen in the iron oxide. This forms a gas that is removed through a stack. With the oxygen removed the product remaining is a relatively pure iron having a spongelike structure. The chemical reaction is:

$$2Fe_2O_3 + 3C \xrightarrow{heat} 3CO_2 \uparrow (gas) + 4Fe$$

(millscale)   (coke)   (pure iron)

Other metals produced commercially by this process include tungsten, molybdenum, nickel, and cobalt.

Various other methods involving precipitation, condensation, and other chemical processes have been developed for producing powdered metals.

SPECIAL POWDER PREPARATION

Prealloyed Powders

Alloyed powder products obtained by blending pure metal powders do not provide some of the properties that are possible with prealloyed powders. Blended-powder products are cheaper to produce and require lower pressures to compact. Prealloyed powders, alloyed in the melting process, provide product properties similar to those possible from the melt composition when maximum density is achieved. This permits the production of alloys such as the stainless steels and other highly alloyed compositions that heretofore have not been successfully obtained by blending. Prealloyed metal powder products may have properties such as corrosion resistance, high strength, or resistance to elevated temperature.

Precoated Powders

Metal powders may be coated with an element by passing the powder through a carrier gas. Each particle is uniformly coated, thus producing a powder product that when sintered has certain characteristics of the coating. This permits a cheap bulk powder to be used as a carrier for the outside active materials. Products made from precoated powders that are sintered are more homogeneous than those produced by blending.

FORMING TO SHAPE

Powder for a given product must be carefully selected to ensure economical production and to obtain the desired properties in the final compact. If only one powder is to be used and the particle size distribution meets specification, additional processing or blending is unnecessary before pressing. Sometimes various sizes of powder particles are mixed together to change the characteristics of flowability or density, but most powder is produced with sufficient particle size variation. Mixing or blending becomes necessary in production when the powders are alloyed or when nonmetallic particles are added. Any mixing or processing of the powder must be done under favorable conditions to prevent oxidation or defects.

Practically all powders have lubricants added in the blending operation to reduce die wall friction and to aid part ejection. Although these lubricants add to the porosity, they increase the production rate and are necessary in presses using automatic powder feed. Lubricants may include stearic acid, lithium stearate, and powdered graphite.

Pressing

Powders are pressed to shape in steel dies under pressures ranging from a few thousand to 200,000 psi (1400 MPa). Because the soft particles can be pressed or keyed together quite readily, powders that are plastic do not require so high a pressure as the harder powders to obtain adequate density. The density and hardness increase with pressure, but in every instance there is an optimum maximum pressure above which little advantage in improved properties can be obtained. Owing to the necessity for strong dies and large-capacity presses, production costs increase with high pressures. The density of the final product varies with compacting pressure as shown in Figure 9.1.

Many commercial presses developed for other materials are adaptable to powder metallurgy. Although mechanically operated presses are generally used because of their high rate of production, hydraulic presses may be employed if the part is large and high

200 POWDER METALLURGY

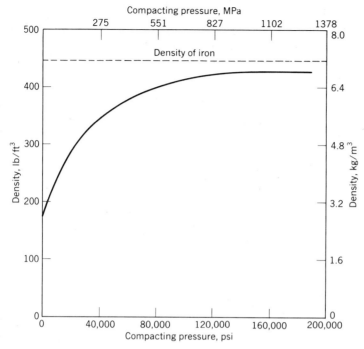

Figure 9.1
Approximate compressibility of iron powder.

pressures are required. The single-punch press and the high-speed rotary multiple-punch press are designed so that their operation, from filling the cavity with powder to ejection of the finished compact, can be either a continuous or a single cycle. Rotary table presses have a high rate of production because they are equipped with a series of die cavities, each provided with top and bottom *punches.* In the course of production the table indexes and the operations of filling, pressing, and ejecting the product are accomplished at the various stations. A simple punch and die arrangement for compacting metal powder is shown in Figure 9.2. Two punches are involved: an upper punch that conforms to the

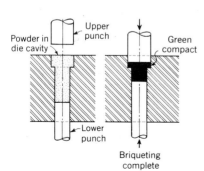

Figure 9.2
Punch and die arrangement for compacting metal powder.

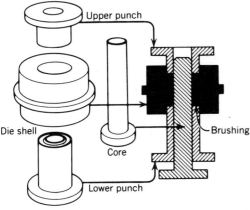

Figure 9.3
Tooling arrangement for briquetting or green compacting bronze powder into a brushing.

top shape of the part and a lower punch that conforms to the lower end of the die cavity. The lower punch also acts as an ejector to remove the **briquetted** part from the die. The die cavity must be smooth to reduce friction and must have a slight draft to facilitate removal of the part. Wall friction prevents much of the pressure from being transmitted to the powder, and, if pressure is exerted only from one side there will be a considerable variation in density from top to bottom. This accounts for using both top and bottom punches in most dies. The travel of the punches depends on the compression ratio of the powder, which for iron and copper is roughly 2.5 : 1. The die cavity is filled to a level about three times the height of the finished compact. The ejected part, known as a **green compact,** resembles the finished part but has only the structural strength derived from interlocking of the powder particles obtained by compression. Final strength is obtained by sintering. A tooling arrangement for compacting loose bronze powder is shown in Figure 9.3.

Figure 9.4 shows a press setup for compacting small pinions from metal powders. Many similar products are completed by the pressing operation and require no further processing other than sintering. Cold-bond pressures of 10 to 35 tons psi (150–500 MPa)

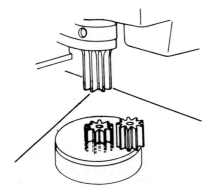

Figure 9.4
Pressing small pinions from metal powder.

are necessary on the part. The sintering operation increases the strength and improves the crystalline structure.

The size of metal powder parts is limited by the capacity of presses used in the compacting operation. Press capacity to 750 tons (7 MN) and more are available. The maximum area of a compact may be readily calculated by the following simple relationship:

$$\text{area } A = \frac{F}{P}$$

where

F = Press capacity, lb or N

P = Required compacting pressure, psi or Pa

The density of a powdered-metal product is one of its most distinguishing characteristics. An increased pressure on a compact causes a higher density part and consequently an increased tensile strength. The density may be increased by using a powder of smaller particle size.

Centrifugal Compacting

A development in powder metallurgy is **centrifugal compacting** heavy metal powders to obtain uniform density. Molds are filled with the powder and then centrifuged to provide pressures around 400 psi (3 MPa). Uniform density is obtained, because the centrifugal force acts independently on each particle of powder. Upon removal from the molds the compacts are processed the same as pressed compacts. This technique is limited to parts made of heavy powders such as tungsten carbide. Parts made in this way should be of nearly uniform section, because small, irregular thicknesses are not successfully compacted.

Slip Casting

Green compacts for tungsten, molybdenum, and other powders are sometimes made by slip casting. The powder, converted to a slurry mixture, is first poured into a plaster of paris mold. Because the mold is porous, the liquid gradually drains off into the plaster leaving a solid layer of material deposited on the surface of the mold. For hollow objects, after sufficient time has been allowed for a desired thickness to accumulate, the remaining slurry is poured out. Upon drying, the green components are sintered in the usual manner. This procedure is simple and permits considerable variation in size and shape.

Increasing emphasis is being directed to this process because parts can be made that are too large or too complex to press. An expensive press is not required, and improved physical properties are additional advantages. The principal disadvantage is time lag in producing parts.

Extruding

Long shapes produced from metal powders must be extruded. Developments in this field make it possible to produce extruded shapes with very high densities and excellent mechanical properties. Methods used for extruding depend on the characteristics of the powder; some are extruded cold with a binder whereas others are heated to a desirable extruding temperature. Generally, the powder is first compressed into a billet and is then heated or sintered in a nonoxidizing atmosphere before being placed in the press. In extreme cases, to avoid oxidation, the billet may be sealed in a metal can before being placed in the press and extruded. Although the greatest use of this process has been toward producing nuclear solid-fuel elements and other materials for high-temperature applications, aluminum, copper, nickel, and many other metals can be extruded.

Gravity Sintering

Porous metal sheets having controlled porosity can be made by a process known as gravity sintering. The process has special application in the fabrication of stainless-steel sheets. A uniform thickness of powder is placed upon ceramic trays and sintered up to 48 h in dissociated ammonia gas at high temperature. The sheets are then rolled to obtain thickness uniformity and a better surface finish. They may then be fabricated into suitable shapes in the same manner as working sheet metal. Porous sheets of stainless steel are corrosion resistant and are used for gasoline, oil, and chemical filters.

Rolling

In a somewhat similar process, powders are fed from a hopper between two rolls that compress and interlock them into a sheet of sufficient strength to be conveyed through a sintering furnace. The sheet can then be passed through another set of rolls and heat treated if necessary. By blending the powders before they enter the rolls, alloy sheets can be made. Metal powders that can be rolled into sheets include copper, brass, bronze, Monel, and stainless steel. Both uniform mechanical properties and controlled porosity can be obtained by this process.

Isostatic Molding

Isostatic molding is a means of obtaining uniform density of the metal powder during the compacting operation. This method is important because high pressures can be used, resulting in a dense product. Isostatic pressing actually refers to situations when the pressure medium is a gas, whereas *hydrostatic pressing* is used when the pressure medium is liquid.

The theory of isostatic molding is based on the fact that a pressure, applied on a static liquid or gas, will cause forces to act equally in all directions on the exposed surfaces. Powder is placed in an elastic container that is subjected to pressure from all sides. Any internal cavity of the molding is made by a metal core or mandrel properly positioned in the container. The entire assembly is confined in a metal pressure chamber as shown in Figure 9.5. Upon completion of the compacting operation, the part is taken from the container and the mandrel withdrawn.

Advantages claimed for this process include uniform product density, uniform strength in all directions, low equipment costs, and a higher green compact strength than is possible

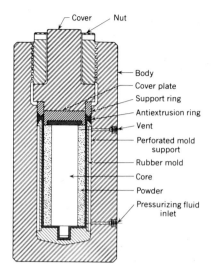

Figure 9.5
Section illustrating isostatic pressing arrangement for dry bag tooling.

to obtain by most other methods. Powder metals that can be compacted by isostatic pressing include aluminum, magnesium, beryllium, iron, tungsten, and stainless steel.

Explosive Compacting

Compacting powders by an explosive charge offers certain advantages for compact-resistant powders. Extremely high pressures are possible, resulting in a high-density product. This reduces the sintering time and limits shrinkage of the compact. Some saving can be made on equipment costs because die designs are relatively simple.

Most designs that have been reported have a closed system. One or more plungers, placed next to the metal powder, are actuated by buffer plates against which the powerful explosive acts. Another design utilizes water in a heavy-walled cylinder. Powders are placed in waterproof bags and put in the cylinder. Hydrostatic pressure is exerted on the compacts by detonating an explosive charge at the end of the cylinder.

Fiber Metal Process

Products made of metal fiber are specially adapted for filters, vibration damping, battery plates, and flame barriers. Most metals or alloys may be formed by this process. Fibers are produced from fine wires or metal wool cut to desired lengths. It is seldom advisable to use straight fibers, so they are usually bent or crimped in a random fashion. Products are then produced by a felting process whereby the metal fibers are mixed with a liquid slurry and poured over a porous bottom. After the liquid has drained off the green mat, which is made up of randomly distributed fibers, the mat is pressed and sintered. The density may be further increased by rolling or coining. Most metals and alloys can be processed by this technique. Figure 9.6 is a cross section of a composite including both tungsten fibers and copper to wet the fiber. Mechanical properties follow Hooke's law.

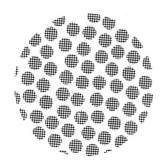

Figure 9.6
Transverse section of 483-5 mil (0.13 mm) tungsten wires imbedded in a copper wetting agent. Unetched magnification ×50.

SINTERING

The operation of heating a green compact to an elevated temperature is known as sintering. As previously stated, it is the process by which solid bodies are bonded by atomic forces. With application of heat the particles are pressed into more intimate contact, and the effectiveness of surface tension reactions is increased. In other words, in sintering the particles are fused together to increase density. During the process grain boundaries are formed, which is the beginning of recrystallization. Plasticity is increased and better mechanical interlocking is produced by building a fluid network. Also, any interfering gas phase present is removed by the heat. The temperatures used in sintering are usually well below the melting point of the principal powder constituent, but may vary over a wide range up to a temperature just below the melting point. Tests have proved that there is usually an optimum maximum sintering temperature for a given set of conditions with nothing to be gained by going above this temperature.

For most metals the sintering temperatures can be obtained in commercial furnaces, but for some metals, requiring high temperatures, special furnaces must be constructed. There is considerable range in the sintering temperature, but the following temperatures have proved satisfactory: 2000°F (1095°C) for iron, 2150°F (1180°C) for stainless steel, 1600°F (870°C) for copper, and 2700°F (1480°C) for tungsten carbide. Sintering time ranges from 20 to 40 min for these metals. The time element varies with different metals, but in most the effect of heating is complete in a very short time and there is no economy in prolonging the operation. Atmosphere is important because the product, made up of small particles, has a large exposed surface area. The problem is to provide a suitable atmosphere of some reducing gas or nitrogen to prevent the formation of undesirable oxide films during the process.

Furnaces for sintering may be either the batch or continuous type. The continuous type, which has a wire mesh belt to carry the compacts through the furnace, is shown in Figure 9.7. Pusher and roller hearth furnaces are also used and are similar in appearance. The dimensional change of growth or shrinkage may occur in sintering. What happens depends on the shape and particle size variation of the powder, the powder composition, sintering procedure, and briquetting pressure. Accurate size is maintained by compensating for the change in making the green compact and then maintaining uniform conditions.

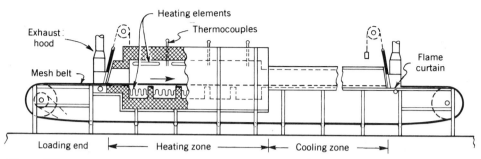

Figure 9.7
Continuous-type furnace for sintering powder metal compacts.

HOT PRESSING

Hot pressing can produce compacted products with improved strength and hardness, improved tolerance, and higher densities than products in which the operations are performed separately. Factors that limit its application include high cost of dies, difficulties in heating and atmospheric control, and the length of time required for the cycle. Some metals, particularly beryllium and cemented carbides, can be hot-pressed in a vacuum using graphite molds and punches. The molds are usually heated by resistance wires wrapped around the mold or else by induction. Vacuum has proved more successful than an inert atmosphere, which means complex and expensive tooling. The compacted powder is cooled to room temperature in the mold, resulting in a slow process. Temperatures adequate for hot-pressing tungsten powders are too great for the use of steel dies, so that graphite dies are required. Hot-pressing is used to some extent in manufacturing cemented carbides, but most attempts to combine sintering with pressing have not been successful. An exception to this is a process known as spark sintering.

SPARK SINTERING

Spark sintering is a method of processing that combines pressing and sintering metal powders to a dense metal part in 12 to 15 s. In operation a high-energy electrical spark, discharged from a capacitor bank, removes within a second or two surface contaminants from the powder particles. This causes the particles to combine as in conventional sintering, forming a solid, cohesive mass. Immediately following the spark the current continues for about 10 s with the temperature well below the melting point of the material, which furthers crystal bonding between the particles. Finally, by hydraulic pressure the mass is compressed between the electrodes to increase its density.

A schematic sketch illustrating a spark-sintering setup for fabricating a bimetallic (carbide and steel) punching die blank is shown in Figure 9.8. The area labeled "sintered carbide" represents the carbide material sinter-bonded to the steel substrate. The powder was originally filled to the top of the graphite mold before being compressed as shown. The vertical graphite centerpiece merely defines the doughnut shape of the punching die.

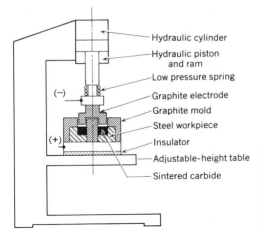

Figure 9.8
Schematic of spark-sintering machine using high-energy spark discharge to stop surface contaminants from powder particles. This is followed by 10- to 12-s spark discharge resistive-heating cycle.

Final densification pressure is applied through the graphite electrodes. In producing a nonmetallic sintered form, the steel substrate shown in Figure 9.8 cannot be used. The graphite electrodes would be machined to the desired dimensions of the final product. This process has been applied to sintering aluminum, copper, bronze, iron, and stainless steel.

FLOWCHART

The flowchart in Figure 9.9 summarizes the various processes and operations used in manufacturing metal powder parts. This chart shows the operational sequence from raw material to the finished parts. The finished parts may undergo additional operations that are not shown on the chart. These include tumbling, welding, brazing, drawing, steam treatment, oil impregnation, or sometimes additional sintering. Tolerances shown are only approximate and apply to small uniform parts.

FINISHING OPERATIONS

Bearings made from porous metal represent one of the important products of powder metallurgy. Porosities ranging from 25% to 35% are generally used, because higher values result in lower bearing strength. Impregnation is accomplished either by immersing the sintered bearing in heated oil for a period of time or by a vacuum treatment, which is much quicker. Such bearings retain oil by capillary force, the oil being released as the bearing is used. Most bearings are of porous bronze or iron composition.

Infiltration is the process of filling the pores of a sintered product with molten metal to decrease porosity or to improve physical properties. In this operation the melting point of the liquid metal must be considerably lower than the solid metal. Prior to the operation a chemical treatment is advisable to increase the extent of infiltration. Liquid metal is

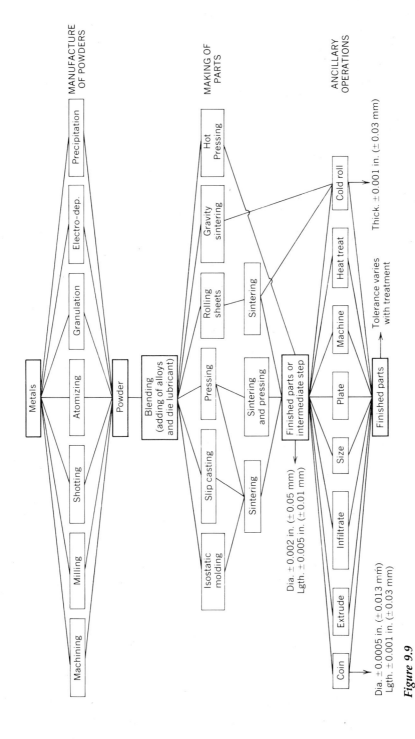

Figure 9.9
Flowchart for fabricating metal powder parts. Tolerances are for uniform parts not exceeding 2 in. (50 mm) in diameter or length.

infiltrated into the part either by allowing it to enter from above or by absorbing it from below. As an example, copper placed upon a piece of presintered iron and heated to 2100°F (1150°C) is drawn into the iron by capillarity.

Sizing and coining operations to close tolerances are regularly performed and may necessitate a final operation such as repressing the part in a die similar to the one used for compacting it. Such *sizing* is a cold-working operation that improves surface hardness and smoothness as well as dimensional accuracy. The density of the part is also increased.

All pressed metal parts may be ***heat-treated*** by conventional methods, although the results do not always conform to those obtained in solid metals. Best results are obtained with dense structures. Porosity influences the rate of heat flow through the part and permits internal contamination if salt bath heat-treating pots are used in the process. For this reason liquid carburizing is not recommended for surface treatment of metal powder parts.

High-density parts can be *plated* by standard procedures, but medium- or low-density parts require some prior treatment to close the pores. Preparations such as peening, burnishing, or plastic resin impregnation will close surface pores and eliminate salt entrapment that causes the plate to blister. Standard plating procedure may then be followed.

One of the characteristics of powder metal products is that they may be die-pressed into dimensionally accurate, finished shapes. However, products requiring such features as threads, grooves, undercuts, or side holes cannot be produced by powder metallurgy methods and must be finish machined. Tungsten carbide tools are recommended, although high-speed steel tools may be used when only a few parts are to be machined. The use of coolants with water is not advisable for iron base parts, as corrosion may occur.

ECONOMIC CONSIDERATIONS

Many products are now being made better and more economically using powder metals. Following are some of the advantages of powder metallurgy.

1. Sintered carbides and porous bearings can be produced only by this process, and bimetallic products can be formed from mold layers of different metal powders.
2. Parts with controlled porosity can be produced.
3. Large-scale production is possible of many small parts that compete favorably with machined parts because of close tolerance and surface finish.
4. Powders that are available in a pure state produce items of extreme purity.
5. The process is economical because there are no material losses in fabrication.
6. Labor cost is low; skilled machinists are not required to operate presses or other necessary equipment.

Powder metallurgy also has the following limitations.

1. Metal powders are expensive and sometimes difficult to store without some deterioration.

2. Equipment costs are high.
3. Some products can be made more economically by other methods, because the size of powder-fabricated parts is controlled by the capacity of the presses available and also by the compression ratio of the various powders.
4. Intricate designs in products are difficult to attain because there is little flow of metal particles during compacting.
5. Some thermal difficulties appear in sintering operations, particularly with low-melting powders such as tin, lead, zinc, and cadmium. Most oxides of these metals cannot be reduced at temperatures below the melting point of metal; hence, if the oxides exist they will have detrimental effects on the sintering process and result in an inferior product.
6. Some powders in a finely divided state present explosion and fire hazards. Such metals include aluminum, magnesium, zirconium, and titanium.
7. A uniformly high-density product is difficult to fabricate by this process.

METAL POWDER PRODUCTS

Among the products for which metal powder is increasingly being used are the machine parts shown in Figure 9.10. Most of these are completed without machining. Following are some of the prominent powder metal products.

Figure 9.10
Variety of machine parts made from metal powders.

Permanent metal powder filters have greater strength and shock resistance then ceramic filters, being made with porosities up to 80%. Fiber metal filters can have a porosity up to 97%. They are used for filtering hot or cold fluids and air. As illustrated in Figure 9.11, one use is in dehydrators for diffusing moisture-laden air around a drying agent such as silica gel. Another common use in gasoline tanks is for separating moisture and dirt from the fuel system. Metallic filters can also be used for flame arresting and sound deadening. Filters used for removing foreign matter are cleaned by reversing the flow of the liquid.

Tungsten carbide particles are mixed with a cobalt binder, pressed to shape, and then sintered at a temperature above the melting point of the matrix metal to produce cemented carbides that are used for cutting tools and dies.

Gears and pump rotors are made from powdered iron mixed with sufficient graphite to give the product the desired carbon content. A porosity of around 20% is obtained in the process. After the sintering operation the pores are impregnated with oil to promote quiet operation.

Brushes for motors are made by mixing copper with graphite to give the compact adequate mechanical strength. Tin or lead may also be added in small quantities to improve wear resistance.

Most bearings are made from copper, tin, and graphite powders, although other metal combinations are used. After sintering the bearings are sized and then impregnated with oil by a vacuum treatment. Porosity in the bearings can be controlled readily and may run as high as 40% of the volume. In Figure 9.12 is a photomicrograph of a bronze bearing, which illustrates the open-pore construction that can be obtained.

Excellent small magnets can be produced from several compositions of iron, aluminum, nickel, and cobalt when combined in powder form. Alnico magnets, made principally from iron and aluminum powders, are superior to those cast.

Electric contact parts are adapted to powder metallurgy fabrication, because it is possible to combine several metal powders and maintain some of the principal characteristics of each. Contact parts must be wear resistant, somewhat refractory, and have good electrical conductivity. Many combinations such as tungsten–copper, tungsten–cobalt, tungsten–silver, silver–molybdenum, and copper–nickel–tungsten have been developed for electrical applications.

Numerous other parts including clutch faces, brake drums, ball retainers, and welding rods are produced by powder metallurgy. Other uses for powder metals include paint

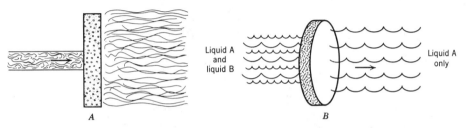

Figure 9.11
Special uses of permanent metal filters. *A*, Diffusing. *B*, Separating.

Figure 9.12
Photomicrograph of oilite bearing with 20% (by volume) voids made from a blend of 10% tin and 90% copper powders.

pigments, missile fuels, and Thermit welding. The addition of powdered metals to plastics increases strength and contributes metallic properties.

QUESTIONS AND PROBLEMS

1. Give an explanation of the following terms:

    Compressibility
    Sintering
    Electrolytic deposition
    Punch
    Briquetting
    Green compact
    Centrifugal compacting
    Hot pressing
    Sizing
    Heat treatment

2. How is the higher cost of powder metal products justified?

3. How would the following metal powder parts be made: porous sheet metal; long, uniformly shaped pieces; beryllium part; cemented carbides; and Alnico magnet?

4. How is sintering accomplished? Why not hot-press all products and eliminate the sintering operation?

5. How and by what method are ferrous alloy powders made?

6. What kind of press is used in slip casting?

7. What is the principal disadvantage of slip casting?

8. What are the purposes of lubricants in powder metallurgy? What are their disadvantages?

9. In the manufacture of metal powder bearings, how may the porosity of the metal be controlled to a desired percentage? What advantage do these bearings have over cast bearings?

10. What is the advantage of irregular-shaped particles in powders?

11. What are the principal advantages of fine powders as compared to coarse powders?

12. What are the two most important characteristics of a powdered-metal part?

13. What accuracy can be expected on a part that has been coined or sized in a press?

14. Why is sintering usually accomplished in a controlled atmosphere?

15. Name three metal powder products that cannot be made by other processes.

16. How are bearings impregnated with oil?

17. Aluminum paint is often manufactured by mixing aluminum powder in a lacquer base. How would you recommend making such powder?

18. What is infiltration and how is it done?

19. If the compacting pressure for stainless steel is 200,000 psi (1333 MPa), what is the maximum-diameter part that can be made in a 300-ton (3-MN) press?

20. Convert the following pressures to SI units: 1000 psi; 5000 psi; 50,000 psi; 500,000 psi.

21. Convert the following densities to g/cm^3: 450 lb/ft^3; 0.260 lb/in.^{3}; 300 lb/ft^3; 260 lb/ft^3.

22. A powdered-metal iron cylinder 1 in. (25.4 mm) in diameter and 1 in. (25.4 mm) long is to be compacted at 50,000 lb/in.^{2}; what is the approximate density of the part?

23. What size press in tons is needed for the part in problem 22?

24. Calculate the approximate volume of raw powder needed for the part in problem 22.

25. Calculate the approximate length of mold that is necessary for the part in problem 22.

26. It is found experimentally that to make a 2-in. (50-mm) cube from powdered iron a press pressure of at least 80,000 psi (551 MPa) is needed on each surface. The vertical press force P is related to the formula $P = 0.6A$, where A is the wall force. What is the force in pounds (newtons) that will be exerted by the press on the powder? Where will the pressed powder be more dense, on the top surface or on the sides?

27. What is the approximate density in pounds per cubic inch of a part that is molded from a powder having a density of 100 lb/ft^3? Also, calculate in SI metric units.

28. How would you define the compressibility of a powder that occupied two-thirds of the space after compaction, as compared to the space it occupies as a free-standing powder?

CASE STUDY

A TO Z PRODUCTS COMPANY

The A to Z Products Company has three hydraulic presses that are little used after the company lost an automotive contract a year ago. The presses are paid for but they do take up plant space. Salespeople have made calls to line up work for the presses but without success. During the past month considerable thought has been given to making a bearing of 90% brass and 10% graphite mixture. These bearings, 1 in. (25.4 mm) long with a 0.50-in. (12.7-mm) OD and a 0.250 ± 0.002 (6.35 ± 0.05 mm) ID can be sold to a lawnmower manufacturer in lots of 10,000 per month. Your company works eight 24-h shifts a month.

You have been asked to make a report to top management outlining the process chart for producing these sintered bearings and advising what additional equipment must be purchased to effect the production line.

The brass powders are known to cost $1.48/lb and the graphite is $3.13/lb. The first estimate at a selling price is four times the material cost. If it is necessary to pay for the additional equipment in 3 years, how much can be invested in new equipment? Densities are 535 lb/ft^3 and 79 lb/ft^3 for brass and graphite.

Looking at cost another way, if it takes an average of 1.7 min per unit to manufacture and the labor cost is $16.20/h, what would the price have to be if 20% profit were desired? Based on either estimate, does this look like a valid economic product for the company based on this order alone?

CHAPTER 10
PLASTIC MATERIALS AND PROCESSES

Plastics are a common material. The term plastics usually refers to *synthetic organic material* that is solid in its final form but is fluid at some stage in the processing and is shaped by heat and pressure. Sometimes the term *polymer* is used, referring to any substance in which several or many thousand molecules or building units are joined into larger, more complex molecules. This chapter considers the types of plastics and the methods of processing them.

MATERIALS

Raw Materials and Properties

The basic plastic molecule is carbon. Resins, necessary for the plastic materials, are produced by chemically reacting *monomers* to form long-chain molecules called polymers. This process is called *polymerization*. Two methods are used to achieve polymerization: *addition polymerization*, where two or more similar monomers directly react to form long-chain molecules, and *condensation polymerization*. This latter method takes two or more dissimilar monomers that react to form long-chain molecules plus the by-product water.

The raw materials for plastic compounds are various agricultural products and numerous minerals and organic materials including petroleum, coal, gas, limestone, silica, and sulfur. During the process of manufacture additional ingredients are added such as color pigments, solvents, lubricants, plasticizers, and filler material. Wood powder, flour, cotton, rag fibers, asbestos, powdered metals, graphite, glass, clays, and diatomaceous earth are the major filler materials. Outdoor chair seats, plastic cloth, garbage cans, machine housings, luggage, safety helmets, and equipment parts are products that utilize fillers. Fillers reduce manufacturing costs, minimize shrinkage, improve heat resistance, provide impact strength, or impart other desired properties to the products. *Plasticizers* or solvents are used with some compounds to soften them or to improve their flowability in the mold. Lubricants also improve the molding characteristics of the compound. All these materials are mixed with the granular resins before molding.

Products can be made from plastic resins rapidly and with close dimensional tolerance. Surface finish is excellent. They often substitute for metals when lightness, moisture or corrosion resistance, and dielectric strength are factors. These materials can be either transparent or colored. They are able to absorb vibration and sound. Often they are easier to fabricate than metals, and the final cost is frequently cheaper than metal parts.

The use of plastics is limited because of comparatively low strength, low heat resistance, low dimensional stability, and material costs that may often be higher than other raw materials. Compared to metals they are softer, less ductile, and more susceptible to deformation under load. Plastics under load are *viscoelastic*, meaning that the material has a viscous and elastic response to applied loads. Unlike metals, which fail by *plastic deformation* or slip of the molecules under load, plastics fail because of viscoelastic deformation. When a load is applied on a plastic there is a combination of rapid elastic change or elastic response, and a slow change, which is the viscous response. The viscoelastic deformation is due primarily to the long-chain molecular structure of plastics. Under an applied load, the long chains slide past each other and the amount of movement is determined by the type of bond. Plastics with weak bonds deform more easily than plastics with strong bonds.

Many plastics are flammable. Unlike metals, plastics are low-density materials. These materials have extensive application as heat and electrical insulators because of their low thermal and electrical conductivity. There are many kinds of plastics in commercial production today and they offer variety in physical properties.

Types of Plastics

Plastics are broadly classified as thermosetting and thermoplastic. ***Thermosetting plastics*** are formed to shape with heat, with or without pressure, resulting in a product that is permanently hard. The heat first softens the granules or resins, but as additional heat or special chemicals are added, the plastic is hardened by a chemical change known as polymerization and cannot be resoftened. Processes used for thermosetting plastics include compression or transfer molding, casting, laminating, and impregnating.

Thermoplastic materials undergo no chemical change in molding and do not become permanently hard with the application of pressure and heat. They remain soft at elevated temperature until they are hardened by cooling, and they may be remelted repeatedly by successive applications of heat, as in the melting of paraffin. Thermoplastic materials are processed by injection, blow molding, extrusion, thermoforming, and calendering. A group of parts molded of thermoplastic resins is shown in Figure 10.1.

THERMOSETTING COMPOUNDS
Phenolics

Phenolic resins are popular for thermosetting applications. The synthetic resin, made by the reactions of phenol with formaldehyde, forms a hard, high-strength, durable material that is capable of being molded under a variety of conditions. The material has high heat and water resistance and can be colored in a variety of ways. It is used in manufacturing coating materials, laminated products, grinding wheels, and metal and glass bonding

Figure 10.1
A sample of injection-molded plastic products.

agents, and can be cast into molded cases, bottle caps, knobs, dials, knife handles, electronic appliance cabinets, and numerous electrical parts. Phenolic resins are used to make wood particle chipboards, and in the foundry as a sand bond for cores and molds. Phenolic compounds may be molded by compression or transfer methods.

Amino Resins

The most important resins are urea-formaldehyde and melamine formaldehyde. These thermosetting compounds can be obtained in the form of molding powders or as a liquid solution for bonding and adhesive applications. Both may be compounded with a variety of fillers to improve mechanical or electrical properties. The good flow characteristics of *melamine* make transfer molding useful for such items as tableware, ignition parts, knobs, electric shavers, and houses. Urea resins suitable for compression and transfer molding have a hard surface and high dielectric strength and can be produced in all colors. Products include electric appliance housings, circuit breaker parts, and buttons. Both resins are used in the application of adhesives and for laminating wood or paper.

Furane Resins

Furane resins are obtained by processing with certain acids, waste farm products such as corncobs, rice hulls, and cottonseeds. The products from the thermosetting resins are dark in color, water resistant, and have good electrical properties. Furane resins are used for core sand binders and as hardening additives for gypsum plaster, as well as bonding agents for floor compositions and graphite products.

Epoxides

Epoxy resins are used for casting, laminating, molding, and potting (encasing electrical parts by pouring the solution into a container), and as paint ingredients and adhesives. Cured resins are low in shrinkage; they have good chemical resistance, excellent electrical characteristics, and strong physical properties. They also adhere well to both glass and metal. As adhesives they are employed to replace other forms of fastening. Epoxies are used in manufacturing laminates, and with glass fibers to make panels for printed-circuit boards, tanks, jigs, and dies. Their resistance to wear and impact makes them applicable for metal-forming dies.

Silicones

The silicone polymers differ materially from most other plastics that are based on the carbon atom. They possess a desirable combination of properties for a large group of industrial products such as oils, greases, resins, adhesives, and rubber compounds. Their outstanding properties include stability, resistance to high temperatures over long periods of time, good low-temperature and electrical resistance characteristics, and water repellency. Some oils and greases operate over a temperature range of $-40°$ to $500°F$ ($-40°$ to $260°C$). The silicone resins may be molded, used in laminates or as coatings; alternatively, they may be processed into foam sheets or blocks. Silicone rubbers are used in molding, extrusion, gaskets, electrical encapsulation of electronic components, glass cloth, electrical connectors, or as shock absorption material. Silicones are available as a liquid for casting, as laminating resins, and as a powdered molding compound for foam products. The high cost of silicone products often limits their use to designs where their unusual properties are the most useful. Silicone-based polymers are processed by compression or transfer molding, extrusion, and casting.

THERMOPLASTIC COMPOUNDS

Cellulosics

These thermoplastics are prepared from various treatments of cotton and wood fibers. They are very tough and can be produced in a variety of colors.

Cellulose acetate is a durable compound having considerable mechanical strength. It can be fabricated into sheets or molded by injection, compression, and extrusion. Display packaging, toys, knobs, flashlight cases, bristle material for paintbrushes, radio panels, film for recording tape, and extruded strips are successfully made of this compound.

Cellulose acetate butyrate molding compound is similar to cellulose acetate, and both are produced in all colors and by the same processes. In general, cellulose acetate butyrate is recognized for its low moisture absorption, toughness, dimensional stability under various atmospheric conditions, and ability to be continuously extruded. Typical butyrate products include steering wheels, football helmets, goggle frames, trays, belts, furniture trim, insulation foil, sound tapes, buttons, and extruded tubing for gas and water.

Ethyl cellulose has the lowest density of the cellulose derivatives. In addition to its use as a base for coating materials, it is employed extensively in the various molding processes because of its stability and resistance to alkalies.

Polystyrene

Polystyrene has the outstanding characteristic of *low specific gravity* (1.07); it is available in colors from clear to opaque, is resistant to water and most chemicals, and has dimensional stability and insulating ability. When loosely packed it is very light. It is an excellent rubber substitute for electrical insulation. Styrene resin is molded into battery boxes, dishes, radio parts, lenses, flotation gears, foundry patterns, ice chests, packaging waste, insulated and disposable cups, and wall tile. Polystyrene is especially adapted to injection molding and extrusion, and is formable in dies. Styrenes will dissolve in gasoline.

Polyethylene

These materials are flexible at room and low temperatures, waterproof, unaffected by most chemicals, and capable of being heat sealed. They can be produced in a variety of colors. Polyethylene, which floats on water, has specific gravity from 0.91 to 0.96. It is one of the inexpensive plastics, and its moisture-resistant characteristics ensure its use for packaging and squeeze bottles. Other products are ice-cube trays, developing trays, fabrics, film for packaging, collapsible nursing bottles, garden hose, coaxial cable, and insulating parts for high-frequency electrical fields. Polyethylene products can be made by injection molding, blow molding, or extruding into sheets, film, and monofilaments.

Polypropylene

Parts can be produced from this material by all thermoplastic molding techniques. It has excellent electrical properties, high impact and tensile strength, and is resistant to heat and chemicals. Monofilaments of polypropylene are used in making rope, fishing line, nets, and textiles. Other products are hospital and laboratory ware, toys, luggage, furniture, film for food packaging, television cabinets, and electrical insulation.

ABS

The initials ABS stand for the combination of acrylonitrile, butadiene, and styrene, which can be compounded to differing hardnesses, flexibilities, and toughnesses. The ABS plastics are used in applications that require abuse resistance, colorability, hardness, electric- and moisture-resistant properties, and limited heat resistance (220°F, 105°C). These plastics are processed by thermoforming and by injection, flow, rotational, and extrusion molding. Applications include household piping, cameras, electrical hand tool housings, telephone handsets, and canoes.

Polyimide

These thermoplastics are produced in the form of solids, films, or solutions, and they have heat resistance to 750°F (400°C), low coefficient of friction, high degree of radiation resistance, and good electrical properties. Products include sleeve bearings, valve seats, tubing, and various electrical components. The films, tough and strong, are used for wire insulation, motor insulation, and printed-circuit backing. The solutions are used in varnishes, wire enamels and coated-glass fabrics. Another grade is the nylons, which are molded and extruded and are employed in the textile fiber and filament field. Molded and extruded products of *nylon* include bearings, gears, valves, tubing, kitchen accessories, and luggage. Nylon monofilaments are used for hosiery, glider tow ropes, and brush bristles.

Acrylic Resin

Acrylic resin has good light-transmitting power, is easy to fabricate, and is moisture resistant. The acrylic resin commonly used is methyl methacrylate, better known by the commercial names *Lucite* or *Plexiglas*. It can be formed by casting, extruding, molding, or stretch-forming for airplane windows, shower doors, toilet articles, and covers where visibility is desirable.

Vinyl Resins

The vinyl resins commercially available include polyvinyl chlorides, polyvinyl butyrates, and polyvinylidene chloride. These thermoplastic materials can be processed by compression or injection molding, extrusion, or blow molding. Vinyl resins are especially suitable for surface coating and for both flexible and rigid sheeting. *Polyvinyl butyrate* is a clear tough resin, used for interlayers in safety glass, raincoats, sealing fuel tanks, and flexible molded products. It has moisture resistance, great adhesiveness, and light and heat stability. *Polyvinyl chloride* has a high degree of resistance to many solvents and will not support combustion. Industrially it is used for rubberlike products, including raincoats, packaging, and blow-molded bottles. *Polyvinylidene chloride* is used for plastic film wraps and pipe. Cellular vinyl foamed products include floats, upholstery, and protective pads for sport uniforms.

Synthetic Rubber

Natural rubber has been synthesized for many years. Many of the industrialized nations had no source of natural rubber. This led to the development that produced many synthetics such as GR-S, nitrile, thiokol, neoprene, butyl, and silicone rubbers. The **synthetic rubber** GR-S is produced in the largest quantity, being particularly adapted for tire use. Similar to natural rubber, it can be substituted in most instances. It is a copolymer of butadiene and styrene and can be cured to any degree of rubber hardness. The strength of GR-S is improved by adding carbon black, and for tire use it is frequently compounded with natural rubber. The butadiene–acrylonitrile copolymers (known as Buna N or nitrile rubbers) are employed principally because of their resistance to oils and are used in oil hose, gaskets, and diaphragms. They also serve as a blending material with phenolics and vinyl plastics. The organic polysulfides, known as *thiokols,* are very resistant to gasoline, oils, and paints as well as to sunlight, and are used in manufacturing hose, shoe heels and soles, coated fabrics, and insulation coatings. Resilient solid objects can be molded in conventional machines used for other plastics.

The chloroprene polymer known as neoprene is produced from coal, limestone, water, and salt. Calcium carbide, a product of coal and limestone, when added to water forms acetylene gas (C_2H_2). This gas, when combined with hydrogen chloride, forms chloroprene, which is changed to neoprene by polymerization. Neoprene has good resistance to oils, heat, and sunlight, and is used for conveyor belts, shoe soles, protective clothing, insulation, hose, printing rolls, tires, tubes, and as a bonding material for abrasive wheels. It has a wider application than other synthetic rubbers and can replace natural rubber. *Butyl,* an isobutylene copolymer, has many of the properties and characteristics of natural rubber. Because of its strength, resistance to abrasion, and low permeability to gases, it is used for inner tubes. Other applications include steam hose, conveyor belting for heated materials, and tank linings. Silicone (*polysiloxane*) rubbers are extremely resistant to both high and low temperatures as well as to lubricating oils, dilute acids, and sunlight. They are used in O-rings and seals for oil and gas lines, sealing doors on airplanes, and wire and cable insulation. Other commercial synthetic rubbers are *polybutadiene,* also used for tires, and *polyacrylate* for oil hose and automotive gaskets. *Urethane* elastomers serve as shock-absorbing pads, forming pads in press work, conveyor rolls, and solid tires.

PROCESSING

The plastics industry consists of, first, the manufacturers who produce the resins and chemicals, and second, the fabricators. The plastic material manufacturers make raw material that includes powders, granules, liquids, and standard forms such as sheets, bars, tubes, flats, rolls, and laminates. The fabricators usually finish the product ready for distribution to industries or the consumer.

Additives

One or more additives, as noted in the following paragraphs, are added to resins for the reasons noted.

Fillers are added to reduce cost and/or increase strength. They include wood flour, quartz, limestone, cotton, rag fibers, powdered metals, graphite, clays, and diatomaceous earth.

Reinforcements are added for physical strength. They include sisal, jute, glass, graphite whiskers, ceramics, and nylon, cotton, and orlon fabrics.

Flame retardants are added to impair burning of the finished product. The principal flame retardant is a phosphate ester.

Stabilizers are added to keep a plastic from oxidizing or degrading during use. Zinc soap is added to vinyls and phenols to the styrenes.

Antistatics are added to prevent a buildup of an electric charge on powders during processing when required amines are added.

Colorants are added to improve consumer acceptance of the product. Colors are added either by mulling with the resin or by metering liquid coloring into the plastic in some injection-molding and extrusion machines. Colors are available as organic and inorganic pigments and dyes as a solid or liquid. Many resins are precolored.

Lubricants are added to improve processing by promoting better flow in molding. They include waxes, zinc, and calcium stearates.

Plasticizers are added to improve flexibility of the final product. Vinyl is very brittle unless phthalate is added.

Ultraviolet protector is added to vinyls, styrenes, polyesters, and fiberglass to improve their life span when subjected to sunshine. Carbon black is one such additive.

Compounding and Preforming

Most industrial products demand a compromise of propertics that require certain ingredients before molding. Much thermoplastic material is purchased as granular material and hence is compounded dry. Thermosetting material, on the other hand, may be purchased as a liquid or as a partially polymerized compound.

Usually it is necessary to mix and prepare the raw materials for a final product. This step is referred to as compounding or preforming. The compounding process is normally carried out in a muller into which any number of additives are mixed. Materials that have been mixed and sometimes melted are placed into the feed hoppers of injection, extrusion, or calendering machines. Some thermoplastic materials are preformed into small pellets of the proper size and shape for a given mold cavity. All preforms are of the same density and weight, and the operation avoids waste of material in loading molds and speeds up production with no possibility of overloading the molds. In the preforming operation the thermosetting powder is cold-molded, and no curing takes place. Preforms are used in compression- and transfer-molding processes.

A *rotary preforming* press used in making disk pellets of various molding compounds is shown in Figure 10.2. The powder is fed by gravity into the mold cells, and excess powder is scraped off. The amount of material fed into each cavity is controlled by regulating the lower punch. As the table revolves, pressure is applied uniformly on both sides compressing the powder charge, and at the end of the cycle the tablet is ejected. Reciprocating machines, which differ from rotary machines, may have more than one cavity, and more than one preform is ejected with each cycle. A formula for the production rate of a reciprocating machine is

$$\text{Minutes per preform} = \frac{1}{(\text{spm} \times \text{no. of cavities in die})}$$

where spm = strokes per minute.

Compression Molding

Compression molding is used mostly for thermosetting material. The material, usually in powder or granular form, is placed in a heated die; the upper half of the die compresses the material, which melts and fills the die cavity. After compression the part solidifies (polymerizes or *cures*), the upper half of the die retracts or opens, and the part is removed. The molding process is illustrated by Figure 10.3. The pressure causes the liquid material to fill and conform to the die shape. Pressures used in compression molding vary from 100 to 8000 psi (0.7–55 MPa), depending on the nature of the material and the size of

Figure 10.2
Rotary preforming press used for making disk pellets of various molding compounds.

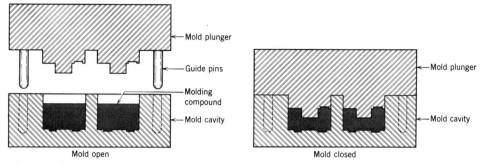

Figure 10.3
The compression-molding process.

product. The temperature range is 250° to 400°F (120°–205°C). Heat is very important for thermosetting resins, because it is used first to plasticize and then to polymerize or make the product permanently hard.

Some thermoplastic materials are processed by compression, but the cycle of rapid heating and cooling of the mold adds to the difficulty in using such material. Unless the mold is sufficiently cooled before ejection, distortion of the piece is likely.

Hydraulic presses are commonly used, although hand-operated ones are available. Automatic presses are available where robots can be adapted to remove the part. Ejector pins force the part from the die. Heat may be transferred from heated platens or applied directly to the metal mold. The heat is supplied by steam, heated liquids, electrical resistance, or ultrahigh-frequency electric currents.

Transfer Molding

This process is used mostly with thermosets. The method is like compression molding in that heat and pressure are used. In *transfer molding,* when the powder or granules are in a semiliquid state, the plastic is forced or transferred to the die cavity through a sprue. After curing the plastic in the mold, the mold is opened and the part is removed. The curing time for transfer molding is usually less than for compression molding. The loading time is also shortened. The method is adaptable for parts requiring small metal inserts. Intricate parts and those having large variations in section thickness can also be produced to advantage by this method. Figure 10.4 is a sectional view of the transfer mold before and after the transfer of the plastic. Losses caused by the runner, sprue, and well are limitations of the method. Die costs may be greater than for compression molding.

Injection Molding of Thermoplastics

Injection-molding machines are somewhat similar to those used for die casting. Thermoplastic material is converted from a granular material to a liquid and then injected into a mold where it solidifies. This material can be repeatedly changed from solid to liquid without chemical change, making it ideal for rapid processing.

224 PLASTIC MATERIALS AND PROCESSES

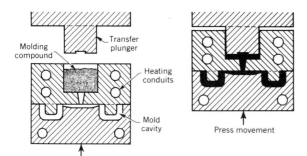

Figure 10.4
The transfer-molding process.

Injecting-molding machines are specified by the tonnage with which the dies may be clamped and the amount of material injected per cycle. Most machines of this type have a 50- to 2500-ton (0.4- to 22-MN) clamping force, and the shot capacity varies from less than an ounce to about 300 oz (9 kg). The machine shown in Figure 10.5 is a 2500-ton (22-MN) hydraulic clamp machine capable of molding 300 oz (9 kg) per cycle. The plastic is plasticized up to 400 lb/h (0.05 kg/s) in the machine before being injected at rates up to 5000 in.^{3}/min (0.081 m^3/s). Figure 10.6 is a schematic sketch showing the operation of injection-molding machines.

The basic steps in *injection molding* are as follows: first, the granular plastic material is loaded into a heating cylinder. The material is then compressed and air is forced out by a ram. Next, the material is transferred to the heated section by the ram where it melts. The moving mold part is now matched to the fixed mold part by a locking device

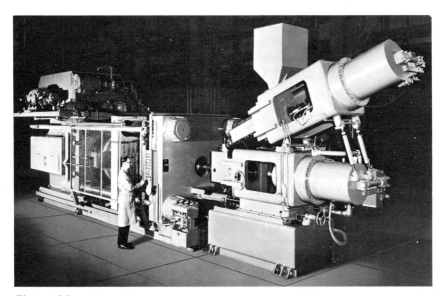

Figure 10.5
Injection-molding machine for plastics.

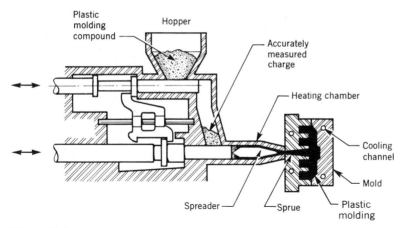

Figure 10.6
Sketch of an injection-molding machine.

on the machine. The ram pressure injects the softened material through the nozzle of the machine, displacing the air content of the mold cavity. The plastic in contact with the cold sidewalls of the mold stiffens first in the gate, sealing the cavity. As cooling takes place, the mass is hardened, permitting the removal of the finished part. After cooling, the mold is opened and the part is removed by the ejector system.

The most important factors of injection molding are the outer and inner pressure and temperature of the material and mold. See Figure 10.7, which is a pressure–time diagram of injection molding. The outer pressure created by the ram pushing the plastic into the mold ensures that the entire cavity is filled. The inner pressure originating in the heating cylinder keeps the product from decreasing in size as it cools. The temperature of the material determines the viscosity of the liquid plastic. A high-temperature, low-viscosity material fills a highly detailed mold quite easily. The temperature of the mold itself affects the time needed for cooling; the cooler the mold, the quicker the material hardens. Most

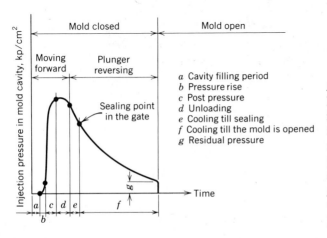

Figure 10.7
Pressure–time diagram of injection molding. *a*, Cavity filling period; *b*, pressure rises; *c*, post-pressure period; *d*, unloading; *e*, cooling till sealing; *f*, cooling till mold is opened; *g*, residual pressure.

226 PLASTIC MATERIALS AND PROCESSES

molds are maintained between 165° and 200°F (74°–93°C) by circulating water; otherwise, after a short time the temperature would be so high that the mold would never harden. The different combinations of these four basic factors determine the type of part produced.

The heating-chamber construction of most injection machines is cylindrical in shape with a *torpedo-like spreader* in the center, so that the incoming material is kept in a layer thin enough to be heated both uniformly and rapidly. The heating-chamber temperature ranges from 250° to 500°F (120°–260°C) depending on the material being charged and mold size. Heat is furnished by a series of electrical-resistance coils. These chambers must be of substantial construction, as injection pressures may reach 30,000 psi (200 MPa).

Also employed in injection molding of many thermoplastics is the in-line reciprocating screw machine. Material is fed from a hopper to the rotating screw, which discharges the material into the front of the extruder barrel. Heat is obtained from the electric band heaters about the screw cylinder and from the frictional forces developed in the material by the rotating screw. The screw rotates until a sufficient amount of material has been plasticized and a limit switch contacted. It then moves forward and injects the plasticized material into the mold, where it remains long enough to solidify. A unidirectional valve prevents any return of the material into the extruder barrel.

Injection molding is faster than compression molding. A production cycle of two to six shots per minute is possible. Mold costs are lower because fewer cavities are necessary to maintain equivalent production. Articles of difficult shape and of thin walls are successfully produced, as illustrated in Figure 10.8. Metal inserts such as bearings, contacts, or screws can be placed in the mold and cast integrally with the product. Material loss in the process is low, as sprues and gates can be reused. Although the capacities of injection machines vary from 2 oz to 8 lb (0.22–3.6 kg), small-parts machines of 8 to 16 oz (0.22–0.45 kg) capacity are most popular.

Figure 10.8
Mold for making berry baskets in an injection machine.

Injection Molding of Thermosets

Thermosetting materials can be injection-molded by a process known as jet molding. With a few changes many of the standard thermoplastic machines can be converted for jet molding. The nozzle, which is the important part of the machine, must be both heated and cooled during the molding cycle. The resin is first heated in the cylinder surrounding the plunger, making it plastic, though not appreciably polymerized. As the plunger forces the resin through the nozzle to the mold, additional heat is applied. When the mold is full the nozzle is rapidly cooled by water to prevent further polymerization.

An alternate to jet molding is the reciprocating screw injection machine (also used for thermoplastic materials) shown in Figure 10.9. Material is fed by gravity to a rotating screw where it is heated by contact with the heated barrel and the frictional heat developed by the rotating screw. As the screw revolves, plasticized material is built up ahead of the screw, being blocked from entering the transfer chamber by the *transfer ram* in upper position until a sufficient amount is accumulated. The ram then returns to the lowered position and the screw, not revolving on the forward stroke, forces the material into the transfer chamber where the plunger pushes it upward into the mold cavities. Precuring of the material is prevented by a water-cooled band around the end of the cylinder. This process is similar to transfer molding except that it is automatic in operation.

Extruding

Thermoplastic materials such as the cellulose derivatives, vinyl resins, polystyrene, polyethylene, polypropylene, and nylon may be extruded through dies into simple shapes of

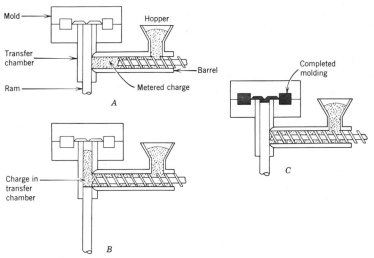

Figure 10.9
Screw injection molding cycle. A, Screw retracts while revolving as molding material feeds into barrel by gravity. B, The screw while not revolving forces material into the vertical plunger chamber. C, Hydraulic plunger forces plasticized material into mold.

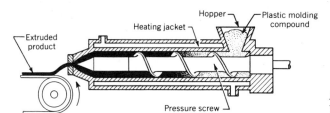

Figure 10.10
Plastic extrusion press.

any length. A schematic diagram of an *extruding* press is shown in Figure 10.10. Granular or powdered material is fed into a hopper and forced through a heated chamber by a spiral screw. In the chamber the material becomes a thick viscous mass, in which form it is forced through the die. As it leaves the die it is cooled by air, by water, or by contact with a chilled surface and fully hardens as it rests on the conveyor. Long tubes, rods, and many special sections are readily produced in this manner. Because thermoplastic extrusions can be bent or curved to various shapes after extrusion by immersion in hot water, such products as conduits for electric conductors and chemicals are made by this process. Thermosetting compounds are not well adapted to this type of extrusion because they harden too rapidly; however, they are used to a limited extent in the production of thick-walled tubes.

A machine for the extrusion of thermosets utilizes a ram instead of a screw to force the material through the die. Material is fed from a hopper at the rear of the cylinder and by repeated strokes of the ram is forced into a long, tapered die that has heated zones. Additional heat results from frictional resistance as the material is forced through the cylinder and die. Curing is complete as it reaches the forward end. Products include tubes, rods, moldings, bearings, brake linings, and gears. Cross-sectional tolerances of 0.005 in./in. (0.4%) can be maintained.

A process known as extrusion coating is used extensively for coating paper, fabrics, and metal foil. A thermoplastic material is extruded through a flat die (Figure 10.11) onto a sheet passing beneath it. The *extrudate*, while soft, blends onto the *substrate* and is contacted by a rubber roll that holds it against the steel roll at a desired pressure. The edges of the sheet are trimmed prior to the windup. Although any thermoplastic material can be extruded as a coating, the ones used most are the vinyls, polyethylene, and

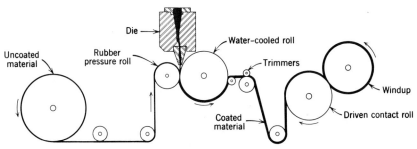

Figure 10.11
Extrusion coating process.

polypropylene. Another important extrusion coating process is that of insulation on wire and cable.

Rotational Molding

Rotational molding employs the simultaneous rotation of thin-walled molds about two axes, primary and secondary, which are perpendicular to one another. After charging with appropriate plastic material, the molds are heated while in rotation, causing the particles to melt on the inner surface of the mold, depositing in layers until all the material is fused. The molds are cooled while still rotating and opened so that the finished article can be removed and the molds recharged. The process is intended primarily for hollow objects from thermoplastic materials. The toy industry uses *plastisols* and polyethylene with rotational molding to make squeeze toys.

The rotational powder method differs from other molding processes in that while the others all require both heat and pressure to plasticize the resin, rotational powder molding requires only that the mold be heated.

Although thin cast-aluminum molds are normally used in rotational molding, electroformed copper or sheet metal is also satisfactory. The mold sections must fit tightly together so that no moisture enters the mold to cause warping. The rotational speeds of the two mold axes are generally controlled by separate motors; normally, a ratio of 3:1 exists between the major and minor axes. The rotational speed of the major axis is generally under 18 rpm, whereas mold temperature ranges from 500° to 700°F (260°–370°C).

The principle of *rotational molding* is schematically shown in Figure 10.12. In one case a single mold is shown; in the other, four molds are assembled on a single-arm unit. In both versions the arm is pivoted so that it can be swung into a heating oven, after which it can be directed to a cooling chamber as shown in Figure 10.13. Some designs have the motors and drive spindles mounted on a track, which permits them to move from the oven to the cooling chamber and the unloading station.

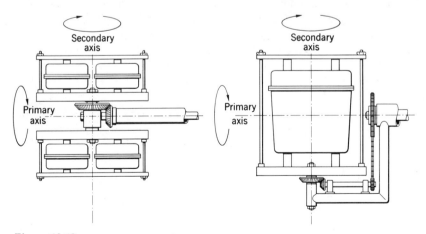

Figure 10.12
Schematic of rotational molding showing two mold mounting systems.

230 PLASTIC MATERIALS AND PROCESSES

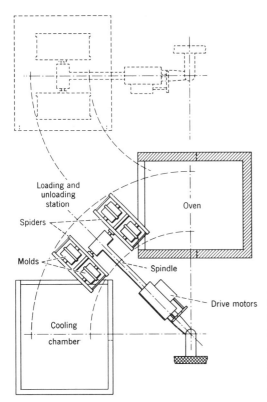

Figure 10.13 Schematic of pivoting-arm mold moving system. Spindle and molds swing in a 90° arc between heating and cooling chambers. Addition of a second cooling chamber, pivot, and spindle (dotted lines) decreases molding cycle time.

Advantages claimed for rotational molding include low initial investment, a flexibility that allows a variety of parts to be made on the same equipment, low tooling costs, totally enclosed and open-end pieces, fine detail, excellent surface finish, and low cost per unit produced. Products made by powder rotational molding are often of considerable size, such as children's chairs, 55-gal (0.2-m^3) drums for food storage, phonograph cases, machinery guards, garbage containers, and gasoline tanks. The same equipment can be used for either thermoplastic powder or plastisol molding.

Potting and Embedding

It is sometimes desirable to protect electrical or mechanical parts in blocks of plastic. *Potting* refers to the insulation and positioning of such parts, whereas embedding (*encapsulation*) is the enclosing of parts in a transparent plastic for preservation and display. These procedures are usually accomplished by casting or dip coating with a liquid material, but other processing methods such as transfer molding can be used.

Blow Molding

Blow molding is used primarily to reproduce thin-walled hollow containers from thermoplastic resins. A cylinder of plastic material, known as a *parison,* is extruded as rapidly

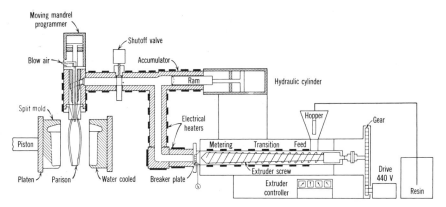

Figure 10.14
Typical blow-molding machine.

as possible and positioned between the jaws of a split mold, as shown in Figure 10.14. As the mold is closed it pinches off the parison and the product is completed by air pressure forcing the material against the mold surfaces. Molds should be adequately vented to eliminate poor surface finish. As soon as the product is cooled sufficiently to prevent distortion, the mold opens and the product is removed. The entire operation is similar to that used for forming bottles in the glass industry.

Figure 10.15 illustrates an eight-station machine for continuously blowing bottles by the *pinch-tube* process. A tube of thermoplastic material is extruded from a plasticizer into an open mold. Each end of the plastic tube is pinched shut by the closing of the mold, and air pressure is fed into the hollow tube by a core tube in the crosshead of the mold. The air pressure expands the plastic to conform with the walls of the mold. After a short cooling cycle, during which air pressure is maintained, the pressure is released,

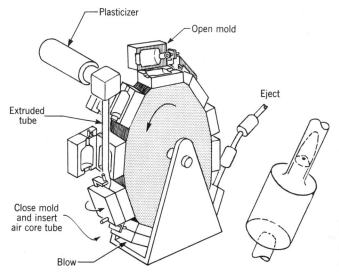

Figure 10.15
Continuous-tube process for making plastic containers.

the mold opens, the bottle is ejected, and the mold is made ready to begin the cycle again. For some plastics the bottle must be cooled to room temperature by a water spray. The top and bottom of the bottle must be trimmed to remove the scrap, but no further processing is necessary. This process is continuously repeated for each of the eight mold stations.

Blow-molded products include cosmetic packaging, bottles, floats, automobile heater ducts, liquid detergent containers, and hot-water bottles. Polyethylene, polypropylene, and cellulose acetate are some of the plastics that can be formed by blowing.

Film and Sheet Forming

The basic methods for producing film or thin sheets are calendering, extruding, blowing, and casting. The one chosen depends on the type of thermoplastic resin selected, which in turn governs the required properties of the product. *Calendering* is the formation of a thin sheet by squeezing a thermoplastic material between rolls (Figure 10.16). The material, composed of resin, plasticizer, filler, and color pigments, is compounded and heated before being fed into the calender. The thickness of the sheet produced depends on the spacing between the rollers that stretch the plastic. Before the film is wound it passes through water-cooled rolls. Vinyl, polyethylene, cellulose acetate films and sheeting, and vinyl floor tile are products of calendering. The same process is used for rolling out uncured rubber stock in tire manufacture.

In making sheets of polypropylene, polyethylene, polystyrene, or ABS, an extrusion process is used. Figure 10.17 is a schematic diagram of this process. After the material has been compounded, it is placed in the feed hopper. The material is heated to not over 600°F (315°C) and forced into the die area at pressures of 2000 to 4000 psi (14–28 MPa) by the screw conveyor. By the combination of the choker bar and die opening, the thickness of the sheet is controlled. After extrusion the sheet passes through oil- or water-cooled chromium-plated rolls before being cut to size. Oil cooling is recommended, because the temperatures should be maintained at approximately 250°F (120°C) for proper curing. Sheet material made in this manner can vary in thickness from 0.001 to 0.125 in. (0.03–3.18 mm).

Blown tubular extrusion produces film by first extruding a tube vertically through a ring die and then blowing it with air into a large-diameter cylinder. The blown cylinder

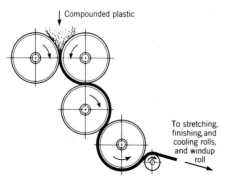

Figure 10.16
Forming film by the calendering process.

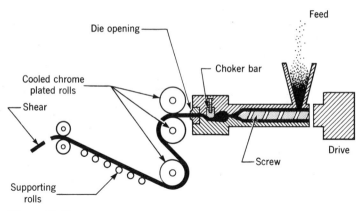

Figure 10.17
Extrusion of thin sheets and film.

is air-cooled as it rises vertically and is ultimately flattened by driven rolls before it reaches the winder. This process permits the extrusion of thin film used for items such as trash bags and packaging materials.

In *film casting* the plastic resins are dissolved in a solvent, spread on a polished continuous belt or a large drum, and conveyed through an oven where they are cured and the solvent is removed. In *cell casting* a cell is made up of two sheets of polished glass, separated according to the sheet thickness desired, and gasketed about the edges to contain the liquid catalyzed monomer. The cell is then raised to the proper temperature in an oven where it remains until curing takes place. Cell casting is used in the production of most acrylic transparent sheet.

Thermoforming

Thermoforming consists of heating a thermoplastic sheet until it softens and then forcing it to conform to some mold by either differential air pressure or mechanical means. Many techniques have been developed for applying pressure to the sheet, such as free-forming by differential pressure (either pressure or *vacuum forming*), *vacuum snapback* forming, *vacuum drawing* or blowing into a mold, *drape* forming, *plug assist* vacuum or pressure forming, and matched-mold forming. Several techniques are shown schematically in Figure 10.18.

Figure 10.18A illustrates what is generally known as *free forming*, a technique that uses air pressure differential and where no male or female form is required. The drawn or blown section retains its shape on cooling. A somewhat similar process known as vacuum snapback forming is illustrated in Figure 10.18B. After the heated sheet is clamped, a vacuum is created in the chamber, which causes the sheet to be drawn down as shown by the dotted lines. The male mold is then introduced into the formed sheet, and the vacuum is gradually reduced causing the sheet to snap back against the mold form. A setup in which sheets are formed to shape by air pressure and are actually blown into the mold is shown in Figure 10.18C. This process is used for more complicated

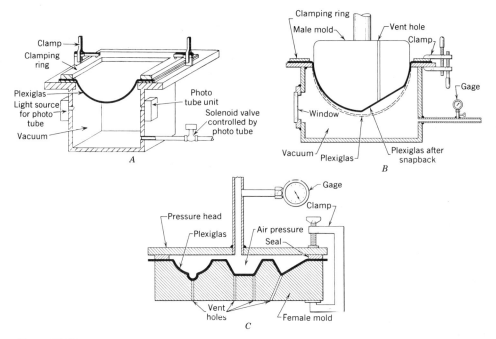

Figure 10.18
Forming methods used for heated thermoplastic sheets. *A*, Free forming. *B*, Vacuum snapback forming. *C*, Positive-pressure molding.

shapes where possible surface defects are not objectionable. By using special synthetic greases in the mold, the tendency for marks to show on the formed part is decreased.

In *drape forming* the plastic sheet is clamped and then drawn over the mold, or the mold is forced into the sheet. Plug assist forming first heats and seals the sheet over the mold cavity. A plug somewhat smaller than the mold form pushes the plastic sheet into a near-bottom position. Vacuum or air pressure is applied to complete formation of the sheet. *Matched-mold* forming is the same as the forming of sheet metal in dies. Molds are made of wood, plaster, metal, or plastics. This technique requires care to avoid marring the sheet surface.

Reinforced Plastics

Reinforced plastics include a wide range of products made from thermosetting resins with random or woven fibers. Although glass fibers predominate, asbestos, cotton, graphite, and synthetic fibers are also used. Polyester resins are low in cost and their properties are good. Epoxies provide extra strength and chemical resistance, whereas silicones find use where heat resistance and electrical properties are important. Other resins are available for special properties and applications.

Fiberglass and other reinforced plastics are made by various processes, but in general all are classified as open and closed molding. The *open-mold* process with a single-cavity

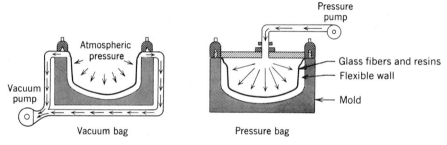

Figure 10.19
Examples of producing reinforced plastic product by the open-mold process.

mold, either male or female, makes a product with little or no pressure. Fiberglass boat bodies are a good example, as the process adapts well to fabricating large objects where only one side is finished. First, the finish coat of paint is sprayed in the mold, and then glass fibers and resin are placed into the mold manually and rolled to compress and remove air. Such molds normally cure in air, but either a vacuum or pressure bag can be used against the layup to provide additional pressure (Figure 10.19). For still more pressure, the assembly can be placed in a steam autoclave at pressures up to 100 psi (0.7 MPa). Other products of the open-mold process include aircraft parts, luggage, truck and bus components, and large containers.

The *closed-mold* or matched-die process uses two-part molds usually made of metal. Both sides are finished and good detail is obtained. The labor cost is low and, because the molds are heated, a high production rate is possible. Products obtained from this process include luggage, helmets, trays, and machinery housing. In general, small-sized products are made by this process because of the high cost of closed molds.

Several other techniques find commercial use in the manufacture of reinforced plastics, one of which is shown in Figure 10.20. In the *sprayup* process, fiberglass and resin are simultaneously deposited in a mold by spray guns. Boats and other large objects are fabricated in this manner. In *filament winding* single strands of fiber are fed through a bath of resin and wound on a mandrel as illustrated in Figure 10.21. This process is used in making pressure cases, tubing, and missile bodies where high strength is a requirement.

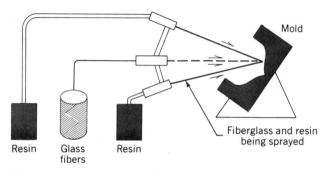

Figure 10.20
Fiberglass and resin being simultaneously deposited in mold by spray guns.

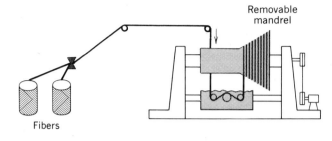

Figure 10.21
Producing high-strength products by filament winding.

Laminated Plastics

Laminated plastics consist of sheets of paper, fabric, asbestos, wood, or similar materials that are first impregnated or coated with resin and then combined under heat and pressure to form commercial materials. These materials, which are hard, strong, impact resistant, and unaffected by heat or water, have desirable properties for numerous electrical applications. The final product may consist of either a few sheets or over a hundred, depending on the required thickness and properties. Although most laminated stock is made in sheet form, rods and tubes as well as special shapes are available. The material has good machining characteristics, which permit its fabrication into gears, handles, bushings, and furniture.

In manufacturing laminated products, the resinoid material is dissolved by a solvent to convert it into a liquid varnish. Rolls of paper or fabric are then passed through a bath for impregnation (Figure 10.22). To facilitate lamination the sheets are then cut into convenient sizes and stacked together in numbers sufficient to make up the thickness of the final sheet. Tubes made by machine-winding strips of the prepared stock around a steel mandrel either are cured by being placed in a circulating hot-air oven or are subjected to both heat and pressure in a tube mold. Paper base laminates are used in electrical products. Fabric base materials, which are stronger and tougher, are better for stressed parts. Gears made of a canvas base are quiet in operation. Fiberglass cloth is recommended for heat-resisting and low water absorption uses. Thin sheets of wood are laminated to produce a light material equal in strength to some metals and resistant to moisture. *Safety glass* is in effect a laminated plastic product, because thermoplastic layers are used between the glass sheets to make it nonshattering. The four resins used most are phenolics, silicones, epoxies, and melamines.

Casting

Thermosetting materials used for *casting* include the phenolics, the polyesters, epoxies, and the allyl resins. The last is especially useful for optical lenses and other applications

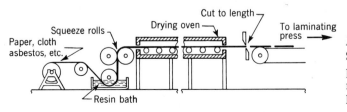

Figure 10.22
Sketch of process for preparing sheet material for lamination.

requiring excellent clarity. These resins have a wider use in casting than the thermoplastics because they have greater fluidity in pouring. Ethyl cellulose and cellulose acetate butyrate, both thermoplastics, are used where impact strength and rigidity are needed for drop hammer and stretch dies. Acrylics are used in casting transparent articles and flat sheets.

Plastics are cast when the number of parts desired is not sufficient to justify making expensive dies. Frequently, open molds of lead are formed by dipping a steel mandrel of special shape into molten lead and stripping the shell from the sides of the mandrel after it solidifies. Cores of lead, plaster, or rubber may be introduced if desired. Hollow castings are produced by the slush-casting method. Solid objects may be made from molds of plaster, glass, wood, or metal. When parts have numerous undercuts the molds are made of synthetic rubber.

MOLDS FOR PLASTICS

Molds for both the compression and injection processes are made of heat-treated steel. The production of these molds demands the same type of machine work and the usual precision required for die casting. There are some differences in construction because of varying characteristics in the materials being processed. Ample draft and fillets should be provided to facilitate removal of the article from the mold. Ejector pins usually provided for this purpose should be located at points where the pin marks are not noticeable. Like metals, plastic materials shrink on cooling and some allowance must be provided. Shrinkage varies according to the type of material and method of processing, but is usually 0.003 to 0.009 in./in. (0.3%–0.9%).

Compression molds are made in hand and semiautomatic types with **positive,** semipositive, and flash designs. The hand molds are charged and unloaded on a bench. Heating and cooling are accomplished by plates on the presses, which are provided with the necessary circulating facilities. The semiautomatic molds are fastened rigidly to the presses and are heated or cooled by adjacent plates. As they open, work is ejected automatically from the molds, which are single or multiple cavity.

Injection molds are made in two pieces, one half being fastened to the fixed platen and the other half to the movable platen. Contact between the halves is made on accurately ground surfaces or lands surrounding the mold cavities. Neither half telescopes with the other, which is the same with many of the compression molds. The cavities should be centrally located with reference to the sprue hole in the fixed half, so as to obtain even distribution of material and pressure in the mold. For locating purposes guide pins, which are similar to those employed on metal press dies, are fastened in the fixed half of the mold and enter hardened bushings in the movable part of the mold.

Mold *cavities* can extend into both halves of the mold. It is best, however, to have the outside of the molded part in the fixed half, provided the shape is suitable for this arrangement. In the cooling process the plastic material tends to shrink away from the cavity walls and is withdrawn from this half as the mold opens. It is retained on the cores of the movable half until the ejector mechanism operates.

Injection molds have cooling channels in both halves to permit the maintenance of a uniform temperature for chilling the molded part, because most materials fabricated by this process are thermoplastic. The material is forced into the mold from the heated

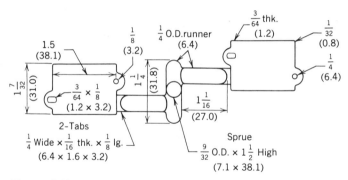

Figure 10.23
Typical molded part showing sprue, runners, and tabs.

cylinder under pressures ranging from 2 to 20 tons/in^2 (30–275 MPa) and is ejected from the mold at a temperature of approximately 125°F (50°C). Ejection of the parts occurs as the mold opens by either ejector pins or **stripper plates**.

Any cores required in injection molding are placed on the movable half of the mold. The normal shrinkage of the molded part tends to cling to the cores, causing it to withdraw freely from the stationary half as the mold opens. Vents to permit the escape of entrapped air are extremely small and are located in such a way as to permit all air to escape quickly.

Molds must be carefully designed to minimize waste and losses of materials. Figure 10.23 is a two-part design that uses a central sprue, runners, and tabs. The finished parts are snapped off the tab during a subsequent cycle of the injection-molding machine, but the sprue, runners, and tabs are essentially lost to the cycle even though a salvage operation can regrind the material for use as a lesser grade.

QUESTIONS AND PROBLEMS

1. Give an explanation of the following terms:

   Thermosetting
   Thermoplastic
   Polystyrene
   Synthetic rubber
   Rotary preforming
   Compression molding
   Torpedo spreader
   Extruding
   Rotational molding
   Potting
   Calendering
   Film casting
   Drape forming
   Filament winding
   Positive mold
   Stripper plates

2. Describe the process of rotational molding and give its advantages.

3. How are the gates and sprues of thermoplastic materials salvaged? Thermosetting materials?

4. Are furane resins organic or inorganic materials? Why?

5. What is meant by polymerization?

6. List the processes used in forming plastics. Give the type of plastic that may be formed in each.

7. How are plastics compounded?

8. What is the advantage of a preform? List its applications.

9. Why is it difficult to process thermoplastic materials by compression molding?

10. Describe the process of compression molding.

11. What advantages does transfer molding have over compression molding?

12. What type of plastic molding is similar to die casting?

13. List the plastic materials generally used in injection molding. What properties make them desirable?

14. Give the purpose in cooling the nozzle in jet molding.

15. Name five products that can be made by extrusion.

16. What processes do you recommend for producing the following products: boat hulls, squeeze toys, garbage containers, film packaging for food, and radio cabinets?

17. Give the purposes for laminated plastics.

18. Describe the vacuum snapback method of forming.

19. Can thermosetting compounds be easily extruded into tubes and rods? Why?

20. How are plastic bottles made?

21. List the reasons that make the pinch-tube method in bottle production attractive.

22. Select a product and construct a flowchart using the calendering process.

23. Give the difference between positive- and flash-type molds.

24. How are plastic molds vented?

25. List the advantages of plastic parts over metal parts, and vice versa.

26. How does film casting differ from the calendering process?

27. What allowances for plastic molds are similar to those for sand castings?

28. Why are gears sometimes made from plastic having a canvas base or filler?

29. How much plastic needs to be plasticized each 24-h day for a 30-oz. (0.85-kg) machine making 2 cycles per minute?

30. If a 1-kg machine has 40 cycles per minute, how many kilograms of plastic will be plasticized each 8-h shift?

31. Convert the following press sizes to SI units: 300 ton, 3000 ton, 30 ton, and 3 ton.

32. Convert the following to SI units: 3 oz, 30 oz, 4 lb, and 20 lb.

33. List the properties of silicone-based polymers that make them desirable as engineering materials.

34. List some plastics that are used as manufacturing materials.

35. Which plastic materials are used in electrical wire insulation, printed-circuit boards, knobs, and chassis? What are the principal properties that enter into your selection?

36. A thermoplastic material having low density and high insulation qualities is required for a design. Suggest several candidate materials.

37. What temperature–pressure combination is required to manufacture a thin-walled, highly intricate molding?

38. What temperature–pressure combinations would be best for rapid manufacture of thick, undetailed objects such as dog food dishes?

39. Find the time in minutes per unit for the production of preforms. Each die has four cavities and the machine is rated at 40 spm.

40. A reciprocating preform production machine is rated at 50 spm. The die unloads four preforms with each stroke. One operator tends three machines that are similar. A part uses 60 g, and each preform weighs about 10 g. The number of parts required is 25,000. Find the production rate for one machine, one operator, and the amount of premixed material for the lot. How many elapsed hours will be required to run the quantity for the lot?

240   PLASTIC MATERIALS AND PROCESSES

41. A part requires 18 g and each preform will be about 8 g. The cavities per stroke are three, and the excess is required for the sprue and runners of the injection machine. One operator tends one machine that makes 80 spm. The order requires 12,000 parts, and the machine that does the molding of the material costs $60/h. The blended material will cost $2/kg. Find the production rate for the operator, machine, and the cost for the part and operator. If the billing rate is two times this cost, what is the price the owner will charge for the lot?

42. A computer part is molded of clear polycarbonate plastic two at a time. A partially dimensioned sketch (Figure 10.23) gives part dimensions, sprue, runners, tabs, and the two parts. Density = 0.0404 lb/in.^{3} (1119 kg/m^3). (a) In customary units find the weight of one part and the shot requirements for the sprue, runners, tabs, and two parts. (b) Determine the yield of the part to total material. (c) The cost of the material is $3.10/lb ($6.89/kg), and waste is recovered at 10% of original value. Find the cost of the part including a fair share of the waste products. (d) The cycle time for one operator and one injection-molding machine is 45 s. For a labor rate of $16.50/h find the labor cost per unit. (e) What is the total cost per unit? (f) Repeat parts (a) to (d) in metric units.

43. List the properties of silicone-based polymers that make them desirable for engineering materials.

44. Which plastic materials are used for electrical wire insulation, printed-circuit boards, knobs, and chassis, and what are the principal properties that dictate their selection?

CASE STUDY

THE GENERAL PLASTICS COMPANY

Rich Hall, die designer for General Plastics, mutters to himself, "What counts in this problem is minimum plastic volume in the runner system." Rich knows that for this plastic mold design it will be impractical to reuse scrap, because the plastic part will be colored and the value of the scrap runners represents a small fraction of the virgin material cost.

The part to be molded is roughly 25 mm in diameter and 10 mm thick, similar to a preform except for the novelty impressions on the surface.

Rich, recently hired in his job, has learned that full-round runners are preferred. They have a minimum surface-to-volume ratio, thus reducing heat loss and pressure drop. Balance runner systems are preferred because they permit uniformity of mass flow from the sprue to the cav-

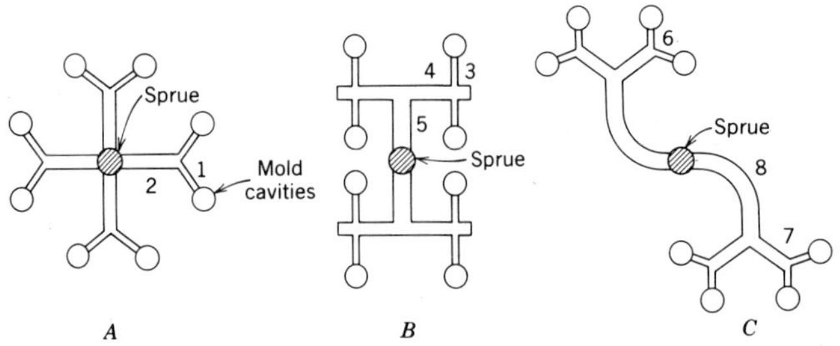

Figure 10.24 Case study. Three design configurations: A, Star pattern. B, "H" pattern. C, Sweep pattern.

ities, because the cavities are at an equal distance from the sprue. Main runners adjacent to the sprue are larger than secondary runners.

Rich has designed three configurations as in Figure 10.24, and he will select the one that uses a minimum of runner material. The time factor is not critical, because the three arrangements provide an identical number of parts per shot. The sprue volume for the three arrangements is equal. Die data are shown in the accompanying tabulation. Determine which arrangement has a minimum of material for the runner system. If the plastic cost $0.60/kg and specific gravity is 1.05, what is the prorated loss per unit? Determine the overall material efficiency if

$$\text{Efficiency} = \frac{\text{Material in parts}}{\text{Material in shot}}$$

Arrangement	Runner section	Diameter, mm	Section length mm
A	1	5	25
	2	6	25
B	3	5	12
	4	6	75
	5	8	25
C	6	5	8
	7	6	175
	8	10	100

CHAPTER 11

METROLOGY AND QUALITY CONTROL

To understand, manipulate, and control production depends, in part, on measurements. Mass production, which is the foundation of modern industry, is based on measured and interchangeable parts. Even small-lot production is inextricably joined to metrology and quality control. Concepts and techniques of measurements and quality control mathematics are introduced in this chapter. Interchangeable production requires that parts be made according to engineering drawings and standards. *Metrology,* the science of measurement, involves inspectors, technologists, and engineers charged with this activity.

MEASUREMENT CONCEPTS

A system of measurements, such as the *English* or *Le Système International d'Unités (SI)* satisfies certain concepts. For each quantity of length, mass, time, or temperature, a unit is necessary. In the English system, the word "pound" is used for a unit of weight or force. The SI system is a decimal system composed of six base units, two supplementary units, and additional derived units as given in Table 11.1. Conversions are provided in the front and back endpapers. In SI the quantities length, mass, and time have the units meter (m), kilogram (kg), and second (s).

A *standard,* such as a 1-lb weight or a yardstick (or a 1-kg mass or meterstick), is necessary. The measurement provides the magnitude of the quantity, but additional knowledge about the correctness of the measurement is necessary. Correctness or *accuracy* is the degree of conformity of a measured or calculated value to some recognized standard or specific value. The difference between the measured and true value is the error of the measurement. *Precision* is the repeatability of the measurement process or how well identically performed measurements agree. This precision concept applies to a set of measurements, not to a single measurement.

The *error* of a set of measurements can be estimated and includes contributions from the randomness of the result about the average and those caused by assignable errors of the process. Random scattering of results about the average measurement can be estimated with confidence using statistics.

When measuring with a micrometer caliper, random scattering could be due to the inspector's "feel," or to progressive differences of the screw thread. Examples of assignable errors would be one-sided wear of the anvil and spindle, nonparallelism or flatness, or failure of the original adjustment to give a zero reading when the faces are tight. *Systematic and constant errors* usually can be predetermined using calibration

Table 11.1 **Units Used in the SI System**

Quantity	Unit	SI Symbol	Formula
Base Units			
Length	meter	m	
Mass	kilogram	kg	
Time	second	s	
Electric current	ampere	A	
Thermodynamic temperature	kelvin	K	
Amount of substance	mole	mol	
Luminous intensity	candela	cd	
Supplementary Units			
Plane angle	radian	rad	
Solid angle	steradian	sr	
Derived Units			
Acceleration	meter per second squared		m/s^2
Angular acceleration	radian per second squared		rad/s^2
Angular velocity	radian per second		rad/s
Area	square meter		m^2
Density	kilogram per cubic meter		kg/m^3
Electric capacitance	farad	F	$A \cdot s/V$
Electric inductance	henry	H	$V \cdot s/A$
Electric potential difference	volt	V	W/A
Electric resistance	ohm	Ω	V/A
Electromotive force	volt	V	W/A
Energy	joule	J	$N \cdot m$
Entropy	joule per kelvin		J/K
Force	newton	N	$kg \cdot m/s^2$
Frequency	hertz	Hz	cps
Luminous flux	lumen	lm	$cd \cdot sr$
Magnetic flux	weber	Wb	$V \cdot s$
Magnetomotive force	ampere	A	
Power	watt	W	J/s
Pressure	pascal	Pa	N/m^2
Quantity of electricity	coulomb	C	$A \cdot s$
Quantity of heat	joule	J	$N \cdot m$
Specific heat	joule per kilogram-kelvin		$j/kg \cdot K$
Stress	pascal	Pa	N/m^2

(*continued*)

Table 11.1 Units Used in the SI System (*continued*)

Quantity	Unit	SI Symbol	Formula
Thermal conductivity	watt per meter-kelvin		W/m − K
Velocity	meter per second		m/s
Voltage	volt	V	W/A
Volume	cubic meter		m^3
Weight (force)	Newton	N	kg − m/s^2
Work	joule	J	n − m

SI Prefixes

Multiplication Factors	Prefix	SI Symbol
1,000,000,000,000 = 10^{12}	tera	T
1,000,000,000 = 10^9	giga	G
1,000,0000 = 10^6	mega	M
1,000 = 10^3	kilo	k
100 = 10^2	hecto*	h
10 = 10^1	deka*	da
0.1 = 10^{-1}	deci*	d
0.01 = 10^{-2}	centi*	c
0.001 = 10^{-3}	milli	m
0.000 001 = 10^{-6}	micro	μ
0.000 000 001 = 10^{-9}	nano	n
0.000 000 000 001 = 10^{-12}	pico	p
0.000 000 000 000 001 = 10^{-15}	femto	f
0.000 000 000 000 000 001 = 10^{-18}	atto	a

*Not preferred engineering factors.

procedures; that is, the device is tested against a more accurate standard and errors are noted.

Sensitivity and readability are used in discussing measurements. Sensitivity and readability are associated with the device, whereas precision and accuracy are associated with the measuring process. *Sensitivity* is the ability to detect differences in a quantity being measured. If one micrometer is able to measure a perceptible difference and another is not, the former is more sensitive. A device is *readable* if the measurements are converted to a readily ascertained number. A vernier on a micrometer makes the instrument more readable.

DIMENSIONING AND TOLERANCE

In dimensioning a drawing, the numbers placed between the arrows represent dimensions that are only approximate. They do not represent any degree of accuracy unless stated by the design. To specify a degree of accuracy it is necessary to add tolerance numbers

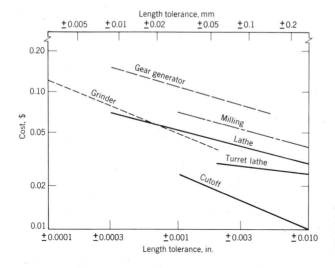

Figure 11.1
Sample cost data for several processes for producing a length tolerance.

to the *dimension*. *Tolerance* is the amount of variation permitted in the part or the total variation allowed in a dimension. A shaft might have an nominal size of $2\frac{1}{2}$ in. (63.5 mm), but for practical reasons this exact dimension cannot be maintained in manufacturing without excessive cost. Hence, a tolerance would be added and, if a variation of ±0.003 in. (±0.08 mm) could be permitted, the dimension would be stated 2.500 ± 0.003 in. (63.5 ± 0.08 mm).

Dimensions given close tolerance infer that the part must fit properly with another part. Both must be given tolerances in keeping with the allowance desired, the manufacturing processes available, and the minimum cost of production and assembly that will maximize profit. The cost of a part goes up as the tolerance is decreased. Figure 11.1 indicates the relative cost of machining a part when the length tolerance is given.

Allowance, which is sometimes confused with tolerance, has an altogether different meaning. It is the minimum clearance space intended between mating parts and represents the condition of tightest permissible fit. If a shaft size $1.498 \, ^{+0.000}_{-0.003}$ is to fit a hole of size $1.500 \, ^{+0.003}_{-0.000}$ the minimum size hole is 1.500 and the maximum size shaft is 1.498. Thus the allowance is 0.002 and the maximum *clearance* is 0.008 as based on the minimum shaft and maximum hole dimensions. The dimensions at this point are dual and could represent either English or metric units.

Tolerances may be either unilateral or bilateral. *Unilateral tolerance* means that any variation is made in only one direction from the nominal dimension. Referring to the previous example,[1] the hole is dimensioned $1.500 \, ^{+0.003}_{-0.000}$, which represents a unilateral tolerance. if the dimensions were given as 1.500, the tolerance would be bilateral, that is, it would vary both over and under the nominal dimension. The majority of manufacturing companies in the United States use the unilateral system. The reason for this can be determined by reference to Figure 11.2, which illustrates both tolerances. The unilateral

---

[1]These dimensions can represent either English or metric units.

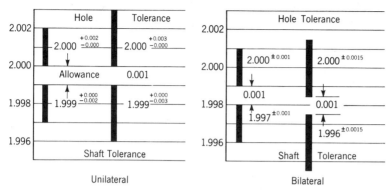

Figure 11.2
Illustrating the application of unilateral and bilateral tolerances.
Dimensions are representative of any English or metric units.

system permits changing the tolerance while still retaining the same allowance or type of fit. With the bilateral system this is not possible without also changing the nominal size dimension of one or both of the two mating parts. In mass production, where mating parts must be interchangeable, unilateral tolerances are customary. To have an *interference or force fit* between mating parts, the tolerances must create a zero or negative allowance.

MEASUREMENT AND INSTRUMENTS

Standard of Measurement

The *standard of measurement* in the United States is the *meter*. For most scientific and technical work SI is superior to the English system. The metric system is more widely accepted than the pound-foot-inch system. The procedure followed in this text shows first the customary units followed by SI units in parentheses.[2]

The advantage of SI is the consistency of one unit for each physical quantity, as seen in Table 11.1. From these elemental units other quantities are derived. In SI, units for force, energy, and power are the same whether the process is mechanical, electrical, or chemical. The case of the decimal relationship between multiples in a number system to the base 10 is an obvious convenience to a weight system that in English units would be represented by 7 lb 2 oz.

The meter is defined as the length of 1,650,763.73 wavelengths of radiation of the atom of krypton 86, the orange-red line. This definition points out the accuracy of metrology but does not provide a standard located in the plant. A standard used for comparison to other measurement devices is the precision gage block set.

Precision Gage Blocks

Gage blocks are square, round, or rectangular in shape, with two parallel sides very accurately lapped to size. In order of hardness and cost, the blocks can be made from

---

[2]Exceptions may be noted where educational confusion will result. Some conversions have no practical significance. ASTM 380-72 *Metric Practice Guide* is used throughout this book as the standard reference.

tool steel, chrome plate steel, stainless, chrome carbide, or tungsten carbide. Tungsten carbide is the hardest and most expensive. Laboratory sets may be obtained with a guaranteed accuracy within two-millionths of an inch per block. These blocks are used mainly for reference in setting gages; for accurate measurements in tool, gage, and die manufacturing; and as master laboratory standards for control of measurements in manufacturing. Their accuracy is valid only at 68°F (20°C). With a standard set of 81 blocks, it is possible to obtain practically any dimension in increments of 0.0001 in. for 0.100 to over 10 in. by combining those of the proper size. Micrometers and vernier-equipped instruments can be used to check tolerance to within one one-thousandth of an inch, but gage blocks are usually necessary when tolerances are in the ten-thousandths. If millionths are to be measured, a constant-temperature laboratory with optical or electronic equipment for gage block calibrations and comparisons is necessary. Special *angle gage blocks* are available with accuracies compatible to the precision gage blocks. A set of 16 angle blocks will permit the measuring of most any angle to an accuracy of 1 s.

Precision gage blocks are assembled by a wringing process. The blocks must first be thoroughly cleaned. One is placed on the other centrally and oscillated slightly. It is then slid partially out of engagement and wrung under light pressure. A slight liquid film between the surfaces of the gages causes them to adhere firmly. Tool steel blocks wear down about five-millionths of an inch in 1000 wringings, so if blocks are to receive great use they should be made from the harder materials. Gage blocks put together in this fashion should not be so assembled for more than a few hours.

In the gage block set of 81 blocks (plus two wear blocks), dimensions are as follows:

9 blocks with 0.0001 in. increment from 0.1001 to 0.1009
49 blocks with 0.001 in. increment from 0.101 to 0.149

TABLE 11.2 **Successive Subtraction Procedure for Using Gage Blocks**

Procedural step		Block used
Desired dimension	= 3.6083	
Ten-thousandth place	= 0.1003	0.1003
Remainder	= 3.508	
Thousandth's place	= 0.108	0.108
Remainder	= 3.40	
Hundredth's place	= 0.30	0.30
Remainder	= 3.10	
Two wear blocks	= 0.10	0.10
Remainder	= 3.0	
Units place	= 3.0	3.0
Remainder	= 0	
TOTAL:		3.6083

248 METROLOGY AND QUALITY CONTROL

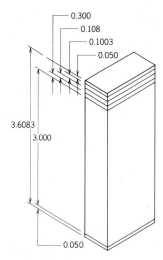

Figure 11.3
Gage blocks assembled to a dimension of 3.6083 in.

19 blocks with 0.050 in. increment from 0.050 to 0.950

4 blocks with 1 in. increment from 1 to 4

2 tungsten steel blocks, each 0.050 in. thick

If a dimensional standard of 3.6083 in. is desired, the procedure for arriving at a suitable combination can be determined by successive subtraction, as shown in Table 11.2. This stack of blocks is shown in Figure 11.3.

LENGTH-MEASURING INSTRUMENTS

Vernier Caliper

Figure 11.4 shows a vernier caliper that can be used for determining inside and outside measurements over a wide range of dimensions. The vernier consists of a main scale graduated in inches and an auxiliary scale having 25 divisions. Each inch on the main scale is divided into tenths and each tenth into four divisions, so that in all there are 40 divisions (each 0.025 in.) to the inch. Vernier and micrometer calipers are also available in metric units. The 25 divisions on the auxiliary or sliding scale correspond to the length of 24 divisions on the main scale and are equal to $^{24}/_{40}$ or $^{24}/_{1000}$ in., which is $^{1}/_{1000}$ in. less than a division on the main scale. Hence, if the two scales were on zero readings, the first two lines would be 0.001 in. apart, the tenth lines 0.010 in. apart, and so on.

The reading on the main scale is first observed and converted into thousandths, and to this figure is added the reading on the vernier. The vernier reading is obtained by noting which line coincides with a line on the main scale. If it is the fifteenth line, 0.015 in. is added to the main scale. These scales are shown in detail in the enlarged view of the vernier scale. As shown, the vernier reads exactly 0.400 in.

Outside measurements are taken with the work between the jaws, and inside measurements with the work over the ends of the two jaws. This method of measurement is

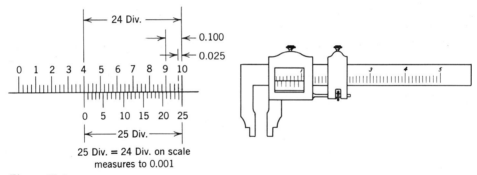

Figure 11.4
Vernier scale.

not so rapid as a micrometer but has the advantage of a wider range with equal accuracy. The principle of the vernier is also used on protractors for angular measurement.

Micrometer

The *micrometer* (Figure 11.5) is used for quick accurate measurements to the thousandth part of an inch. Micrometers are available that read in metric units. This tool requires an accurate screw thread in obtaining a measurement. The screw is attached to a spindle and is turned by a thimble at the end. The barrel that is attached to the frame acts as a nut to engage the screw threads, which are accurately made with a pitch of 40 threads per inch. Each revolution of the thimble advances the screw $1/40$ of an inch or 0.025 in. The outside of the barrel is graduated in 40 divisions, and any movement of the thimble down the barrel can be read next to its beveled end. When the spindle is in contact with the anvil on a 1-in. micrometer, the zero readings on barrel and thimble should coincide. The graduations on a metric micrometer caliper in the 0- to 25-mm range are 0.002 mm with a vernier reading.

The scale on the barrel and thimble edge can best be understood by reference to the enlarged view of Figure 11.5. On the beveled edge of the thimble are 25 divisions, each division representing 0.001 in. To read the micrometer the division on the thimble coinciding with the line on the barrel is added to the number of exposed divisions on the

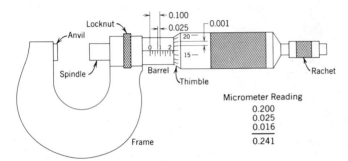

Figure 11.5
One-inch micrometer caliper.

barrel converted into thousandths. Thus, the reading shown in Figure 11.5 is made up of 0.200 plus 0.025 on the barrel, or 0.225 in., to which is added 0.016 on the thimble to give a total reading of 0.241.

Because most micrometers read only over a 1-in. range to cover a wide range of dimensions, several micrometers or else different-length spindles are employed. The micrometer principle of measurement is also applied to inside measurements, to depth reading, and to the measurements of screw threads. Micrometers are available with optical and electronic readouts.

For accurate shop measurements to 0.0001 in. a bench micrometer may be used. This machine is set to correct size by precision gage blocks, and readings may be made directly from a dial on the headstock. Constant pressure is maintained on all objects being measured, and comparative measurements to 0.00005 in. are possible. Precision measuring machines utilizing a combination of electronics and mechanical principles are capable of an accuracy of 0.00001 in.

Optical Instruments

Because of their extreme accuracy and ability to measure parts without pressure of contact, numerous optical instruments have been devised for inspecting and measuring. A microscope for toolroom work is shown in Figure 11.6. An object viewed is greatly enlarged,

Figure 11.6
Toolmaker's microscope.

and the image is not reversed as in the ordinary microscope. To be measured a part is first clamped in proper position on the cross slide stage. The microscope is focused and the part to be measured brought under the crossline seen in the microscope. The micrometer screw is then turned until the other extremity is under the crossline, the dimension being obtained from the difference in the two readings. The micrometer screws operate in either direction and read to an accuracy of 1 part in 10,000.

The *optical gage,* shown in Figure 11.7, can be used to measure heights to 1 part in 10,000 in the production shop or gage laboratory. The gage is equipped with interchangeable spindles to accommodate dimensions to 9 in. (230 mm). The spindle is lowered by an electric motor until it contacts the workpiece with a predetermined load between 7 and 10 oz (2–3 N). The actual dimension can be read on an illuminated, knob-adjusted vernier scale. Used like a micrometer or visual gage, it is faster than a manual micrometer.

Caliper and Divider

The caliper is used for approximate measurements, both external and internal. It does not measure directly but must be set to size with a steel rule or some form of gage being used. Most shop *calipers,* known as spring calipers, consist of two legs with a flat spring head plus a nut and screw to hold them in position. The legs on *outside calipers* curve inward so that the caliper may pass over cylindrical work, while on *inside calipers* the legs are straight with the ends turned outward. *Hermaphrodite calipers* are used principally

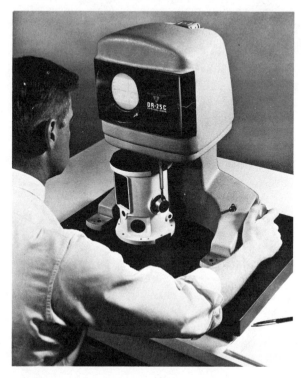

Figure 11.7
Optical gage.

for locating centers and layout work. They have one leg similar to the leg on an outside caliper, whereas the other is a straight point. A *divider* is similar in construction to a caliper except that both legs are straight with sharp, hardened points at the end. This tool is used for transferring dimensions, scribing circles, and doing general layout work.

ANGULAR MEASUREMENT

Angular measurement is standardized by the *radian*. This is the unit of measure of a plane angle with its vertex at the center of a circle and subtended by an arc equal in length to the radius. Usually common angular measuring instruments read degrees directly from a circular scale scribed on the dial or circumference. There are also devices that require the aid of other measuring instruments and calculations to obtain the result.

The plain or universal *bevel protractor* measures directly in degrees and is adapted to all classes of work in which angles are to be laid out or established.

Sine Bar

A *sine bar* is a simple device used either for accurately measuring angles or for locating work to a given angle. Mounted on the center line are two buttons of the same diameter at a known distance apart. For purposes of accurate measurement the bar must be used in connection with a true surface.

The operation of the sine bar is based on the trigonometric relationship that the sine of an angle is equal to the opposite side divided by the hypotenuse. Measurement of the unknown side is accomplished by a height gage or precision blocks.

In Figure 11.8 is a sine bar set to check the angle on the end of a machined part. In this case,

$$\sin \theta = \frac{(h_1 - h_2)}{L}$$

where L is a known distance, usually 5 or 10 in. (127 or 254 mm) depending on the sine bar used. The heights h_1 and h_2 are built up to correct heights in corresponding units of linear dimension with precision gage blocks, and their difference in elevation over L gives the sine of the angle θ.

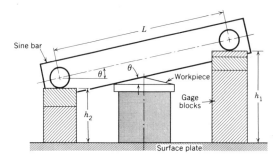

Figure 11.8
Sine bar setup on gage blocks for measuring an angle on workpiece.

SURFACE MEASUREMENTS

Surface checking instruments are for finding a measure of the *accuracy of a surface* or the condition of a finish. Much of this work is done on a flat, which is an accurately machined casting or lapped-granite block known as a *surface plate*. It is the base upon which parts are laid out and checked with the aid of other measuring tools. These plates are carefully made and should be accurate to within 0.0001 in. (0.003 mm) flatness from the mean plane to any point on the surface. Small plates, known as *toolmakers' flats*, are lapped to a greater degree of accuracy. Their field of application is limited to small parts and they are normally used with precision gage blocks.

Surface Gage

The *surface gage,* shown in Figure 11.9, checks the accuracy or parallelism of surfaces. It also transfers measurements in layout work by scribing lines on a vertical surface. When in use it is set in an approximate position and locked. The spindle can be finely adjusted by turning the knurled nut that controls the rocking bracket. When used with the scriber, it is a line measuring or locating instrument. If the scriber is replaced by a dial indicator or a transducer, it becomes a precision instrument for checking surfaces.

Optical Flat

Measurements to the millionth part of an inch (~ 25 nm) are made by *interferometry,* the science of measuring with light waves. Measurements by this principle are made with

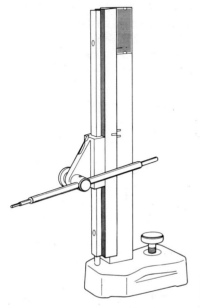

Figure 11.9
Height transfer (surface) gage.

small instruments known as *optical flats*. They are flat lenses with very accurately polished surfaces that have light-transmitting quality. Optical flats are usually made from natural quartz because of its hardness, low coefficient of expansion, and resistance to corrosion. Optical flats are available from 1 to 12 in. (25–300 mm) in diameter with a thickness about one-sixth the diameter. It is not necessary that the two surfaces of a flat be absolutely parallel.

One of the common uses for optical flats is the testing of plane surfaces. The optical flat is placed on the flat surface to be tested, and light is reflected both from the optical flat and the surface being tested through the very thin layer of air between the two surfaces. When the light waves are in phase, there is a light band; when they are out of phase, a dark band is created. If the thickness of the air layer measures one-half a wavelength of light or more, an interference effect occurs. The interference between the rays reflected from the bottom of the flat and from the top of the work causes dark bands called newton's rings to appear.

If the surface is irregular, the appearance is similar to a contour map. The position and number of lines show the location and extent of the irregularities. When the bands are straight, evenly spaced, and parallel to the line of contact as shown in Figure 11.10, the surface is perfectly flat. If the bands were straight but not evenly spaced of if the bands curved, the surface would not be flat. If the wavelength of the light source is known, any deviation from this pattern indicates an error in the surface, the amount of which can be measured. A monochromatic light of one wavelength, such as fluorescent helium, is used for the bands to be sharply defined. Each fringe or band in such a light indicates a difference in height of 0.0000116 in. (295 nm), the one-half wavelength of helium.

The fringe patterns are interpreted differently according to whether the optical flat is pressed hard against the surface or slanted slightly with a wedge-shaped air space between the block and the workpiece as in Figure 11.11 *A* and *B*. Because contact fringes are more difficult to interpret, the wedge method is used whenever possible. Figure 11.11 shows two parts being inspected, one concave and one convex. The number of bands or fringes that appear depend on the wedge thickness, but the curvature is the measure of surface flatness. Figure 11.11*A* shows a condition in which each band curves an amount slightly over two band intervals. The 2.2 bands of curvature indicate that the workpiece

Figure 11.10
Straight interference fringes shown by this optical flat indicate a flat gage block.

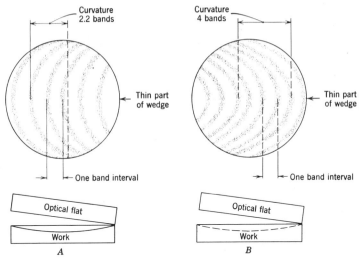

Figure 11.11
Use of the optical flat to determine curvature. *A*, Convex surface with near and far edges 11.6 μin. (295 nm) low. *B*, Concave surface with center 46 μin. (1168 nm) low.

is 2.2 × 0.0000116 in. = 0.0000255 in. = 25 μin. (625 nm) high in the center, because the bands curve toward the thin part of the wedge. In Figure 11.11*B* the bands curve away from the thin side of the wedge, so that the surface of the workpiece is concave. The four-band curvature means that the surface is 46 μin. (1150 nm) lower in the center than on the edges.

Surface Roughness

Several devices have been developed to measure *surface roughness*. The simplest procedure is a visual comparison with an established standard. Other methods include microscopic comparison, direct measurement of scratch depth by light interference, and the measurement of the magnified shadows cast by scratches on a surface or even touch comparison of tactile standards. The usual procedure is to employ a diamond stylus to trace over the surface being investigated and to record a magnified profile of the irregularities.

To measure roughness and other surface characteristics, a standard was developed by the American Standards Association (ASA B46.1-1962) that deals with such surface irregularities as height, width, and direction of the surface pattern. These surface irregularities, as well as the symbols for specifying surface roughness on a drawing, are shown in Figure 11.12.

An instrument used in making surface roughness measurements is shown in Figure 11.13. This is a direct-reading instrument that measures the number of roughness peaks per inch above a preselected height by passing a fine tracing point over the surface. The unit consists of a tracer that converts the vertical movements of the tracing point into a

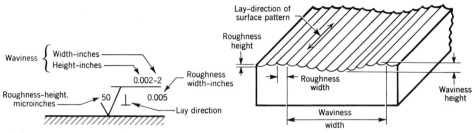

Figure 11.12
Surface characteristics and symbols for indicating their maximum values.

small, fluctuating voltage that is related to the height of the surface irregularity, a motor-driven device (profilometer) for operating the tracer, and the amplimeter. The amplimeter receives the voltage from the tracer, amplifies it, and integrates it so that it may be read as digital values or shown on a strip chart recorder. The process is a continuous one, and the instrument shows the variation in **average roughness** from a reference line as illustrated in the magnified profile of a surface in Figure 11.14. Readings may be either *arithmetical* (AA) or *root mean square* (RMS) average deviation height from the reference line CD. The difference in result of the two methods of calculation is indicated in the example worked out in connection with the figure. The instrument may be operated either manually or mechanically, and readings can be taken on both plain and curved surfaces.

Surfaces with the same average roughness height can be very different, because the height and number of the peaks and valleys and the roughness width can be dissimilar. The unit shown in Figure 11.13 will also determine the number of peaks per inch above a preselected height in order that surfaces can be more thoroughly identified and inspected. Surface roughnesses available by common production methods are indicated in Figure 11.15.

Designers and production personnel tend to seek very smooth finishes on parts for

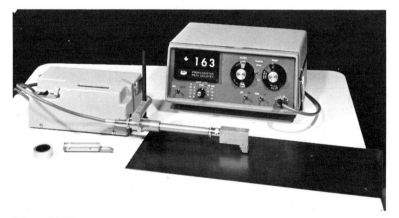

Figure 11.13
Surface gage including transducer tracer, amplifier, and indicator for measuring surface roughness.

SURFACE MEASUREMENTS   257

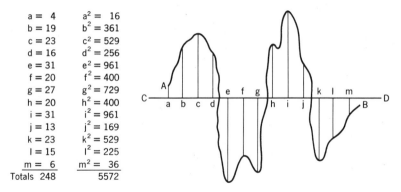

Figure 11.14
Relationship between arithmetic average and root-mean-square values used in determining surface roughness.

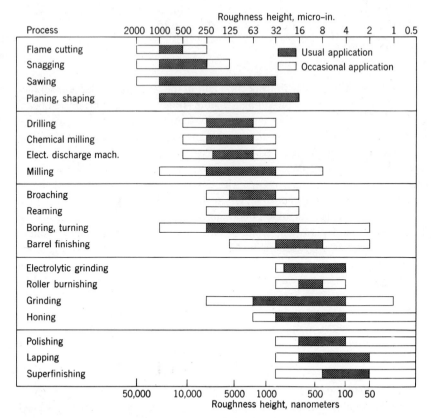

Figure 11.15
Surface roughness available from common production methods.

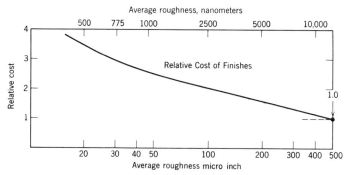

Figure 11.16
Relative cost of surface finishes.

aesthetic purposes and as a mark of pride. Fortunately, highly machined and polished surfaces are not required on most parts; if they were, the cost would be prohibitive. Figure 11.16 illustrates the relative cost of producing finishes. The costs are increased for products requiring low roughness because of equipment costs and additional labor and inspection. Protecting a highly polished surface until assembly or delivery can be more expensive than the production cost of the surface itself.

Laser

Helium–neon *gas lasers* are popular in inspection and in the assembly of large machines because they are the only method of providing a visible straight line. The bright-red beam does not sag or bend. There is no other inspection system with this accuracy over large distances. Lasers are used in production shops and in inspection laboratories for checking straightness, flatness, squareness, and levelness.

A laser is a device capable of generating coherent electromagnetic radiation at wavelengths shorter than those of the microwave. The gas laser contains a gas whose atoms or molecules are raised to higher energy levels in the gas. The material portion of the laser consists of a long cylindrical tube containing the gaseous medium, a means for exciting a discharge in the medium, and a pair of mirrors facing each other that constitute the resonator for the laser energy. Generally, this laser may be summarized as a gaseous medium excited by electric discharge and containing within it a closed optical path in which optical energy can be contained for relatively long periods of time. Figure 11.17A is a sketch of the essentials of the laser gun.

The most accurate measurement and calibration techniques involve a process called interferometry. This method is simply measuring distances making use of a known coherent source (most often a helium–neon laser). Laser light from a coherent source is divided into two beams as illustrated in Figure 11.17B. A light from a mirror mounted on one surface interferes with reflected light from a second surface. One mirror remains in a fixed position while the other is moved to cause alternately constructive and destructive interference. This alternation produces variations of wave intensity that can be detected by a photoelectric detector and counted with a digital counter.

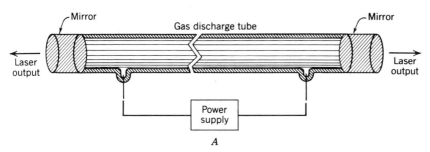

Figure 11.17A
Essentials of a gas laser.

This method (accurate to two millionths of the desired distance) can measure lengths from 0.2 mm to 120 m and is used primarily for testing and measuring machine tools, precision measurement, comparisons with standards, and calibration.

The *laser interferometer* consists of three parts: a power supply, a combination laser and interferometer, and a retroflector. Simply stated, beam splitters send one-half the laser light to the retroflector, and the other half is directed to a photo detector. The light going to the retroflector is reflected back to the interferometer and the fringes created by the interference of the two sources of light are a measure of distance. Such units can be equipped with digital or graphical readout. So fast is the interpretation of distance that the retroflector may move at rates up to 200 in./min (0.1 m/s).

Figure 11.18 shows a *tooling laser,* centering detector, and a readout unit used for checking the flatness of surface plates. The laser emits a 10-mm-diameter columnated and visible-red laser beam lined up with its mounting to within 30 arc seconds. The detector senses the centroid of the beam, and any offset between the centroid and the center of the detector is indicated on the panel meter readout. The beam displacement is indicated for both orthogonal axes perpendicular to the beam. Such a unit has a resolution of about 10 μin./ft (80 × 10^{-6}%) of laser beam up to 250 ft (75 m).

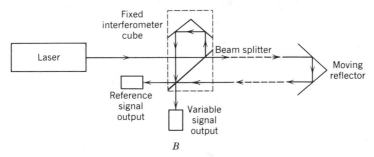

Figure 11.17B
Principle of operation of laser interferometer. (Adapted from Harry, *Industrial Lasers and Their Applications.*)

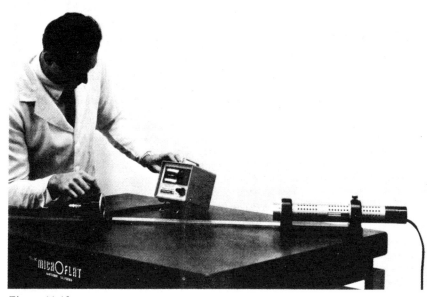

Figure 11.18
Laser, center detector, and readout meters.

GAGES

Gages are used to measure fixed shape or size in production work. They represent a standard for comparing manufactured parts. Their use is limited to one or several dimensions on a part because adjustments of a gage are not normally made during a production cycle. Inspecting a part requires a minimum of time with gages. Much gaging is done by the operators in the shop while their equipment is in operation, thus causing no delay of production time. This device determines whether the part has been made to the design tolerance. The gage does not usually indicate a specific dimension. *Inspection gages* are used by inspectors in the final acceptance of the product. They ensure that the product is made in accordance with the tolerance specification on the design. *Working or manufacturing gages* are frequently made to slightly smaller tolerances than the inspection gages, to keep the size near the center of the limit tolerance. Parts manufactured around limit sizes will still pass the inspector's gage.

The success of a gage is measured by its accuracy and service life, which in turn, depends on the workmanship and materials used for the gage. Gages are subject to abrasive wear during service, and the selection of material is important. High-carbon and alloy tool steels have been the principal materials, although glass is also used. These materials can be accurately machined and heat-treated for additional hardness if necessary. Heat-treating operations increase hardness and abrasives resistance. Steel gages are subject to some distortion because of the heat-treating operation. Low-carbon steels have limited hardness and are not always suitable.

These objections are largely overcome after chrome-plating the surface or in application of cemented carbides. Chrome plating permits the use of steel having inert qualities, because wear resistance is obtained with a hard chromium surface. Chrome resurfacing is used in reclaiming worn gages. But cemented carbides applied on metal shanks by the powder metallurgy technique provides the hardest wearing surface.

Snap Gages

A *snap gage,* used in the measurement of plain external dimensions, consists of a U-shaped frame having jaws equipped with suitable gaging surfaces. A plain gage has two parallel jaws or anvils that are produced to some standard size and cannot be adjusted. This type of gage can be replaced by adjustable gages to allow a means for changing tolerance settings or adjusting to wear. Most gages are provided with the *"go" and "no-go"* feature in a single jaw, such a design being both satisfactory and rapid. The general design shown in Figure 11.19 has been selected because it incorporates most of the advantages of similar gages. It is light in weight, sufficiently rigid, easy to adjust, provided with suitable locking means, and is designed to permit interchangeability with many parts.

The tolerance for the settings, as shown in Figure 11.19, must account for the total *gage allowance,* which is customarily taken as 10% of the tolerance of the part (i.e., 5% of the part tolerance for each button, and the wear allowance, which is 5% of the part tolerance). The allowance for wear is usually made only for "go" gages, since "no-go" gages have little wear.

The usual practice is to allocate both gage tolerance and wear allowance entirely within the tolerance limits of the part to be inspected. An occasional part may be rejected even though the part may have been made within its tolerance.

Figure 11.19 indicates the appropriate sizes for the "go" and "no-go" dimensions when the gage is set to measure $1.000 \, ^{+0.000}_{-0.004}$ in. $(25.40 \, ^{+0.00}_{-0.10}$ mm).

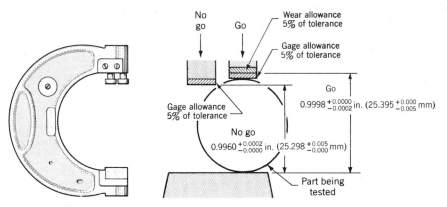

Figure 11.19
Adjustable limit snap gage set for inspecting a dimension of $1.000 \, ^{+0.000}_{-0.004}$ in. $(25.40 \, ^{+0.00}_{-0.10}$ mm).

Plug Gages

A plain *plug gage* is an accurate cylinder used as an internal gage for the size control of holes. It is provided with a suitable handle for holding and is made in a variety of styles. These gages may be either single- or double-ended. Double-ended plain gages have "go" and "no-go" members assembled on opposite ends, whereas progressive gages have both gaging sections combined on one end.

The allowance for manufacturing snap gages and the allowance of the part to be inspected must be considered in the design. Figure 11.20 shows a "go" and "no-go" gage with the appropriate dimensions for checking a hole size of $0.750 \;^{+0.000}_{-0.004}$ in. ($19.05 \;^{+0.00}_{-0.10}$ mm).

Other gages include ring, taper, thread, and thickness. *Ring gages*, for outside diameters, are used in pairs, a "go" and "no-go." *Taper gages* are not dimensional gages but rather a means of checking in terms of degrees. Their use is a matter more of fitting rather than of measuring. A *thickness* or *feeler gage* consists of a number of thin blades and is used in checking clearances and for gaging in narrow places.

Dial Indicator

A *dial indicator* is composed of a graduated dial, spindle, pointer, and a satisfactory means for supporting or clamping it firmly. Most indicators have a spindle travel equal to $2^1/_2$ revolutions of the hand. Between the test point and the hand is interposed an accurate multiplying mechanism that magnifies on the dial any movement of the point. This tool may be considered either a measuring device or a gage. As a measuring device, it measures inaccuracies in alignment, eccentricity, and deviations on surfaces supposed to be parallel. In gaging work it gives a direct reading of tolerance variations from the exact size.

A dial test indicator equipped with a permanent magnet base is sown in Figure 11.21. This base operates in the same fashion as a permanent magnet chuck. With the handle in the base turned to the "on" position, the indicator is held securely on the horizontal, vertical, or overhead flat surface of any machine. Other methods of support are a suitable clamp or a heavy base as used on a surface gage.

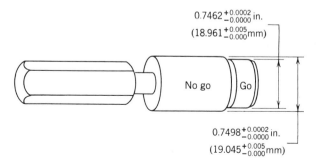

Figure 11.20
Plug gage dimensioned for checking a hole size of $0.750 \;^{+0.000}_{-0.004}$ in. ($19.05 \;^{+0.00}_{-0.10}$ mm).

GAGES 263

Figure 11.21
Dial indicating gage with permanent magnet base.

Projecting Comparators

Projecting comparators like Figure 11.22 are designed on the same principle as a projection lantern. An image is placed before a light source and the shadow of the profile is projected on the screen at some enlarged scale. Usual magnifications are ×20 and ×50, although other magnifications up to ×100 can be used.

The outline of the object is reflected to the screen. Outline inspection is important for many tools, dies, gages, and formed products. These tools and their outline surfaces are repeated on the article. Projecting comparators assure that the outline is accurate. Additionally, it is employed in the inspection of small parts such as needles, saw teeth, threads, forming tools, taps, and gear teeth. Because it checks work to definite tolerances, it is useful for studying wear on tools or distortion caused by heat treatment.

Pneumatic, Electric, and Electronic Gaging

Because of speed, accuracy, and adaptability to automatic inspection, *pneumatic, electric,* and *electronic gaging* are used as inspection and production devices. These gages can

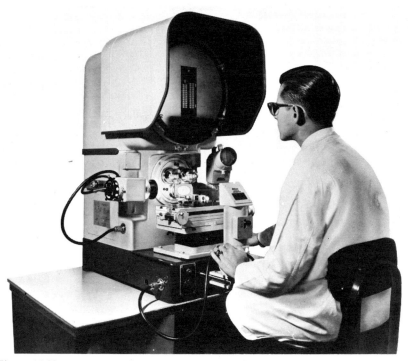

Figure 11.22
Horizontal profile projector.

be used for checking other gages and dimensional standards, piece-by-piece inspection, automatic inspection, and machine control.

Pneumatic gaging employs compressed air and measures back pressure of the air as it exits from the gage by metering its flow. An air spindle, shown in Figure 11.23, has two small diametrically opposed holes for air flow. The amount of air flow is controlled by the size of the annulus space between the air spindle and the work. This change in flow is registered on the dial, which is calibrated to read in fractions of a thousandth of

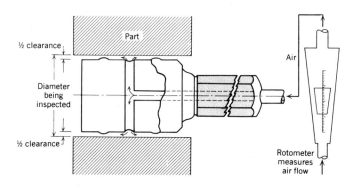

Figure 11.23
Schematic drawing of a spindle used to measure internal diameters.

an inch (0.02 mm). The relationship between the rate of flow and the size of hole is true only for small clearances, and the maximum range is around 0.003 in. (0.08 mm). High amplification permits reading in fractions of a tenth of a thousandth (0.002 mm).

Pneumatic gages can check internal and external dimensions, and multiple checking of several dimensions can be done simultaneously. Air gages will reveal hole or shaft taper, out-of-roundness, and tool gouges that are difficult to detect with a plug gage. Pneumatic gages have the following advantages: speed and simple operation, accuracy to about 0.0005 in. (0.013 mm), relatively low cost, nonscratching, even of the finest or softest finish, low gage wear, and minimum skill requirement. Dimension can be magnified and displayed. Many dimensions can be read at once, which facilitates inspection and selective assembly. These gages can be fitted with electric or electronic gages and amplifiers.

ELECTRIC GAGING

Two types of electric gaging employ either sensitive microswitches (contact gages) or a dial indicator with two limit switches. Microswitch-type units are usually employed for inspection of very large parts that have tolerances of at least 0.002 in. (0.05 mm). Large die castings, for example, can have many dimensions inspected simultaneously by deploying microswitches previously set with gage blocks on a surface plate. Each dimension to be checked is provided with pairs of red and green lights that indicate whether the dimensions meet the specified tolerance. These gages are designed for simultaneous checking of several dimensions. Master parts and their dimensions are employed in setting the gage to the correct limits.

The dial indicator type of electric gage has two switching limits. One limit is set at the upper and the other at the lower tolerance limit. Production work can be divided into three categories. Parts too large, too small, and within tolerance can be electrically shuttled if a control circuit is incorporated. This type of gage is fast, is accurate to 0.0001 in. (0.003 mm), and can be used in automatic inspection and control.

ELECTRONIC GAGING

Electronic gaging systems are popular in production, inspection, and control. Most measurement systems are composed of a generalized three-stage arrangement as seen in Figure 11.24. The first stage is the detector-transducer, followed by an intermediate modifying stage and a terminating stage. The purpose of the detector-transducer is to sense the input signal while being insensitive to other inputs. A strain gage should be sensitive to strain but unaffected by temperature. The purpose of the second stage is to modify or sometimes amplify the transduced information to have it acceptable to the third stage. It may amplify power to drive the terminating device. The terminating stage provides the information in a form understandable to one of the human senses or to a controller, minicomputer, digital readout, or recorder.

The first contact that a measuring system has with the quantity to be measured is

266 METROLOGY AND QUALITY CONTROL

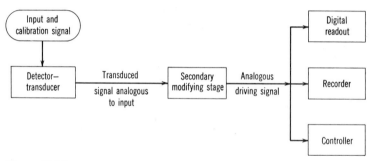

Figure 11.24
Three stages of a generalized measurement signal.

through the input information accepted by the transducer. The transducer must be changed by the quantity to be measured. In Figure 11.25A is a variable-inductance transducer or, more commonly, the linear variable differential transformer (**LVDT**). The LVDT provides an a-c voltage output proportional to the displacement of a core passing through the windings. It is a mutual inductive device using three coils. The LVDT has advantages over other transducers. It converts length displacement into a proportional electric voltage and it cannot be overloaded mechanically, because the core is separable from the remainder of the device and is relatively insensitive to temperature.

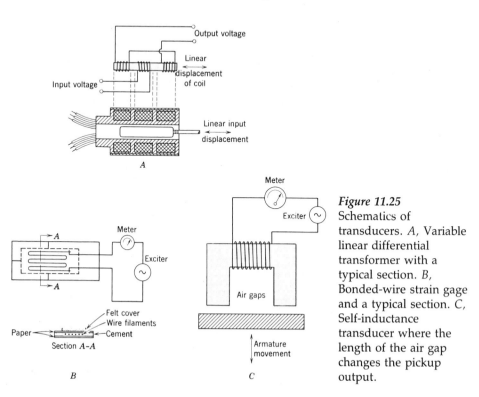

Figure 11.25
Schematics of transducers. A, Variable linear differential transformer with a typical section. B, Bonded-wire strain gage and a typical section. C, Self-inductance transducer where the length of the air gap changes the pickup output.

Figure 11.25*B* is a resistance strain gage bonded to a surface. Once the block is loaded—say, by a compressive force—the force indirectly strains the wires of the gage. The elongation of the gage wires reduces their diameter and a longer length results in increased resistance. Figure 11.25*C* is a self-inductance transducer in which the length displacement changes thereby changing inductance.

Figure 11.26 is an example of a solid-state modular electronic column being used for gaging three dimensions. It uses an LVDT sensing element to produce a light-emitting diode display. A reject light signals an unacceptable part. Multiple- or single-dimension inspection using spindle, snap, or ring gages is possible. Electronic gaging is superior to mechanical or pneumatic gaging as it requires less response time and has improved transducer linearity over the range of measurement. Accuracies to 0.00005 in. (0.0001 mm) are reported.

Figure 11.27 is a photograph of a coordinate measuring machine that provides a decimal readout of a probe position. The part is loaded and leveled on the gaging table, and a zero point is located by a reference point. As the probe is moved about, the readout console displays the *x, y,* and *z* coordinate positions from the reference point. Bore location and size can be determined. Movements in all axes are measured by a reading head as it travels over a steel grating having 1000 lines per inch (40 lines per millimeter). A corresponding segment of grating mounted on the reading head creates a *moire fringe pattern* as it passes over the grating. As the fringe patterns are counted, output signals from the head provide a digital readout of movement and position. Accuracies of 0.001 in. (0.03 mm) are common.

Figure 11.26
LVDT transducer and light-emitting diode used in three electronic columns for gaging of three dimensions.

Figure 11.27
Coordinate measuring machine.

The roundness and geometrical gage shown in Figure 11.28 measures roundness, concentricity, and alignment of inside and outside diameters, squareness, and flatness. The work is positioned on the table, which has a precision spindle that will rotate at 0.4 to 12 rpm. The stationary gaging head reflects the movement of the table to accuracies within $1^1/_2$ millionths of an inch (0.00004 mm). Measurements are recorded on a 10-in. (250-mm) polar chart that is synchronized with the spindle. Figure 11.28 shows two gage heads being used for determining a differential reading for measuring wall thickness and the concentricity of inside and outside diameters. Figure 11.29 shows two kinds of faults with bar stock (roughness and ovality), which are measured with a roundness meter.

The electronic gage, like the electric type, can be "tailored to fit" large, unusually shaped parts. The advantages include excellent accuracy and sensitivity, fast readings, zero adjustments, high magnification, and versatility.

MACHINE GAGING

Electronic measurement devices can be mounted on the machine to monitor the dimensions dynamically. Gaging can control the amount of metal removed, for example. Consider a dynamic balancing machine integrated into a production line where engines are being produced continuously. The crankshaft, which is sensitive to imbalance, requires a monitoring activity to determine the amount of metal to remove to satisfy performance spec-

MACHINE GAGING 269

Figure 11.28
Roundness and geometry gage.

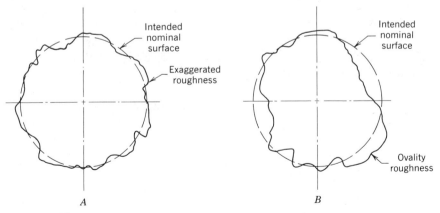

Figure 11.29
A, Transverse variations in roughness. *B*, Ovality variations in roughness.

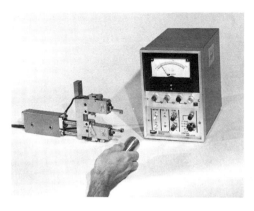

Figure 11.30
LVDT and controlled fingers for on-line gaging and control of ODs.

ification for vibration control. Once the nominal dimension is achieved, a control system produces a feedback signal to stop machining.

Figure 11.30 is a unit adaptable to continuous monitoring of ODs. In this simple device hydraulic retractable fingers move the gage out of the way during the load and unload cycle. These transducer-monitor instruments are used in several ways for machine gaging, and the example of the dynamic balancing machine for the rotary crankshaft will have a feedback computer control system capable of continuously interpreting results and making adjustments in the metal removal.

Automatic Inspection Machines

Automatic inspection machines are usually one-of-a-kind units designed to gage or inspect a mass-produced part. The gaging units are sequentially employed in a timed cycle, and the results are either displayed for an inspector's decision, or more often, the parts are classified or rejected by mechanisms within the machine. Figure 11.31 shows a machine employed to inspect the diameter of V-8 cylinder bores at the rate of 150 per hour. Each bore is measured, classified, and has its size stamped beside the bore by the machine. The accuracy is such as to make classifications according to variations in diameter of 0.004 in. (0.010 mm). Because each bore is known, selective pistons can be used in assembly. An out-of-tolerance bore activates a circuit that ejects the block from the assembly line.

The machine in Figure 11.31 uses pneumatic electronic gages. The machine can record the number of each classification of bore size, and this information can be transmitted to piston inventory records for selective fitting of pistons to bores. This matching produces engines with optimum efficiency.

When pneumatic and electric or electronic gaging is used, the advantages of each system are realized. The more automated a process the more likely that the combination will be used. The response speed of a combination of these gages is no less than that of an individual pneumatic gage; hence, inspection speeds of 6000 parts per hour are possible. Dynamic balancing of crankshafts, and appropriate metal removal to assure rotational vibration performance specifications are often at a rate of 100 units per minute.

Figure 11.31
Pneumatic electronic automatic inspection of V-8 cylinder blocks.

NONDESTRUCTIVE INSPECTION

The purpose of *nondestructive testing* (NDT) is simply to determine flaws or defects without damage to the object. Preservation of the part and economics are the motivating considerations.

Hardness Measurements

In Chapter 2 the methods of determining material hardness are discussed. It is common practice to have hardness specified by engineering for heat-treated parts. Because most parts are near their final stage of production when hardness is verified, equipment must be selected that will cause a minimum impression and distortion of the workpiece. Figure 11.32 is a photograph of a portable, hand-held penetration hardness tester suitable for determining the hardness of mild steel and nonferrous alloys. Most hardness-testing units are semiportable and require movement of the machine to the production area, or else the parts must be taken to the hardness tester.

Magnetic Particle Inspection

In *magnetic particle inspection* an intense magnetic field is set up in the part to be inspected. Cracks, voids, and material discontinuities cause the lines of magnetic flux to

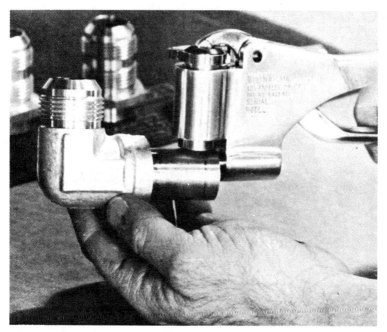

Figure 11.32
Portable hardness tester.

be distorted and they break through the surface as in Figure 11.33. Ferromagnetic powders applied to the part build up at the point where the defect occurs. This method is used to indicate surface imperfections in any material that can be magnetized. Fluorescent magnetic particles that glow in ultraviolet light can be used to intensify the effect.

Figure 11.34 indicates the way in which cracks in a truck axle king pin are revealed using this technique. Trade terms that describe the process are *"magnaflux"* for magnetic particle inspection and "magnaflow" for fluorescent particle inspection.

In some alloys hardness measurements cannot be correlated with strength, but certain magnetic properties do correlate with strength and other physical characteristics of the metal. Testing of magnetic properties is gaining importance.

Radiographic Inspection

Radiographic inspection is accomplished by exposing a part of either X rays, gamma rays, or radioisotopes, and viewing the image created by the radiation on a fluoroscope

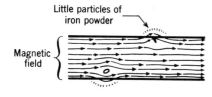

Figure 11.33
Principle of magnetic particle inspection.

Crank pin as it appears in visual inspection.

Cracks revealed with magnetic flux inspection— Magnaflux process.

Cracks revealed with fluorescent particles— Magnaglow process.

Figure 11.34 Cracks in crank pin revealed with magnaflux and magnaglow inspection.

or film. To examine a piece of steel 5 in. (127 mm) thick with X rays requires a machine of over 1000 kVA capacity. X Rays are very sensitive and can be used to inspect any thickness of almost any material. Because of the powerful nature of gamma rays, much radiographic work is done with radium or "cobalt-60" sources. The principal advantages of gamma-ray inspection are the low cost and portability of the source. Electrical power and cooling water are unnecessary. Special techniques shield the capsule of radioactive material when it is not in use. Radiography is employed to examine for internal defects and check for alignment and operation of assembled parts.

Gamma rays may be used to inspect a casting or welds. Flaws that may be found by radiography in welds include porosity, slag, incomplete fusion, and undercutting.

Fluorescent Penetrants

Various fluorescent penetrants can find surface defects in almost any material. The penetrants are normally oil based and may be applied by dipping, spraying, or brushing. The penetrant is later washed off the surface and a powder is applied to absorb the penetrant remaining in cracks and voids. The part is then examined under special light, and the colored powder reveals the flaws.

Ultrasonic Testing

In ultrasonic testing a high-frequency vibration or supra-audible signal is directed into the part to be tested. A quartz crystal that changes electrical signals to ultrasonic inaudible sound waves is pressed against the part. When the sound waves reach the other side of the part or reach a discontinuity, they are reflected back and the crystal generates a signal upon receiving them. A cathode ray tube measures the time lag between the initial signals and the returning ones; hence, metal thicknesses or distances to discontinuities may be measured with precision, and the metallurgical characteristics may be monitored with an unusual degree of precision in some alloys.

Eddy Current Testing

This NDT method is useful for flaw detection, sorting by metallurgical properties such as hardness, and thickness measurement. This method induces an eddy current from a coil adjacent to the surface. Discontinuities in the part change the amplitude and direction of flow of the induced current. The changes of magnitude and phase difference can be used to sort parts according to alloy, temper, and other metallurgical properties.

QUALITY CONTROL

Manufacturing processes lead to variation in output. This is natural and expected. **Quality control** is a professional field that deals with these variations in an effort to provide quality production at minimum cost. In quality control this acceptance of inevitable variation is a central concept. The point of quality control is to study ongoing processes to assure they are proceeding in control. This involves analysis of the characteristics of the population output by inference of the sample output. Thus trends are detected in the output that ultimately lead to finding rejects; then the necessary corrections are made to keep the process in control, which means avoiding rejects. The trends that are detected result from *assignable causes* as opposed to *random causes* that are inherent in manufacturing processes. This is the task of quality control: to identify assignable causes of variation. The main tools are the control charts, of which the $\bar{x}$ chart and the p-charts are prominent.

The purpose of the *control chart* is to distinguish between random fluctuations and the fluctuations attributed to assignable causes. This is achieved with an appropriate choice of control limits calculated using the laws of probability. If these limits are exceeded, the process is out of control. The chart (called the x-bar chart) uses *measurements* such as the values read from a micrometer. The p-chart deals with variables that are called *attributes* or characteristics that are evaluated on qualitative data. This is in contrast to information evaluated on actual measured values. The information for the p-chart would be determined from a "go, no-go" plug gage, for example. The go, no-go plug gage determines whether or not the diameter is satisfactory or not satisfactory. It does not indicate any magnitude of measurements. If actual measurements were required, a micrometer would be used. The plug gage is faster than a direct-reading micrometer, and it requires a different evaluation for quality control.

When parts must be inspected in large numbers, 100% inspection of each part is slow and costly and may not eliminate all of the defective pieces or uncover faulty dimensions. Mass inspection using manual methods tends to be careless, operators become fatigued, and inspection gages become worn or out of adjustment more frequently. The risk of passing defective parts is variable and of unknown magnitude, whereas in a planned sampling procedure the risk can be calculated. Many products such as fuses or food cannot be 100% inspected, because any final test results in the destruction of the product. Inspection is costly, and nothing of value is added to the product because of inspection if the product is produced to specifications.

The term *interchangeable manufacture* implies that the parts that go into an assembly may be selected at random from large numbers of parts. In such a system of manufacturing selective fitting is unnecessary except where special allowances are encountered. Extreme accuracy is not necessary or desirable, because manufacturing costs increase as working limits become closer. No part should be made with a greater degree of accuracy than is required. A balance must be established between the cost of manufacturing and ease of assembly.

The inspector will sample the parts being produced in a mathematical manner and determine whether or not the entire stream of production is acceptable, provided the company is willing to allow a certain known number of defective parts. This number of acceptable defectives is usually taken as 3 out of 1000 parts produced.

To use quality control techniques in inspection the following steps are followed:

1. Sample the stream of products.
2. Measure the desired dimensions or use a gage for qualitative judgments of the sample.
3. Perform the statistical evaluation using probability mathematics.
4. Construct a control chart.
5. Plot succeeding data on the control chart.

A sample must be chosen with impartiality. Dimensions in a manufactured part may vary as a result of either random causes or assignable causes. Variations with *random causes* are inevitable and, for a given process and machine, cannot be eliminated or reduced. ***Assignable causes*** can be eliminated because they include factors such as worn-out equipment or out-of-specification machine adjustment, improper tooling, material defects, or poorly trained workers. The control chart or quality control technique accepts the chance of normal dispersion of dimensions but signals the inspector when a determinable defect occurs.

It is important to choose the parts for inspection on an impartial basis, so they will be representative of the parts being made. Various techniques are used to choose parts, such as by programmed electronic signals or at fixed times, but the best one mathematically speaking is on a ***random basis***.

Control Limit Calculations for the $\bar{x}$ Chart

The standard deviation σ of the dimensions of the parts inspected must be calculated to construct the control chart, which is the fundamental tool of the quality control procedure. The ***standard deviation*** is a measure of dispersion of a dimension about a mean dimension. Once sufficient data, free of assignable causes of variation, have been obtained and the standard deviations calculated, the $\bar{x}$ control chart may be constructed. Further calculation of σ is necessary only to check its value if the process is changed even slightly.

Most inspectors group the data in subgroups of 4 to 10 and call this a ***sample*** even if the parts are not gathered at the same instance. The parts must be grouped in the order taken. The standard deviation σ is found as follows:

1. Calculate the average size $\bar{x}$, called x-bar, of the dimensions measured for each group.
2. Calculate the standard deviation σ for each group by

$$\sigma = \sqrt{\frac{(x_1 - \bar{x})^2 + (x_2 - \bar{x})^2 + \ldots + (x_n - \bar{x})^2}{n}}$$

where

$x_1, x_2, \ldots, x_n$ = Individual dimensions in the sample, in. (mm)
$\bar{x}$ = Average dimension of the sample, in. (mm)
n = Number of parts in each sample

3. Calculate the average standard deviation $\bar{\sigma}$ by using the number of subgroups, N.

$$\bar{\sigma} = \frac{\Sigma \sigma}{N}$$

The *control chart*, such as shown in Figure 11.35, is constructed by plotting the average dimension A_1 of a sample against time, and the upper and lower control limits are drawn at a distance equal to

$$A_1 \bar{\sigma} = 3\sigma_{\bar{x}}$$

above and below the mean dimension line. The value of $3\sigma_{\bar{x}}$ is an arbitrary limit that has found acceptance in industry. Thus control limits are set so that only 3 pieces out of 1000 will be defective.

The value of A_1, calculated by probability theory, is dependent on the number of pieces in the sample and is as follows:

Sample Size (units)	A_1
2	3.76
3	2.39
4	1.88
5	1.60
10	1.03

The *drawing tolerance limits* for a part are larger than the control limits, and are never inside of them. Once the control chart has been established, data are recorded on it and it becomes a record of the variation of an inspected dimension over a period of time. The data plotted should fall in random fashion between the control lines if all assignable

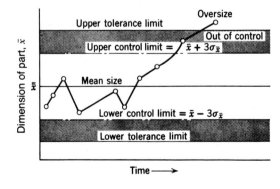

Figure 11.35
Control chart characteristics.

causes for variation are absent. When the data fall in this manner, it can be assumed that the part is being made correctly 99.73% of the time; that is, no more than 3 out of 1000 will leave the process incorrectly made.

So long as the points fall between the control lines, no adjustments or changes in the process are necessary. If five to seven consecutive points fall on one side of the mean, the process should be checked. When points fall outside of the control lines, the cause must be located and corrected immediately. Observe Figure 11.35 where the dimension as measured moves above the upper control line and the upper tolerance limit. This dimension is out of control, and action would be taken to correct the problem.

If all of the dimensions taken over a very long period of time are used and there are no assignable causes of variation, the frequency with which each size occurs can be plotted against size. A curve, known as a *normal curve* and similar to Figure 11.36, results. The significance of the standard deviation may be realized more fully by noting that 68.27% of the data fall within $\pm 1\sigma$, 95.45% within $\pm 2\sigma$, and 99.73% within $\pm 3\sigma$.

To illustrate the manner in which control limits are calculated, assume that round pieces of stock called gear blanks have been inspected. The dimensions, which may be viewed as numbers representing either English or metric units, are grouped in sample sizes of three and are shown in Table 11.3. The process was carefully controlled in that no assignable causes of errors were known to exist. The standard deviation σ is calculated according to the steps shown previously.

$$\bar{\bar{x}} = \frac{27.495}{11} = 2.500$$

$\bar{\bar{x}}$ = grand average or average of sample averages

and

$$\bar{\sigma} = \frac{0.02286}{11} = 0.00208 \text{ (say 0.002)}$$

The **upper control limit, UCL,** and the **lower control limit, LCL,** are each equal to

$$A_1\bar{\sigma} = 3\sigma_{\bar{x}}$$

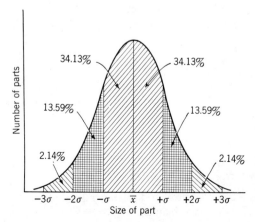

Figure 11.36
Normal distribution and percentage of parts that will fall within sigma limits.

METROLOGY AND QUALITY CONTROL

Table 11.3 **Control Limit Calculations for $\bar{x}$ Chart**

Sample number	Dimension			Average Size in Sample $\bar{x}$	Standard deviation σ
	x_2	x_2	x_3		
1	2.495	2.501	2.499	2.498	0.00252
2	2.501	2.500	2.496	2.499	0.00216
3	2.501	2.495	2.498	2.498	0.00245
4	2.497	2.500	2.503	2.500	0.00245
5	2.497	2.503	2.501	2.500	0.00252
6	2.502	2.500	2.498	2.500	0.00163
7	2.499	2.499	2.496	2.498	0.00141
8	2.500	2.503	2.505	2.503	0.00216
9	2.500	2.497	2.499	2.498	0.00141
10	2.499	2.503	2.501	2.501	0.00163
11	2.503	2.497	2.501	2.500	0.00252
				27.495	$\Sigma \bar{x} = 0.02286$

Therefore,

$$3\sigma_{\bar{x}} = 2.39(0.002) = 0.0048$$

$$= 0.005 \text{ (approx.)}$$

Figure 11.37 represents the control chart, and future data can be plotted on this chart to determine the control of the process. Usually more data are used in determining σ. This example indicates the method only.

Bilateral tolerance limits must be at least 2.500 ± 0.005 and would almost always be at least 2.500 ± 0.006 in order that there be some leeway between control and tolerance limits.

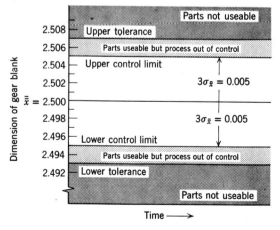

Figure 11.37 Control and tolerance limits for gear blanks.

Control Limit Calculations for the p-Chart

The characteristic being plotted on these charts is p, the fraction of defective parts found in a sample. Sometimes p is called a *proportion*. This type of quality control analysis evaluates characteristics that are obtained on a "go" or "no-go" basis. Therefore, the p-chart provides an overall picture of quality, and the products are thus divided into two categories only, either acceptable or unacceptable. For this fraction of defectives chart, the distinction is whether the lot of product is a conforming one, or it is defective. The lot is accepted on the basis of the p-value, that is if the p-value falls within the control limits. Otherwise the lot is rejected and returned to the supplier.

In the formation of p-charts the observations are classified into subgroups called samples. We will consider samples of equal size. The sample proportion defective is

$$p = d/n$$

where

d = Number of defectives found in a sample

n = Pieces inspected, or sample number

The overall average proportion of defectives is the sum of sample proportion defectives divided by the combined samples size or $\bar{P}$ (called P-bar). This will be the central line on the p chart.

$$\bar{P} = \frac{\Sigma p}{N}$$

where N is the number of samples.

The standard deviation of the number of defectives derived from probability mathematics is

$$\sigma_{\bar{p}} = \sqrt{\frac{\bar{P}(1 - \bar{P})}{n}}$$

Instead of operating in terms of the number of defectives, the control limits for the p chart are established in terms of the mean proportion of defectives p. Control limits may then be established. The choice of 3σ control limits (three standard deviations from the central line p) is an economic choice based on experience in the field that has proved satisfactory for detecting assignable causes of variation. Therefore, control limits are as follows:

$$\text{LCL} = \bar{P} - 3\sigma_{\bar{p}}$$

$$\text{UCL} = \bar{P} + 3\sigma_{\bar{p}}$$

If a negative lower limit is obtained, the limit is replaced by zero. On a p-chart, plot the sample number along the x-axis direction and the proportion defective p in the y-axis direction.

Consider the following example. A producer of silicon computer chips randomly selects 1000 chips per day ($n = 1000$) for inspection. The results of the inspection are to accept the chip as either OK or not OK. This type of inspection results in a *p*-chart. Data for 20 days of inspection ($N = 20$) are in Table 11.4.

$$\text{Average proportion defective} = \overline{P}$$
$$= 1025/(20 \times 1000)$$
$$= 0.0513$$

The standard deviation for the *p*-chart is found using *P*-bar, and is

$$\sigma_{\overline{p}} = \sqrt{\frac{0.0513(0.9488)}{1000}} = 0.0070$$

$$\text{UCL} = 0.0513 + 3 \times 0.007 = 0.072$$
$$\text{LCL} = 0.0513 - 3 \times 0.007 = 0.030$$

Observe Figure 11.38 where the points are plotted on the *p* charts, and all the points fall within the upper and lower control limits, and the process is considered in control.

Table 11.4 Data for 20 Days of Inspection

Date	Number of defects	Proportion defective
1	44	0.044
2	57	0.057
3	44	0.044
4	68	0.068
5	58	0.058
6	36	0.036
7	52	0.052
8	44	0.044
9	69	0.069
10	63	0.063
11	35	0.035
12	69	0.069
13	36	0.036
14	51	0.051
15	51	0.051
16	64	0.064
17	42	0.042
18	49	0.049
19	43	0.043
20	50	0.050
TOTAL:	1025	

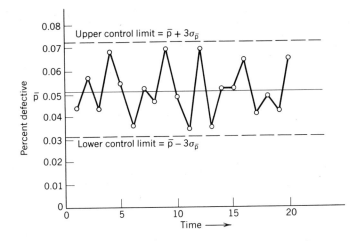

Figure 11.38
p-Chart of the fraction of good units to sample number of units.

The inspection will continue and when sample fraction defective points fall outside of the limits, action is required.

QUESTIONS AND PROBLEMS

1. Give an explanation of the following terms:

   Metrology
   Dimension
   Tolerance
   Allowance
   Unilateral tolerance
   Gage blocks
   Micrometer
   Sine bar
   Surface measurement
   Surface plate
   Interferometry
   Surface roughness
   Average roughness
   Inspection gages

   Snap gage
   "Go," "no go"
   Plug gage
   Electronic gage
   LVDT
   Nondestructive test
   Magnaflux
   Radiography
   Quality control
   Assignable cause
   Random choice
   Standard deviation
   Sample
   Upper limit

2. What is meant by nominal size and tolerance?

3. Discuss the allowances in the manufacture of a snap gage.

4. Why is unilateral tolerance preferred over bilateral tolerance?

5. Sketch a micrometer barrel showing 0.862 in. (21.89 mm). Also in metric dimension.

6. Why is 100% inspection impractical for most purposes? When is it necessary?

7. Discuss the relationship of product cost to accuracy.

8. Discuss pro and con the statement that quality does not add to product value.

9. What is meant by interchangeable manufacture?

10. Write a method for certifying the accuracy of a micrometer.

11. What is the difference between assignable and random causes of variations?

12. Give 10 examples of assignable and random causes of variation.

13. What percentage of the data would fall outside control limits naturally if $\pm 2\sigma$ were used as control limits?

14. What percentage of the data should fall within $\pm 1\sigma$ on the normal distribution curve? Within $\pm 2\sigma$?

15. Sketch a vernier caliper reading of 0.702 in.

16. Explain the setting of tolerance limits using a control chart.

17. How could a laser determine if a machine were set so that its bed was horizontal?

18. Discuss the sample size and its effect on control chart construction and the use of a control chart for inspection.

19. Discuss the principal advantage of making a plug gage using glass.

20. Sketch the barrel of a 4-in. (100-mm) micrometer that is reading 3.762 in. (95.56 mm).

21. How would you measure the lead of a small screw using a toolmaker's microscope?

22. Sketch a ground scissor blade and give inspection procedure for its acceptance or rejection.

23. Indicate ways of ensuring long linear measurements, 3 m or more, that are correct to 1 part in 100.

24. Define, sketch, or describe the differences between a surface gage, profilometer, toolmaker's flat, optical flat, and surface plate.

25. Sketch the fringes on an optical flat if the surface of the part being inspected is high in the center and low on all edges; if the part has a "valley" going one direction down the center and a "hill" going perpendicular to it; and if the part is perfectly level.

26. List the advantages of the principal types of nondestructive testing.

27. Give the advantages of pneumatic, electric, and electronic gages.

28. Why is the electronic gage preferred over the electric gage?

29. Laser light has unique advantages for inspection. What are they?

30. Referring to Figure 11.15, estimate the surface roughness of the following: scissor blade, window glass, carburetor die casting, airplane wing, master gage block, surface plate, structural steel, paper clip, drawing triangle, and the bottom of a cast-iron frying pan.

31. What general type of dimensional inspection equipment do you recommend for "mass-production" inspection of bottle caps, gears for a watch, large forging, extruding tube, and plastic steering wheels?

32. Find the average and the standard deviation for the following samples: (1.501, 1.504, 1.503); (1.501, 1.504, 1.503, 1.503); and (2.501, 2.503, 2.506, 2.506).

33. Find the upper and lower control limits for an average 1.503, standard deviation of .005, and sample size of 4. Reconstruct the limits for a sample size of 5.

34. Calculate the standard deviation of the following dimensions if they are grouped with a sample size of 3. Use horizontal groupings.

4.188	4.186	4.190
4.186	4.187	4.189
4.184	4.183	4.187
4.191	4.189	4.189
4.183	4.190	4.186
4.186	4.182	4.186
4.188	4.183	4.184
4.182	4.188	4.190
4.184	4.185	4.189
4.189	4.187	4.183
4.189	4.191	4.183
4.183	4.184	4.185

Calculate the standard deviation.

35. Calculate from problem 34 the control chart limits and draw the control chart with upper and lower control limits.

36. If, after constructing the control chart in problem 35, the following average sample dimensions are found by inspection, give your opinion as to whether the process is in control:

4.189	4.191	4.189	4.196
4.192	4.193	4.194	4.191

37. Recalculate the standard deviation in problem 34 if the sample size is 2. Each column provides six samples. Draw a control chart for data in problem 34 using a sample size of 2.

38. Recalculate the standard deviation and the average for problem 34 with a sample size of 4. Each column provides three samples. Draw the control chart.

39. Construct a p-chart if the number of total defectives is 1350 for 20 samples of 1000 units.

40. What typical measurement devices will require a p-chart or an x-bar chart?

41. During 30 days 1907 defectives were determined from equal-sized samples of 1000 units. Determine the upper and lower control limits for the p-chart.

42. Examining Figure 11.11, what is the irregularity if the fringes are made by using monochromatic light that has a wavelength of 0.000017 in. (432 nm)?

43. Calculate the resolution of a laser beam in metric units. How much does this amount to in a 5-m length?

44. Four parts with the following dimensions are to be assembled in tandem. What is the tolerance on the assembled length?

Part A = $6.191^{+0.005}_{-0.001}$

Part B = 4.752 ± 0.003

Part C = $13.352^{+0.006}_{-0.008}$

Part D = 7.273 ± 0.005

The tolerance of an assembled length T_s equals $\sqrt{(T_A)^2 + (T_B)^2) + \ldots}$ where T_A, etc. equals the tolerance of the individual parts. "Redesign" the dimensions and make them (1) all unilateral and (2) all bilateral. In SI also.

45. A 3-in. gage block is set up under one end of a 10-in. sine bar. (a) What height would have to be used on the other end to check an angle of 60°? For 48°? (b) Repeat for 30° (c) Repeat (a) if $L = 20$ in.

46. Find the average and root-mean-square for the following measurements: $a = 0.0042$, $b = 0.0043$, $c = 0.0039$, $d = 0.0047$, $e = 0.0045$.

47. A surface is measured and the following deviations are noted. Calculate the arithmetic and root-mean-square average roughness.

$a = 0.0039$	$g = 0.0049$	$m = 0.0040$
$b = 0.0045$	$h = 0.0058$	$n = 0.0031$
$c = 0.0057$	$i = 0.0061$	$o = 0.0042$
$d = 0.0046$	$j = 0.0052$	$p = 0.0045$
$e = 0.0030$	$k = 0.0045$	
$f = 0.0035$	$l = 0.0053$	

48. Determine the out-of-flatness constant error for the following calibration data of a micrometer caliper. Using a 0.2500-in. ball gage, five different locations on the anvil and spindle face were tested. The locations are shown in Figure 11.39. The average measurement is determined from a series of readings made of each location.

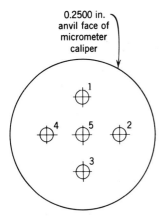

Figure 11.39
Figure for problem 48.

Ball gage size		1	2	3	4	5
0.2500	reading	0.2501	0.2500	0.2500	0.2501	0.2501
0.3125	reading	0.3125	0.3122	0.3122	0.3122	0.3125

(a) Give the out-of-flatness for the 0.2500-in. ball. (b) Parallelism for the anvil and spindle face can be found using a second, 0.3125-in., ball gage that is placed in the same positions, and the data are recorded as above. The 0.3125-in. dimension results in a 180° difference of the angular position of the spindle. Find the average difference to indicate any non-parallelism.

49. Find the gage blocks for a 0.3125-in. dimension, and for a 4.0245-in. dimension.

50. Determine gage block set for a 3.5687 in. dimension using the fewest number of blocks. Repeat for 2.178 in.

CASE STUDY

MINIMUM COST TOLERANCE

McBride Production makes gear trains for industry. Whaley Enterprises has approached McBride with a requirement. Bob Oliver, after studying the specifications, provides a design similar to that in Figure 11.40. The important components and their critical functional dimensions are as follows:

Part number	Critical dimension	Basic dimension	Tolerance
Spacer	A	0.250 in. (6.35 mm)	±0.010 in. (±0.25 mm)
Spur gear	B	1.000 in. (25.40 mm)	±0.010 in. (±0.25 mm)
Bevel pinion	C	3.250 in. (82.55 mm)	±0.010 in. (±0.25 mm)
		4.500 in. (114.30 mm)	±0.030 in. (±0.75 mm)

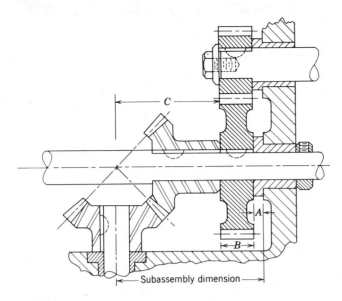

Figure 11.40
Case study. Gear subassembly.

Design criteria dictate that the subassembly length be 4.500 ± 0.030 in. (114.30 ± 0.75 mm). Oliver is confident that if three maximum (or minimum) length parts happen to come together in the same assembly, the assembly will function.

Metrologist Pat Toole, after seeing the design, asks engineering to reconsider increasing the tolerances.

"All right," says Bob, "but don't increase the cost or make rejects."

Pat Toole reasons that a model given by

$$T_{sum} = 0.030 \text{ in.}$$
$$= [(T_a)^2 + (T_b)^2 + (T_c)^2]^{1/2}$$

will serve, and as the first approximation he lets $T_a = T_b = T_c$. Determine what these balanced tolerances are. Determine a process and cost to produce Oliver's tolerance and compare this cost to your tolerance cost as determined by the formula. Use several trials, where $T_a \neq T_b \neq T_c$, and determine a lower cost. Continue in this manner until you have lowest cost and a practical process to produce the dimension. Use Figure 11.1 to indicate cost.

Once you have specified the minimum tolerances, establish a testing procedure to ensure quality compliance to ±0.030 in. (±0.75 mm) subassembly tolerance. What inspection methods do you recommend?

CHAPTER 12
HOT WORKING OF METAL

An ingot of steel has little economic application until it has been formed into shape for subsequent manufacturing. A cold ingot cannot be economically converted into other shapes. Hot working covers the process of shaping an ingot into structural shapes, bar stock, or sheet form. With the ingot hot, it can be hammered, pressed, rolled, or extruded into other shapes. Because of scaling and oxidation at high temperatures, most ferrous metals are finish processed by cold working to obtain an improved surface finish, higher dimensional accuracy, and better mechanical properties.

This chapter involves forming materials at a temperature above the *recrystallization temperature*. This class of processes is different from shaping materials by casting, molding, machining, or chip removal.

PLASTIC DEFORMATION

The two principal types of mechanical work in which material undergoes plastic deformation and is changed in shape are hot working and cold working (discussed in Chapter 13).

Like many metallurgical concepts, the difference between hot and cold working is not easy to define. When metal is hot worked, the forces for deformation are less and the mechanical properties are relatively unchanged. When a metal is cold worked, greater forces are required and the strength of the metal is increased. In hot working the thickness of the material is substantially changed, but in some cold-working operations, such as the finish rolling of sheet metal, the thickness remains approximately the same.

In the manufacture of metal components, basic alternatives available for the production of a desired shape include casting, machining, consolidating small pieces (welding), and deformation processes. Hot working is a deformation process. Metal deformation processes exploit an interesting property of metals: their ability to flow plastically in the solid state without accompanying deterioration of properties. Moreover, in forcing the metal to the desired shape there is little or no waste.

Hot working is the plastic deformation of metals above their recrystallization temperature. It is important to note that the recrystallization temperature varies with different materials, and hot working does not necessarily imply high absolute temperature. For instance, lead and tin are hot worked at room temperature. Some recrystallization temperatures of common metals are listed here:

Metal	Recrystallization temperature °F (°C)
Aluminum	300 (150)
Copper	390 (200)
Gold	390 (200)
Iron	840 (450)
Lead	Below room temperature
Magnesium	300 (150)
Nickel	1100 (590)
Silver	390 (200)
Tin	390 (200)
Zinc	At room temperature

Although hot working causes plastic deformation because it is done above the recrystallization temperature, it does not produce strain hardening. The hot-worked metal does not possess a greater elastic limit or become stronger and usually undergoes a decrease in yield strength, that is, a point where additional strain occurs without any increase in stress load on the material. Ductility, which is the ability of a material to be deformed plastically without fracture, is impaired. Thus it is possible to alter the shape of the metals drastically with moderated forces by hot working and without causing fracture.

The recrystallization temperature of a metal determines whether or not hot or cold working is being accomplished. For steel, recrystallization starts around 950° to 1300°F (500°–700°C), although most hot work on steel is at temperatures considerably above this range. There is no tendency for hardening by mechanical work until the lower limit of the recrystalline range is reached. Some metals such as lead and tin have a low recrystalline range and can be hot worked at room temperature, but most commercial metals require some heating. Alloy composition has an influence on the proper working range, the usual result being to raise the recrystalline range temperature. This range may also be increased by prior cold working.

During all hot-working operations the metal is in a plastic state and is readily formed by pressure. In addition, hot working has the following advantages:

1. Porosity in the metal is largely eliminated. Most ingots contain many small blow holes. These are pressed together and eliminated.
2. Impurities in the form of inclusions are broken up and distributed throughout the metal.
3. Coarse or columnar grains are refined. Because this work is in the recrystalline range, it should be carried on until the low limit is reached to provide a fine grain structure.
4. Physical properties are generally improved, principally as a result of the grain refinement. Ductility and resistance to impact are improved, strength is increased,

and greater homogeneity is developed in the metal. The greatest strength of rolled steel exists in the direction of metal flow.
5. The amount of energy necessary to change the shape of steel in the plastic state is far less than that required when the steel is cold.

All hot-working processes present a few disadvantages that cannot be ignored. Because of the high temperature of the metal there is a rapid oxidation or *scaling* of the surface with accompanying poor surface finish. As a result of scaling, close tolerances cannot be maintained. Hot-working equipment and maintenance costs are high, but the process is economical compared to working metals at low temperatures.

The term *hot finished* refers to steel bars, plates, or structural shapes that are purchased in the "as-rolled" state from the hot-working operation. Some descaling is done, but otherwise the steel is ready for use in bridges, ships, railroad cars, and other applications where close dimensional tolerances are not required. The material has good weldability and machinability, because the carbon content is less than 0.25%.

ROLLING

Steel that is not remelted and cast into molds is converted to useful products in two steps:

1. The ingot is rolled into intermediate shapes: blooms, billets, and slabs.
2. These blooms, billets, and slabs are further rolled into plates, sheets, bar stock, structural shapes, or foils.

The ingot remains in molds until the solidification is about complete and the molds are removed. While still hot, the ingots are placed in gas-fired furnaces called **soaking pits,** where they remain until they have attained a uniform working temperature of about 2200°F (1200°C) throughout. The ingots are taken to the rolling mill where, because of the large variety of finished shapes to be made, they are first rolled into intermediates shapes as blooms, billets, or slabs. A **bloom** has a square cross section with a minimum size of 6 by 6 in. (150 by 150 mm). A *billet* is smaller than a bloom and may have any square section from 1 $1^{1}/_{2}$ in. (38.1 mm) up to the size of a bloom. *Slabs* may be rolled from either an ingot or a bloom. They have a rectangular cross-sectional area with a minimum width of 10 in. (250 mm) and a minimum thickness of 1 $1^{1}/_{2}$ in. (38.1 mm). The width is always three or more times the thickness, which may be as much as 15 in. (380 mm). Plates, skelp, and thin strips are rolled from slabs.

One effect of a hot-working rolling operation is the grain refinement brought about by *recrystallization*. This is shown diagrammatically in Figure 12.1. The coarse structure is broken up and elongated by the rolling action. Because of the high temperature, recrystallization starts immediately and small grains begin to form. These grains grow rapidly until recrystallization is complete. Growth continues at high temperatures, if further work is not carried on, until the low temperature of the recrystalline range is reached.

Arcs AB and $A'B'$ are the contact arcs on the rolls. The wedging action on the work

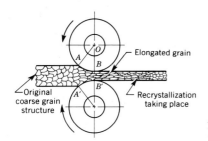

Figure 12.1
Effect of hot rolling on grain structure.

is overcome by the frictional forces that act on these arcs and draw the metal through the rolls. In the process of rolling, stock enters the rolls with a speed less than the peripheral roll speed. The metal emerges from the rolls traveling at a higher speed than it enters. At a point midway between A and B, metal speed is the same as the roll peripheral speed. Most deformation takes place in thickness, although there is some increase in width. Temperature uniformity is important in all rolling operations, because it controls metal flow and plasticity.

In rolling, the quantity of metal going into a roll and out of it is the same, but the area and velocity are changed. Thus,

$$Q_1 = Q_2 = A_1 V_1 = A_2 V_2$$

where

Q_1 = Quantity of metal going into roll

Q_2 = Quantity of metal leaving roll

A_1 = area (ft^2) of an element in front of roll

A_2 = Area (ft^2) of an element after roll

V_1 = Velocity (ft/s) in element before roll

V_2 = Velocity (ft/s) in element after roll

$$\frac{A_1}{A_2} = \frac{V_2}{V_1}$$

As the cross-sectional area is decreased the velocity increases, as do lengths of the material.

Most primary rolling is done in either a two-high reversing mill or a three-high continuous rolling mill. In the *two-high reversing mill* (Figure 12.2A) the piece passes through the rolls, which are then stopped and reversed in direction, and the operation is repeated. At frequent intervals the metal is turned 90° on its side to keep the section uniform and to refine the metal throughout. About 30 passes are required to reduce a large ingot into a bloom. Grooves are provided on both the upper and the lower rolls to accommodate the various reductions in cross-sectional area. The two-high rolling mill is versatile, because it has a range of adjustment as to size of pieces and rate of reduction. It is limited by the length that can be rolled and by the inertia forces that must be overcome each time a reversal is made. These are eliminated in the **three-high mill** (Figure 12.2C),

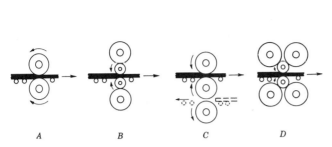

Figure 12.2
Various roll arrangements used in rolling mills. A, Two-high mill, continuous reversing. B, Four-high mill with backing-up rolls for wide sheets. C, Three-high mill for back-and-forth rolling. D, Cluster mill using four backing-up rolls.

but an elevating mechanism is required. The three-high mill is less expensive to manufacture and has a higher output than the reversing mill.

Billets could be rolled to size in a large mill for blooms, but usually this is not done for economic reasons. Frequently, they are rolled from blooms in a continuous billet mill consisting of six or more rolling stands in a straight line. The steel makes but one pass through the mill and emerges with a final billet size, approximately 2 by 2 in. (50 by 50 mm), which is the raw material for many final shapes such as bars, tubes, and forgings. Figure 12.3 illustrates the number of passes and the sequence in reducing the cross section of a 4 by 4 in. (100 by 100 mm) billet to round bar stock.

Other arrangements of rolls used in rolling mills are shown in Figure 12.2. Those that have four or more rolls use the extra ones for backing up the two that are doing the rolling. In addition, many special rolling mills take previously rolled products and fabricate

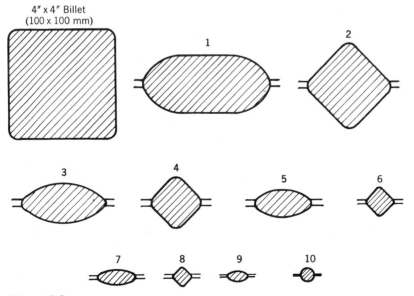

Figure 12.3
Diagram illustrates number of passes and sequences in reducing the cross section of a 4-by-4 in. (100 by 100 mm) billet to round bar stock.

them into such finished articles as rails, structural shapes, plates, and bars. Such mills usually bear the name of the product being rolled and, in appearance, are similar to mills used for rolling blooms and billets. A mill specializing in rolling rails is able continuously to roll a rail almost a quarter of a mile long.

FORGING

Hammer or Smith Forging

Hammer or *smith forging* consists of hammering the heated metal either with hand tools or between flat dies in a steam hammer. *Hand forging* as done by the blacksmith is the oldest form of forging. The nature of the process is such that accuracy is not obtained nor can complicated shapes be made. Forgings ranging from a few pounds to over 200,000 lb (90 Mg) are made by smith forging.

Forging hammers are made in the single- or open-frame type for light work, whereas the double-housing type is made for heavier service. A typical steam hammer is shown in Figure 12.4. The force of the blow is closely controlled by the hammer operator, and considerable skill is required.

Drop Forging

Drop forging differs from hammer forging in that closed-impression rather than open-face or flat dies are used. The dies are matched and separately attached to the movable ram and the fixed anvil. The forging is produced by impact or pressure, which compels the hot and pliable metal to conform to the shape of the dies, as in Figure 12.5. In this operation there is drastic flow of the metal in the dies caused by the repeated blows on

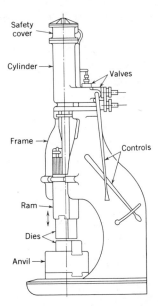

Figure 12.4
Open-frame steam hammer.

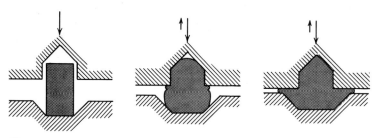

Figure 12.5
Drop forging with closed dies.

the metal. To ensure proper flow of the metal during the intermittent blows, the operation is divided into a number of steps. Each step changes the form gradually, controlling the flow of the metal until the final shape is obtained. The number of blows required varies according to the size and shape of the part, the forging qualities of the metal, and the tolerances required. For products of large or complicated shapes, a preliminary shaping operation using more than one set of dies may be required. Approximate forging temperatures are as follows: steel, 200° to 2300°F (1100°–1250°C); copper and its alloys, 1400° to 1700°F (750°–925°C); magnesium, 600°F (315°C); and aluminum, 700° to 850°F (370°–450°C). Closed-die steel forgings vary in size from a few ounces to 22,000 lb (10 Mg).

The two principal types of drop-forging hammers are the *steam hammer* and the *gravity drop* or **board hammer.** In the former the ram and hammer are lifted by steam, and the force of the blow is controlled by throttling the steam. These hammers operate at over 300 blows a minute. The capacities of steam hammers range from 500 to 50,000 lb (2–200 kN). They are usually of double-housing design, with an overhead steam cylinder assembly providing the power for activating the ram. For a given weight ram a steam hammer will develop twice the energy at the die as can be obtained from a board or gravity drop hammer.

In the gravity-type hammer the impact pressure is developed by the force of the falling ram as it strikes upon the lower fixed die. A piston lift gravity drop hammer is shown in Figure 12.6. It utilizes air or steam to lift the ram. This type of hammer permits the preselection of a series of short- and long-stroke blows. The operator is relieved of the responsibility of regulating stroke heights, and greater uniformity in finished forgings results. Hammers of this type are procurable for ram weights of 500 lb (225 kg) up to and including ram weights of 10,000 lb (4500 kg). The board drop hammer has several hardwood boards attached to the hammer for lifting purposes. After the hammer has fallen, rollers engage the boards and lift the hammer up to 5 ft (1.5 m). When the stroke is reached the rollers spread and the boards are held by dogs until they are released by the operator. The force of the blow is entirely dependent on the weight of the hammer, which seldom exceeds 8000 ℓb (35 kN). The board hammer is not so quick lifting as the air or steam unit. Gravity hammers find extensive use in industry for such articles as hand tools, scissors, cutlery, and implement parts. Operator protection for hands, eyes, and ears is required for both types of hammers.

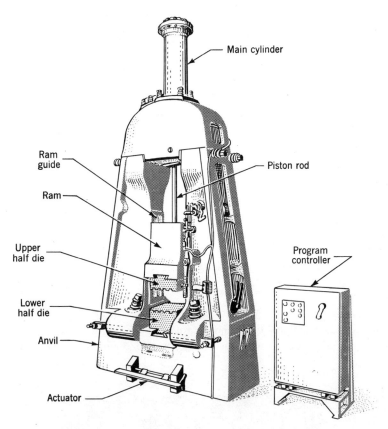

Figure 12.6
Piston lift gravity drop hammer.

The *impacter forging hammer* (Figure 12.7) has two opposing cylinders in a horizontal plane which actuate the dies toward each other. Stock is positioned in the impact plane in which the dies collide. Its deformation absorbs the energy, and there is less shock or vibration in the machine. With this process the stock is worked equally on both sides, there is less time of contact between stock and die, less energy is required than with other forging processes, and the work is held mechanically.

A forging will have a thin projection of excess metal extending around it at the parting line that is removed in a separate trimming press. Small forgings may be trimmed cold, although care must be taken in the trimming operation not to distort the part. The forging is usually held uniformly by the die in the ram and pushed through the trimming edges. Punching operations may also be done while trimming is taking place.

Figure 12.8 shows the dies for forging the main landing gear outer cylinders for a large aircraft. The dies weight over 31 tons (28 Mg). Some forging operations require reheating of the part between die stations.

Upon completion all forgings are covered with scale and must be cleaned. This can

294 HOT WORKING OF METAL

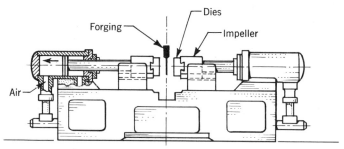

Figure 12.7
Horizontal-impact forging machine.

be done by pickling in acid, shot peening, or tumbling, depending on the size and composition of the forgings. If some distortion has occurred in forging, a sizing or straightening operation may be required. Controlled cooling is usually provided for large forgings and, if certain physical properties are necessary, provision is made for heat treatment.

Advantages of the forging operation include a fine crystalline structure of the metal, closing of any voids, reduced machining time, and improved physical properties. Forging is adaptable to carbon and alloy steels, wrought iron, copper, and aluminum and mag-

Figure 12.8
Forging die for main landing gear outer cylinder.

nesium alloys. Disadvantages include scale inclusions and the high cost of dies, which prohibits short-run jobs. Die alignment is sometimes difficult to maintain, and care is required in die design to prevent cracks from occurring in the forging as a result of the metal's folding over during the operation. Closed-impression die forgings have better utilization of material than open, flat dies, better physical properties, closer tolerances, higher production rates, and less requirement of operator skill.

Press Forging

Press forging employs a slow squeezing action in deforming the plastic metal as contrasted to the rapid impact blows of a hammer. The squeezing action is carried completely to the center of the part being pressed, thoroughly working the entire section. These presses are the vertical type and may be either mechanically or hydraulically operated. The mechanical presses, which are faster operating, can exert 500 to 10,000 tons (4–90 MN) force.

The pressure necessary to form steel at forging temperature varies from about 3000 to 27,000 psi (20–190 MPa). These pressures are based on the cross-sectional area of the forging when measured across the surface of the die at the parting line.

The press capacity is expressed as

$$F = \frac{P}{A \times 2000} = \frac{5 \times P \times 10^{-4}}{A}$$

where

F = Press capacity, tons
P = Pressure required, psi (usually about 15,000 psi for mild steel)
A = Area of the forging at the parting line, in.^{2}

For small press forgings *closed-impression dies* are used, and only one stroke of the ram is normally required to perform the forging operation. The maximum pressure is built up at the end of the stroke, which forces the metal into shape. Dies may be mounted as separate units and one or two or more cavities may be cut into a single block. There is some difference in the design of dies for different metals. Copper alloy forgings can be made with less draft than steel; consequently, more complicated shapes can be produced.

In the forging press a greater proportion of the total energy input is transmitted to the metal than in a drop hammer press. Much of the impact of the drop hammer is absorbed by the machine and foundation. Press reduction of the metal is faster, and the cost of operation is consequently lower. Most press forgings are symmetrical in shape with surfaces that are smooth, and they provide a closer tolerance than is obtained by a drop hammer. However, many parts of irregular and complicated shapes can be forged more economically by drop forging. Forging presses are often used for sizing operations on parts made by other processes.

Upset Forging

Upset forging entails gripping a bar of uniform section in dies and applying pressure on the heated end, causing it to be upset or formed to shape (Figure 12.9).

The maximum length of stock to be upset is expressed as

$$L = \frac{kP}{\pi}$$

where

L = Maximum length of stock to be upset, in. (mm)

P = Perimeter of cross section, in. (mm)

k = Constant with values of 2 or 3 (usually about 2.6 for steel)

The length of the stock to be upset cannot be more than two or three times the diameter or else the material will bend rather than bulge out to fill the die cavity.

For some products the heading operation may be completed in one position, although in most cases the work is progressively placed in different positions in the die. The impressions may be in the punch, in the gripping die, or in both. In most instances the forgings do not require a trimming operation. Machines of this type are an outgrowth of smaller machines designed for cold-heading nails and small bolts.

Progressive piercing, or internal displacement, is the method frequently employed on upset forging machines for producing parts such as artillery shells and radial engine cylinder forgings. The sequence of operations for cylinder forging is shown in Figure 12.10. Round blanks of a predetermined length for a single cylinder are first heated to forging temperature. To facilitate handling the blank, a *porter bar* is pressed into one end. The blank is upset and is progressively pierced to a heavy-bottom cup. In the last operation a taper-nosed punch expands and stretches the metal into the end of the die, frees the porter bar, and punches out the end slug. Large-cylinder barrels weighing over 100 lb (45 kg) can be forged in this manner. Parts produced by this process range from small to large products weighing several hundred pounds. The dies, not limited to upsetting, may also be used for piercing, punching, trimming, or extrusion.

To produce more massive shapes by this method, a continuous upsetting machine has been developed. This machine can feed induction-heated bar stock to the die cavity, where rapid blows of the upsetting die build up the part. Some of these machines have a hollow upsetting die, so that long lengths of constant cross section shape can be produced.

Another variation to upset forging is ***metal gathering.*** Rather than form an opening

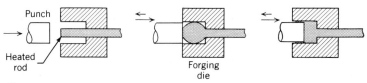

Figure 12.9
Upset forging.

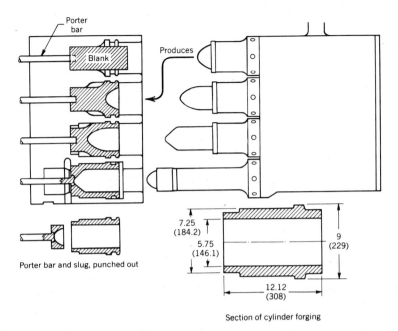

Figure 12.10
Sequence of operations for a cylinder forging on an upset forging machine.

in heated bar stock, an operation of forging a conical shape, similar to Figure 12.9 is followed.

Roll Forging

Roll forging machines are primarily adapted to reducing and tapering operations on short lengths of bar stock. The rolls on the machine (Figure 12.11) are not completely circular but are 25% to 75% cut away to permit the stock to enter between the rolls. The circular portion of the rolls is grooved according to the shaping to be done. When the rolls are in open position the operator places the heated bar between them, retaining it with tongs.

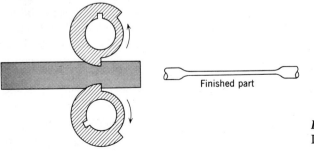

Figure 12.11
Principle of roll forging.

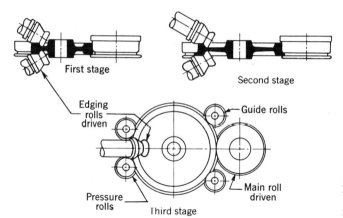

Figure 12.12
Wheels formed by hot-roll forging.

As the rolls rotate, the bar is gripped b the roll grooves and moved forward. When the rolls open, the bar is pushed back and rolled again or is placed in the next groove for subsequent forming work. By rotating the bar 90° after each roll pass, there is no opportunity for flash to form.

In rolling wheels, metal tires, and similar items a *roll mill* of somewhat different construction is used. Figure 12.12 shows how a rough forged blank is converted into a finished wheel by the action of the various rolls circumventing the wheel. As the wheel rotates, the diameter is gradually increased while the plate and rim are reduced in section. When the wheel is rolled to its final diameter, it is transferred to a press and given a dishing and sizing operation.

Roll forging is sometimes used for axles, blanks for airplane propellers, crowbars, knife blades, chisels, tapered tubing, and ends of leaf springs. Parts made in this fashion have a smooth finished surface and tolerances equal to other forging processes. The metal is hot worked thoroughly and has good physical properties.

EXTRUSION

Metals that can be hot worked can be extruded to uniform cross-sectional shape by the aid of pressure. The principle of *extrusion,* similar to the act of squirting toothpaste from a tube, has long been utilized in processes ranging from the production of brick, hollow tile, and soil pipe to the manufacture of macaroni. Some metals, notably lead, tin, and aluminum, may be extruded cold, whereas others require the application of heat to render them plastic or semisolid before extrusion. In the actual operation of extrusion, the processes differ slightly, depending on the metal and application, but in brief they consist of forcing metal (confined to a pressure chamber) through specially formed dies. Rods, tubes, molding trim, structural shapes, brass cartridges, and lead-covered cables are typical products of metal extrusion.

Most presses used in conventional extruding of metals are a horizontal type that is hydraulically operated. Operating speeds, depending on temperature and material, vary from a few feet a minute up to 900 ft/min (4.6 m/s).

The advantages of extrusion include the ability to produce a variety of shapes of high strength, good accuracy, and good surface finish at high production speeds with a relatively low die cost. More deformation or shape change can be achieved by this process than by any other except casting. Almost unlimited lengths of a continuous cross section can be produced, and because of low die costs production runs of 500 ft (150 m) may justify its use. The process is about three times as slow as roll forming and the cross section must remain constant. There are several variations of this process.

Direct Extrusion

Direct extrusion is illustrated diagrammatically in Figure 12.13. A heated round billet is placed into the die chamber and the dummy block and ram placed into position. The metal is extruded through the die opening until only a small amount remains. It is then sawed off next to the die and the butt end removed.

Indirect Extrusion

Indirect extrusion (Figure 12.13) is similar to direct extrusion except that the extruded part is forced through the ram stem. Less force is required by this method because there is no frictional force between the billet and the container wall. The weakening of the ram when it is made hollow and the difficulty of providing good support for the extruded part constitute limitations of this process.

Impact Extrusion

In *impact extrusion* a punch is directed to a slug with such a force that the metal from the slug is pushed up and around it. Most impact extrusion operations, such as the manufacture of collapsible tubes, are cold-working ones. However, there are some metals and some products, particularly those in which thick walls are required, in which the slugs are heated to elevated temperatures. Impact extrusion is covered in the chapter on cold working (Chapter 13).

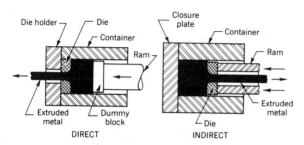

Figure 12.13
Diagram illustrating direct and indirect extrusion.

PIPE AND TUBE MANUFACTURING

Pipe and tubular products may be made by butt or electric welding, formed skelp, piercing, and extrusion. Piercing and extrusion methods are used for seamless tubing, which is found in high-temperature, high-pressure applications, as well as for transporting gas and chemical liquids. Seamless steel pipe up to 16 in. (400 mm) in diameter has been manufactured. Extruded tubes are also used for gun barrels, because the process can be adapted to internal configurations such as rifling and grooves. Butt-welded pipe is the most common and is used for structural purposes, for posts, and for conveying gas, water, and wastes. The electric-welded pipe is used primarily for pipelines carrying petroleum products or water.

Butt Welding

Intermittent and continuous butt-welding methods are used. Heated strips of steel known as *skelp*, which have the slightly beveled edges, are used so that they will meet accurately when formed to a cylindrical shape. In the intermittent process, one end of the skelp is trimmed to a V shape to permit the entry into the *welding bell,* as shown in Figure 12.14A. When the skelp is brought up to welding heat, the end is gripped by tongs that engage a draw chain. As the tube is pulled through the welding bell, skelp is formed to a cylindrical shape and the edges are welded together. A final operation passes the pipe between sizing and finishing rolls for correct sizing and scale removal. Continuous butt welding of pipe is accomplished by supplying the skelp in coils and providing a means for flash welding the coil ends to form a continuous strip. As the skelp enters the furnace, flames impinge on the edges of the strip to bring them to welding temperatures. Leaving the furnace, the skelp enters a series of horizontal and vertical rollers that form it into pipe. A schematic view of the rollers showing how the pipe is formed and sized is shown in Figure 12.14B. As the pipe leaves the rollers, it is sawed into lengths that are finally processed by

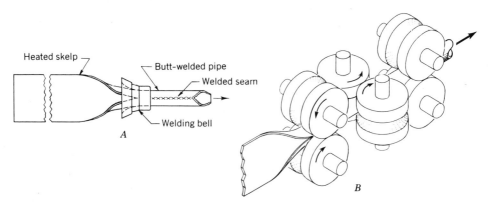

Figure 12.14
Producing butt-welded pipe. A, Drawing skelp through a welding bell, B, Skelp being formed into a continuous butt-welded pipe.

descaling and finishing operations. Butt-welded pipe is made by this method in sizes up to 3 in. (75 mm) in diameter.

Electric Butt Welding

The electric butt welding of pipe necessitates cold forming of the steel plate to shape prior to the welding operation. The form is developed by passing the plate through a continuous set of rolls that progressively change its shape. This method is known as *roll forming*. The welding unit, placed at the end of the roll forming machine consists of three centering and pressure rolls to hold the formed shape in position and two electrode rolls that supply current to generate the heat. Immediately after the pipe passes the welding unit shown in Figure 8.14, the extruded flash metal is removed from both inside and outside the pipe. Sizing and finishing rolls then complete the operation by giving the pipe accurate size and concentricity. This process is adapted to the manufacture of pipe up to 36 in. (915 mm) in diameter with wall thicknesses varying from $1/8$ to $1/2$ in. (3.2–12.7 mm). Pipes of larger diameter are usually fabricated by submerged-arc welding after being formed to shape in large, specially constructed presses.

Lap Welding

In the *lap welding* of pipe the edges of the skelp are beveled as it emerges from the furnace. The skelp is then drawn through a forming die or between rolls, to give it cylindrical shape with the edges overlapping. After being reheated, the bent skelp is passed between two grooved rolls as shown in Figure 12.15. Between the rolls is a fixed mandrel to fit the inside diameter of the pipe. The edges are lap welded by pressure between the rolls and the mandrel. Lap-welded pipe is made in sizes 2 to 16 in. (50–400 mm) in diameter.

Piercing

To produce *seamless tubing*, cylindrical billets of steel are passed between two conical-shaped rolls operating in the same direction. Between these rolls is a fixed point or mandrel that assists in the **piercing** and controls the size of the hole as the billet is forced over it.

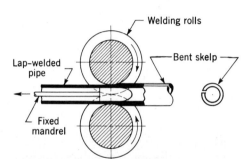

Figure 12.15
Method of producing lap-welded pipe from skelp.

302   HOT WORKING OF METAL

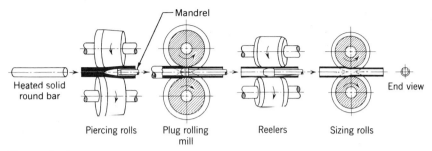

Figure 12.16
Principal steps in the manufacture of seamless tubing.

The entire operation of making seamless tubing by this process is shown in Figure 12.16. The solid billet is first center-punched and then brought to forging heat in a furnace before being pierced. It is then pushed into the two piercing rolls, which impart both rotation and axial advance. The alternate squeezing and bulging of the billet open up a seam in its center, the size and shape of which are controlled by the piercing mandrel. As the thick-walled tube emerges from the piercing mill, it passes between grooved rolls over a plug held by a mandrel and is converted into a longer tube with specified wall thickness. While still at working temperature the tube passes through the *reeling machine,* which further straightens and sizes it and in addition gives the walls a smooth surface. Final sizing and finishing are accomplished in the same manner as with welded pipe.

This procedure applies to seamless tubes up to 6 in. (150 mm) in diameter. Larger tubes up to 14 in. (355 mm) in diameter are given a second operation on piercing rolls. To produce sizes up to 24 in. (610 mm) in diameter, reheated, double-pierced tubes are processed on a rotary rolling mill as shown in Figure 12.17, and are finally completed by reelers and sizing rolls, as described in the single-piercing process.

In the continuous method shown in Figure 12.18, a $5^{1}/_{2}$-in. (139.7-mm) round bar is pierced and conveyed to the 9-stand mandrel mill, where a cylindrical bar or mandrel is inserted. These rolls reduce the tube diameter and wall thickness. The mandrel is then removed and the tube reheated before it enters the 12-stand stretch-reducing mill. This mill reduces not only the wall thickness of the hot tube but also the tube diameter. Each successive roll is speeded up to produce a tension sufficient to stretch the tube between stands. The maximum delivery of this mill is 1300 ft/min (6.6 m/s) for pipe around 2 in. (50 mm) in diameter.

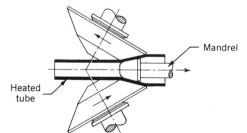

Figure 12.17
Rotary seamless process for large tubing.

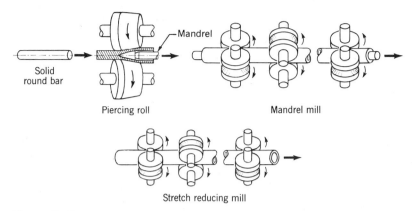

Figure 12.18
Principal steps in the manufacture of continuous tubing.

Tube Extrusion

The usual method for **extruding tubes** is shown in Figure 12.19. It is a form of direct extrusion but uses a mandrel to shape the inside of the tube. After the billet is placed inside, the die containing the mandrel is pushed through the ingot. The press stem then advances and extrudes the metal through the die and around the mandrel. The entire operation must be rapid, and speeds up to 10 ft/s (3 m/s) have been used in making steel tubes. Low-carbon steel tubes can be extruded cold, but for most alloys the billet must be heated to around 2400°F (1300°C).

DRAWING

For products that cannot be made with conventional seamless rolling mill equipment, the process illustrated in Figure 12.20 is used. A bloom is heated to forging temperature and, with a piercing punch operated in a vertical press, the bloom is formed into a closed-end hollow forging. The forging is reheated and placed in the hot drawbench consisting of several dies of successively decreasing diameter mounted in one frame. The hydraul-

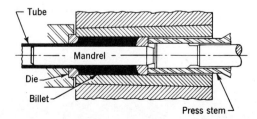

Figure 12.19
Extruding a large tube from a heated billet.

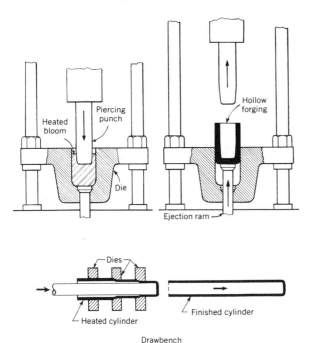

Figure 12.20
Drawing thick-walled cylinders from heated bloom.

ically operated punch forces the heated cylinder through the full length of the drawbench. For long thin-walled cylinders or tubes, repeated heating and *drawing* may be necessary. If the final product is a tube, the closed end is cut off and the tube is processed through finishing and sizing rolls similar to those used in the piercing process. To produce closed-end cylinders similar to those used for storing oxygen, the open end is swaged to form a neck or reduced by hot spinning.

SPECIAL METHODS

Hot Spinning

Hot spinning of metal is used commercially to dish or form heavy circular plates over a rotating form and to neck down or close the ends of tubes. In both cases a form of lathe is used to rotate the part rapidly. Shaping is done with a blunt-pressure tool or roller that contacts the surface of the rotating part and causes the metal to flow and conform to a mandrel of the desired shape. Once the operation is started, considerable frictional heat is generated, which aids in maintaining the metal at a plastic state. Tube ends may be reduced in diameter, formed to some desired contour, or completely closed by the spinning action.

Warm Forging

A process known as ***Thermo-Forging*** utilizes a temperature between that normally used for cold and for hot working. There are no metallurgical changes in the metal and no surface imperfections often associated with metal worked at elevated temperatures. Figure 12.21 is a photograph of a cross section of an acid-etched socket-head cap screw. The continuous-fiber structure indicative of high strength is visible. Because the flow lines follow the contour of the part, stress concentrations are reduced. The temperature of the metal and the forging pressures and speeds must be accurately controlled, because the metal is below the recrystallization temperature.

Additional Methods

To obtain thinner sections in forgings, heated dies can be employed. If the proper lubricant is used, additional surface oxidation is minimized, closer tolerances can be obtained, the work remains pliable for a longer period of time, and the production rate is increased. Die life is decreased, however, and there is a cost associated with heating the die. Unless thin sections are desired, the process is seldom justified.

High–energy rate forming is usually associated with cold-working operations, but some high-velocity presses are driven by various mechanisms, explosive charges, or capacitor discharges. Most parts formed in this manner are completed with one blow. Because the operation is fast, thin sections can be forged before the heat is lost. Because of the impact load and the rapid temperature increase of the die associated with this type of operation, die life is relatively short. The process is useful in forging high-temperature, difficult-to-form alloys.

Figure 12.21
Thermo-Forged socket head cap screw.

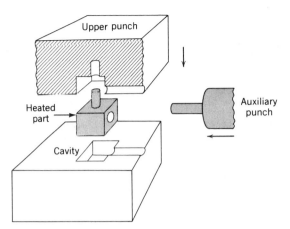

Figure 12.22
Use of an auxiliary punch in die forging.

Because of the specialized problems encountered in mass-produced parts, some forging presses are fitted with auxiliary rams or punches that move within or through conventional ones. Figure 12.22 shows the use of an *auxiliary punch* to produce a hole in the forging. Usually punches of this kind are delayed in their operation until the dies have either almost or fully completed their work. Because of the complexity of such operations, only mass-production runs can be considered with this process.

Metals that are difficult to forge (e.g., titanium) can be cast in a press surrounded by inert gas. This process, known as *environmental hot forming,* eliminates most oxidation and scaling and tends to prolong die life. For very large forgings, the inert gas is introduced into the forming area alone, but in case of small presses they are totally enclosed by a cabinet into which argon is admitted.

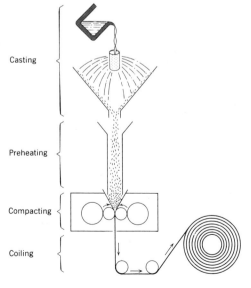

Figure 12.23
Rolling aluminum sheet from pellets.

Small aluminum pellets, smaller than grains of rice, can be rolled into sheets. Figure 12.23 shows how molten aluminum is poured into a revolving perforated cylinder. It is transported by air to a preheating chamber, hot rolled into sheet, and coiled. This process is adaptable to high-volume production with a minimum outlay of equipment. Theoretically, sheets of unlimited length can be formed by this process.

QUESTIONS AND PROBLEMS

1. Given an explanation of the following terms:

   Recrystallization
   Soaking pits
   Blooms
   Three-high mill
   Smith forging
   Board hammer
   Impact forging
   Press forging
   Upset forging
   Progressive piercing
   Metal gathering
   Roll forging
   Skelp
   Piercing
   Tube extrusion
   Drawing
   Hot spinning
   Thermo forging

2. Contrast hot- and cold-working methods in a table of attributes.

3. Why is porosity largely eliminated when metal is hot worked?

4. What is the recrystallization temperature?

5. Draw a general curve indicating how temperature affects the energy required for rolling, plasticity of metal, and the rate of oxidation.

6. Describe the difference between hot working and hot finishing.

7. Describe the difference between bars of steel finished by cold-working and hot-working processes.

8. Give general statements about the physical properties of steel as a result of hot working.

9. Why are steel ingots not cooled and then reheated before rolling?

10. A board hammer receives its name because of what feature?

11. Describe the following shapes used in connection with the rolling of steel: ingot, bloom, slab, and billet.

12. How would the following size rolled shapes be classified: 15 by 15 in. (375 by 375 mm); 20 by 2 in. (500 by 50 mm); 3 by 1 in. (75 by 25 mm); and 3 by 3 in. (75 by 75 mm)?

13. List the advantages of impact forging.

14. Sketch the shape of the rolls and describe the operations used in rolling an I-beam from an ingot.

15. Compare a forging press to a drop hammer press and discuss the energy transmitted to the work.

16. Why is it necessary to cool the rolls of a rolling mill with water when this causes scale to form on the metal and also reduces its temperature?

17. Why can open-die forgings be made larger than closed-die forgings?

18. Describe progressive piercing and why it is done.

19. What advantages does press forging have over drop forging?

20. Why does the press forging process give greater efficiencies so far as input–output work effort is concerned?

21. Why can metals like aluminum and tin be extruded cold?

22. Describe why continuous upsetting is a competitive process with extrusion.

23. List products that are shaped by roll forging.
24. Sketch a method used to extrude a lead sheath on wire cable.
25. Describe the process for hot rolling a metal wheel.
26. Sketch five methods for the manufacture of pipe.
27. Why is it necessary in forming tubing that each successive set of rollers run at a higher speed?
28. Describe the rollers in a stretch-reducing mill.
29. Describe the continuous method of making seamless tubing.
30. List the advantages and disadvantages of heating dies for making forgings.
31. Design a process for producing square tubing. Sketch the roll system.
32. Give the purpose of forging in an inert-gas atmosphere.
33. Why is it necessary to have the extruded shape's cross section remain constant?
34. What is the reason for double-piercing large-size tubes?
35. If a toothpaste tube is manufactured by impact extrusion, how is it filled?
36. Compare the processes of forming aluminum sheet by pellet rolling, by the continuous-casting process, and by the conventional method.
37. If about 30 passes are required to reduce a large ingot to a bloom, about what percentage reduction per pass is attained?
38. Using Figure 12.3, estimate (by plotting on graph paper) the percentage reduction for each pass.
39. At what percentage of the melting temperature are the following usually forged: aluminum, steel, and copper?
40. Given a material with the same recrystallization temperature as nickel, and a very high yield strength, what recrystallization temperature and method of hot working would you use?
41. What size press is needed to forge an open-end wrench that has a "flat size" cross section of 9.3 in.^{2} (6000 mm^2)?
42. A 2-in.-square steel cross section is to be upset. What is the maximum length that can be upset?

CASE STUDY
THE YUNGK COMPANY

Steel forgings are the product of this company. Material cost is an important part of the price and is estimated carefully. Opportunities exist for the improvement of material yield, and George Yungt recognizes that die design and construction, choice of stock size, and minimizing of tong hold, flash, and scrap make for an attractive price and the desirable profit. If George does not have a good yield, he loses money.

The Yungk Company, in business for over 60 years, has developed its own approach in estimating material cost. The shape-volume contained within the forging drawing, including openings, is found. A cut-weight is the weight of the material at the press to produce one forging. It is equal to the shape weight plus allowances for flashing, tong hold, and scale losses. Additionally, crop end losses, resulting from the possibility that the required forging raw material bar stock will be an uneven multiple of pur-

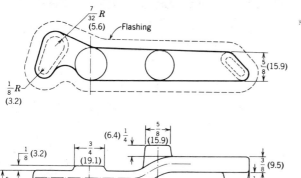

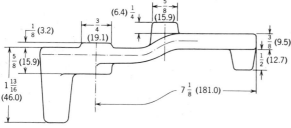

Figure 12.24
Case study.

chased bar stock length, are added to the scrap estimate.

George knows that using the proper bar size to make a forging is important. If the bar OD is too large, excessive fullering must be done to reduce the size. This results in low production and excessive die wear. When the bar size is too small, it becomes impractical to fill the heavier sections of the part.

An arm forging print (Figure 12.24) has been received. Shape-volume has been determined to be 3.55 in.³ (55,000 mm³); the density = 0.283 ℓb/in³ (7765 kg/m³). Studying the print, George believes that the forging will require a $^{15}/_{16}$-in. (50.8-mm) OD and will have flashing 0.075 in. (3.8 mm) thick by 1 in. (25.4 mm) wide for a periphery of 20 in. (508 mm). "A tong hold of one inch is sufficient," George says. "End losses for cropping should be about 2% and scrap will be 3%."

Determine the total weight for one forging. What is the forging material cost if bar stock costs $0.425/lb (0.946/kg)? What is the percentage yield of finished forging? What percentage scrap can George tolerate if he makes 25% profit on the material?

CHAPTER 13
COLD WORKING OF METAL

When a metal is rolled, extruded, or drawn at a temperature below the recrystallization temperature, the metal is cold worked. Most metals are cold worked at room temperature. Even though the forming does cause a temperature rise, it is not as much as hot working. Hot working, on the other hand, is performed on metal in the plastic state and refines the grain structure. Cold work distorts the grain and does little toward reducing the size of the material. Cold work improves strength, machinability, dimensional accuracy, and surface finish of metal. Because oxidation is less for cold working, thinner sheets and foils can be rolled than by hot working. Many of the same processes and equipment are used for both hot and cold work, but the forces required and the results are diffcrent.

PRINCIPLES OF COLD WORKING

To understand the action of cold working, one must have knowledge of the structure of metals. All metals are crystalline in nature and are composed of irregularly shaped grains of various sizes. These grains may be seen using a microscope if the metal has been properly polished and etched. Each grain is constructed of atoms in an orderly arrangement known as a lattice. The orientation of the atoms in a given grain is uniform but differs in adjacent grains. When material is cold worked, the change in material shape brings about marked changes in the grain structure. Structural changes that occur are grain fragmentation, movement of atoms, and lattice distortion. *Slip planes* (Figure 13.1) develop through the lattice structure at points where thc atom bonds of attraction are the weakest, and whole blocks of atoms are displaced. When slip occurs, the orientation of the atoms is not changed. In cases where there is reorientation a phenomenon called *twinning* occurs. In twinning the lattice on one side of the plane is oriented in a different fashion from the other, but the atoms have shapes identical to adjacent atoms. Slip is the more common result of deformation.

Much greater pressures are needed for cold working than for hot working. The metal, being in a more rigid state, is not permanently deformed until stresses exceeding the elastic limit are passed. Because there can be no recrystallization of grains in the cold-working range, there is no recovery from grain distortion or fragmentation. As grain deformation proceeds, greater resistance to this action results in increased strength and hardness of the metal. The metal is said to be *strain hardened* and, for some metals that will not respond to heat treating, it is the only known method of changing such physical properties as hardness and strength. Several theories have been advanced by metallurgists

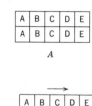
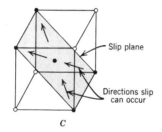

Figure 13.1
Representation of slip in a body-centered lattice system.

to explain this occurrence. In general, they refer to resistance built up in the grains by atomic dislocation, fragmentation, or lattice distortion, or a combination of all three phenomena.

The amount of cold work that a metal will stand is dependent on its ductility. The higher the ductility of a metal the more it is able to be cold worked. Pure metals withstand a greater amount of deformation than metals having alloying elements, because alloying increases the tendency and rapidity of strain hardening. Large-grain metals are more ductile than small ones. When metal is deformed by cold work, severe stresses known as *residual stresses* are set up in the metal. To remove these undesirable stresses, the metal must be reheated to slightly below the recrystalline range temperature. In this range the stresses are rendered ineffective without appreciable change in physical properties or grain structure. Heating into the recrystalline range eliminates the effect of cold working. Sometimes it is desirable to have residual stresses in the metal. The fatigue life of small parts may be improved by shot peening, which causes the surface metal to be in compression and the material below the surface to be in tension (see later section for details).

Advantages and Limitations

Many products are *cold finished* after hot rolling to make them commercially acceptable. Hot-rolled strips and sheets are soft and have surface imperfections. They lack dimensional accuracy and certain desirable physical properties. The cold-rolling operation reduces size only slightly, permitting accurate dimensional control. Surface oxidation does not result from the process and a smooth surface is obtained. Strength and hardness are increased. For metals that do not respond to heat treatment, cold work is a possible method to increase hardness. Ductile materials can be extruded at temperatures below the recrystallization range. Higher pressure and heavier equipment are needed for cold-working than for hot-working operations. Brittleness results if the metal is overworked, and an annealing operation then becomes necessary. In general, cold working produces the following effects:

1. Stresses are set up in the metal that remain unless they are removed by subsequent heat treatment.
2. Distortion or fragmentation of the grain structure is created.
3. Strength and hardness of the metal are increased with a corresponding loss in ductility.

312 COLD WORKING OF METAL

4. Recrystalline temperature for steel is increased.
5. Surface finish is improved.
6. Close dimensional tolerance can be maintained.
7. The process is economical and rapidly produces parts in high-volume applications.

COLD-WORKING PROCESSES

The effects just discussed are not achieved by all cold-working processes. Operations involving bending, drawing, and squeezing metal result in grain distortion and changes in physical properties, whereas shearing or cutting operations change only form and size. Cold-working processes pertain primarily to rolling, drawing, or extrusion.

Tube Finishing

Tubing, which requires dimensional accuracy, smooth surface, and improved physical properties, is finished by either *cold drawing* or a *tube reducer*. Tubing that has first been hot rolled is treated by *pickling* and washing to remove all scale. Before the cold tube-finishing operation, a lubricant is applied to prevent galling, reduce friction, and increase surface smoothness. Drawing is done in a ***drawbench*** (Figure 13.2). One end of the tube is reduced in diameter by a swaging operation to permit it to enter the die, then it is gripped by tongs fastened to the chain of the drawbench. In this operation the tube is drawn through a die smaller than the outside diameter of the tube. The inside surface and diameter are controlled by a fixed mandrel over which the tube is drawn.

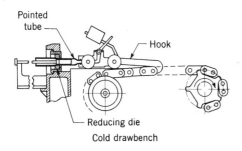

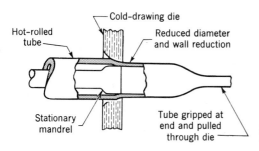

Figure 13.2
Process of cold-drawing tubing.

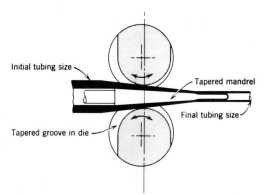

Figure 13.3
Schematic of a tube reducer.

This mandrel may be omitted for small sizes or for larger sizes if the accuracy of the inside diameter is not important. Drawbenches require a pulling power ranging from 50,000 to 300,000 lb (0.2–1.3 MN) and may have a total length of 100 ft (30 m).

The operation of drawing a tube is severe. The metal is stressed above its elastic limit to permit plastic flow through the die. The maximum reduction for one pass is around 40%. This operation increases the hardness of the tube so much that if several reductions are desired, the material must be annealed after each pass. This method also produces tubes having smaller diameters or thinner walls than can be obtained by hot rolling. Hypodermic tubing is produced in this manner with an outside diameter of less than 0.005 in. (0.13 mm).

The *tube reducer* has semicircular dies with tapered grooves through which the previously hot-rolled tubing is alternately advanced and rotated. The dies (Figure 13.3) rock back and forth as the tubing moves through them. A tapered inside mandrel regulates the size to which the tube will be reduced. The tube reducer can make the same reduction in one pass that might take four or five passes in a drawbench, but its chief advantage is the much longer lengths of tubing that can be produced.

Tubes that are finished by either method have all the advantages found in cold-worked metals and can be made in longer lengths and with thinner walls than is possible by hot working.

Wire Drawing

Wire is made by cold-drawing hot-rolled wire rod through one or more dies, as shown in Figure 13.4, to decrease its size and increase the physical properties. The wire rod, about $7/32$ in. (6 mm) in diameter, is rolled from a single billet and cleaned in an acid bath to remove scale, rust, and coating. A coating is applied to prevent oxidation, to neutralize any remaining acid, and to act as a lubricant and a coating to which a later-applied lubricant may cling.

Both *single-draft* or *continuous-drawing* processes may be used. In the first method a coil is placed on a reel or frame and the end of the rod pointed so that it will enter the die. The end is grasped by tongs on a drawbench and pulled through to such length as may be wound around a *drawing block* or reel. From there on the rotation of the draw

314   COLD WORKING OF METAL

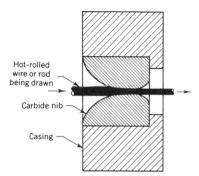

Figure 13.4
Section through a die used for drawing wires.

block pulls the wire through the die and forms it into a coil. These operations are repeated with smaller dies and blocks until the wire is drawn to its final size. In continuous drawing (Figure 13.5) the wire is fed through several dies and draw blocks arranged in series. This permits drawing the maximum amount in one operation before annealing is necessary. The number of dies in the series will depend on the kind of metal or alloy being processed and may vary from 4 to 12 successive drafts. The dies are usually made from tungsten carbide, although diamond dies can be used for drawing small diameters.

Wire drawing is often determined by the following relationships:

$$\% \text{ Reduction in area} = \frac{(A_o) - A_f \times 100}{A_o}$$

Figure 13.5
A continuous wire-drawing machine.

$$\% \text{ Elongation} = \frac{L_f - L_o}{L_o} \times 100$$

where A_o, A_f, L_o, and L_f are original and final area and lengths. The continuity equation also holds for wire drawing.

$$Q_o = Q_f = A_o V_o = A_f V_f$$

where

Q_o, Q_f = Quantity of metal entering and leaving a die

V_o, V_f = Velocity of wire entering and leaving a die

Foil Manufacture

Foils are made from a broad variety of pure metals and alloys by cold rolling to thicknesses as thin as 0.00008 in. (0.0020 mm). Most foil manufacturers find it necessary to control the raw material so closely that they employ their own vacuum-melting equipment. Only the purest metals are charged into the melting furnaces and alloys are added when required. In most instances the foil is continuously cast, cooled, and rolled as it comes from the furnace. One producer of aluminum foil casts the material by the *Hunter process,* in which the molten metal is forced up through an asbestos-based nozzle or tip onto water-cooled rolls where it solidifies 0.250 in. (6.35 mm) thick in less than 3 s. In one continuous series of operations the aluminum is rolled to about 0.006 in. (0.15 mm) in thickness at velocities approaching 2000 ft/min (10 m/s). Thinner foil is produced by further rolling. Foil thickness is produced by a combination of roll pressure and controlled tension on the material. Most foils can be produced with two sides bright or one side bright and the other a satin finish. The latter is produced by pack or double-thickness rolling. When two sheets are passed through the rollers at the same time, the faces contacting the rollers are bright and the mating faces have the satin finish when stripped apart.

Metal Spinning

Metal spinning is the operation of shaping thin metal by pressing it against a form while it is rotating. The nature of the process limits it to symmetrical articles. This type of

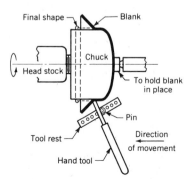

Figure 13.6
Metal spinning operation.

Figure 13.7
Metal spinning a 120-in. (3-m)-diameter head for missile.

work is done on a speed lathe, similar to the wood lathe except that in place of the usual tailstock, it is provided with some means of holding the work against the form as shown in Figure 13.6. The forms are usually turned from hardwood and attached to the face plate of the lathe, although smooth steel chucks are recommended for production jobs.

Practically all parts are formed with the aid of blunt hand tools that press the metal against the form. The cross slide has a hand or compound tool rest in the front for supporting the hand tools and some means for supporting a trimming cutter or forming roll in the rear. Parts may be formed either from flat disks of metal or from blanks that have been previously drawn in a press. The latter method is used as a finishing operation for many deep-drawn articles. Most spinning work is done on the outside diameter as shown in the figure, although inside work is also possible. Figure 13.7 shows spinning a $5/8$-in. (3.6-mm) thick, 140-in. (3.6-m) diameter plate that, when finished, is a 120-in. (3.0-m) diameter, elliptical-shaped head.[1]

Bulging work on metal pitchers, vases, and similar parts is done by having a small roller, supported from the compound rest, operate on the inside and press the metal out against a form roller. The part must first be drawn and often given a bulging operation beforehand, as spinning cannot be done near the bottom.

---

[1] When a dimension such as wire diameter or thickness of sheet or plate is given in customary units, the practice of providing SI units is followed, although there may not be SI commercial sizes available. Exceptions are numerous, however; for example, 2 by 4 in. lumber or $1/4$-20 UNC thread which are not converted to SI units.

Lubricants such as soap, beeswax, white lead, and linseed oil reduce the tool friction. Because metal spinning is a cold-working operation, there is a limit to the amount of drawing or working the metal will stand, and one or more annealing operations may be necessary. Spinning lends itself to short-run production jobs of about 5000 pieces or less, although it has many applications in quantity production work. Spinning has several advantages over press work in that tooling costs are lower, a new product can be brought to the production stage sooner, and for very large parts the high cost of a press capable of doing the job may be prohibitive. Labor costs are higher for spinning than for press work and the production rate may be much less. Simple shapes can be formed from soft nonferrous metals up to $1/4$ in. (6 mm) in thickness and from low-carbon steel up to $3/16$ in. (5 mm) in thickness. Tolerances up to $\pm\ 1/32$ in. ($\pm\ 0.8$ mm) can be easily maintained for diameters under 18 in. (460 mm). This process is frequently used in making bells on musical instruments and also for light fixtures, kitchenware, reflectors, funnels, and large processing kettles.

Shear Spinning

In spinning thick metal plates, power-driven rollers must be used in place of conventional hand spinning tools. This operation is called *shear spinning*. In Figure 13.8 are the progressive steps of a shear-spinning operation where a conical shape is formed from a flat plate. The plate is initially held securely against the mandrel by a holder. The roll formers force the plate to conform to the mandrel, maintaining a uniform wall thickness from the starting point until completion. The wall thickness obtained is equal to the plate thickness times sin $\alpha/2$, α being the cone included angle. Parts having a cone angle less than 60° require a conical preform. Reduction in wall thickness up to 80% is possible, although, in some cases, the reductions are much smaller. This relationship can be given as

$$t_f = t_s (\sin \alpha/2)$$

where

t_f = Final thickness, in.

t_s = Starting thickness, in.

α = Included angle of cone, °

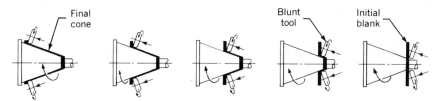

Figure 13.8
Progressive forming in a shear spinning operation in which a conical shape is formed from a flat plate.

318 COLD WORKING OF METAL

In conventional spinning the wall thickness remains about the same throughout the operation. Spinning tools merely bend or flare the metal into a new contour and do not cause a plastic flow or reduction in wall thickness. In shear spinning, the metal is uniformly reduced in thickness over the mandrel by a combination of rolling and extrusion. Advantages claimed for shear spinning include increased strength of part, material savings, reduction in cost, and good surface finish.

Most metal can be formed by this process. Although heat is sometimes applied throughout the cycle, it is not required for steel alloys and most nonferrous metals.

Stretch Forming

In forming large sheets of thin metal involving symmetrical shapes or double-curve bends, a metal stretch press can be used effectively. Figure 13.9A shows one of the simpler hydraulically operated presses. A single die mounted on a ram is placed between two slides that grip the metal sheet. The die moves in a vertical direction and the slides move horizontally. Large forces of 50 to 150 tons (0.5–1.3 MN) are provided for the die and slides. The process is a stretching one and causes the sheet to be stressed above its elastic limit while conforming to the die shape. This is accompanied by a slight thinning of the sheet, and the action is such that there is little springback to the metal once it is formed. Adapted to both production and short-run jobs, inexpensive dies of wood, kirksite, plastic, or steel can be used. Large double-curvature parts, difficult by other methods, are easily made with this process. The process can be used with many hard-to-form alloys, there is little severe localized cold working, and the problem of unequal metal thinout is minimized. Scrap loss is fairly high because material must be left at the ends and sides for trimming, and there is a limitation to the shapes that can be formed.

Stretch forming requires the metal to be stretched to a point greater than its yield strength and less than its tensile strength. For estimating the pressure required for stretch forming the following formula can be used:

$$P = 1.25 Y_s A$$

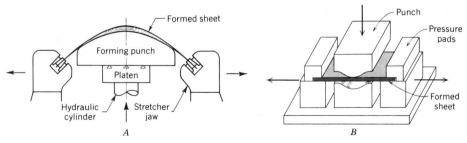

Figure 13.9
Stretch processes. A, Stretch forming. B, Stretch draw forming.

where

P = Stretch-forming pressure, lb
Y_s = Yield strength of metal, psi
A = Cross-sectional area, in.^{2}
1.25 = Empirical constant

A combination of stretch and draw forming (Figure 13.9B) can be employed. Manufacturers of stretch–draw forming presses maintain that all metals and alloys become unusually ductile when stretched from 2% to 4% and can be formed with about one-third the force normally required. Despite scrap losses caused by the necessity of gripping the material in jaws during stretching, the process is used not only for short-run aircraft parts of aluminum but is also employed in the automotive industry to make steel roof panels, hood covers, rear deck lids, and doorposts. Titanium and stainless-steel sheet can also be formed in this manner.

Swaging and Cold Forming

These terms refer to methods of cold working by a compressive force or impact that causes the metal to flow in some predetermined shape according to the design of the dies. The metal conforms to the shape of the dies, but it is not restrained completely and may flow at some angle in the direction to which the force is applied. *Sizing,* the simplest form of cold forging, is the process of slightly compressing a forging, casting, or steel assembly to obtain close tolerance and a flat surface or a flash removal operation. The metal is confined only in a vertical direction. Small pinions, less than 1 in. in diameter, are cold extruded. *Rotary swaging* (Figure 13.10) is a means of reducing the ends of bars and tubes by rotating dies that open and close rapidly on the work, so that the end of the rod is tapered or reduced in size by a combination of pressure and impact. Mechanical pencils, metal furniture legs, and umbrella poles are examples of parts made by this process. Because swaging action is rather severe, the material hardens and an annealing operation is necessary if much reduction is desired.

The *cold forming,* formerly called **cold heading,** or *upsetting* of bolts, rivets, and other similar parts done on a cold-forming machine, is another form of swaging. Because

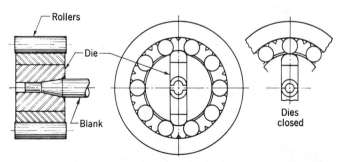

Figure 13.10 Illustrating the operation of dies in a swaging machine.

the product of the cold header is made from unheated material, the equipment must withstand the high pressures that develop. Also, alignment of the upsetting tool with the dies must be accurate so that the work turned out will be free from defects. A solid die machine of this type is illustrated in Figure 13.11. The rod is fed by straightening rolls up to a stop, and it is then cut off and moved into one of the four types of header dies shown in Figure 13.12. The heading operation may be either single or double, and upon completion the part is ejected from the dies. Production rates of 600 parts per minute are not uncommon.

Nails, rivets, and small bolts are made from coiled wire and forged cold, whereas large bolts require heating of the end of the rod before the heading operation. In nail making, the head is formed before shearing the wire. The wire is fed forward, clamped, headed, and pinched or sheared off to form the point, and is finally expelled. The nails are tumbled together in sawdust to remove the lubricants and "whiskers" of metal before packaging. Figure 13.13 shows some examples of cold-headed parts.

Bolt-making machines are available that completely finish the bolt before it leaves the machine. The operation consists of cutting off an oversize blank, extruding the shank,

Figure 13.11
Single-stroke, cold-header machine producing 36,000 parts per hour from $^3/_{16}$-in. (4.8-mm) coiled steel.

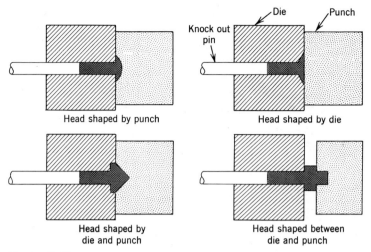

Figure 13.12
Types of cold-header dies.

heading, trimming, pointing, and roll threading. All operations are carried on simultaneously, and outputs range from 50 to 300 pieces per minute.

Intraforming is a process in which metal is squeezed at a pressure of about 300 tons (4000 MPa) or less, onto a die or mandrel to produce an internal configuration. A mandrel, workpiece, and finished part are shown in Figure 13.14. Ferrous and nonferrous forgings, powdered metal parts, and tubing can be produced. Splines, internal gears, special-shaped holes, and bearing retainers are produced by this method. Tooling is inexpensive, die

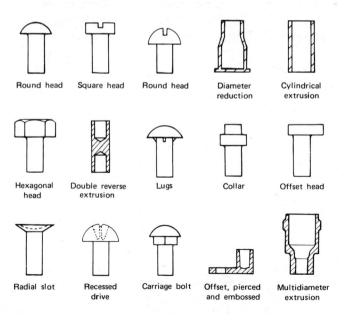

Figure 13.13
Typical parts made by cold heading from wire stock.

Figure 13.14
Intraforming workpiece, finished part, and mandrel.

life is from 1000 to 100,000 parts, and good surface finishes and high accuracy may be obtained. The maximum ID of the workpiece that may be accommodated is 4 in. (100 mm).

Hobbing

Mold cavities (Figure 13.15) are produced by forcing a hardened steel form or hob into soft steel. The *hob*, machined to the exact form of the piece to be molded, is heat treated to obtain the necessary hardness and strength to withstand the tremendous pressures involved. Pressing the hob into the blank requires much care, and frequently several alternative pressings and annealings are necessary before the job is complete. During the

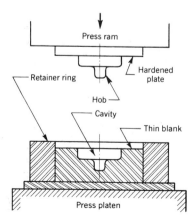

Figure 13.15
Die hob producing a mold cavity by pressing into soft steel.

hobbing operation the flow of metal in the blank is restrained from any appreciable lateral movement by a heavy retainer ring placed around it. The actual pressing is done in hydraulic presses having capacities ranging from 250 to 8000 tons (2–70 MN).

The advantage of hobbing is that multiple identical cavities can be produced economically. The surfaces of the cavities have a highly polished finish, and machine work is unnecessary other than to remove surplus metal from the top and sides of the blank. This process is used in producing molds for the plastic and die-casting industries.

Coining and Embossing

The operation of *coining* (Figure 13.16) is performed in dies that confined the metal and restrict its flow in a lateral direction. Shallow configurations on the surfaces of flat objects such as coins are produced in this manner. Because special-type presses that develop high pressures are required in this operation, its use is limited to relatively soft alloys.

Embossing is more of a drawing or stretching operation and does not require the high pressures necessary for coining. The punch diameter is usually reduced in diameter so that it touches only the part of the blank that is being embossed and avoids the surfaces of the mating die. The major use for embossing is making nameplates, medallions, identification tags, and aesthetic designs on thin sheet metal or foil. The embossed design is raised from the parent metal. The mating die (Figure 13.16) conforms to the same configuration as the punch, so that there is very little metal squeezing in the operation and practically no change in the thickness of the metal. Rotary embossing using cylindrically shaped dies is extensively used on thin sheet metal and foils. The metal is fed through the rolls as shown in Figure 13.17.

Riveting, Staking, and Stapling

These processes are used to fasten parts together as illustrated in Figure 13.18. In the usual *riveting* operation a solid rivet is placed through holes made in the parts to be fastened together, and the end is pressed to shape by a punch. Hollow rivets may have the ends secured by curling them over the edges of the plate. Explosive rivets have a powder charge that expands the nonhead end.

Staking is a similar operation in that the metal of one part is upset to cause it to fit tightly against the other part. A staking punch may have one or more projections, as shown in the figure, or it may be in the form of a ring with sharp chisellike edges. Both

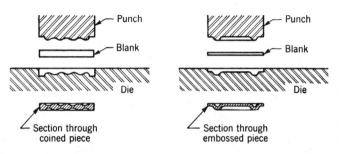

Figure 13.16
Illustrating the difference between coining and embossing.

324 COLD WORKING OF METAL

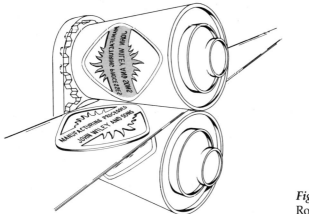

Figure 13.17
Rotary embossing.

operations can be performed on small presses, because not much pressure is required. Stapling is used to join the two or more sheets of metal and to join sheet to wood.

Roll Forming

Cold roll forming machines are constructed with a series of mating rolls that progressively form strip metal as it is fed continuously through the machine at speeds ranging from 50 to 300 ft/min (0.3–1.5 m/s). A machine is shown in Figure 13.19 in which tubular sections are being produced by five pairs of rolls. The tubular section enters a resistance welder after being formed and is continuously welded. The number of roll stations depends on the intricacy of the part being formed. For a simple channel four pairs may be used, whereas for complicated forms several times that number may be required. In addition to the mating horizontal rolls, these machines are frequently equipped with guide rolls mounted vertically to assist in the forming operation and straightening rolls to "true up" the product as it emerges from the last forming pass.

Figure 13.20 shows a variety of typical parts that have been roll formed. Note also the sequence of forming operations for a window screen section. In forming the sequence the vertical center or pass line is first established so that the number of bends on either side is about the same. Forming usually starts at the center and progresses out to the two

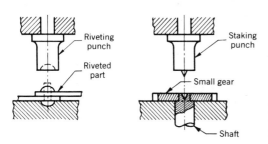

Figure 13.18
Illustrating the difference between riveting and staking processes.

Figure 13.19
Cold roll tube-forming machine. Strip enters machine from coil (not shown) and is bent to tubular shape by five pairs of rolls before being welded.

edges as the sheet moves through the successive roll passes. The amount of bending at any one roll station is limited. If the bending is too great, it carries back through the sheet and affects the section at the preceding roll station. Corner bends are limited to a radius equal to the sheet thickness.

In terms of capacity for working mild steel, standard machines form strips up to 0.156 in. (3.96 mm) thick by 16 in. (400 mm) wide. Special units have been made for much heavier and wider strip steel. The process is rapid and is applicable to forming products having sections requiring a uniform thickness of material throughout their entire length. Unless production requirements are high, the cost of the machine and tooling cannot be justified.

Tolerances of roll forming are affected by the size of the section, material, product, and the gage and gage tolerances of the material. In addition, the length of the piece must be tolerated, and it is influenced by the speed of operation, length of the piece to be cut, and accuracy of the cutting device. Attainable tolerances are as follows: piece length, $\pm 1/64$ to $1/8$ in.; length, $\pm 1/64$ to $1/8$ in.; straightness and twist, $\pm 1/64$ to $1/4$ in. in 10 ft; cross-sectional dimension, $\pm 1/64$ to $1/16$ in; and angles, $\pm 1°$ to $2°$.

There are several guidelines to follow when designing a product to be cold roll formed. A slight angle is more favorable than long vertical side walls, and blind corners and radii should be avoided to prevent inaccuracy resulting on rolls without control features. Smaller-bend radii are easier and less expensive to make than those that are larger.

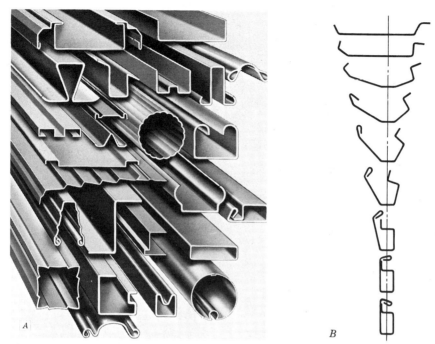

Figure 13.20
Cold-rolled formed parts. *A*, Miscellaneous parts formed from coiled strip.
B, Sequence of forming operation for window screen section.

Plate Bending

Another method of bending metal plates and strips into cylindrical shapes is by a roll-bending machine as illustrated in Figure 13.21. This machine consists of three rolls of the same diameter. Two of them are held in a fixed position and one is adjustable. As a metal plate enters and goes through the rolls, its final diameter is determined by the position of the adjustable roll. The closer it is moved to the other rolls, the smaller the diameter. Machines of this type are made in capacities ranging from those that form small-gage thicknesses to others that form heavy plates up to $1^{1}/_{4}$ in. (30mm).

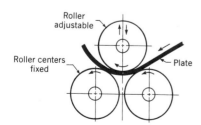

Figure 13.21
Plate-bending rolls.

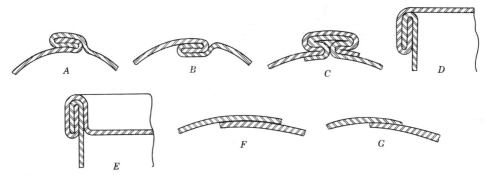

Figure 13.22
Seams used in container manufacture: A, Longitudinal outside lock seam. B, Longitudinal inside lock seam. C, Compound longitudinal seam. D, Double seam for flat containers. E, Double seam for recessed-bottom containers. F, Adhesive-bonded top seam. G, Resistance-welded seam.

Seaming

In the manufacture of metal drums, pails, cans, and numerous other products made of light-gage metal, several types of *seams* are used. The most common are shown in Figure 13.22. The lock seam used on longitudinal seams is adapted for joints that do not have to be absolutely tight. After the container is formed the edges are folded and pressed together. The *compound seam,* sometimes called the Gordon or *box seam,* is much stronger and tighter than the lock seam and is suitable for holding fine materials. Both these joints may be formed and closed on either hand or power seaming presses.

Bottom seams, which are somewhat similar to the longitudinal seams, are made in either flat or recessed styles. Flat-bottom seaming is limited to one end of a container, as the container must be open to make the joint. Double seaming with recessed bottoms can be done on both ends of a container. *Edge flanging, curling, and flattening,* the operations necessary to make a recessed double seam, are shown in the figure.

Double-seaming machines may be hand operated, semiautomatic, or automatic. Semiautomatic machines must be loaded and unloaded by the operator, but the operation of the machine is automatic. In automatic machines the cans are brought to the machine by conveyor, and ends are supplied by magazine feed. The cans are fed from the conveyor to a star wheel, which transfers them to an automatic delivery turret. The delivery turret feeds them into position with the seaming heads, and the closing seam is made.

HIGH–ENERGY RATE FORMING

High–energy rate forming (HERF) includes a number of processes in which parts are formed at a rapid rate by extremely high pressures. Although the term inaccurately describes the process, it is accepted by industry for what would be better called high-

COLD WORKING OF METAL

Table 13.1 **Approximate Deformation Velocities**

Process	Velocity	
	ft/s	m/s
Hydraulic press	0.10	0.03
Brake press	0.10	0.03
Mechanical press	0.1–2.4	0.03–0.73
Drop hammer	0.8–14	0.24–4.3
Gas-actuated ram	8–270	2.0–82
Explosive	30–750	9–230
Magnetic	90–750	27–230
Electrohydraulic	90–750	27–230

velocity forming. The deformation velocities of several processes are noted in Table 13.1. By imparting a high velocity to the workpiece, the size of equipment to form large parts can be reduced, and certain materials that may not lend themselves to conventional forming methods can be processed. The die costs are low, good tolerances can be maintained, and the production costs can be minimized. While development of these processes has been centered on forming relatively thin metal, applications of high–energy rate forming include compacting metal powders, forging, cold welding, bonding, extruding, and cutting.

Explosive Forming

A variety of methods for applying energy at a high rate has been developed. Several are diagrammed in Figure 13.23. **Explosive forming** has proved to be an excellent method of utilizing energy at a high rate, because the gas pressure and rate of detonation can be controlled. Both low and high explosives are used in the various processes. With *low explosives,* known as cartridge systems, the expanding gas is confined and pressures may build up to 100,000 psi (700 MPa). *High explosives,* which need not be confined and which detonate with a high velocity, may attain pressures of up to 20 times that of low explosives. Explosive charges, whether exploded in air or a liquid, set up intense shock waves that pass through the medium between the charge and the workpiece, but decrease in intensity as the waves spread over more area. Springback is minimized in explosive forming but does exist. Less springback occurs with the use of sheet explosives close to the workpiece, high clamping forces on the hold-down areas, and, in the absence of lubricants, thick materials exhibit less springback than do parts made from thin blanks. Aside from the generation of gas pressure by powder, high gas pressures may also be attained by the expansion of liquefied gases, explosion of hydrogen–oxygen mixtures, spark discharges, and the sudden release of compressed gases. A device known as Dynapak[2]

---

[2] A machine developed by General Dynamics used primarily for forging and extrusion operations.

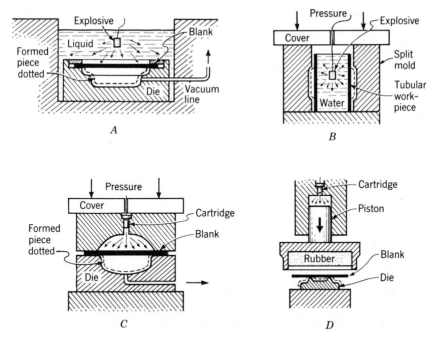

Figure 13.23
Methods in high–energy rate forming. A, Direct forming with fluid pressure. B, Bulging operation. C, Direct forming by gas pressure. D, Gas-actuated drop hammer.

has been developed that utilizes this last procedure. Dynapak uses compressed nitrogen, which upon release accelerates a heavy ram to a speed of 2000 in./sec (50 m/s) or less. Attaining very high pressures, it is used for open-end extrusion, impact extrusion, forging, and for compacting powders.

Figure 13.23C and D show examples of expanding-gas methods. In Figure 13.23C the gas presses against the workpiece and forces it to conform to the die. In Figure 13.23D the gases act against a piston that forces the confined rubber punch over the blank and die. This method is similar to that of a drop hammer but is much more rapid. Thin-wall tubing may be formed by "slow" explosive forming using a powder that deflagrates rather than detonates. The expanding gases are trapped inside a boot within the tubing, and the expanding boot forces the tubing into the configuration of the die. Figure 13.24 is an example of such forming using a shotgun shell.

Electrohydraulic Forming

Electrohydraulic forming, also known as ***electrospark forming,*** is a process whereby electrical energy is directly converted into work. The forming equipment for this process is similar to Figure 13.23A or B, but pressure is obtained from a spark gap instead of an explosive charge. A bank of capacitors is first charged to a high voltage and then discharged

330   COLD WORKING OF METAL

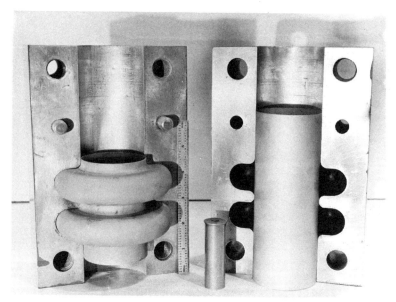

Figure 13.24
Bellows explosively formed using a 12-gage shotgun shell.

across a gap between two electrodes in a suitable nonconducting liquid medium. This generates a shock wave that travels radially from the arc at high velocity, supplying the necessary force to form the workpiece to shape. This process is safe to operate and has low die and equipment cost. The energy rates can also be closely controlled.

Magnetic Forming

Magnetic forming is another example of the direct conversion of electrical energy into useful work. At first it served primarily for swaging-type operations such as fastening fittings on the ends of tubes and crimping terminal ends of cables. More recent applications are embossing, blanking, forming, and drawing, all using the same power source but differently designed work coils.

Figure 13.25 is a simple diagrammatic sketch illustrating how electromagnetic forming works. The charging voltage E is supplied by a high-voltage source into a bank of

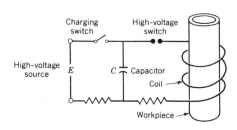

Figure 13.25
Schematic of an electromagnet-forming circuit.

capacitors connected in parallel. The amount of energy stored can be varied either by adding capacitors to the bank or by increasing the voltage. The latter is limited by the insulating ability of the dielectric material on the coils. The charging operation is rapid and, when complete, a high-voltage switch triggers the stored electrical energy through the coils establishing a high-intensity magnetic field. This field induces a current into the conductive workpiece placed in or near the coil, resulting in a force that acts on the workpiece. This force, when it exceeds the elastic limit of the material being formed, causes permanent deformation.

Three different forming possibilities are shown in Figure 13.26. In Figure 13.26A the coil surrounds a tube that when energized forces the material tightly around the fitting. The same principle would apply if a conducting ring were placed around a number of wire ends. If the coil is placed inside an assembly, as indicated in Figure 13.26B, the force will expand the tube into the collar. By changing the design of the coil as in Figure 13.26C, flat plates may be embossed or blanked. The process can be used to assemble fragile parts, such as the swaging of an aluminum dial on a plastic knob.

Both permanent and expandable coils are used in this process. Because the forces act on the coils as well as the work, the coils and insulation must be capable of withstanding the forces or else they will be destroyed. Dielectric insulators fail at around 10 kV and are expensive, whereas the cost of expendable coils is less. As a result, expendable coils are generally used when a high energy level is required. In magnetic forming, the conducting metal being formed is rapidly accelerated so that much of the forming takes place after the magnetic impulse. Because the magnetic field will not be restricted by nonconducting materials, it is possible to form a part within a container. The pressure on the work is uniform, production rates are rapid, reproducibility is excellent, lubricants are unnecessary, there are no moving mechanical parts to the machine, and relatively unskilled labor is required. The process is limited in that complex shapes may be impossible to form, pressures cannot be varied over the workpiece, and present units are limited to 60,000 psi (400 MPa) pressure.

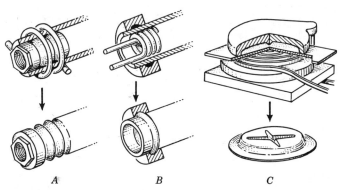

Figure 13.26
Different applications of magnetic forming.
A, Swaging. B, Expanding. C, Embossing or blanking.

OTHER METHODS
Impact Extrusion

An interesting example of *extrusion by impact* is in the manufacture of collapsible tubes for shaving cream, toothpaste, and paint pigments. These extremely thin tubes are pressed from slugs, as illustrated in the upper half of Figure 13.27. The punch strikes a single blow of considerable force causing the metal to squirt up around the punch.

The outside diameter of the tube is the same as the diameter of the die, and the thickness is controlled by the clearance between the punch and die. The tube shown in Figure 13.27 has a flat end, but any desired shape can be made by properly forming the die cavity and the end of the punch. For toothpaste tubes a small hole is punched in the center of the blank and the die cavity is shaped to form the neck of the tube. On the upstroke the tube is blown from the ram with compressed air. The entire operation is automatic, with a production rate of 35 to 40 tubes per minute. The tubes are then threaded, inspected, trimmed, enameled, and printed. Zinc, lead, tin, and aluminum alloys are worked in this fashion. Some tubes are lined with foil of a different material that has been previously clad to the blank. Both the foil coating and base metal are formed contiguously. Impact extrusions are low in cost, have excellent surface quality, and are adapted to production rates of from 100,000 to 20 million parts per year. The process is used to make shell cases, soft drink cans, and innumerable hollow mechanical parts that have one end totally or partially closed.

The lower half of Figure 13.27 illustrates a variation of what is known as the *Hooker process* for extruding small tubes or cartridge cases. Small slugs or blanks are used as in the impact extrusion process, but in this case the metal is extruded downward through the die opening. The size and shape of the extruded tube are controlled by the space between the punch end and die cavity wall. Copper tubes having wall thicknesses of

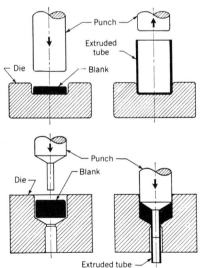

Figure 13.27 Methods of cold-impact extrusion for soft metals.

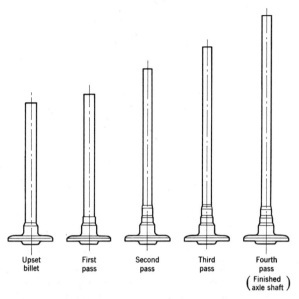

Figure 13.28
High-speed cold extrusion operations used to manufacture axle shafts.

0.004 to 0.010 in. (0.10–0.25 mm) can be produced in lengths of about 12 in. (300 mm). The process may work-harden the material to an extent that intermediate annealing must be done.

High-speed cold extrusion is used to manufacture axle shafts, shown by Figure 13.28. Upset billets are hand-loaded onto a three-station, cascade-type loading magazine. From this station on, the feeding through the four work stations is automatic. The cold-forming extrusion process used to elongate the shaft is accomplished by forcing a ring die over the axle shaft; the diameter of the die is less than the shaft. The total overall diameter reduction is accomplished in the four stages. Some advantages of this method of manufacture are as follows: (1) An improved surface finish extends the fatigue life of the material; (2) less stock removal for final finishing is required; and (3) work hardening of the surface improves the physical characteristics of the material.

Shot Peening

This method of cold working has been developed to improve the fatigue resistance of the metal by setting up compressive stresses in its surface. This is done by blasting or hurtling a rain of small shot at high velocity against the surface to be peened. As the shot strikes, small indentations are produced causing a slight plastic flow of the surface metal to a depth of a few thousandths of an inch. This stretching of the outer fibers is resisted by those underneath, which tend to return them to their original length, thus producing an outer layer having a compressive stress while those below are in tension. In addition, the surface is slightly hardened and strengthened by the cold-working operation. Because fatigue failures result from tension stresses, having the surface in compression greatly offsets any tendency toward such a failure.

Figure 13.29
Surface character of 45 Rockwell C steel that has been shot peened with steel shot. A $10^{1}/_{2}$-in. (490-mm)-diameter Wheelabrator unit was used at a speed of 2250 rpm.

Shot peening uses an air blast or some mechanical means such as centrifugal force for hurling steel shot onto the work at a high velocity; Figure 13.29 is an example of the surface obtained by shot peening. Surface roughness or finish can be varied according to the size of shot. Stress concentrations caused by the roughened surface are offset because indentations are close together and sharp notches do not exist at the bottom of the pits. Intense peening is undesirable as it may cause weakening of the steel.

This process adds increased resistance to fatigue failures of working parts and can be used on parts of irregular shape and on local areas that may be subject to stress concentrations. Surface hardness and strength are also increased, and in some cases the process is used to produce a commercial surface finish. It is not effective for parts subjected to reversing stresses nor is its effect appreciable on heavy metal sections.

QUESTIONS AND PROBLEMS

1. Give an explanation of the following terms:
   Residual stresses
   Tube reducer
   Drawbench
   Sizing
   Swaging
   Cold heading
   Coining
   Embossing
   Staking
   Explosive forming
   Electrospark
   Magnetic forming
   Impact extrusion
   Shot peening

2. Refer to the body-centered space lattice structure described in Chapter 2. Show how a material made of this structure might develop a slip plane.

3. Give reasons that prevent hot rolling of foils.

4. What effect does the elastic limit of a material have on its ability to be cold worked?

5. Explain the process of strain hardening.

6. How are residual stresses removed from cold-worked metals?

7. Why are pure metals more easily cold worked than alloys?

8. Give the history of cold-drawn tubing from ore to finished product.

9. List the principal advantages for the cold-drawing and tube reducer methods of tube production.

10. Explain the production of hypodermic tubing.

11. Discuss the speed of travel of wire feeding through a group of successive dies.

12. Compare spinning over press work for sheet metal products.

13. Explain how bright and and satin finishes are produced on foil.

14. What are the charactcristics that give advantages in stretch-draw forming?

15. Describe the process of making a nail in a cold-header machine. Sketch a set of dies and cutters.

16. Describe the physical properties of a metal that has been cold swaged.

17. What is debossing and why is it sometimes used in preference to embossing?

18. Sketch a mandrel to intraform a small internal gear in a forged tube.

19. Prepare a series of sketches for form rolling house gutters.

20. What other processes are competitive with die hobbing?

21. Design a small tube for holding mashed potatoes for space flights. Describe the steps in its manufacture and filling.

22. How does coining differ from embossing? Which requires more energy at the point of impact?

23. What is the difference between die casting and intraforming?

24. Sketch the following and detail the types of seams that might be used: oil drum, bucket, and food can.

25. Describe how the fatigue resistance of metal can be improved by cold working.

26. If a $1/8$-in.-thick stock is shear spun and the cone angle is 70°, what is the final thickness?

27. List the high–energy rate forming operations and state the type of work for which each is adapted.

28. A metal is shear spun to a thickness of 0.098 in., and a cone angle of 65° is necessary. What original stock thickness is required?

29. Discuss the word "springback" by using concepts from Chapter 2 on stress versus strain.

30. Design a die and explosive-forming system that would make a small metal drinking cup.

31. What stretch-forming pressure is required for a material having a yield strength of 105,000 psi on a part that has an area of 194 in.^{2}?

32. Why do thick materials exhibit less springback than thin ones when explosively formed?

33. For what purposes is shot peening used? List the advantages and disadvantages.

34. An aircraft panel that has a compound form has an area of 63 in.^{2} This material has a yield strength of 140,000 psi. How many pounds force are required?

35. A 4-in. (102-mm) round pipe with a 1-in. (25.4-mm) inside diameter is to be drawn into a 2-in. (50.8-mm) inside diameter pipe. About how many passes are required?

36. Approximately how many feet of 0.005-in. (0.13-mm) foil 3 ft. (0.9 m) wide could be made from 3 lb (1.4 kg) of lead? The density of lead is given in Chapter 2.

37. Calculate the wall thickness of a large nose cone made from 25.4-mm plate. The included angle of the cone is 60°.

38. Convert the following pressures to SI units: 40,000 psi, 10 psi, 750 psi.

39. If it is desired that steel shot leave the wheel at 500 ft/s, what size wheel is needed if it operates at 3600 rpm? What is 500 ft/s in SI units?

CASE STUDY

CENTRIFUGAL FAN FATIGUE PROBLEM

The Wayne Company is in the business of fabricating high-velocity, high-pressure industrial blowers. Until recently, assembly of the steel fan impellers was completed by welding the blades to the impeller sides (Figure 13.30). Welding causes the blades to deform slightly, and dy-

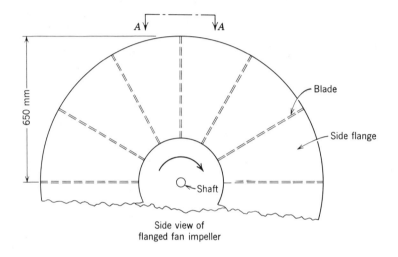

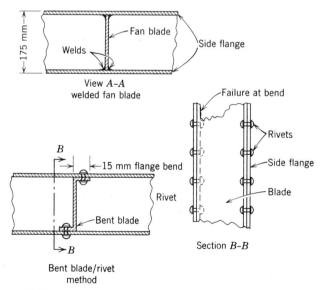

Figure 13.30
Case study.

namic balancing of the impeller is difficult because of weld grinding. To eliminate the balancing operation, manufacturing superintendent Dave Hock proposes to the company owner that he buy a press to bend a 15-mm flange on both edges of the blades to allow riveting to the sides, thus eliminating welding. Dave claims that the bending and riveting operations would avoid the dynamic problems experienced with welding.

The process to bend the flange on the impeller blades was installed, and after the usual amount of machine adjustment the process seemed to operate as well as Dave had claimed. The binding and riveting process proved to yield consistent parts that required no dynamic balancing.

Two fans were tested on the test stand for performance. After 1500 h of operation both fans developed severe vibrations and were shut down for troubleshooting.

Dave, being interested in the success of the project, takes the lead in vibration analysis. He finds that the vibrations are caused by the absence of large pieces that have broken off from the ends of a few of the fan blades. After studying the crack pattern, Dave determines that the failure was caused by fatigue cracks that originated at the flange bends on the blades and that the failure must be related to the new bending process.

Help Dave with this problem. You must determine, given a fatigue crack originating at the bend, what the possible alternatives are for altering the manufacturing process or material preparation to eliminate the cause of fatigue failure. What additional provision should you consider? How about reinstituting the balancing operation? Consider the solution of a different thickness. Examine library sources for stress patterns that result from different radii used for bends.

CHAPTER 14
PRESS WORK AND TOOLING

The machine used for most cold-working and some hot-working operations is known as a *press*. It consists of a machine frame supporting a bed and a ram source of power, and a mechanism to move the ram at right angles to the bed.

A press is equipped with dies and punches designed for producing parts in *press-working* operations. These tools are necessary for forming, ironing, punching, blanking, slotting, and the many operations that use press-working equipment.

Although some presses are better adapted for certain classes of work than others, most forming, punching, and shearing operations can be performed on any standard press if the proper dies and punches are employed. This versatility allows use of the same press for different jobs and operations, which is a desirable feature for short-run production.

Presses are capable of rapid work, because the operation time is only one stroke of the ram plus the time necessary to feed the stock. Production costs are low, and the process is suitable for mass-production methods, as is noted by its application in the manufacture of automotive and aircraft parts, hardware specialities, and kitchen appliances.

TYPES OF PRESSES

A classification of press machines is difficult to make, as most presses are capable of varied work. It is not entirely correct to call one press a bending press, another an embossing press, and still another a blanking press, because these three operations can be done on one machine. However, some presses are especially designed for one type of operation, and are known by the operation name. Examples are a punch press and a coining press. A simple classification is according to the source of power, either manual or power operation. Manual-operated machines are used on thin sheet metal, particularly in jobbing work, but most production machines are power operated. Presses can be grouped according to the number of rams or method of operating the rams. Most manufacturers name them according to the general design of the frame, although they may also be designated according to the means of transmitting power.

Several factors are considered in selecting a press. The operation, size of part, power required, and the speed of operation are important. For most punching, blanking, and trimming operations, crank or eccentric-type presses are generally used. In these presses the energy of the flywheel may be transmitted to the main shaft either directly or through a gear train. For coining, squeezing, or embossing operations, the knuckle joint is ideal.

It has a short ram and is capable of exerting a tremendous force, especially at the most extended position of the ram. Hydraulic presses for drawing operations have slower speeds than those employed for punching and blanking. The standard practice is not to exceed 50 ft/min (0.25 m/s) when drawing mild steel; aluminum and other nonferrous metals may be worked up to 150 ft/min 0.75 m/s).

The presses have a tonnage rating, which is expressed as follows: "The energy expended on the work at each stroke of the press must equal the tonnage required times the distance through which this tonnage must act." This energy is available from that stored in the flywheel and is given to the work as the flywheel slows down. This energy may be calculated by

$$E = \frac{N^2 D^2 W}{5.25 \times 10^9}$$

where

E = Energy (ton-in.) available at 10% slowdown of the flywheel
N = rpm of flywheel
D = Flywheel diameter, in.
W = Weight of flywheel, lb

Tonnage may also be limited by the motor restrictions.

Inclined Press

An *inclinable open-back press* with a gap frame is shown in Figure 14.1. This press, though shown in the vertical position, can be tilted backwards to permit the parts and

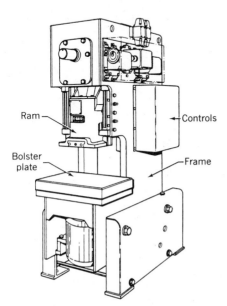

Figure 14.1
Open-back inclinable (OBI) press.

scrap to slide off the back side. Parts can slide by gravity into a *tote box,* or material may be fed by chute into the dies. Most presses of this type are adjustable and vary from vertical to a steep-angle position. Inclinable presses are often used in the production of small parts involving bending, punching, blanking, and similar operations.

Gap Press

Gap or *C-frame* presses are named because of the open arrangement of the press frame as shown in Figure 14.1. This is also shown in Figure 14.2. Gap presses provide excellent clearance around the dies and permit the press to be used for long or wide parts. Stamping operations can be performed on a gap press, and frequently the inclinable feature is used.

Arch Press

The arch press is shown in Figure 14.2 and is named for the particular shape of its frame. The lower part of the frame near the bed is wide to permit the working of large-area sheet metal. The crankshafts are small in relation to the area of the slide and press bed, as these presses are not designed for heavy work. They are used for blanking, bending, and trimming.

Straight-Side Press

As the capacity of the press is increased, it is necessary to increase the strength and rigidity of the frame. Straight-side presses are stronger because the heavy loads are taken up in a vertical direction by the massive side frame, and there is little tendency for the punch-and-die alignment to be affected by the strain. These presses are available for capacities in excess of 1250 tons (11 MN).

Straight-side presses are manufactured with various means of supplying power and different methods of operation. For the smaller presses a single crank or eccentric is usually employed, but as the size of the press increases, additional cranks are needed to distribute the load on the slide uniformly. The slide can be suspended in place by either

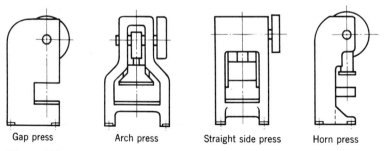

Gap press   Arch press   Straight side press   Horn press

Figure 14.2
Typical frame designs used in presses.

one, two, or four guides or points of suspension. ***Double-acting presses***, used extensively in drawing operations, have an outer ram that precedes the punch and clamps and the blank before the drawing operation. The outer ram is usually driven by a special link motion or cams, whereas the inner ram carrying the punch is crank driven.

A large straight-sided, enclosed, double-action toggle press is shown in Figure 14.3. Pressure is applied on the slide in four places. There is a distinct advantage in large-area presses because such construction prevents tilting of the slide with unbalanced loads. The toggle mechanism in this machine controls the motion of the blank holder. A ***toggle mechanism***, shown in Figure 14.11, may be described as a grouping of two or more bars so that although joined together end to end, they are not inclined except when the "knee" is straightened. Then great force is achieved on the ends; the time that the force is applied and when there is no motion on the blank holder is known as the *dwell* period. Dwell is necessary for blank holding on drawing operations, and it is frequently advisable to have a slight dwell on the punch to allow the metal to adjust itself properly under pressure. Straight-side frames are also used on hydraulic presses where heavy loads are encountered such as forming heavy-gage material, press forging, coining, and deep drawing.

Horn Press

Horn presses (Figure 14.2) ordinarily have a heavy shaft projecting from the machine frame instead of the usual bed. Where a bed is furnished, provision is made to swing it to one side when the horn is used. This press is used principally on cylindrical objects involving seaming, flanging edges, punching, riveting, and embossing.

Figure 14.3
Turret top completely formed with one stroke of enclosed toggle press.

Knuckle Joint Press

Presses designed for coining, sizing, and heavy embossing must be massive to withstand the large concentrated loads imposed upon them. The press shown in Figure 14.4 is designed for this purpose and is equipped with a ***knuckle joint*** mechanism for actuating the slide. The upper link or knuckle is hinged at the upper part of the frame at one end and fastened to a wrist pin at the other. The lower link is attached to the same wrist pin and the other end to the slide. A third link is fastened to the ends of the wrist pin and acts in a horizontal direction to move the joint as illustrated in Figure 14.11. As the two knuckle links are brought into a straight-line position, tremendous force is exerted by the slide.

This press is widely used for striking coins. According to tests made at the United States Mint in Philadelphia, a force of 98 tons (0.9 MN) is required to bring out clear impressions on half dollars made in a closed die.

Aside from striking coins, many other parts such as medals, key blanks, car tokens, license plates, watch cases, and silverware are cold pressed in this type of machine.

Figure 14.4
Knuckle joint press, 600-ton (5.3-MN) capacity, where frame is cast iron.

Sizing, cold heading, straightening, heavy stamping, and similar operations can also be performed. As the stroke of this press is short and slow, it is not adapted to drawing or bending operations.

Press Brake

Press brakes are used to brake, form, seam, emboss, trim, and punch sheet metal. These presses have a maximum width of 30 ft (9 m) and will form thickness to $^5/_8$ in. (16 mm). Capacities range from ten to several thousand tons of bending capacity. The bed is stationary and is the mounting surface for the lower die. Figure 14.5 is a mechanical brake that will form, punch, and cut to length steel parts.

The pressure capacity required of a press brake is determined by the length of work it will take, the thickness of the metal, and the radius of the bend. The minimum inside radius of a bend is usually limited to a radius equal to the thickness of the material. For bending operations the pressure required varies in proportion to the tensile strength of the material. Press brakes have short strokes and are generally equipped with an eccentric-type drive mechanism.

Figure 14.6 illustrates two unusual presses used with heavier gage material for the

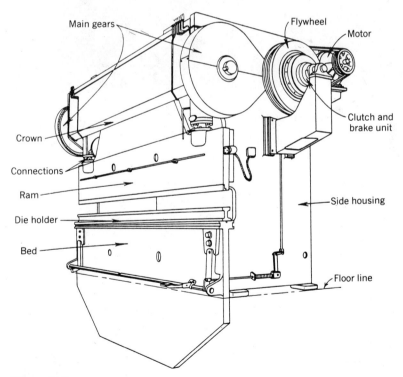

Figure 14.5
Power-press brake.

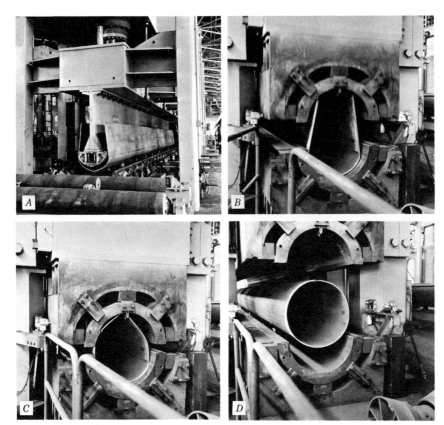

Figure 14.6
Steps in press-forming large-diameter pipe.

production of 30- and 36-in. (760-mm, 915-mm) pipe for gas transmission systems. The press-type brake bends the huge plate into a U shape as a first step in the operation. From this shape it is squeezed in an "O" press at pressures up to 18,000 psi (125 MPa) into a tubular form. Following this series of forming operations, the pipe is resistance welded, cleaned, and inspected.

Squaring Shears

This machine is used entirely for shearing sheets of steel and is made in both manual- and power-operated types. Sheets to a width of 10 ft (3 m) can be accommodated. Hydraulic hold-down plungers are provided every 12 in. (300 mm) to prevent any movement of the sheet during the cutting. In operation the sheet is advanced on the bed so that the line of cut is under the shear. When the foot treadle is depressed, the hold-down plungers descend and the shearing blade cuts progressively across the sheet.

Turret Press

Turret presses are especially adapted to the production of sheet metal parts having varied hole patterns of many sizes. In conventional presses of this kind a template is prepared to guide the punch, and the hole size is selected by rotating a turret containing the punches. Figure 14.7 illustrates a 30-ton (0.3 MN), tape-controlled turret punch press that will handle plate or sheet metal sizes to 48 by 72 in. (1200 by 1830 mm). The sheet metal is positioned under a punch with a table speed of 300 in./min (0.1 m/s). Holes up to $4^{3}/_{4}$ in. (120 mm) can be punched in $^{3}/_{8}$-in. (9.4-mm)-thick steel at a rate of more than 30 per minute to an accuracy of 0.005 in. (0.13 mm). Thirty-two different punches will fit in the turret.

Hydraulic Press

Hydraulic presses have longer strokes than mechanical presses and develop full tonnage throughout the entire stroke. The capacity of these presses is readily adjustable, and only a fraction of the capacity may be used. The length of the stroke is adjustable and the presses are adapted to deep-drawing operations because of their slow, uniform motion. They are also used for briquetting powdered metals, extruding, laminating, plastic molding, and press forging. They are not recommended for heavy blanking and punching operations, as the breakthrough shock is detrimental to the press. Maintenance is higher than for mechanical presses, even though the operation of the press is much slower. Small hydraulic presses resemble straight-side presses. For large-area work the post or four-column type of construction is used.

The hydraulic press (Figure 14.8) is especially designed for making deep draws in all kinds of sheet metal. The main draw punch mounted on the upper slide moves in tandem

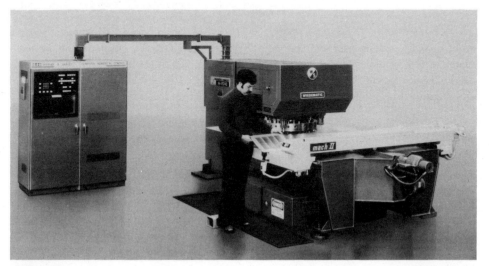

Figure 14.7
Tape-controlled turret punch press.

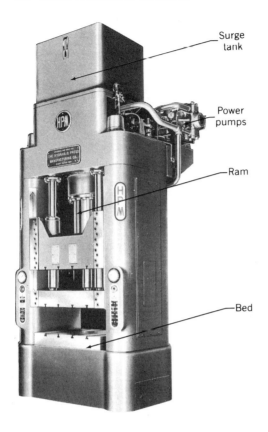

Figure 14.8
Double-action metal-drawing press.

with the blank holder slide and ring below it until the blank is contacted. The die rests upon the bolster plate; below it is a die cushion that can assist in maintaining pressure on the blank or ejecting the formed part. By locking the blank holder to the main slide and the die cushion idle, the press acts as a single-action hydraulic press.

Transfer Press

Transfer presses, being fully automatic, are capable of performing consecutive operations simultaneously. Material is fed to the press by rolls or as blanks from a stack feeder. In operation the stock is moved from one station to the next by a mechanism synchronized with the motion of the slide. Each die is a separate unit and is provided with a punch that may be independently adjusted from the main slide. Figure 14.9 is a 250-ton (2.2-MN) unit and produces 1600 starter end plates per hour.

The economical use of transfer presses depends on quantity production, as their usual production rate is 500 to 1500 parts per hour. Sheet metal products made on this equipment

TYPES OF PRESSES 347

Figure 14.9
Transfer press with capacity of 250 tons (2.2 MN) produces 1600 starter end plates per hour.

have a sequence of shearing, blanking, and forming operations. High-volume production is common.

Fourslide

The *fourslide* machine has many advantages over the punch press for complex forming operations on small parts made of sheet metal. The basic machine has four power-driven slides set 90° apart that are separately controlled by cams to move progressively through a cycle. Figure 14.10 shows the sequence of operations for forming a spring clip.

The fourslide can be equipped with punches, cutting-off and blanking tools, lifting and shifting devices, drills, and, in some cases a vertical punch. The tools can be made to pivot or open and close if the operation so requires. If either wire or strip, coiled stock is fed into the machine, it is first straightened by passing it through rollers attached to the frame. Hoppers and precision locators can be used if the parts are preblanked; spot

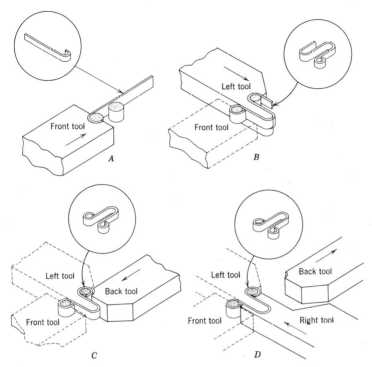

Figure 14.10
Sequence of operations on a fourslide machine.

and butt welders can also be employed. The process can be almost automatic and lends itself to mass-produced parts.

DRIVE MECHANISMS FOR PRESSES

Most of the drive mechanisms for transmitting power to the slide are shown in Figure 14.11. The most common drive is the *single crank,* which gives a slide movement approaching simple harmonic motion. On a downstroke the slide is accelerated. Reaching its maximum velocity at midstroke, it is then decelerated. Most press operations occur near the middle of the stroke at maximum slide velocity. The *eccentric drive* gives a motion like that of a crank and is often used where a shorter stroke is required. Its proponents claim it has greater rigidity and less tendency for deflection than a crank drive. *Cams* are used where some special movement is desired such as a swell at the bottom of the stroke. This drive has some similarity to the eccentric drive except that roll followers are used to transmit the motion to the slide.

Rack and gear presses are used for applications requiring a very long stroke. The movement of the slide is slower than in crank presses, and uniform motion is attained.

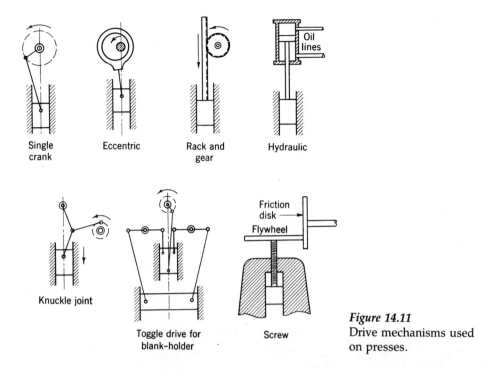

Figure 14.11 Drive mechanisms used on presses.

These presses have stops to control the stroke length and may be equipped with a quick-return feature to raise the slide to starting position. The *arbor press* is a familiar example.

Hydraulic drive is used in presses for a variety of work. It is especially adapted to large pressures requiring slow speed for forming, pressing, and drawing operations.

In the *screw drive,* the slide is accelerated by a friction disk that engaged the flywheel. As the flywheel moves down, greater speed is applied. From beginning to end of the stroke the slide motion is accelerated. At the bottom of the stroke the amount of stored energy is absorbed by the work. The action resembles that of a drop hammer, but it is slower and there is less impact. Presses of this type are known as *percussion presses*.

Several link mechanisms are used in press drives because of the motion or because of their mechanical advantage. The *knuckle joint* mechanism (Figure 14.11) is common. It has a high mechanical advantage near the bottom of the stroke when the two links approach a straight line. Because of the high load capacity of this mechanism, it is used for coining and sizing operations.

Eccentric or *hydraulic* drives may be substituted for the crank shown in the figure. *Toggle* mechanisms used primarily to hold the blank on a drawing operation are made in a variety of designs. The auxiliary slide in the figure is actuated by a crank, but eccentrics or cams may be used. The principal aim of this mechanism is to obtain a motion having a suitable dwell so that the blank can be held effectively.

An application for the microprocessor in control of a press is in stretch forming. About 75% of aircraft frames are stretch formed. A microprocessor that has memory capabilities can be programmed by using an acceptable part that has been formed manually. Under a playback mode any number of parts can be produced. Another advantage of the microprocessor is that programs can be memorized and transferred to tapes or disks that serve as permanent records.

FEED MECHANISMS

Safety is a paramount consideration in press operation, and every precaution must be taken to protect the operator. Wherever possible, material should be fed to the press by a means that eliminates any chance of the operator having his or her hands near the dies. In the long-run production jobs such features can be designed into the operation. Feeding devices applied to medium-sized and small presses have the advantage of rapid, uniform machine feeding in addition to the safety features.

One common feeding mechanism is the *double-roll* feed utilizing coiled stock and scrap reels. The operation of the feed rolls is controlled by an eccentric on the crankshaft through a linkage to a ratchet wheel that pulls the material across the die. Each time the ram moves up the rolls turn and feed the correct amount of material for the next stroke.

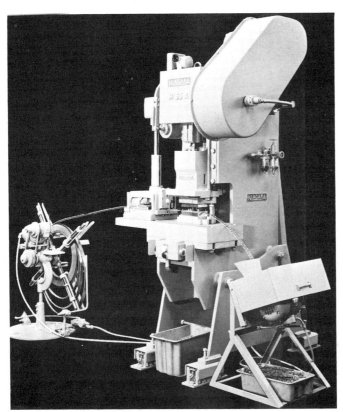

Figure 14.12
Automatic high-speed press with propeller shaft-driven slide feed. Capacity 35 tons (0.3 MN).

By providing the machine with a variable eccentric, the amount of stock fed through the rolls can be varied. An automatic high-speed press equipped with a slide feed is shown in Figure 14.12. The scrap from this press, instead of being rerolled on a scrap reel, is sheared to small lengths for easy handling. For heavy materials, straightening rolls also act as feeding rolls.

Another feeding device is the *dial station* feed. This method is designed for single parts previously blanked or formed in some other press. The indexing is controlled by an eccentric on the crankshaft through a link mechanism to the dial. For each stroke the dial indexes one station. All feeding by the operator takes place at the front of the machine away from the dies.

Light parts can be stacked in a magazine and placed in the die position by a *suction* device. A blank is lifted off the top of the stack by suction fingers and placed against a stop gage on the die. *Magazine* feeds may also be used with a reciprocating mechanism that feeds blanks from the bottom of the stack. *Gravity* feed is sometimes used on inclined presses, the blank sliding into a recess at the top of the die.

PRESS OPERATIONS AND TOOLING

The tools in most presses come under the general title of *punches* and *dies*. The punches of the assembly is attached to the ram of the press and is forced into the die cavity. The die is usually stationary and rests on the press bed. It has an opening to receive the punch, and the two must be in perfect alignment for proper operation. Punches and dies are not interchangeable. A single press may do a large variety of operations depending on the dies used.

Shearing

Cutting metal involves stressing it in shear above its ultimate strength between adjacent sharp edges as shown in Figure 14.13. As the punch descends upon the metal the pressure first causes a plastic deformation to take place as in Figure 14.13*B*. The metal is highly stressed adjacent to punch-and-die edges, and fractures start on both sides of the sheet as the deformation continues. When the ultimate strength of the material is reached, the

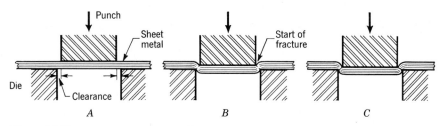

Figure 14.13
Process of shearing metal with punch and die. *A*, Punch contacting metal. *B*, Plastic deformation. *C*, Fracture complete.

fracture progresses, and, if the clearance is correct and both edges are of equal sharpness, the fractures meet at the center of the sheet as shown in Figure 14.13C. The amount of clearance, which plays an important part in die design, depends on the hardness of the material. For steel it should be 5% to 8% of the stock thickness per side. If improper clearance is used, the fractures do not meet, and instead must cross the entire sheet thickness, using more power.

Flat punches and dies as shown in the figure require a maximum of power. To reduce the shear force the punch or die face should be made at an angle, so that the cutting action is progressive. This distributes the shearing action over a greater length of the stroke and can reduce the power required by up to 50%.

Blanking, as shown in Figure 14.14, is the operation of cutting out flat areas to a desired shape. It is usually the first step in a series of operations. In this case the punch should be flat and the die given some shear angle so that the finished part will be flat. Punching or piercing holes in metal, notching metal from edges, or perforating are all similar operations. For these operations the shear angle is on the punch and the metal removed is scrap. *Trimming* is the removal of "flash" or excess metal from around the edges of a part and is essentially the same as blanking. *Shaving* is similar except that it is a finishing or sizing process. **Slitting** is making incomplete cuts in a sheet as illustrated in Figure 14.15. If a hole is partially punched and one side bent down as a louver, it is called **lancing.** All these operations may be done on presses of the same type and differ little except in the dies that are used.

The formula for the pressure P (in pounds-force) that is required to blank or punch a material, assuming there is no shear angle on the punch or die, is expressed as

$$P = SLt$$

and for round holes,

$$P = \pi DSt$$

where

S = Shear strength of materials, psi (Pa)
L = Sheared length, in. (mm)
D = Diameter, in. (mm)
t = thickness of materials, in. (mm)

A blanking punch and die for stamping stainless-steel razor blades is shown in Figure 14.16. Because of the severe service encountered, die parts are made from carbide. When the punch becomes dull, usually after about 150,000 punches, it is reground. The small

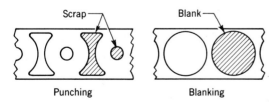

Figure 14.14
Illustrating the difference between punching and blanking operations.

PRESS OPERATIONS AND TOOLING 353

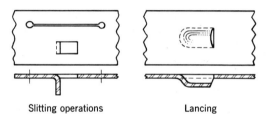

Slitting operations   Lancing

Figure 14.15
Examples of slitting and lancing operations.

high-speed presses used with these dies employ coiled stock and operate at 300 spm. After the press operation the punched material is recoiled and hardened. Then follows grinding, honing, and stropping. Finally, the blades are snapped apart, inspected, and, depending upon brand, are coated with Teflon and packaged.

Bending and Forming

Bending and forming may be performed on the same equipment as that used for shearing—namely, crank, eccentric, and cam-operated presses. Where bending is involved the metal is stressed in both tension and compression at values below the ultimate strength of the

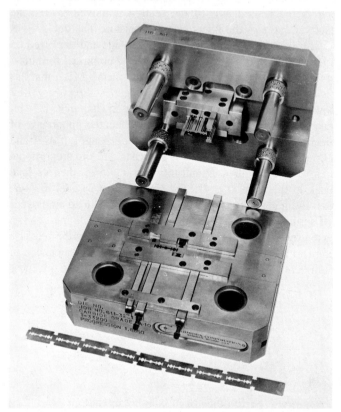

Figure 14.16
Punch and die for razor blades.

material without appreciable changes in its thickness. As in a press brake, simple bending implies a straight bend across the sheet of metal. Other bending operations, such as curling, seaming, and folding, are similar, although the process is slightly more involved. Bending pressures can be determined using the following empirical relationship:

$$F = \frac{1.33LSt^2}{W}$$

where

F = Bending force required, tons (MPa)
L = Length of bend, in. (mm)
S = Ultimate tensile strength, tons/in.^{2} (Pa/mm^2)
t = Metal thickness, in. (mm)
W = Width of V-channel, or U lower die, in. (mm)

A *V-channel* or a *U-die* is a punch-and-die arrangement. These punch and dies often form a 90° internal angle between the faces of the metal. The empirical constant of 1.33 is a die-opening factor proportional to metal thickness.

In designing a rectangular section for bending, one must determine how much metal should be allowed for the bend, because the outer fibers are elongated and the inner ones shortened. During the operation the neutral axis of the section is moved toward the compression side, which results in more of the fibers being in tension. The thickness is slightly decreased, and the width is increased on the compression side and narrowed on the other. Although correct lengths for bends can be determined by empirical formulas, they are influenced considerably by the physical properties of the metal. Metal that has been bent retains some of its original elasticity, and there is some elastic recovery after the punch is removed as shown in Figure 14.17. This is known as ***springback***.

Figure 14.18 shows a generalized stress–strain curve that explains the importance of the yield point in bending operations. If a part *A–B* is stretched to a length of *A–C* and released, it will return to its original length *A–B*. If *A–B* is stretched past the proportional limit *L* on the stress–strain curve to a length *A–D* and the load removed, then its final length is *A E*. The springback in this case is *E–D*. The line *M–E* is parallel to *L–B*. Thus, depending on the stress–strain characteristics of the metal and the load employed, an indication of the extent of springback is possible.

Springback may be corrected by overbending an amount such that when the pressure is released, the part will return to its correct shape. Springback is more pronounced in large-bend radii. The minimum-bend radius varies according to the ductility and thickness of the metal.

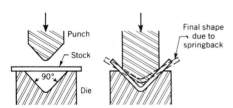

Figure 14.17
Springback in bending operations.

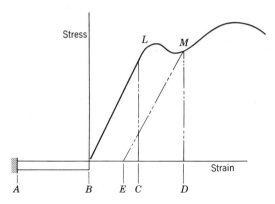

Figure 14.18
Springback and its relationship to stress–strain.

A forming die, designed to bend a flat strip of steel to a U-shape, is shown in Figure 14.19. As the punch descends and forms the piece, the knockout plate is pressed down, compressing the spring at the bottom of the die. When the punch moves up, the plate forces the work out of the die with the aid of the spring. Such an arrangement is necessary in most forming operations because the metal presses against the walls of the die, making removal difficult. Parts that tend to adhere to the punch are removed by a knockout pin that is engaged on the upstroke.

Drawing

Three bent flanges are shown in Figure 14.20. The first one (Figure 14.20A) is the simple straight bend. The stretch flange and shrink flange (Figure 14.20 and B and C, respectively) involve a plastic flow of metal that does not take place in a straight-bend flange. This plastic flow or adjustment of metal is characteristic of all drawing operations. Stresses are involved that exceed the elastic limit of the metal so as to permit the metal to conform to the punch. However, these stresses cannot exceed the ultimate strength without developing cracks. If the stretch flange (Figure 14.20B) is considered to be a section of a

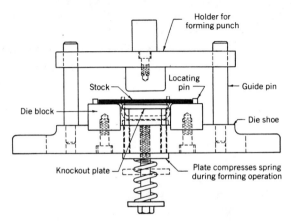

Figure 14.19
Forming punch and die.

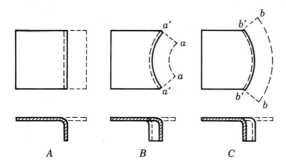

Figure 14.20
Types of flanges. A, Straight. B, Stretch. C, Shrink.

circular depression that has been drawn, the metal in arc *aa* must have been stretched to $a'a'$. The action is a thinning one and must be uniform to avoid cracks. In the **shrink flange** (Figure 14.20C) the action is just the opposite, and the metal in the flange is thickened. Most drawn parts start with a flat plate of metal. As the punch is forced into the metal, severe tensile stresses are induced into the sheet being formed. At the same time the outer edges of the sheet that have not engaged the punch are in compression and undesirable wrinkles tend to form. This must be counteracted by a blank holder or pressure plate, which holds the flat plate firmly in place.

In a simple drawing operation of relatively thick plates the plate thickness may be sufficient to counteract wrinkling. This may be done in a single-acting press as shown in Figure 14.21. Additional draws may be made on the cup-shaped part, each one elongating it and reducing the wall thickness.

Most drawing, involving the shaping of thin metal sheets, requires double-acting presses to hold the sheet in place as the drawing progresses. Notice Figure 14.8. Presses of this type usually have two slides, one within the other. One slide controlling the blank-holding rings moves to the sheet ahead of the other to hold it in place. This action is illustrated in Figure 14.22. The motion of the blank-holding slide is controlled by a toggle or cam mechanism in connection with the crank. Hydraulic presses are well adapted for drawing because of their relatively slow action, close speed control, and uniform pressure. Figure 14.23 shows a sectional diagram of an *inverted* drawing die. The punch is stationary and is mounted on the bed of the press. As the die descends, the blank is contacted; then as its downward movement continues, the blank-holding ring maintains contact with the blank during drawing. By the use of a die cushion to control the holder the pressures on the blank can be increased and controlled.

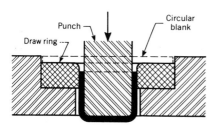

Figure 14.21
Arrangement of punch and die for simple drawing operations.

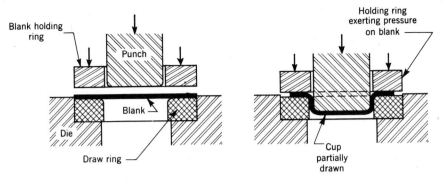

Figure 14.22
Action of blankholder and punch in a drawing operation.

The pressure applied to the punch necessary to draw a shell is equal to the product of the cross-sectional area and the yield strength Y_S of the metal. A constant that covers the friction and bending is necessary in this relationship. The pressure P for a cylindrical shell may be expressed by the empirical equation

$$P = \pi dt S(D/d - 0.6)$$

where

D = Blank diameter, in.
d = Shell diameter, in.
t = Metal thickness, in.
S = Tensile strength, psi

The amount of clearance between a punch and die for blanking is determined by the thickness and kind of stock. For thin material the punch should be a close-sliding fit. For

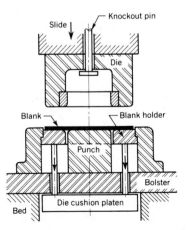

Figure 14.23
Drawing operation using an inverted punch.

heavier stock the clearance must be larger to create the proper shearing action on the stock and to prolong the life of the punch.

There is a difference of opinion as to the method designating clearance. Some claim that clearance is the space between the punch and die on one side or one-half difference between the punch and die sizes. Others consider clearance as the total difference between the punch and die sizes. For example, if the die is round, clearance equals die diameter minus punch diameter. The advantage of designating clearance as the space on each side is particularly evident in the case of dies with irregular form or shape. Whether clearance is deducted from the diameter of the punch or added to the diameter of the die depends on the work. If a blank has a required size, the die is made smaller. When holes of a certain size are required, the punch is ground to the required diameter and the die is made larger.

SPECIAL DIES AND FORMING PROCESSES

The die sets shown in Figures 14.19 and 14.21 are classified as *simple* dies. Only one operation is performed with each stroke of the ram. *Compound* dies combine two or more operations at one station as illustrated in Figure 14.24. Strip stock is fed to the die, two holes are punched, and the piece is blanked on each stroke of the ram. When the operations are not similar, as in the case of a blanking and forming operation, dies of this type are frequently known as *combination* dies.

A *progressive* die set performs two or more operations simultaneously but at different stations. A punch-and-die set of this type is shown in Figure 14.25. As the strip enters the die, the small square hole is punched. The stock is then advanced to the next station, where it is positioned by the pilot as the blanking punch descends to complete the part. This general type of design is simpler than the compound dies, because the respective operations are not crowded together. Regardless of the number of operations to be performed, the finished part is not separated from the strip until the last operation. A progressive die set that performs 15 operations on a can opener, completing one at each stroke, is shown in Figure 14.26. Production is rapid and close tolerances are difficult to maintain.

In the *perforation* of metal one or two rows of punches are employed. The metal sheet

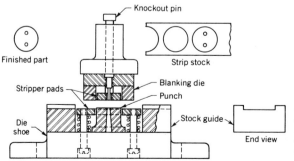

Figure 14.24
Compound punch and die.

SPECIAL DIES AND FORMING PROCESSES   359

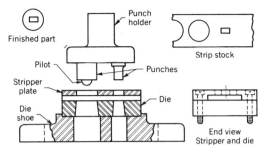

Figure 14.25
Progressive punch and die.

is moved incrementally through the press each time the punches are withdrawn. The punched holes may be almost any shape. All nonbrittle metals can be perforated.

Misalignment of punch and die causes excessive pressures, shearing or chipping of die edges, or actual breaking of the tools. Such action may occur through shifting even though the setup was originally correct. To prevent these occurrences, proper alignment is ensured by guide rods at two or four corners of the die that fit into holes in the punch holders. These dies are known as *pillar* dies. This arrangement of having the punch and die held in proper alignment facilitates the setting up of the tools. A similar arrangement, known as a ***subpress die*** and occasionally used on small work, employs a punch and die mounted in a small frame so that accurate alignment is always maintained. Pressure is applied by a plunger that extends out of the top of the assembly.

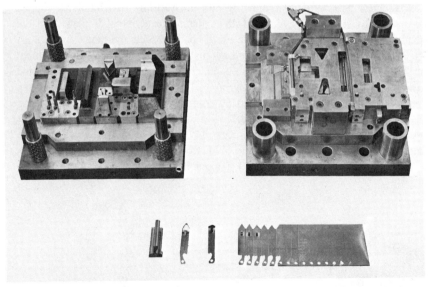

Figure 14.26
Progressive die set that performs 15 operations on can openers and completes one on each stroke.

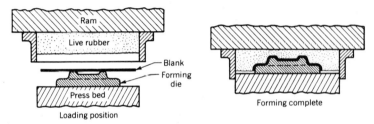

Figure 14.27
Method for forming sheet metal using single die and rubber pad.

Rubber Pad Processing

In *rubber die* processing a rubber or urethane pad confined in a container replaces the die or punch. Under pressure, the rubber pad flows around a form block or punch, and the blank is punched or formed. Advantages of these methods are lower cost tooling for short-run production.

In the **Guerin process** the rubber pad is in a boxlike container mounted on the ram side. As the **platen** moves down, the force of the ram is exerted evenly in all directions resulting in the sheet metal being pressed against the die block as shown in Figure 14.27. *Cutting die blocks* are merely steel templates of the part to be made and need not be over $3/8$ in. (10 mm) thick. Forming dies may be made of hardwood, aluminum, or steel. Aluminum sheet can be cut in thickness up to 0.051 in. (1.30 mm); for bending and forming the usual limit is approximately $3/16$ in. (5 mm) thick.

A *Marform* process permits deeper drawing and the forming of irregularly shaped parts. In the operations shown by Figure 14.28, a flat piece of metal is placed on the blank holder plate. As the platen descends the rubber pad contacts the blank and clamps it against the top of the punch and surrounding plate. As the downward movement continues, the blank is formed over the end of the punch and sufficient pressure is exerted over the unformed portion to prevent metal wrinkle. During the drawing operation the downward movement of the blank is opposed by controllable pressure pads in forming aluminum sheet thicknesses up to 0.675 in. (17.15 mm) have been processed.

Deep drawing can be achieved by another process known as *Hydroform*, which employs a rubber cavity in the ram containing hydraulic fluid. Lowering the ram clamps the blank between the flexible rubber die member and a blank holder on the bolster. A punch

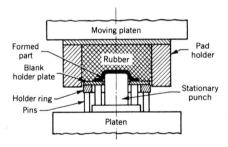

Figure 14.28
Arrangement of the components in a forming operation with the Marform process.

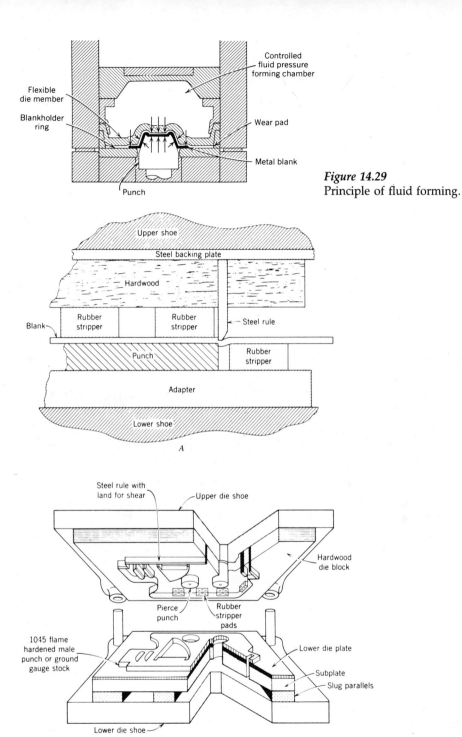

Figure 14.29
Principle of fluid forming.

Figure 14.30
Steel rule die process. *A*, Cross section of shearing action of steel rule. *B*, Assembled steel rule die.

attached to a hydraulic cylinder assembly moves upward and increases the diaphragm pressure on the blank, drawing it around the punch. Figure 14.29 shows the elements of the process.

Steel Rule Dies

This process employs steel ribbons mounted in hardwood as the die. The ***steel rule die*** is similar to a cookie cutter. The process eliminates the solid-metal die section used in blanking sheet metal stock. The steel rules are shaped to the outer configuration of the part to be blanked and are inserted in a hard or specially laminated wood backing. The shearing action and a steel rule are shown in Figure 14.30. The punch may be of 1045 flame-hardened steel, or ground-gage stock. The strip of stock used in steel rule dies varies from 0.056 to 0.166 in. (1.42–4.22 mm) thickness, and the shearing-edge side is ground at a 45° bevel. Steel rule dies are useful for low-quantity blanking of up to $3/8$-in. (10-mm) steel or 0.55-in. (15-mm) aluminum.

QUESTIONS AND PROBLEMS

1. Give an explanation of the following terms:

   Double-acting press
   Toggle mechanism
   Horn
   Knuckle joint
   Turret press
   Transfer press
   Fourslide
   Slitting
   Lancing
   Springback
   Shrink flange
   Perforating
   Subpress die
   Guerin process
   Platen
   Steel rule die

2. Explain the differences between a die and a punch.

3. Give the principal advantage of an inclined press over a vertical one, and also over one with a rigid frame.

4. Define embossing and how it differs from other processes.

5. Sketch an arch press and describe the work for which it is designed.

6. What press is used for making auto license plates? Why is it used?

7. When would you recommend a turret press?

8. List the differences between hydraulic and mechanically-driven presses?

9. Describe a transfer press. List 10 products that can be produced on a transfer press.

10. Sketch the operation of a knuckle joint mechanism. On what type of press is this used?

11. Select a part to be made economically on a transfer press and sketch the sequence of operations.

12. When is the hydraulic press adopted for making parts?

13. Consider the ultimate strengths of aluminum and mild steel, and then discuss the probable results of shearing them.

14. Discuss the clearance between a punch and die for shearing operations. Why is it important?

15. Construct a flow diagram for manufacture of razor blades.

16. List the advantages of the five types of press drives.

17. Explain the purpose of double-acting presses.

18. Briefly describe the following operations: blanking, punching, shaving, slitting, and lancing.

19. A press flywheel has the diameter of 28 in. (710 mm) and a weight of 690 lb (310 kg); it operates at 360 rpm. What is the available energy of the flywheel?

20. List the differences between compound and progressive dies.

21. Find the total energy of the flywheels in ton-inches if each of two wheels is 36 in. in diameter, has a thickness of 4.25 in., operates at 420 rpm, and the density is 0.29 lb/in.^{3}

22. Sketch a process to form a small metal $1/4$-cup (0.12-liter) measuring cup.

23. Find the pressure in pounds-force to punch the strip stocks shown by Figure 14.31. Cold-rolled steel shear strength is 42,000 psi.

24. Assume that the material of Figure 14.31 is stainless steel and has a shear strength of 122,000 psi. Find the pressure in pound-force.

25. What role does the rubber pad play in process using only a single die?

26. A press brake is required to bend a cold-rolled steel sheet along a 48-in. dimension. A V-shape width is $1/2$ in. The ultimate tensile strength of the material is 95,000 and the thickness is $1/4$ in. What is the bending force? What is the importance of this number?

27. What press do you recommend for each of the following jobs: forming steel tops for automobiles, stamping coins, screw tops for glass jars, nose cones for small missiles, and file-case drawer fronts?

28. A $1/4$-in. plate is 96 by 36 in. and will have a brake along both directions. A V-channel width = $1/4$ in. Corners are notched to permit double bending. The ultimate tensile strength is 135,000 psi. Find the bending forces.

29. Find the punching pressure for 2024-T grade of aluminum having a shear strength of 40,000 psi. The blank perimeter is 17 in. and thickness is 0.070 in.

30. Find the drawing pressure for a beverage can shown in Figure 14.33. The tensile strength of this grade of aluminum is 27,500 psi.

31. Find the drawing pressure for the following material and design. Assume a free draw with clearance sufficient to prevent ironing, and a maximum reduction of 50%, a deep

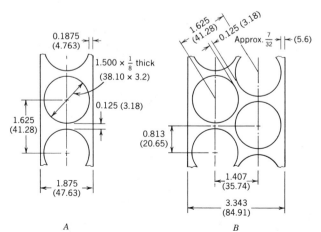

Figure 14.31
Strip stock layouts.

draw of steel stock $1/8$ in. thick with a tensile strength of 50,000 psi into a shell 10 in. in diameter. The blank diameter is 12 in.

32. Consider the following: metal having a tensile stress of 40,000 psi, mean diameter of shell 4 in., blank diameter of 9.5 in., and thickness of $1/16$ in. Find the pressure applied to the punch.

33. Describe a steel rule die. Why are they limited to short-run work?

34. Convert the following press sizes to SI units: 400 ton, 1200 ton, and 3 ton.

35. What is the maximum area of the blank that can be formed in a 700-ton (6.2-MN) press if the pressure necessary to form the material is 14,000 psi (96.5 MPa)?

36. Calculate the minimum inside radius that usually will apply if a 1-in. (25.4-mm) plate is being bent. Also calculate for a $1/4$-in. (6.4-mm) plate.

37. Sketch the manner in which a ribbon of thin metal 0.020 in. (0.51 mm) thick and $1/4$ in. (12.7 mm) wide can be bent into the shape of a paper clip on a fourslide machine.

38. Consider a single-crank drive and plot a velocity diagram of the slide as a function of the crank position.

39. What size press in tons is needed if a 6-in. (150-mm) diameter is to be formed at a pressure of 20 tons/m² (275 MPa)?

40. Suppose 1-in. (25.4-mm) disks are punched from a 1.25-in. (31.8-mm) flat coiled stock. What is the minimum percentage of waste?

41. According to tests made at the U.S. Mint in Philadelphia, a force of 98 tons (0.9 MN) is required to bring out clear impressions in half dollars made in a closed die. What pressure does this force create?

42. Consider Figure 14.31 again. The material cost is $0.75/ℓb ($1.65/kg). In addition to blanking losses of the skeleton, add a 5% loss for overall waste. (a) Find the losses for both designs. Density = 0.278 ℓbshin.³ (7692 kg/m³). (b) Salvage of the waste and skeleton is recovered at 10% of the original value. Find the economic yield of both designs. Which one is cheaper? (c) Also answer the preceding in metric units.

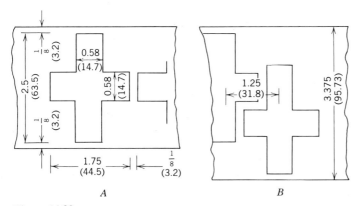

Figure 14.32
Layouts of blanks.

43. Consider Figure 14.31. Material cost = $0.57/lb ($1.27/kg). In addition to blanking losses, there is an additional loss of 4% for waste. Density = 0.278 lb/in.³ (7692 kg/m³). Salvage of the material skeleton and waste is sold back at 5% of the original price of the material. (a) Find the cost of one blank for both designs in English units. (b) Find the cost of one blank for both designs in SI.

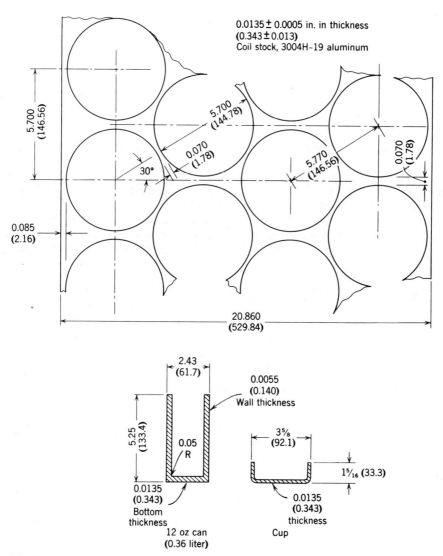

Figure 14.33
Case study.

44. Two potential blank layouts are given in Figure 14.32. Material is cold-rolled steel having a tensile strength of 40,000 psi. Thickness is $1/8$ in. Determine punching pressure. Density = 0.278 lb/in.^{3} (7692 kg/M^3). Cost of the material is $0.75/lb ($1.65/kg). Waste loss in addition to the skeleton is 4%. Find the efficiency of strip stock layouts A and B. Which design is preferred?

CASE STUDY

TWELVE-OUNCE BEVERAGE CONTAINER COMPANY

This company manufactures beverage containers in the popular 12-fl oz (0.36-liter) size. With production exceeding 7.5 million cans daily, material efficiency is absolutely necessary. The can is composed of three pieces: the body, top, and pull ring. The rivet connecting the pull ring to the top is formed from the top during the mechanical joining process. The container body is blanked from 3004-H19 aluminum coils, and the layout is shown in Figure 14.33. An intermediate cup is formed without any significant change in thickness. The cup is then drawn in a horizontal drawing machine, and metal is "ironed" to a sidewall thickness of 0.0055 in. (0.140 mm). The bottom thickness remains unchanged. The can is trimmed to a final height of 5.25 in. (133.4 mm) to give an even edge for rolling to the lid.

(a) Determine the "metal efficiency," that is, the final metal in the can as related to the original metal in the coil. What is the metal efficiency of the can to the 5.700-in. (144.78-mm) blank? [Disregard the 0.05-in. (1.3-mm) radius.] If the 3004-H19 aluminum coil stock costs $0.5864/lb ($1.293/kg) and the density = 0.0981 lb/in.^{3} (2715 kg/m^3), find the cost of the can. Assume that waste is recovered and recycled at $0.275/lb ($0.606/kg). What is the material cost lost in waste per can? For a production of 1 million cans daily, what are the material costs lost in waste? Describe some alternatives for improving the yield. Repeat in SI units.

CHAPTER 15
BASIC MACHINE TOOL ELEMENTS

Most machine tools are constructed by using two or more components. These components, although they may have different functions in such machines as a lathe, mill, or drill press, have common characteristics.

Because of the demand for metal removal machines such as lathes, machining centers, milling machines, grinders, and the many others shown throughout this book, there has been continuous development in flexible *machining centers*. The mass-produced and the special machine tools are constructed of basic elements. This chapter describes those elements.

Important requirements for machine tool structures include *rigidity, shape, operator* and *part accessibility,* ease of *chip removal,* and *safety*. In terms of machine tool performance, static and dynamic stiffness is necessary for accuracy and precision. Stability of the machine structure is required to prevent machining chatter. Understanding the basic machine elements is necessary to appreciate the breadth of modern machining methods in the manufacture of products.

STRUCTURES FOR CUTTING MACHINES

Castings, forgings, and hot- or cold-formed shapes usually require machining. The variety of sizes, shapes, and materials calls for diversity in machining.

Machine tools differ not only in the number of cutting edges they employ, but also in the way the tool and workpiece are moved in relation to each other. In some machines (vertical machining centers, drill presses, boring machines, milling machines, shapers, and grinders) the workpiece remains virtually motionless and the tool moves. In others (planers, lathes, and boring mills) the tool is virtually fixed and the workpiece moves. But it should be pointed out that seldom are these simple elements applied without modification. Study Figure 15.1 for the traditional processes used for machining parts.

The single-point tool-shaping machines are the easiest to visualize. The lathe and the boring machine are *kinematic inversions* employing the single-point tool. In Figure 15.2A the work rotates in the lathe, but the cutting tool is stationary. In the boring machine (Figure 15.2B) the tool rotates while the work is stationary. Although the lathe tool and the boring machine worktable are not truly stationary, this is overlooked for the moment. To feed a tool carriage past rotating work is usually more acceptable than to feed rotating work with headstock and supports past a stationary tool post.

The shaper and planer use *single-point cutting tools*. Figure 15.2C and D point out

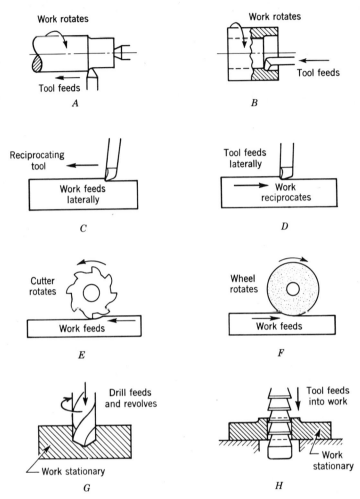

Figure 15.1
Traditional processes used for machining parts to specified dimensions. A, Turning. B, Boring. C, Shaping. D, Planing. E, Milling. F, Grinding. G, Drilling. H, Broaching.

that size of the workpiece is a factor in machine structural design. The smaller workpiece is more efficiently machined on the shaper than on the planer. The general appearance of the machine is changed by reversing the kinematic relationship of work and tool. However, the cutting action principle is identical.

With the introduction of the milling cutter by Eli Whitney in the early 1800s, the rotating tool was used only as a boring tool. But Whitney gave it a new application. The milling cutter was no longer used for cutting circular bores exclusively, but was used for cutting keyways, slitting, sawing, slab and face milling, gear cutting, and shaping irregularly formed pieces. Use of the rotating tool combined with traversing work was introduced in the milling machine as shown in Figure 15.2E. The kinematic inversion of the

STRUCTURES FOR CUTTING MACHINES 369

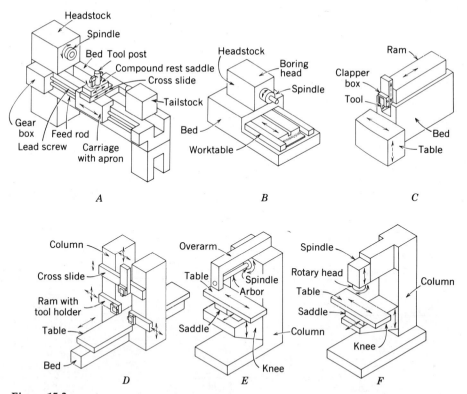

Figure 15.2
Basic structural elements in conventional machine tools. A, Lathe. B, Horizontal boring machine. C, Shaper. D, Planer. E, Horizontal milling machine. F, Vertical milling machine.

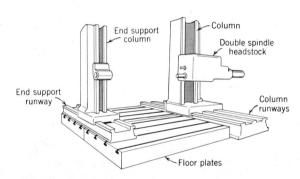

Figure 15.3
Basic elements in a floor-type, horizontal milling, drilling, and boring machine.

370   BASIC MACHINE TOOL ELEMENTS

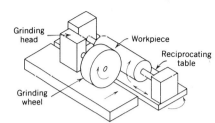

Figure 15.4
Basic elements in a bench-type grinding machine.

standard milling machine is the floor-type horizontal boring, drilling, and milling machine illustrated in Figure 15.3

The cylindrical grinder adopts motions of the lathe and boring machine except for the substitution of rotating tools (the grinding wheel) for single-point tools. The work and cutting tool rotate in the grinder as shown in Figure 15.4.

The characteristics of these basic cutting machines are listed in Table 15.1.

Table 15.1 **Cutting and Feed Movement for Conventional Machines**

Machine	Cutting Movement	Feed Movement	Types of Operation
Lathe	Workpiece rotates	Tool and carriage	Cylindrical surfaces, drilling, boring, reaming, and facing
Boring machine	Tool rotates	Table	Drilling, boring, reaming, and facing
Planer	Table traverses	Tool	Flat surfaces (planing)
Shaper	Tool traverses	Table	Flat surfaces (shaping)
Horizontal milling machine	Tool rotates	Table	Flat surfaces, gears, cams, drilling, boring, reaming, and facing
Horizontal boring	Tool rotates	Tool traverses	Flat surfaces
Cylindrical grinder	Tool (grinding wheel) rotates	Table and/or tool	Cylindrical surfaces (grinding)
Drill press	Tool rotates	Tool	Drilling, boring, facing, and threading
Saw	Tool	Tool and/or workpiece	Cut off
Broaching	Tool	Tool	External and internal surfaces

MACHINE FRAME

Machine frames either are cast from gray cast iron or steel or else are manufactured by welding steel plate. The frame is usually cast if it is too intricate for welding fabrication or if the weight of the frame is relatively important. Gray cast iron is preferred for most moderate-size machines in which vibration forces may be a problem because it has a very high damping capacity. Large, heavy-duty frames that must withstand impact loads are often made of cast steel.

Welded or fabricated frames are increasing in popularity as compared to cast frames for the following reasons:

1. Savings in weight may amount to 25%.
2. Repairs are relatively easy for damaged frames.
3. Various grades of steel may be used in the same frame depending on the design requirements of a machine member.
4. Design changes are less costly because there is no investment in patterns or cores.
5. Errors in machining or design are easier to correct.
6. Additional material can be located near the stress zones to control vibration and deflection.

The disadvantages of the fabricated frame are:

1. Gray cast iron absorbs vibrations better than welded steel frames.
2. Cast material is homogeneous, and hence, chemical reactions are negligible.
3. Casting process may be more adaptable to high production rates.
4. Frames for heavy-duty machines may need considerable weight to absorb the loads.

BASIC ELEMENTS

Metal-cutting machines are composed of self-contained elements, each having a special function. The elements consist of *headstock, column, table, saddle, bed, base or runways,* and *cross* or *slide rails*. Machines receive their specialized identity from the combination of units. For example, there are four distinct types of horizontal boring, drilling, and milling machines: table, floor, planer, and multiple head. The table-type machine has a table and a saddle, and the workpiece is placed on the table. The floor-type machine combines floor plate sections and runways. The planer-type machine derives its name from a reciprocating table; and the multiple-head machine incorporates additional headstocks, a crossrail, and column supports. These basic units can be employed in a number of ways to machine a part.

The headstock drives and feeds the cutting tool or rotates the part. Figure 15.5 is a cutaway view of a headstock showing gearing. This headstock is used on boring, drilling, and milling machines. The spindle speed is infinitely variable from 25 to 1120 rpm, and it is driven by a 25- to 40-hp (18–30 kW) d-c drive.

Spindle rotation may be reversed to accommodate threading and tapping operations. For certain machines spindles are provided with power feed and adjustments in several

Figure 15.5
Open view of a single-spindle headstock showing gearing.

directions. The revolutions per minute or the feed in inches per revolution (millimeter per revolution) is usually variable unless the machine is designed for only one specific production operation as in a turning operation of railroad car wheels.

A cutaway of a grinding-wheel spindle is shown in Figure 15.6. This spindle is $1\frac{1}{2}$ hp (1100 W), 3600 rpm, and is direct motorized. The bed or base supports other basic elements as in Figure 15.7A. In this case the bed allows for linear motion of the table, and the table is guided by *double V-ways*. In the lathe the bed supports the headstock, tailstock, cross slide, and carriage. In a boring machine the bed supports a rotating table, whereas in the planer the bed supports a reciprocating table.

The column (Figure 15.7B) provides vertical support and guides the headstock for a

Figure 15.6
Grinding spindle.

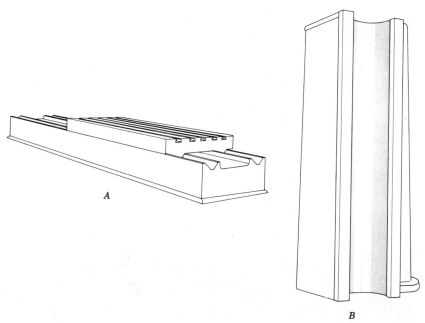

Figure 15.7
A, Planer-type table with T-slots on cast-iron, double-V way bed. B, Cast-iron column with scraped flat ways.

certain class of machines. For large machines the column is ribbed, heavily constructed, and has provision for counterweighting heavy headstocks to reduce horsepower. The counterweight may be a chain over a pulley, and the counterweight is hidden inside the column, as shown in Figure 15.7B.

The function of a table is to support the workpiece and to provide for locating and clamping. In some machines the table is provided with *power feeds* in one to three directions. Figure 15.8A illustrates a table with a saddle that has two-axis movement in the horizontal plane. Carriages are found in lathes and provide movement along the axis of the bed.

Runways carry a floor planer-type machine column and rotary tables. These are primarily used with larger machines. When a column base with a column and headstock traverse as a complete unit and are supported on an element, the element is called a runway and not a base. When a column base, column, and headstock are an integral unit, the supporting element is called a bed and not a runway, although both are similar in principle (Figure 15.8B).

End supports or tailstocks, as shown by Figure 15.9, serve as an outboard support for cutting tools or the workpiece.

The fixed and movable elements that form a machine tool structure locate and guide each other in accordance with the relative position between workpiece and cutting tool. Whenever basic elements slide over one another, as a carriage over a bed, it is necessary

374 BASIC MACHINE TOOL ELEMENTS

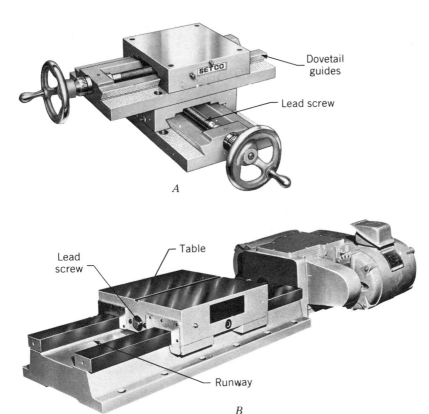

Figure 15.8
A, Feed table and saddle for compound movement using hand feed.
B, Motorized feed system using lead screws on runway.

to provide accurate movements. This is provided with either *flat* or *V-ways* or a combination of the two. The V-shape acts as a guide in two directions. The ways are oiled by force-feed lubrication to avoid scoring or seizing. A V-way has the advantage of not becoming loose as wear occurs, although the sliding unit may ride up one surface when the side thrust is large. Ways may be constructed with roller or ball bearings. Some machines use hard plastic inserts to reduce friction, and wipers are necessary to avoid embedding chips.

When cutting forces or loads are not severe, ways and contacting members have scraped cast-iron sliding members. *Scraping* is performed manually with a sharp chisel-shaped carbide tool point much like a flat file. A rectangular carbide tip is brazed on the end. The surface is scraped leaving small 0.0002-in. (0.005-mm) pockets randomly arranged. During scraping the ways are compared to a true flat *master surface plate*. The plate is coated with red lead or Prussian blue and by rubbing the master on the ways, the high spots show dye. These high spots are removed by hand-scraping away minute amounts of metal. The process of scraping and checking with a master and compound is repeated

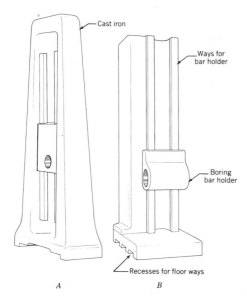

Figure 15.9
End-support columns, enclosed and open.

until a true plane is obtained. Figure 15.7 A and B illustrate flat and V-ways, which may have scraped or ground surfaces.

The power capacity and the performance of the machine determine static and dynamic stiffness. The size and shape of the workpiece, together with the cutting processes and operating and loading conditions, affect the shape and design of the structure. Load-carrying capacity is limited by the allowable stresses in the material and the shapes and sizes of the various cross sections of the members. In a structure subjected to impact loads, excessive stiffness may be unwanted and elasticity desired. An open C-shaped frame has less stiffness than a closed frame, as illustrated in Figure 15.10. Most machine tool structures can be resolved into elements that may be modeled as beams subject to transverse bending and torsion.

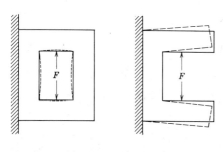

Figure 15.10
Deformations of open and closed frame under axial load.

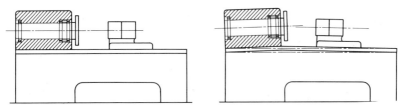

Figure 15.11
Misalignment effect of temperature changes on the bearings of lathe headstock.

The mass of the structure affects its performance. One example is the planer machine bed, which is "held on" to its base by gravity alone. As the effect of the planer's mass is greater than the upward component of the cutting forces, it is not lifted from the sliding ways during metal removal.

When a machine is idle, deviations from defined straightness and flatness are caused by inaccuracies in the machine manufacture. During machining, static and dynamic deformations and changes in oil film thickness can cause conditions as shown by Figure 15.11.

FLEXIBLE MACHINE TOOL SYSTEMS

The basic structure of machine tools has changed little since the early days of metal cutting, but automation has altered machine tools into several distinct types designed for specialized processes. Recently, however, the trend has been away from dedicated special machines and toward highly adaptable self-contained systems.

Flexible machine tool systems may have several power units, each of which can drive any number of multiple-spindle machining heads. These special-purpose heads can be set up ahead of time and changed quickly. Workpieces are transported automatically between workstations, and the entire head-changing and workpiece transport operation is under the control of a central process computer.

Machining centers not only allow different ways to make parts but a different way to do business. The modular construction of the system makes it possible to expand when necessary. Design changes and prototypes are easily accommodated because of the low setup cost, and inventory can be precisely controlled, In general, the flexibility of the system permits changes in the production schedule, design specifications, and even the product, all at a much lower cost than with conventional systems.

A flexible machining system is capable of machining cast-iron transmission cases at the rate of about five per hour. Operations include drilling, tapping, spot facing, reaming, boring, milling, chamfering, and recess grooving. The system has machining units that can be used in any combination with 77 machining heads containing from 1 to 23 spindles. Heads not in use are stored overhead ready to be transferred to the machining units as needed and can be changed in about 15 s.

DRIVES

A detailed study of electrical motors is beyond the scope of this text. The types of motors are so great that only a few of the principal machine tool drives can be discussed.

The *alternating-current* motors are generally of the single-phase of induction type. The series motor has the motor field winding and the armature connected in series. The connections to the armature are made through brushes to the segments of a commutator. The speed of the motor is controlled by a variable series resistance, and it has a high starting torque. Arcing between the brushes and the commutator and between the commutator segments themselves is a disadvantage of this motor. It must be totally enclosed if it is used around inflammables.

The a-c *induction* motor can be operated from single-, two-, and three-phase current. The windings on the armature form closed circuits known as a "squirrel cage." The low starting torque is a disadvantage of this type of motor.

Direct-current motors are often used with numerically controlled equipment. Although industrial plants seldom have an extensive d-c supply, motor generator sets and static converters are employed to provide the power. There are two basic types of d-c motors: the *series motor* and the *shunt motor*.

The series-wound motor is controlled by a variable resistance in series with the field coils. It has a high starting torque, but speed is decreased with an increased load. The shunt motor can maintain a more constant speed when the load is increased, but it has low starting torque.

The motors are either directly connected through gearing to work and tool-holding devices or else by one or more "V" belts. High-production machines are usually "geared head," whereas small laboratory-type machine tools are often belt driven. Fluid drives employing hydraulic motors are used when power requirements cause sharp power surges.

METHODS OF HOLDING WORKPIECE

The method for holding a workpiece depends on size and shape, its machine, and requirements for rapid production. With quantity production machines such as turret lathes, holding devices are usually actuated hydraulically, pneumatically, electrically, or by cam action to minimize clamping time. With automated or numerically controlled machines, holding devices are programmed to release the workpiece at the conclusion of machining and to clamp the next part fed to it automatically.

Supporting Work between Centers

The common way to support a rotating workpiece is to mount the part *between centers*. This method is able to support heavy cuts and is convenient for long parts. Because the work is mounted between two tapered centers, it will not rotate with the spindle unless it is attached. Attachment is made through a pear-shaped forging known as a ***dog***, which consists of an opening to accommodate the stock being turned, a setscrew to fasten the work securely, and an elongated portion at the top (known as the tail) that is bent at a right angle—parallel with the stock—so that it may engage a slot in the face plate. After the setscrew is tightened, the tail of the dog is fitted into the face plate and the work is ready to be turned. The center in the headstock turns with the work; consequently, no lubrication of that center is necessary. The tailstock center or **dead center** acts as a conical bearing and for this reason must be kept clean and lubricated or else must be a ball bearing type. Allowance for expansion of the workpiece as a result of the heat generated by the machining operation and the rotation of the stock is necessary.

In turning long slender shafts or boring and threading the ends of spindles, a **center rest** gives additional support to the work. The center steady rest is attached to the bed of the lathe and supports the work with three jaws or rollers. Another somewhat similar rest is known as a follower rest. It is attached to the saddle of the carriage and supports work of small diameter that is likely to spring away from the cutting tool. This rest moves with the tool, whereas the center rest is stationary.

In machines employing high rotational speeds for carbide tool turning, the hydrostatic steady rest is used. This steady rest has three jaws that are held against the workpiece with a constant pressure.

Mandrel

Cylindrical work that has been bored or reamed to size may be held between centers by one of the several *mandrels* shown in Figure 15.12. Solid mandrels have hardened, ground surfaces and are available in standard sizes. The surface is ground with a taper of about 0.0006 in./in. (0.06%) in length, the small end being 0.0005 in. (0.013 mm), undersized to facilitate insertion into the work. The work must be pressed on the mandrel in an arbor press, because considerable force is required. The other mandrels are adapted for holding work that may not have holes accurate to size or where several parts are to be machined in one setup.

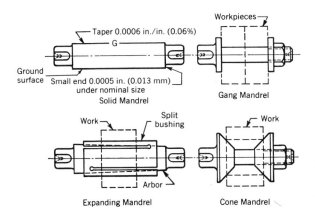

Figure 15.12
Various types of mandrels used for holding stock between centers.

Face Plate

The workpiece may be held to the face plate by means of clamps, bolts, or else in a fixture or special holding device attached to it. Figure 15.13 illustrates work supported by being bolted to a face plate. Such mounting is suitable for flat plates and parts of irregular shape. Figure 15.13 illustrates a boring bar for internal turning.

Chucks

Chucks are used for holding large and irregularly shaped parts and are either bolted or screwed to the spindle making a rigid mounting. Chucks are made in several designs and may be classified as follows:

1. *Universal chuck.* All jaws maintain a concentric relationship when the chuck wrench is turned.
2. *Independent chuck.* Each jaw has an independent adjustment as shown in Figure 15.14.
3. *Combination chuck.* Each jaw has an independent adjustment and in addition has a separate wrench opening that controls all jaws simultaneously.
4. *Drill chuck.* A small universal screw chuck used principally on drill presses but frequently used on lathes for drilling and centering.

Power chucks operated pneumatically, hydraulically, or electrically relieve the operator of the effort involved in tightening and loosening the work. Additional advantages of the power chuck are that it is quick acting; the chucking pressure can be regulated; and it can be used for both bar and chucking work. "Chucking" in this case refers to an irregular-shaped casting, forging, or previously machined part; bar refers to **bar stock,** which is usually round, square, or hexagonal in cross section.

Figure 15.13
Boring an eccentric hole on the face plate of a lathe.

Figure 15.14
Independent jaw lathe chuck.

Because it is difficult to mount all types of work on standard equipment, many special chuck jaws or holding fixtures must be devised. Standard face plates are frequently used for mounting such fixtures. The holding device is held to the face plate either by bolting or by T-slots on the face of the plate.

Collets

Collets, commonly used for bar stock material, are made with jaws of standard sizes to accommodate round, square, and hexagonal stock. Collets of the *parallel*-closing type are sometimes used for large stock, but ordinary collets of the *spring* type are more common. These collets are solid at one end and split on the tapered end. The tapered end contacts a similarly tapered hood or bushing and, when forced into the hood, the jaws of the collet tighten around the stock. Spring collets are made in three types: push-out, draw-back, and stationary.

The *push-out* and *draw-back* collets operate in a similar way. The push-out collet shown in Figure 15.15 operates in the following manner. When the plunger is moved to the right, the tapered split end of the collet is forced into the taper of the head, which causes the collet to tighten about the stock. The *hood* is screwed on a threaded spindle. The draw-back collet operates in the same way, except that the collet is drawn back against a tapered hood for tightening action. Push-out collets are recommended for bar work, because a resulting slight movement of the stock pushes it against the **bar stop**. The bar stop assures that all material protrudes from the collet to the same length. Draw-back collets are not widely employed for bar stock but are useful when the collets are of extra-capacity size and are utilized for holding short pieces. The slight back motion in closing forces the work against the locating stops.

A simple draw-in collet used on engine lathes is shown in Figure 15.16. The proper-sized collet is placed in the sleeve and screwed to the **draw bar** extending through the spindle. Work can be placed in the collet and held by turning the hand wheel on the end of the draw bar. This forces the collet back against the tapered surface of the sleeve and causes the collet jaw to grip the work.

With both the push and draw collets there is a slight movement of the workpiece, because the collet moves as it is tightened. In most cases this is not a disadvantage, but when it is, the *stationary* collet can be used. A shoulder on the collet comes to rest against

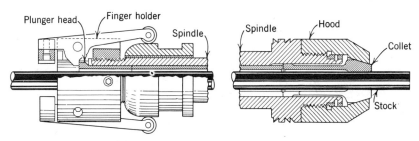

Figure 15.15
Push-out type of collet.

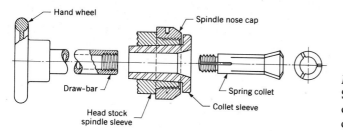

Figure 15.16 Section showing construction of draw-in collet attachment.

the head to provide this endwise accuracy. Because there are more sliding surfaces in the stationary collet, the rotational concentricity is not as accurate as for other types of collets.

There are a number of special collets, including those employing hydraulic force on a flexible steel sleeve in which the workpiece is held. Such collets give a high-torque connection with close accuracy.

Arbor

Expanding or *threaded arbors* hold short pieces of stock that have in them a previously machined, accurate hole. The action in holding the work is controlled by a mechanism similar to that used with collets. An expanding plug-type arbor is shown in Figure 15.17. The work is placed on the arbor against the stop plate and, as the draw rod is pulled, the tapered pin expands the partially split plug and grips the work. The threaded arbor operates in a similar fashion except that the work is screwed on the arbor by hand until it is forced back against a stop or flange.

Both collets and arbors may be power operated by pneumatic, hydraulic, or electrical devices located at the end of the spindle. Such an arrangement is frequently used on high-production work to provide quicker and easier operation.

T-Slots and Vises

Worktables on millers, planers, and shapers are constructed with *T-slots* on their surfaces to provide means for holding and clamping down parts that are to be machined. On planers most work is held by clamping directly to the table, and a wide variety of clamps,

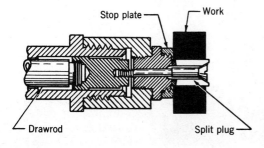

Figure 15.17 Expanding plug-type arbor.

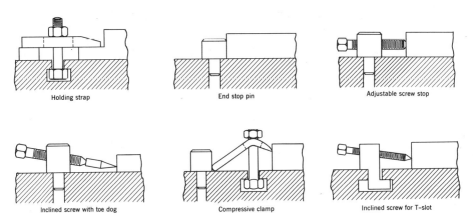

Figure 15.18
Methods of holding work on planer table.

stop pins, and holding devices are shown in Figure 15.18. Several of these arrangements are adapted to holding down plates such that the entire surface may be machined.

Vises can be clamped to a worktable provided with T-slots. Delicate or complex-shaped parts are often held in a ***pocketed chuck*** by casting them in place with a low-melting-point lead alloy or a hard tooling wax.

Magnetic Chucks

Ferrous work can be held on surface grinders and other machine tools by magnetic chucks. This method is both simple and rapid but is only adaptable to situations in which the forces on the workpiece are low. Parts are placed on the chuck, which is energized by the turning of a switch.

Two types of chucks used are the *permanent magnet* chucks and those magnetized by direct current. The *direct-current* chucks are made in both rectangular and circular shapes. The rectangular style is suitable for reciprocating grinders or for light milling machine work. Rotary chucks, designed for lathes and rotating table grinders, are shown in Figure

Figure 15.19
Concentric-gap and radial-pole rotary chucks.

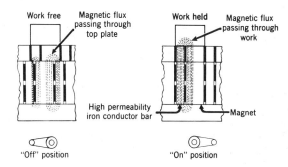

Figure 15.20
Schematic showing how work is held on permanent magnet chuck.

15.19. The problem of energizing the current to those chucks is overcome by collector rings and a brush unit mounted at the back end of the chuck or spindle. The pulling power varies according to the type of winding used and may be as high as 165 psi (1.1. MPa). The equipment for furnishing the direct current consists of a motor generator set and demagnetizing switch or silicon-controlled rectifiers.

Parts held on a magnetic chuck should be *demagnetized* after the work is finished. Several types of demagnetizers are available, operating on either alternating or direct current, which successfully removes the residual magnetism.

A permanent magnet chuck does not require any electric equipment. The operation of this chuck is by a lever. Figure 15.20 shows the operation when the operating lever is shifted. In the "off" position the conductor bars and separator are shifted in such a way that the magnetic flux passes through the top plate and is short-circuited from the work. When the handle is turned to the "on" position, the conductor bars and nonmagnetic separators line up so that magnetic flux in following the line of least resistance goes through the work in completing the circuit. The holding power, obtained by the magnetic flux passing through, is sufficient to withstand the action of grinding wheels and other light machining operations. This chuck and the d-c chucks may be used for either wet or dry operations.

METHODS OF HANDLING WORKPIECE

For small quantities the customary method of handling a workpiece is manual if the mass if less than 30 to 50 lb (10–25 kg) or by crane or conveyor if heavier. The Occupational Safety and Health Act (OSHA) and management practice encourage operator safety and welfare, which provides for robots, automatic equipment, and other handling devices if the quantity is large or the weight is prohibitive.

If the quantity of production is sufficient, mechanical loaders have an economic advantage over manual loading. A variety of mechanisms is available to load, position, control the cycle, and unload the workpiece. Systems are available that can completely process and assemble the item. See Chapter 17 for an example.

Mechanical loading reduces operator fatigue. If a workpiece has a mass of 35 lb (16 kg) at a production rate of 60 pieces per hour, the operator will load and unload about 32,000 lb (7600 kg) in a working day.

Figure 15.21 is a loader–unloader that can pick up and place small parts or pieces

384 BASIC MACHINE TOOL ELEMENTS

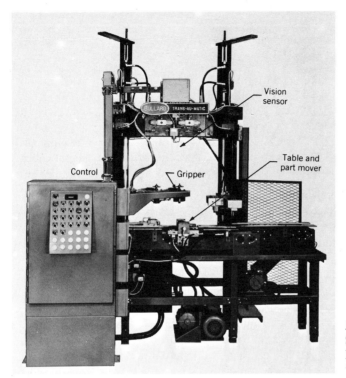

Figure 15.21
Semiautomatic loading–unloading "robot" machine.

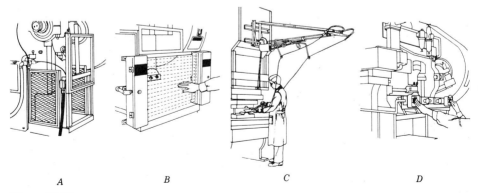

Figure 15.22
Methods of point-of-operation protection. *A*, Gate barrier. *B*, Photoelectric light curtain. *C*, Pull-back. *D*, Two-hand controls.

weighing several hundred pounds. It loads and unloads vertical chucking machines from conveyors. Such robots may include press feeders, mechanical hands, vibrating positioners, and assembly systems.

METHODS OF CONTROL

Few production machines are controlled entirely by human intervention, although some small general-purpose machines are. Machines having only hand feed, which implies that muscle power replaces the motor, are found with small drill presses known as *sensitive* machines. Other machines such as engine lathes, though controlled by hand, have power feeds and speeds. Mechanical systems assist in controlling equipment like the shaper where the machine is adjusted to control the operation mechanically from beginning to end. The reversal of the tool and the feed of the table are done mechanically during the machining operation.

Machines using *cam* action are semiautomatic. After they are loaded and started, each successive operation in the cycle begins when the previous operation is completed. The cams actuate changes in speeds or feeds or in the cutting tool itself. Hydraulic drives and controls are used primarily when a machine has a reciprocating member, on presses, and where constant velocity is difficult to maintain by mechanical means. Planer tables and reciprocating tables on grinders are two examples of the latter.

Timing cycles are frequently used in automatic machine tools. The timers actuate either microswitches or solenoids that in turn control machine movements.

A chapter is devoted to numerically controlled equipment. Simply stated, this is a system where commands that control the machine are stored on a tape or disk. General-purpose equipment like milling machines that are provided with numerical control may be used for a "one-of-a-kind" job or in semi-mass production by simply changing the programming and the tooling. These machines are not automatic because they may not have loading and unloading capabilities.

SAFETY

Many methods of safety are considered by the manufacturers of machine tools. Training programs are implemented by suppliers of parts, and schools are involved in educating students about the importance of safety in the manufacturing laboratories. Using safety glasses and guards; avoiding wearing of loose articles of clothing, carelessness, and worn tools; and observing sensible practice are stressed by management, unions, and state and federal laws. Federal laws require designs that meet minimum standards. *Point-of-operation guards* prevent the entry of hands or fingers reaching through, over, under, or around the guard.

Figure 15.22A shows a machine guard that surrounds the tool working area of a punch press. A gate barrier device must be closed before machine operation can begin. This is in addition to machine guards, in that the hands or arms will be removed before the gate can be closed, all of which precedes the machine cycle.

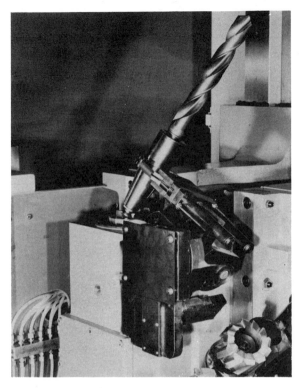

Figure 15.23
Magazine in lower-right corner and tool gripper with tapered shank drill.

Another safety measure employed is a *sensing* device that normally uses a photoelectric light curtain or a radiofrequency sensing field. These light curtains are installed to prevent or stop normal stroking of the press or operation of the machine if the operator's hands are placed in or near the point of operation.

Pull-back devices as shown in Figure 15.22C are another method. These are designed to prevent the operator from reaching into the point of operation. These devices have wristlets for each operator's hand and are attached to a pulling cable.

Other popular devices are two hand buttons or palm buttons and trips that require both trips to be operated jointly before the machine cycle can begin.

TOOL CHANGING

Many computer-controlled machines employ tool changers. Selection and changing is controlled with numerical control and high-speed mechanisms. These tool changers move *dimension-preset tools* from their storage area to the spindle. In the more basic approach the spindle can move to a tool storage rack or carousel located on the machining envelope and swap tools directly. After the spindle descends into contact with the new tool and inserts it into its taper, a power-driven draw bar actuates a split collet that closes around the tool retention knob opposite its working end. The tool magazine may be some distance from the spindle. The most popular method for tool changing is the two-hand arm. The desired tool is collected from a drum in one end of the special arm. At the moment of the tool change the empty end of the arm grips the tool in the spindle, removes it, indexes 180°, and inserts the new tool into the spindle taper.

If the magazine is located permanently adjacent to the spindle, the changer arm can stay open, remove the working tool, and replace it with the next tool. This is the arrangement shown by Figure 15.23. In the photograph, the gripper is moving a drill from the magazine to the spindle. Note the other tools on the drum that are garrisoned ready for work. The gripper takes a tool from the spindle, rotates 180° to swing down, around, and up to deposit the tool in an injection-molded pocket in a 20-tool rotary magazine. The magazine indexes to the predetermined position according to the program, thus being ready for the next tool for the arm to grip and carry back to the spindle.

QUESTIONS AND PROBLEMS

1. Give an explanation of the following terms:

   Kinematic inversion   Bar stock
   Saddle   Bar stop
   Machining center   Draw bar
   Dog   Arbor
   Dead center   T-slot
   Center rest   Pocketed chuck
   Mandrel

2. Write the advantages and disadvantages for a welded machine tool frame.

3. What is a multiple-head machine?

4. What are the advantages of gray cast iron for a machine frame?

5. Name two ways to traverse a spindle or table.

6. Discuss torque for both a-c and d-c motors.

7. Name features that provide for machine table versatility.

8. Describe methods for supporting long slender shafts that are to be machined.

9. Describe the difference between a bed and a runway.

10. Why are mandrels tapered?

11. What is dynamic stiffness? What is its purpose for basic machine tool elements?

12. What are the advantages of a universal chuck over an independent chuck, and vice versa?

13. Discuss the differences between an open or C-frame and a closed frame.

14. Why is a stationary collet less accurate in concentricity but more accurate for stock location, so far as endwise dimensions are concerned?

15. What is the purpose of a tailstock?

16. Sketch cams to start a machine, engage the feed, and stop the machine.

17. What are the principal advantages of V-type ways?

18. Sketch a simple machine using basic machine tool elements that will simultaneously drill all of the holes to hold a cylinder head to the block. Assume that there are four holes on each side and one per end.

19. Sketch the components for a single machine to turn and thread a hexagonal bolt and slot it for a screwdriver. Label the parts.

20. What is the customary taper on a mandrel when expressed in millimeter per meter?

21. Determine the total force holding a 4-in. (102-mm)-diameter plate on a direct-current chuck.

22. What is the maximum force holding a 40 in. (102-mm)-diameter plate to a rotary magnetic chuck?

23. Convert the following feeds to SI units: 3 in., per revolution, 0.25 in. per revolution, 3 in./min, and 0.25 in./min.

24. If a workpiece weighs 20 lb and a part is loaded every 6 min, how many pounds must a worker handle each 8-h shift? Calculate also in SI units.

25. What is the fundamental difference between conventional and flexible machining systems?

26. List advantages of flexible systems from an operator's point of view. From a management point of view.

CASE STUDY

PORTABLE MACHINE TOOL

An impeller turbine blade (Figure 15.24) is attached to a water turbine wheel. The turbine is located at a hydroelectric plant. The turbine blades are bolted to the wheel through three holes. After many years of on-line service, the holes in the wheel and blade have elongated. The maintenance department believes that reboring and rebolting the blades to the wheel are necessary. Special steel bolts will be used. The maintenance supervisor, Carlos Ramirez, says, "The 'buckets' can be rebored in a machining center. No need to do that here, but the wheel is a different matter." Down time of the unit, disassembly of the wheel, and transportation to a repair facility

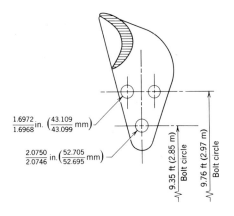

Figure 15.24
Case study. Turbine blade.

are costly, because the hydroelectric plant is in a remote location.

Devise a system of machine components to allow reboring the wheel to the dimensions given by Figure 15.24. Assume that the wheel remains mounted on the turbine generator shaft and that it can be rotated and braked after the shroud is removed. Specify and sketch the machine component building blocks, accessories, platform, and machine that will do the work. A crane is available for assembly of the machine components at the site. Give your special machine tool a name and identify the component parts by arrows. Show two view of the new machine tool.

CHAPTER 16
NUMERICAL CONTROL

Numerical control refers to the operation of machine tools from numerical data stored on paper or magnetic tape, tabulating cards, computer storage disks, or direct computer information. A historical example of using instructions punched on paper tape is the player piano. Notes to be played (instructions) are defined as a series of holes on a piano roll (punched paper tape), then sensed by the piano (using a pneumatic system powered by a foot-operated bellows), which plays the notes (executes the instructions).

Because mathematical information is used, the concept is called *numerical control (NC)*. NC is the operation of machine tools and other processing machines by a series of coded instructions. Perhaps the most important instruction is the relative positioning of the tool to the workpiece. An *organized list of commands* constitutes an *NC program*. The program may be used repeatedly to obtain identical results. Manual operation of machine tools may be unsurpassable in producing fine-quality work, but such qualities are not consistent. NC is not a machining method; it is a means for machine control. NC is considered as one of the most dramatic and productive developments in manufacturing in this century.

NUMERICAL CONTROL MACHINE TOOLS

Early NC design placed control units on existing machine tool structures to accomplish numerical control. As experience was gained, it was apparent that NC machines were more efficient in the overall operation than conventional machines. But the controls became more adaptable to production equipment, and most of the equipment shown in this book has the potential for numerical control. Inspection, pipe bending, flame cutting, wire wrapping, circuit board stuffing of electronic chips, laser cutting of fabric, drafting machines, and production processes have proven applications. In some instances the added controls cost more than the basic machine tool, but it is often a worthwhile expense. Solid-state circuitry has provided more reliable control at lower cost than previous electronic technology.

NC should be considered whenever there is similar raw material and work parts are produced in various sizes and complex geometries. Applications are in low-to medium-quantity batches, and similar processing sequences are required on each workpiece. Those production shops having frequent changeovers will benefit.

NC machine tools incorporate many advances such as programmed optimization of cutting speeds and feeds, work positioning, tool selection, and chip disposal. The adoption

Figure 16.1
NC controlled slant-bed turret combination turning-chucking lathe.

of NC has altered existing designs to the point that NC machine tools have their own characteristics separate from the machine tools described in the various chapters. For example, modifications to the turret lathe have resulted in a turret slanted on the back side rather than placed on the front horizontal ways. A greater number of tools can be mounted on the turret as a result of the structural adaptation. This can be seen in Figure 16.1.

Development of the *machining center* with *tool storage* resulted from NC. Figure 16.2 shows an NC machining center with a storage of 24 tools in a magazine. Each tool can be selected and used as programmed. These machining centers can do almost all types of machining such as milling, drilling, boring, facing, spotting, and counterboring. Some machining operations can be programmed to occur simultaneously. The NC program selects and returns cutting tools to and from the storage magazine, if equipped, and also inserts them into a spindle. Parts can be loaded and moved between pallets, manipulated by rotation, and inspected after the work is finished. Robotic operation is possible, also being accomplished by NC.

OPERATIONAL SEQUENCE

NC starts with the parts *programmer* who, after studying the engineering drawing, visualizes the operations needed to machine the workpiece. These instructions, commonly called a program, are prepared before the part is manufactured and consist of a sequence of symbolic codes that specify the desired action of the tool workpiece and machine. Even in computer aided design and manufacturing this interpretation is necessary. The engineering drawing of the workpiece is examined, and processes are selected to transform the raw material into a finished part that meets the dimensions, tolerances, and specifications. This process planning is concerned with the preparation of an operations sheet

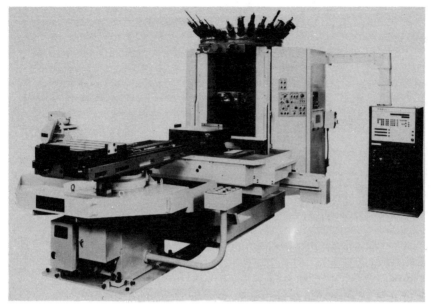

Figure 16.2
NC machine with tool carrousel on top of column and pallet for part loading in front.

or a route sheet or traveler. These different titles describe the procedure or the sequence of the operations, and it lists the machines, tools, and operational costs. The particular order is important. Once the operations are known, those that pertain to NC are further engineered in that detail sequences are selected. Chapter 25 discusses operations and cost estimating.

A program is prepared by listing codes that define the sequence. A part programmer is trained about manufacturing processes and is knowledgable of the steps required to machine a part, and documents these steps in a special format. There are two ways to program for NC, by *manual* or *computer assisted* part programming. The part programmer must understand the ***processor language*** used by the computer and the NC machine.

If manual programming is required, the machining instructions are listed on a form called a part program *manuscript*. This manuscript gives instructions for the cutter and workpiece, which must be positioned relative to each other for the path instructions to machine the design. Computer assisted part programming, on the other hand, does much of the calculation and translates brief instructions into a detailed instruction and coded language for the control tape. Complex geometries, many common hole centers, and symmetry of surface treatment can be simply programmed under computer assistance, which saves programmer time.

Tape preparation is next, as the program is "typed" onto a tape or punched card. If the programming is manual, the 1-in. (25-mm)-wide perforated tape is prepared from the part manuscript on a typewriter with a standard keyboard but equipped with a punch device capable of punching holes along the length of the tape. If the computer is used, the internal memory interprets the programming steps, does the calculations to provide

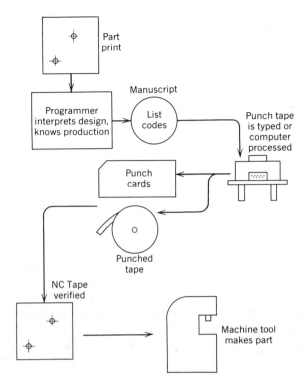

Figure 16.3
Flowchart of numerical control steps.

a listing of the NC steps, and additionally will prepare the tape. Some tapes contain electronic or magnetic signals; other systems use disks or direct computer inputs.

Verification is the next step, as the tape is run through a computer, and a plotter will simulate the movements of the tool and graphically display the final paper part often in a two-dimensional layout describing the final part dimensions. This verification uncovers major mistakes.

The final step is production using the NC tape. This involves ordering special tooling, fixtures, and scheduling the job. A machine operator loads the tape onto a program reader that is part of the ***machine control unit,*** often called a MCU. This converts coded instructions into machine tool actions. The media that the MCU can sense may be perforated tape, magnetic tape, tabulating cards, floppy disks, or direct computer signals from other computers or satellites. ***Perforated paper tape*** is the predominant input medium, but the concepts are the same whatever the input. These general steps are shown by Figure 16.3.

TYPES OF CONTROL

Controls are either open or closed loop. ***Open-loop control*** is defined as a system where the output or other system variables have no effect, or *feedback* on the control of the input. In the open-loop system an operational device such as a machine slide is instructed to move to a certain location, but whether or not the slide reaches the predetermined location is not ascertained by the control unit. It may be possible to achieve the desired results by means other than a control system. Figure 16.4 illustrates an open-loop, two-axis system. The axis coordinates are in the X–Y plane, and a third axis, Z, is possible. The input medium, usually paper tape, is scanned in a unit called a reader. Discrete signals feed into the control unit, and instructions proceed to the stepping motor drive unit. Each machine slide or movement that is to be controlled has its own *stepping motor* and drive. The stepping motor is usually electric, but hydraulic units are sometimes found. The drive to the machine element may be conventional leadscrews, ball bearing screws, or pinion and rack arrangements. Open-loop control is simple and less costly, but it is not as accurate as closed-loop control systems. On the other hand, some systems do not require the additional closed-loop feature.

A *closed-loop control* system for a single-axis control is shown in Figure 16.5. The machine motion as actuated by the *servomotors* is recorded or monitored by a feedback unit that may be electronic, mechanical, or optical. It may be a *transducing* device that indicates the position the machine table, slide, or tool has reached in response to the tape command. The feedback unit transmits position signals through the feedback signal circuit to the control unit where the signals are continually compared with program signals. The command signal is fed through an amplifier to actuate the drive motor until the difference between the command signal and the actual slide position reaches zero error. When the error signal is zero or null in a closed-loop system, the machine movements are at the exact position commanded. Most NC systems are closed loop. The control unit may initiate one or more of the following actions:

1. Record the accuracy of the command—that is, the position of the tool or selection of the component.
2. Automatically compensate for error and create an additional requirement for the tool to move to a new position.
3. Stop the motion when the input and feedback signals are the same.

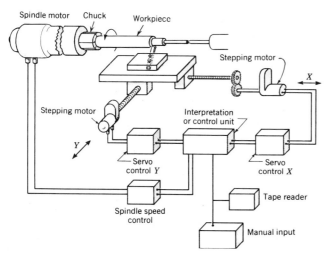

Figure 16.4
Two-axis, open-loop numerical control system.

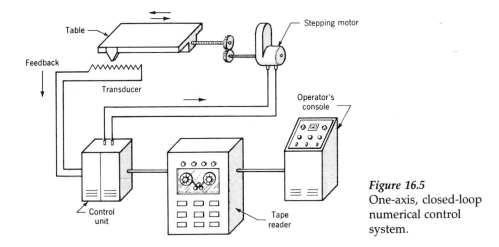

Figure 16.5
One-axis, closed-loop numerical control system.

Although Figure 16.5 illustrates this principle for one axis, it is also pertinent to three-axis machines.

RECTANGULAR COORDINATES

NC uses *rectangular or Cartesian coordinates* to define a point in space. This is the common system where distances are equal; for example, between 2 and 3 is the same distance as between 8 and 9. Through this coordinate system a point in space can be described in mathematical terms from any other point along three mutually perpendicular axes. Each machine tool has a standard coordinate-axis system, which allows the part programmer to define unambiguously the sequence of operations and movements of the machine tool, cutting tool, and part. Machine tool construction is based on two or three perpendicular axes of motion and an axis of rotation.

Generally, the Z axis of motion is parallel to the principal spindle of the machine, whereas the X axis of motion is horizontal and parallel to the workholding surface. Once the X axis is oriented, the remaining planes fall into place. The Y axis of motion is perpendicular to both X and Z. The location of the router in Figure 16.6 is $X = -2$, $Y = +3$, and $Z = +1$. The axes designations for typical machine tools are given in Figure 16.7. A similar program of notation exists for rotary motion. Separate axes motions may include tilt and swivel of gimballed heads. Milling an elliptical part with sloping walls may use five axes of the machine simultaneously.

Once the coordinate axis is known, the part programmer may have the option of specifying the tool position relative to the origin of the coordinate axes. NC machines may specify the zero point as a *fixed zero* or *floating point*. In fixed zero the origin is always located at the same point on the machine table and is the lower left-hand corner. Locations are defined by positive X and Y coordinates. A floating point allows the zero point to be set at any position on the machine table. The workpiece may be symmetrical and the zero point would be at the center of symmetry. The floating-point method is the more common.

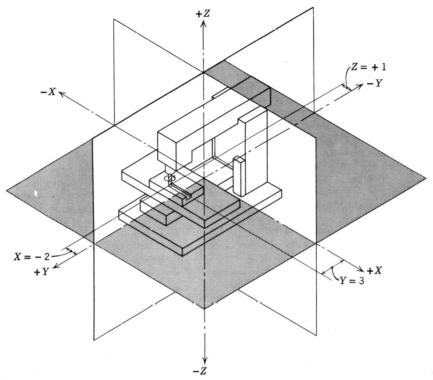

Figure 16.6
Principal X, Y, and Z axes for a milling machine.

Absolute dimensions are ones that always start from a fixed-zero reference point and span the distance. For NC the tool locations are defined in relation to the zero point. An *incremental dimension* is one that always starts from a prior location that is not a zero reference. Figure 16.8 represents a part requiring two holes to be drilled. The holes are labeled 1 and 2 and their coordinate axes numbers are given. For absolute dimensioning, the tool will move to hole 1 with dimensions $x = 4$ and $y = 3$, and after drilling, the tool will move to hole 2 with dimensions $x = 7$ and $y = 7$. For incremental positioning, the tool will move from hole position 1 to 2 along the X axis 3 units ($\Delta x = +3 = 7 - 4$) and 4 units along the Y axis ($\Delta y = +4 = 7 - 3$).

PUNCHED TAPE

Material for the tape can be punched or perforated paper, mylar-reinforced paper, mylar-coated aluminum, or certain plastics materials. Mylar-reinforced paper tape is the most widely used.

Although various tape formats are available, emphasis continues to be with those NC

PUNCHED TAPE 397

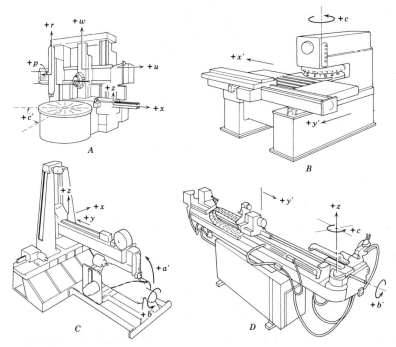

Figure 16.7
Principal numerical control machine axes. A, Vertical turret lathe or vertical boring mill. B, Turret punch press. C, Welding machine. D, Right-hand tube bender.

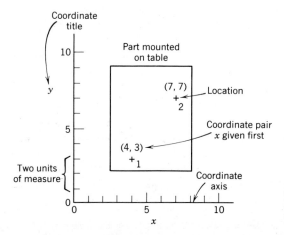

Figure 16.8
Example of absolute and incremental locations.

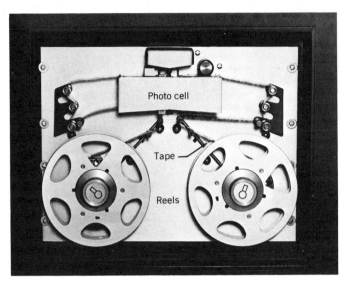

Figure 16.9
High-speed photoelectric tape reader.

systems manufactured to standards[1] that define variable block tape formats for positioning and contouring controls. The tape is 1 in. (25.4 mm) in width and has eight channels. Holes are punched in the channels in code patterns. A tape reader senses the hole patterned by photoelectric cells, fingers, brushes, or a vacuum method. Figure 16.9 is an example of a reading head. Reading speeds of 100 to 1000 characters per second are available using photoelectric methods. Magnetic tape, though less frequently used, is 1 in. (25.4 mm) wide with 14 channels.

In manual part programming a typewriter machine will punch holes upon the strike of a key. With computer assisted tape preparation, not only does the computer do supplementary calculation, but a tape is also produced and for all practical purposes is identical to tape produced by an NC typewriter. During part production the tape is fed through the tape reader once for each workpiece. While the machine is performing one machining sequence, the tape reader is feeding the next instruction into the controller's data buffer. Machine operation is more efficient in this manner, as the machine tool is not waiting for the next instruction to be fed into its active memory. After the last instruction is recorded into the data buffer, the tape is rewound ready for the next part.

Figure 16.10 illustrates punched paper tape. Holes are punched in the tape in coded form; with magnetic tape or floppy disks a charge is placed on the media and on audio tape an audio signal is recorded. Tabulating cards use discrete holes punched in the card to indicate commands.

Figure 16.10A illustrates the EIA standard coding for 1-in. (24.5-mm)-wide, eight-channel punched tape for the ***binary-coded decimal*** (BCD) system. An ASCII standard

---

[1]Electronics Industries Association (EIA) RS 273-A and RS 274-B for positioning and contouring.

is also widely used. The coding of a tape is by the absence or presence of a hole. This absence or presence of a hole is binary, that is, two. The base number 2 system can represent any number in the more familiar 10 or decimal system. The binary code uses only 0 or 1. The 0 or 1 is referred to as a bit. Computers operate on a form of binary arithmetic, where a number can be expressed by a combination of "on" and "off" circuits. This concept is suitable for NC, because a number may be expressed by a hole or no-hole in a tape. Letters can also be expressed by a combination of binary bits. The meaning of successive digits in the binary system is based on the number 2 raised to successive powers.

This system is used in almost all numerically controlled operations. There are eight

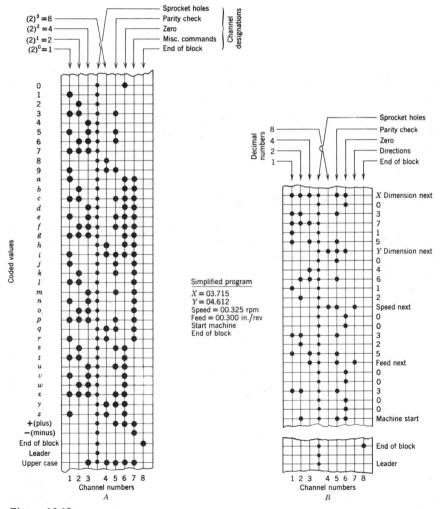

Figure 16.10
A, Binary coded decimal tape. B, BCD program.

designated channels and one line of sprocket holes on the tape. When numerical data are the input, channels 1, 2, 3, 4, and 6 are employed. The first four channels represent the numbers 1, 2, 4, and 8, which are powers of 2; that is, $2^0 = 1, 2^1 = 2, 2^2 = 4, 2^3 = 8$. Hence, referring to the figure, the number 7 is read into the tape by punching holes in the first, second, and third channels, which total $1 + 2 + 4 = 7$. To indicate the number 5, a hole is punched in channels 1 and 3. The tape reader makes elementary checks on the accuracy with which the tape has been punched. This is called a parity check. There must be an odd number of holes in each row or the tape reader and, hence, the machine stops. Therefore, each time a command calls for an even number of holes to be punched, an additional one must be punched in channel 5, the parity check channel. The sprocket drive holes are not considered in the parity check.

Figure 16.10B is a schematic of the way a short strip of tape might appear for the simplified program shown. All numbers are usually depicted with either five or six digits, and the first two refer to whole inches (millimeter or centimeter) and the latter ones to the decimal fraction.

The coding in Figure 16.10 will not necessarily be the same for different manufacturers' products. Such things as "coolant on" and "coolant off" are regularly punched into the tape and meet machine tool builders' specifications. Various commands may be coded as illustrated, although programming of a given tape is characteristic of a particular machine tool and control arrangement, and universally used standards for all commands are not available.

POINT-TO-POINT PROGRAMMING

NC programming is often segregated into point-to-point (PTP) or continuous path. Although many NC controls have capability in both methods and distinctions are obscured, the concepts are sufficiently different for learning purposes.

The *point-to-point* or positioning method is characterized by punching, spot welding, or drilling machines. It is used extensively on machines that can move in only one direction. PTP locates the working spindle or workpiece in a specific relative position, and the tool operates either by tape instruction or manually. The tool does not contact the work then moving between coordinate positions. For example, the holes for Figure 16.8 would be drilled in successive stops. Whether the tool moves along the x axis first and then the Y axis is immaterial to PTP. Actually, some machines move simultaneously along both axes. Many positioning control systems use the eight-channel, 1-in. (25.4-mm)-wide perforated tape with standard codes and formats. Some NC PTP machines have only the X and Y axes controlled, whereas others will be programmed on three or more axes as well as have tool selection, feed, speed, spindle rotation, coolant flow, and other functions controlled.

In some cases the PTP method may be programmed to machine a *straight line and a contour*. To machine the surface *FG* in Figure 16.11A, the tool could be positioned by tape instruction at two different x–y locations. If 18 different positions were programmed, the actual machine path would be more accurate. For Figure 16.11B nine program steps are shown, but it is usually necessary to have 100 programmed commands per quadrant.

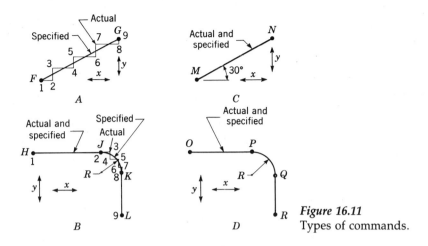

Figure 16.11
Types of commands.

For these operations the tape length may become excessive in machining complex contours involving PTP commands.

Some machines are basically positioning ones, but they do have some contouring features such as straight-line milling as shown in Figure 16.11C. Turret lathes and turret drills are machines that are positioning ones with some contouring ability.

A simple positioning program to EIA RS 273 standards for drilling four holes is given in Figure 16.12. The term *format* refers to the general arrangement of codes placed on the tape. The *tab code* (holes in tracks 2, 3, 4, 5, and 6 in a transverse row) precedes each command. The instruction for the x coordinate precedes that for the y coordinate.

CONTINUOUS-PATH PROGRAMMING

In *continuous-path programming* (CPP) the cutting tool contacts the workpiece as co-ordinate movements take place (Figure 16.13). *Contouring operations* include milling, turning, and flame cutting. Contouring differs in movement between program points. An *interpolation* routine differentiates CPP from PTP programming. The problem is to provide control for the tool continuously, which requires frequent changes in two or more axes simultaneously. During this movement, the tool touches the workpiece. There are three interpolation methods used to connect defined coordinate points: linear, circular, and parabolic.

In *linear interpolation* the machine shape is the result of a series of straight-line machining moves programmed in sufficient quantity to give an acceptable comparison between the drawing contours and the finished shape. Linear interpolation allows movement of two or more axes of the cutter at the same time. If a circle is machined, perhaps several thousand finite and discrete points connected by straight lines would be required. Inasmuch as the points appear on the program tape, it can be seen that the tape may become lengthy.

In *circular interpolation* the programing for a circle would be the endpoints of the

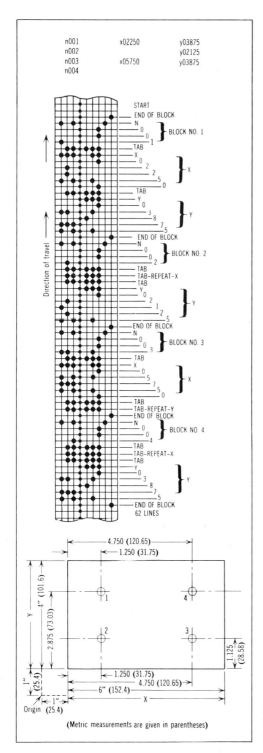

Figure 16.12
EIA RS 273 tape format for drilling four holes. Channel No. 1 left side of tape.

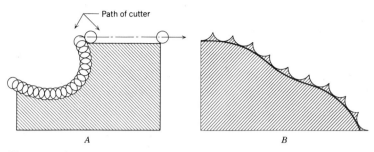

Figure 16.13
A, Cutter path for continuous programming. B, Cusps resulting in nonparallel axes machining.

arc, the radius, and center, and the direction of the cutter as shown by Figure 16.11D. The circular interpolator in most machine control units is a computer component that breaks up the span into the smallest straight-line *resolution units* available in the control, 0.0001 to 0.0002 in. (0.003–0.005 mm). The control computes and generates the controlling signal for the tool. For 1000 blocks of linear interpolation information on a tape, only 5 blocks are necessary with circular information. Parabolic interpolation has application in free-form designs such as molds or die sculpturing.

In continuous programming, feed rates, tool geometry and offset, depth of cut, and materials are entered into the program logic. When a surface not parallel to one of the machine axes being machined, a smooth surface is usually unobtainable, as shown in Figure 16.13B. The choice of cutter, ball-nosed or square-end mill, does not eliminate the problem. There may be cusps that are ground away in a later operation. The deepest part of the cut represents the finished dimension.

OTHER NUMERICAL CONTROL SYSTEMS

Systems having a computer controlling more than one machine tool are known as *direct numerical control* (DNC). One or more NC machine tools is connected to a common computer memory to receive "on-demand" or real-time distribution of data. In this system there is no punched tape. The tape reader, which is not considered a reliable component, is omitted in DNC. DNC may also comprise a management information retrieval package, where information is returned to the central computer and a variety of reports and actions are presented for management interpretation. The DNC system includes telecommunication lines or other methods of transmitting the instructions. DNC was established to overcome the expensive nature of large computers that were dedicated to one machine tool. It allows storage of extremely long programs that will not fit into the memory of a computer NC machine.

Computer numerical control (CNC) systems use a dedicated stored-program minicomputer to perform NC functions in accordance with control programs stored in computer memory. It provides basic computing capacity and data buffering as a part of the control unit. Part programs are entered in the common way to the tape reader, but the tape reader is not necessary for subsequent parts, as the computer is doing the directing. CNC is also known as "soft wired," implying that the program can be changed along with built-in control features. The computer may be used as a terminal to accept information from another computer or telephone data.

As a rule the programmer establishes feeds and speeds. If during the machine operation some unforeseen problem such as hard spots or worn or broken tools occurs, an *adaptive control* may be employed to slow or stop the machine. Conversely, adaptive control may sense machining conditions and can be used to increase speeds or feeds as the situation dictates. The sensed variables include torque, heat, deflection, or vibration.

There are additional methods of machine tool control sometimes associated with NC. *Dial control* is the ability to dial each axis dimension for the workpiece.

HIGHER LEVEL LANGUAGES

Computer assisted NC programming, in particular the popular APT language, is the de facto high-level programming language for NC of machine tools. Space does not permit its full development, because an entire book would be required. Nor should the student believe that programming is punching of holes on a tape. The authors have used the "paper tape" as a substitute for programming instructions because of space limitations.

There are basically two types of languages: machine language and programming language. When programming in machine language, the parts programmer provides the data required by the machine to perform each and every operation in the machining sequence. When coding in a programming language, the NC programmer uses a combination of statements, symbols, and digital information. A computer language such as ***APT*** (Automatic Programmed Tools) is a ***high-level language*** especially designed for NC. A post processor then converts the program into machine language required by the particular NC tool. The final program used by the machine tool will be identical whether it is prepared in machine language or by the computer in a programming language. But different machine tools require their information differently. APT is almost identical for these different machines.

Most APT statements are composed of two parts separated by a slash. Words such as GORGT/TO, LIN2, PAST, LIN3, directs the tool to move right from its present position until its periphery is tangent to LIN2, and its center is past LIN3 on the far side. Normally a parts programmer will define the lines, points, circle, and so forth, to describe the part near the beginning of the program prior to the first motion statement. The moving geometry of a cutter must correspond to the engineering drawing requirements.

BASIC (Beginners' All-purpose Symbolic Instruction Code) is perhaps the easiest language to learn for programming a computer. But BASIC is not considered a special language like APT. There are many NC languages. When counting public and proprietary languages, the number is well above 100. The bibliography provides additional sources of study.

NC offers economic advantages in moderate production and job lot industries, including the following:

1. The amount of nonproductive time is reduced and "chip time" is increased.
2. The number of jobs and fixtures, particularly those used to define positioning, is reduced because the tape does the positioning.
3. A complex component may be machined almost as easily as a simple one once a tape has been prepared.
4. There is usually a reduction in machining setup and cycle times, although the machining requirements for cutting are identical to those machines that have no NC. Scheduled throughput time is reduced.
5. It is adaptable to short runs as compared to special-purpose production machines.
6. There are fewer rejects because reliability and quality are consistent.
7. The program may be changed to allow for machining modifications.
8. Inspection costs are usually reduced.

Disadvantages to NC machines include the following:

1. The capital cost for buying a machine is high, and sometimes the marginal investment saving over non-NC general-purpose machines is not warranted.
2. There is a loss in machine flexibility if a tape or the control malfunctions.
3. Control systems are costly but costs have been coming down.
4. Maintenance costs increase because of the sophistication of the control systems.

QUESTIONS AND PROBLEMS

1. Give an explanation of the following terms:

   Numerical control machine
   Machining center
   Tool storage
   Programmer
   Processor language
   Program reader
   Punched tape
   Control unit
   Open loop
   Closed loop
   Rectangular coordinates
   Binary-coded decimal
   Point to point
   Format
   Continuous path
   Interpolation
   ATP
   Higher level language

2. What is meant by feedback?
3. How does a player piano tape differ from numerical control tape?
4. Make a schematic diagram of a two-axis, closed-loop control system.
5. How does a machining center differ from a numerically controlled turret lathe?
6. List the media for NC. Which is most common?
7. What is the purpose of the parity check?
8. Write a paragraph about the machine control unit.
9. What are the characteristics of a stepping motor?
10. What types of operations are best suited for point-to-point commands?
11. Why is a coordinate system necessary for numerical control?
12. What is the importance of fixed zero and floating point?

13. What is adaptive control and what is its purpose?

14. Why does the first numerically controlled machine tool located in a plant have disproportionate costs associated with it?

15. Show the coordinate location in a three-dimensional place of $x = 5$, $y = 3$, and $z = 4$.

16. Sketch the following machines and show the X, Y, and Z axes for each: drill press, lathe, and milling machine.

17. Develop a flow chart showing the steps in NC metal machining where you start with the part drawing.

18. Make a schematic of tape and with the binary-coded decimal system show the numbers 0 to 9, end block, and six miscellaneous operations identified by the first six letters of the alphabet.

19. Using the Electronic Industries Association standard program on a tape, facsimile the following: "Manufacturing Processes"; "The Year of 2421"; "6 + 7 = 13."

20. Prepare a schematic of BCD tape for the following: $x = 30.720$, $y = 2.000$, $z = 0.500$, feed = 0.025 in./min, coolant on, coolant off, end block. (*Hint:* Coolant on, channels 1, 5, and 7. Coolant off, channels 2, 5, and 7.)

21. Prepare a schematic of the BCD tape for the following: $x = 22.750$, $y = 1.250$, $z = 2.117$, feed = 0.015 in./min, coolant on, coolant off, end block.

22. Prepare a tape schematic for the following: $x = 15.005$, $Y = 0.005$, $Z = 1.275$, feed = 0.009 in. per revolution, coolant on, coolant off, end of block.

CASE STUDY

DRILLED PLATE

"Well, what do you think about the new request for quotation from Baltimore Machine Tool for machining their drilled plates?" Ralph Digiamco, president of Ace Tool and Die, a company that subcontracts work for larger companies, was addressing his production engineer, Bob Quinn.

"I don't know if we can do it," answered

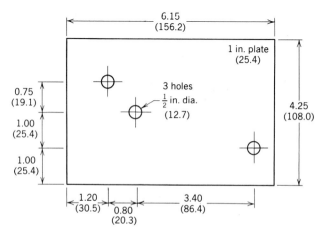

Figure 16.14
Case study.

Bob, looking at the drawing. "We can flame cut the blank from plate stock oversize and then mill the external dimensions. Perhaps while it is on the miller we can drill the holes, but the drilled plate may not meet quality specs. There probably won't be more than a few that meet quality control specs."

Ralph nodded in agreement. "It will be definitely too expensive to do it that way. What if we use the NC to drill the holes instead of the vertical miller?"

"A good idea," said Bob, sitting up in his chair. "That way we would only have to change the NC tape to get a new hole layout for other jobs that might come in. I'll look into this and get back to you in a few days."

Help Bob with his program (see Figure 16.14) using the following hints: Assume that the outline has been completed by the milling machine. Use the axis system suggested by the drawing, where the origin is defined in the lower left-hand corner of the part. Let the drilling feed = 0.025 in./min, speed = 90 rpm, and use start and stop machine and EOB. Prepare a schematic of the tape for this job.

CHAPTER 17
MANUFACTURING SYSTEMS

A system of manufacturing can be described as the factory organized to produce parts and products. Signals to the system are the purchase order and the delivery of the product. Along the way, people, plant, design, production processes, material, energy, and money are managed to achieve these objectives in the best way. Collectively, the people, plant, . . . , and money are called the manufacturing system.

The organization of the system, for the purposes of this text, begins with planning for production. Machines, material, money, and manpower (the four M's of the system) must be matched to the quantity and design requirements to produce the part or product to cost and on-time delivery. A complete manufacturing system involves more than the equipment, for example. In systems engineering we include everything from the peripheral equipment required such as the jigs and fixtures to control software, tooling design, programming, debugging, maintenance, training, and spare parts.

Robots are a consideration in manufacturing systems. Robots are adopted if the job is dull, dirty, or dangerous and the economy makes sense.

Quantity requirements can vary significantly, and the manufacturing engineering service staff is requested to prepare plans. Batch production or automation, robotics, cellular or flow layout of equipment, and special sequencing of the lots are typical plans that may be undertaken. These diverse plans depend on a number of considerations covered in this chapter.

SYSTEM ECONOMICS

A manufacturing system coordinates elements of machines, materials, manpower, and money to return a *profit to the shareholders*. The benefits of the enterprise make it worthwhile to the people to sustain it profitably. The tangible output of manufacturing is the product. The enterprise responds to the sale of the product with respect to the variety and quantity.

Manufacturing of durable goods can be broadly classified as mass, moderate, or job lot production. In *mass production,* sales volume is established and production rates are independent of individual orders. Mass-production examples include automated transfer machines for discrete products, partial and fully automated assembly lines, industrial robots for spot welding, machine loading, and spray painting, automated materials-handling systems, and computer production monitoring.

In *moderate production,* parts are produced in variable quantities and perhaps irreg-

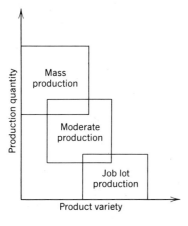

Figure 17.1
Concepts of manufacturing systems depend on product variety and quantity.

ularly over the year. Output is more dependent on single sales orders. The lots may be produced once or periodically. Examples include books, clothing, and industrial machinery.

Job lot industries are the most flexible, and their output is directly connected to the customer order. Most often the job lot industry will not build the product unless the order from a customer is assured. In mass production the manufacturer will frequently build for inventory and may be unaware of the customer. Figure 17.1 describes these interconnections. Notice those zones that are shared between two kinds of product sales. The Figure 17.1 classification depends on product sales between a buying and a producing organization.

Production can be classified according to another conceptual model. Notice Figure 17.2, in which product variety and quantity suggest differing modes of production systems.

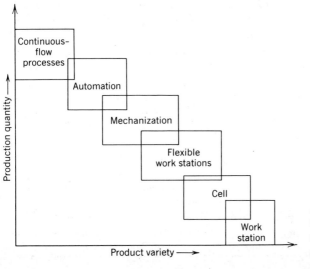

Figure 17.2
Concepts of production systems for product variety and quantity.

The meaning of a *production system* is more limiting than a manufacturing system. A production system implies converting equipment such as machines, processes, benches, and tools. Production systems can be divided into six categories:

1. Work station
2. Cell
3. Flexible work stations
4. Mechanization
5. Automation
6. Continuous-flow processes

These systems are defined in Table 17.1.

Bench work or **manual assembly** is an example of a simple **work station**. Manual assembly is still an important application for small-quantity production. "Ones and twos" are not unknown quantities in prototype work, and these circumstances dictate human assembly. These work stations may have little to no numerical control or computer involvement. With larger production quantities there is a danger of human error in production, and costs of labor can be significant. NC machine tools then become the

Table 17.1 **Characteristics of Production Systems**

Type of System	Characteristics
Work station	Individual bench, machine, or discrete process; craft work, single stations, or NC machine tools working independently; the most primitive production system. Many chapters in this book consider individual work stations.
Cell	Simplest organized production effort; may or may not be computer controlled or robot assisted; composed of two or more work stations.
Flexible work stations	Volume of production has less effect, and use of the computer is characteristic; examples include low quantities of engine blocks.
Mechanization	Dedicated to production of large quantities of one product, with little model variation; examples include high-volume and automobile parts. Computers and robots do not have a significant role, although tools and pneumatic, electric, and electronic controls are important.
Automation	Examples include transfer lines, which may or may not be computer controlled. More recent automation has included robots, which are used for arc welding and parts handling, for example.
Continuous-flow processes	Examples include production of bulk product, such as chemical plants and oil refineries. Features are: flow process from beginning to end, sensor technology available to measure important process variables, use of sophisticated control and optimization strategies, and full computer control.

work station. Because of the low overall efficiency of single work stations, there are numerous attempts to connect them to a system.

Computer technology is important in automation and continuous-flow processes, but not in mechanization. The production systems of today use computers. Association between the digital computer and manufacturing automation may seem logical to the student, but this relationship has not always existed. Historically, automation technology preceded computer technology.

Competition in high-quantity production requires the greatest productive efficiency. Automation has traditionally concentrated on the manufacturing operation and its equipment for high volume. Today, however, computer aided design and computer aided manufacturing (CAD/CAM) includes the manufacturing operation, design, and planning functions that precede the actual production of the product; thus the need for computer integrated manufacturing systems. Although high volume was the first recipient of this technology, mini, personal, and mainframe computers are involved with cell and flexible work stations.

If a company automates one of its assembly lines, the savings in assembly time must pay for the expense of the new machinery. Assembly processes can be classified into primary and secondary assembly operations. *Primary operations* are those that add value to the assembly such as painting, fitting parts to the assembly, and packaging. All other necessary operations, such as rotating and moving, that do not add value to the assembly are secondary. Using primary and secondary assembly operations, an overall efficiency for an assembly process can be rationalized as follows:

$$E = \frac{P}{P + S} \times 100$$

where

E = *Process efficiency,* percentage
P = Total time to complete primary operations
S = Total time to complete secondary operations

By analyzing the efficiency of a given process, it is possible to examine the output for that process.

In planning for *CAM* it is necessary to outline the essential requirements that must be met. *Designing the product* with CAM and robots concentrates on part geometry for easier automatic handling, feeding, locating, holding, loading, and unloading, and many management functions. In addition, the *basic components of CAM* are described in detail and are called *specifications*. These include both special and commercially available feeders, selectors, loaders, unloaders, transfer devices, indexing equipment, vision systems, and numerous other pieces of equipment.

Is CAM really necessary? If a manufacturer hopes to stay in business over the long term, the firm must steadily improve manufacturing methods. Whenever a manufacturer can produce a comparable or better product at lower cost than competitors, the firm will prosper. The selection of the CAM system depends on planning and economics.

In the economic evaluation step, a number of strategies are possible. Consider the situation where there is a minimum product variety but a large quantity of output. Automation may be a choice but full automation may not be economical. Instead, **partial automation** with the addition of automatic devices until either *full automation* or the point of diminishing economic returns has been reached is preferred. It is possible to describe levels of automation that range from an automatic fixture at a single machine, or *mechanization,* to full automation. Part variety may be a dominant requirement and, with low volume, flexible manufacturing systems (FMS) may be the choice.

The choice of the production system depends on many factors. Very important, however, is the objective of *profit maximization or cost minimization.* Figure 17.3 is a layer model of the cost and price structure for a product or part. These ideas are discussed further in Chapter 25, but the motive of profit is strong in the selection of a production system.

Planning at this stage includes the economic considerations of the design, quantity, equipment, plant layout of equipment, computers, programming tool change time, maintenance, load and unload cycles, and other normal production interruptions. Each system can be evaluated using the layer chart model of Figure 17.3. Cycle time per station required, type and number of machines employed, and the cost of both direct and indirect labor for both automated and conventional methods are among the considerations made.

The economic objective of FMS is to approach the efficiency of mass production for batch production. *Batch production* exists whenever the number of parts manufactured occurs in lots ranging from several units to many. Annual demand may require several lots or batches. There is a setup and a teardown for the general-purpose equipment that prepares for the design, then readies the equipment for the next lot. *Mass production* occurs when high annual volumes of production utilize special-purpose equipment. Setup approaches a zero time per unit for mass-production quantities.

Economic experience with Japanese FMS report paybacks of about 4 years. Payback is usually defined to be gross investment divided by net annual savings or

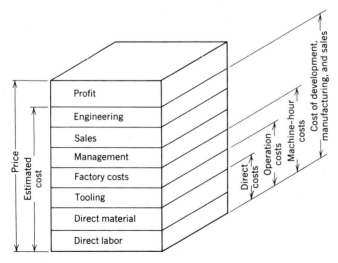

Figure 17.3
Elements in a cost–price model for a manufacturing enterprise.

$$\text{Years payback} = \frac{\text{Investment in manufacturing system}}{\text{Net annual savings}}$$

This is a crude measure of desirability and is used because of its ease of understanding. Engineering economic methods that find the rate of return or net present worth are preferred, and that subject is deferred to those textbooks.

In some experiences as reported on Japanese FMS, many of the parts are *nonrecurring* (i.e., parts that are done only once and there little chance of repeated production). This kind of work is sometime called "special." Reduction in direct labor cost is an objective for FMS. *Untended night shift* or reduced operation for the night shift means that much work is necessary to prepare the parts for later production.

The integration of the robot into individual manufacturing processes requires systems engineering. Whereas robots are available off the shelf, a robotic system is not. It must be carefully planned and customized to each application. In a typical application the robot represents about 30% to 40% of total system cost. Engineering and peripheral equipment might constitute 60% to 70% of the total cost. Yet these nonrobot costs may yield a greater return on investment than the robot itself.

CELL PRODUCTION

Cell production can be thought of as "garage-style" production located in a bigger plant. Although the comparison is not always apt, it suggests several characteristics. First, it is the next higher level of manufacturing work stations as compared to a single work station. Cell production may involve several machines connected by a conveyor system. If it is a specialized cell, it will make only certain classes of parts. In some situations cell design may approach the efficiency of an automated transfer line.

Cell production encourages the following benefits: for simple product design fewer and more straightforward tooling and setups, shorter material-handling distances, and less complicated production and inventory control. Improved process planning procedures are possible.

High Technology

Unlike "garage-style" production, there is often a high-technology core within the cell. Several CNC machine tools are arranged within the reaching arm radius of a robot. A *robot slave* is responsible for part handling. The cell computer supervises and coordinates the various operations, which makes it compatible to the direct numerical control (DNC) computer. In developmental experiences, the core runs 24 h continuously and requires workers' participation only during the day. The day shift loads the computer program enabling the production scheduling, and may do secondary or noncell operations such as heat treatment, mounting part on special pallets, unloading pallets, and cleaning and removal of chips.

During a part-processing activity by the machine in a cell, the robot can perform housekeeping functions such as chip removal, staging tools in the tool changer, and

Plant Layout

As cells are a higher order over a single machine or processing unit, plant layout becomes a consideration. Plant layout is the physical arrangement of all facilities within the factory. Like all planning acts, *plant layout* decisions depend on product design and specifications, production volume, manufacturing operations for the product, and assembly sequence for the product. The best layout will attempt to use the least space consistent with safety, comfort, and product manufacture. It will consider operations, inspections, delays, transportation (material handling), and storage. Figure 17.4 provides basic plant and *cellular layouts* and uses one of the following patterns or straight line: **S-shaped, U**-shaped, circular, and random angle. The type employed depends on many factors, and the existing shape of the building is a prominent one.

Cells need not be in a circle, as a variety of layouts are possible. Nor do cell layouts require that the cells be far from each other. The cells may be stationed along an in-line material transfer system such as a conveyor. Raw and intermediate-finished parts move along the conveyor. A "ready for workpiece" signal from the control unit of the first machine in a manufacturing cell instructs the robot to look for the required workpiece on the conveyor. The robot picks up the workpiece, loads it onto the machine, and sends a signal to the machine control to begin its operations on the workpiece. A "part finished" signal from the last machine tool to the robot requests that the completed part be unloaded and transferred to the outgoing conveyor. The cycle would then be repeated.

Group Technology Layout

One variation of group technology (GT) includes the concept of GT machine cells, groups of machines arranged to produce similar part families. A cellular arrangement of pro-

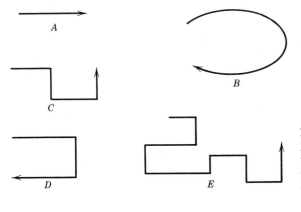

Figure 17.4
Basic layout plans. *A*, Straight line. *B*, Circular. *C*, S-shaped. *D*, U-shaped. *E*, Random.

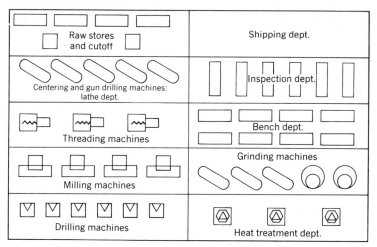

Figure 17.5
Typical plant layout by machine type classification.

duction equipment achieves an efficient work flow within the cell. Labor and machine specialization for the particular part families produced by the cell are possible, raising the total productivity of the cell.

Figure 17.5 is a typical plant layout by *machine type* classification. If a part is moved between these departments, the shop path would appear as spaghetti or as a random walk. The logic to this type of layout is that the skills to operate the equipment are clustered and the operators can, perhaps, interchange among the equipment. A GT layout for common parts is given in Figure 17.6. This pattern reduces in-process storage, the distance parts are being moved, and indirect labor costs of handling.

One limitation of the *flow line layout* requires that all parts in the family be processed through the machines in the same sequence. While some processing steps can be ignored,

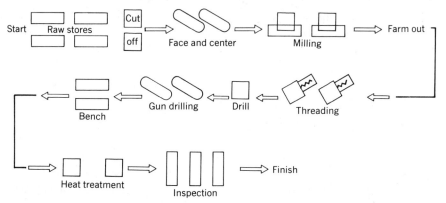

Figure 17.6
Group technology machining center layout.

416   MANUFACTURING SYSTEMS

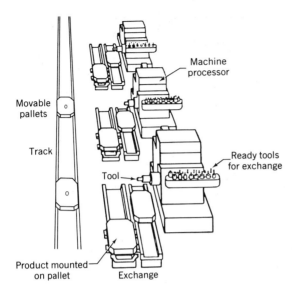

Figure 17.7
In-line manufacturing cell with movable pallets.

it is necessary that the flow of work through the system be unidirectional. Reversal of work flow is accommodated in the more flexible group machine layout, but not conveniently in the flow line configuration as shown by Figure 17.7.

A machine cell is able to provide machining operations. The machines, fixtures, and tools are arranged for efficient flow of work parts through the cell. A single-machine approach can be used for work parts whose design allows them to be made using a one type of process such as a turning or milling machine center. The group machine layout is a cell design in which several machines are used together with no provision for conveyorized parts movement between the machines. Manual operations may be substituted for the robot-assisted conveyors.

Planning for production in a manufacturing cell is different than for a single machine tool. In a manufacturing cell, only one machine can operate with a cutting speed derived by the conventional minimum production time criterion or with an adaptive control system that maximizes the material removal rate. Such a cell drifts down to the slowest work station in the production process.

FLEXIBLE MANUFACTURING SYSTEMS

Flexible manufacturing systems (FMS) are arrangements of individual work stations, cells, machining cells, and robots under the control of a computer. Sometimes FMS means flexible *machining* systems. Workpieces are mounted on pallets that move through the system transferred by towlines, conveyors, or drag chains. Operator interdiction is discouraged by FMS. As jobs are changed the computer is reprogrammed to handle new requirements. FMS is closely related to cellular systems.

Equipment and manufacturing cells are located along the material transfer highway.

Different parts move on the conveyor and generally the quantity is small. The workpieces are complex, however, and can require complicated manufacturing steps. The parts along the highway will not need all the equipment or cells. Production of the various parts requires processing by different combinations of manufacturing, but FMS is versatile and can perform different operations on a variety of products. Often an FMS machine can perform many processing steps. The process begins with a robot or operator loading or unloading a CNC machine in FMS. Notice Figure 17.8. After processing in FMS the robot will return the semifinished or finished part to the conveyor.

Pallet transfer to and from machines is by automatically guided vehicles (AGVs). These carriers are often rated by the tonnage capacity they are able to move. The moving pallets play synthesized tunes through loud speakers to warn pedestrians as they move about the factory floor, and tunes are changed from time to time. These unmanned pallets can move the part to a central area where special work can be conducted. Some loading stations prepare work for the night shift, and these are inventoried for accessibility on the pallets by robots. The fixtures aboard the pallet are designed for a specific class of parts such as prismatic, shafts, or plate. Automatic loading palletizers can be used to mount and unmount the part on the vehicle.

Not only are parts moved, but tools are moved between the stations if a part requires tooling that is not available on a specific machine. The robots mount the tool into the carriage where it is available for the part at the time it is needed. The robot arms are able to reach the tool and extend it the required distance.

For *unmanned machines* or factories it is necessary to control chips because cleanliness is important. Various means are used to keep pallets clean. Designs are available to dump

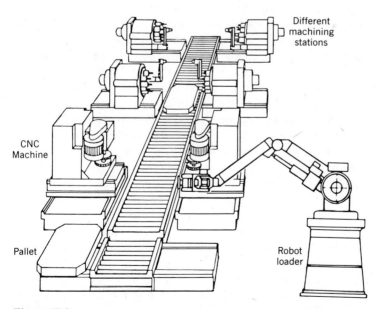

Figure 17.8
Flexible manufacturing system with robot handling system.

the pallets, blow off the chips, or use vacuum systems to suck up even the smallest chips. Chips are often conveyed in underfloor tubes, where they are eventually gathered in a large cyclone for collection, cleaning if economical, and compacting.

Tool breakage for the automatic machines is controlled by several means. Probes on each machine check for tool breakage, and spare tools are contained in the magazine if necessary. In some cases acoustic emission monitoring checks for broken tools. When tools need to be changed they are sometimes supplied from a central location where they have been present. Changing may be automatic and follow tool life criteria rather than a breakage alarm. This tool life is programmed, and then maintenance follows the tool life requirements as instructed by the central computer.

If the tool breaks while machining, the tool and part are removed for later inspection. The operation continues, however, with replacement of a new part on the machine.

Guidance of the vehicles is by several methods. A hidden wire, tracks, or electronic targeting may be used to guide the vehicles.

FMS is integrated with CAD/CAM. CAD, for example, will limit the number of tools to a preset number, such that the factory will not store more than a specific number. Another approach finds the number of tools and then reduces that number by cost control methods. Standardization of tools, their kind and quantity, and specifications are a natural development of FMS.

MECHANIZATION

Mechanization is a well-developed production system. Traditionally, it has not involved the computer or adaptive controls, although sensors, electronic, and electrical controls are common. Significant mechanization is found, for example, with the automatic screw machines, which may be cam and drum driven. These machines have been available for a century. Transfer dies in press work is a highly developed form of mechanization.

Many examples of small-batch mechanization are found in job shops, in die-making shops, and in the aerospace and machine tool industries. The accuracy of the final product depends not only on the machine but also on the skill of the operator who has to set up the workpiece. The task is facilitated by *jigs,* which hold the workpiece and incorporate a guide for the tool, and *fixtures,* which firmly hold one or more workpieces in the correct position in relation to the machine tool bed. Machining or processing then proceeds. These tasks are described in other chapters.

The use of several tools in a machine to perform several operations (either simultaneously or in sequence) saves time over individual machining elements and is a form of mechanization. Mechanization requires automatic workpiece handling and tooling. Handling will sequence a workpiece through a work cycle, perhaps "index and travel" the part from one location to another. Clamping and locating of the workpiece occur at the work station, and machining follows.

A variety of design configurations is possible. The movement of the workpiece may be circular around the machine or linear along the machine if one machine is used. Circular movement uses less floor area. Straight-line type machines allow addition and subtraction of stations, thus facilitating interchangeability and continuous chip or waste

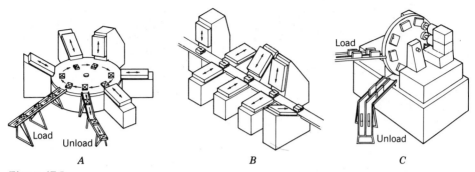

Figure 17.9
Transfer machines. *A*, Circular table. *B*, In-line. *C*, Drum.

removal. Indexing is most suited for drilling, tapping, and turning operations. In constant-travel mechanization the workpiece is advanced with indexing in either a circular or straight-line path, and locating and clamping the workpiece is required only once. Constant-travel mechanization is preferred for milling, broaching, and grinding operations. Figure 17.9 shows three types of automation.

Some assembly operations lend themselves to simple mechanical methods. Thus, screws or bolts can be driven and parts placed, crimped, or riveted with mechanical devices. Complicated machines are available, for example, to manufacture hospital syringes completely and aseptically in large daily volume. The cost is reduced while productivity and consistency of the product increases, but only if the reliability of mechanization can be assured. The cost of downtime repairs can cancel all savings. In-line inspection to pinpoint and remove imperfect assemblies, either during or at the end of the assembly operation, is an important element for complicated mechanization operations. Many of the elements used in automatic assembly are the same as in mechanized production and workpiece handling, and may be mechanical, electromechanical, fluidic, or numerically or computer controlled.

AUTOMATION

Automation involves automatic handling between machines and continuous automatic processing at the machines. The elements *continuous* and *automatic* are necessary to separate automation from mechanization. For example, a single machine can be mechanized so it is completely automatic, but this is not the concept of automation considered here. According to our definition, automation exists only when a group of related operations are tied together mechanically or electronically or with the assistance of computers or with robots. Computers and robots are not necessary to have automation.

Notice Figure 17.10, which is a specially designed automatic machine. This electric armature-balancing machine produces 650 parts per hour, and provides a dynamic balance correction in two planes to a limit of 0.001 oz-in. (0.07 g-cm) per plane. It features a microprocessor-based balanced computation system with adaptive feedback to compensate

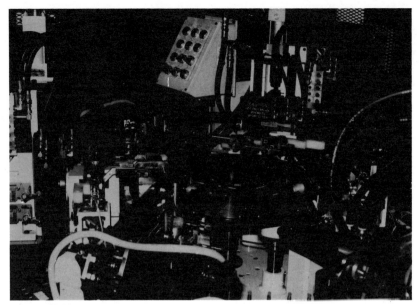

Figure 17.10
Electric armature-balancing machine producing 650 parts per hour.

for tool wear. A specially designed high-life carbide mill will machine 2 million armatures without correction. Armatures are loaded and unloaded automatically to the balance correction machine.

History

Automation is not a new technology. Food and beverage processing, petroleum and chemical industries, and telephone services have been "automated" for decades. Grain milling is a processing operation that approached complete automation about two centuries ago. Grain was fed into a flour mill by means of a bucket conveyor, and water power moved it over a series of endless belt and screw conveyors through coarse and fine grinding operations until flour emerged. The Jacquard loom has been controlled by punched cards since 1805. Automation has existed in varying degrees in the metalworking industry for a long time.

The word automation was coined at Ford Motor Company in 1945 to describe "a logical development" in technical progress where automatic handling *between* machines is combined with *continuous processing* at machines. Having the combination of two or more automatic operations on a standard machine, such as found with automatic bar machines, vertical turret lathes, and others is not what is meant by automation. These machines are highly mechanized. These machines may be considered to be automated, however, when they are mechanically joined for continuous automatic handling and processing. If an automatic bar machine is connected mechanically to a material feed or to the conveyor that advances parts, it is called automation.

Purposes for Automation

Reduction in direct labor costs is the major objective. Automation is also necessary to assure product quality. Large volume and its attendant boredom causes quality problems. Companies use automation to obtain uniform quality with machines instead of employing the variable skills of labor. Operator fatique, carelessness, or other human frailties are reduced.

Repetitive manual operations may cause conscious or unconscious *job dissatisfaction*. In some situations these jobs can be improved from the operator's point of view by the installation of at least partial automation.

Safety is another reason for automation. Safety is important for positive employee morale and good operating practice. Automation means reduced industrial accidents.

Shop efficiency is improved as loaders, unloaders, feeders, automatic inspection, and other automation devices require efficient distribution of parts from one machine to another. This smooths out the delivery of parts between machine operations. Automation of a single machine improves processing efficiency, because feeders and loaders cannot accept parts that fail to meet specifications.

Automation improves efficiency by segregating short runs from high-volume production. Flip-flop between short and long run results in extended setups; time is spend changing over from one to the other, and it is difficult to maintain systematic work scheduling. Automated setups, of course, must be left that way within practical limits. Keeping long-run jobs progressing through one line and low-volume work through another is a planning necessity to maintain the economic efficiency.

Tooling

The use of standard tools, instead of specially designed tooling, may help reduce tool change time to a minimum. Tools are located in carousels, and can be retrieved automatically or under the control of robots or attending operators. There is a specific place for each tool in these tool bins, and each is identified and computer keyed to the position of the tool in the machine. Moreover, each tool in the bin is *preset,* so that it can replace a worn tool in the machine without adjustments. A clock or counter is connected to the tool on the machine and indicates when a predetermined number of parts have been machined. Setting these clocks, of course, is based on the known life of a tool. When any particular tool has machined a predetermined number of pieces, the machine automatically stops, and the preset tool on the board is installed. Out-of-tolerance parts made by worn tooling are minimized, and fewer tools are broken.

Prevention of Shutdown

Automatic production lines engineered according to the principles of **modular automation units** can be backed up with conventional machines in case of a serious failure. Prevention of shutdown and reliability of operation is important. Sections of modular engineered lines can be removed and replaced with standard units.

Other planning can consider the possibility of unexpected failure. Uninterrupted production at intermediate stations of a transfer line is essential. Inventory banks of partially finished parts at each station are one answer. Shuttle conveyors to normally idle standard machines are another. Emergency work stations included as part of an integrated line are also valuable. But a *policy dictum* may be more important; that is, have no shutdowns and engineer the design to avoid it.

Preventive maintenance is important to uninterrupted production. This may involve keeping records of lubrication and the life of perishable tools, motors, bearings, relays, and switches, and their replacement before they fail and stop production. *Cleanliness* is important to avoid shutdown. Automatic removal of chips and scrap pieces is necessary. A chip disposal system for a large machining line may be necessary. A device for tilting a part to drain coolants may prevent problems in the subsequent operations.

Part Design

A basic principle in automation is that the design of the part and the design of the process should be related as closely as possible. Close cooperation between product and equipment designers is essential, not only to avoid excessive costs in automated machines and to ensure maximum efficiency in processing, but also to provide the highest degree of flexibility possible. Without some idea of what future changes may be made in part, the equipment builder cannot provide for them. And, of course, one of the first questions asked by the machine designer is, "Will the produce design remain stable for the life of the machine or shall we make allowances for future changes?"

Automation in Mass Production

Handling and moving the part was solved first for mass production. Some mechanization examples are progressive and transfer dies for automotive sheet metal operations, upsetters and cold headers in material deformation, and automatic screw machines in machining. Bottle-making machines for glass and plastic containers are the automation extensions of these mechanization examples. These are large-volume production of identical parts, which simplifies the problem of holding the part firmly in a fixed position relative to the tool. The part is moved from station to station by mechanical means, and the only manual movement remaining is that of loading a new length or batch of raw materials. But even these are integrated. In plastic bottle making, the plastic raw material is conveyed in pipes from the train boxcar or truck directly to bulk storage, where it is automatically conveyed to the machine.

Transfer Equipment

Mechanical loading and transfer devices are used to move components of varied geometry from machine to machine. Special jaws grip the part, life, move, and turn it on arms, and place it into the new work position. Travel distance, direction, sequence, and speed are controlled mechanically, electromechanically, or with fluidic controls. Robots and computer control are used. Dead stops, *mechanical arms,* or *iron hands* can be reprogrammed, but not very readily. But robotic manipulators overcome the inflexibility of mechanical manipulators. Figure 17.9 illustrates transfer equipment.

Electric, Air, and Hydraulic Power

The basic machine movements are linear, as in the travel of a drill or rotary such as an indexing table. The power for these movements can be electrical, pneumatic, or hydraulic. The selection of the power source depends on availability, relative cost, amount of power, space, and speed requirements.

Parts Orientation

A step in the assembly process is the orientation of parts. This work can be done by operators who will pick, orient, and place the part in another location. The parts may be jumbled from a previous operation or they may be stacked horizontally, vertically, leaved, in order, random, or size arranged. The choice here is almost limitless. Parts orientation work is popular, especially in low-volume work. *Direct-labor employees,* a classification of labor for the factory, will handle the parts. As volume increases, it pays to consider other ways. Numerous mechanical, pneumatic, vibratory, and ingenious devices exist to orient parts.

A robot or a part feeding system is required to orient an individual part and present the component consistently for assembly. Parts can be manually oriented either directly into a feeding system on the robot arm or into a magazine tray from which the robot removes them. In contrast a computerized visual detection system can orient and present the part without the need for manual labor.

A computerized *visual feeding* system is composed of a feeding and an orientation section. The feeding section consists of a bowl and track feeding unit that separates the disoriented parts and passes them along to the orientation section. One method for separation is vibratory motion, which first reduces the pile of parts to a single layer, and then aligns them into a single row. At this point single parts are removed from the row and fed for orientation.

The part (see Figure 17.11) must be checked to see if it is inverted, and turned so that all pieces face the same direction. Determining which side of the part is up is first. Notice in Figure 17.12A that the isolated part is fed into a holding unit and held between a horizontal light source and a light detection sensor. The sensor reads the position of the part and the computer determines if any orientation is necessary. If the component

 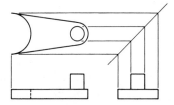

Figure 17.11
Design of part for orientation.

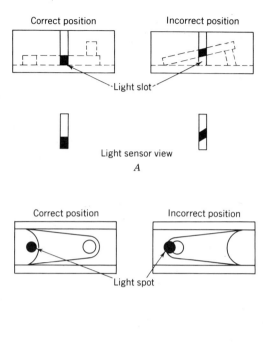

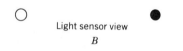

Figure 17.12
Vision-determining system.
A, Inversion from side view.
B, Direction from top view.

is positioned wrong side up, it is vertically rotated 180°. The part is returned to the orientation sensing unit, where it is checked using vertical light sensors, as shown in Figure 17.13. If part orientation is wrong, it is rotated horizontally 180°. All these actions are shown in Figure 17.13.

Bowl feeders (Figure 17.14) are popular for small parts, including nuts, screws, shafts, pins, and washers. Some devices allow vibration from an electromagnet supported on the base. When components are placed in the bowl, the vibratory motion causes them to climb up a track to the outlet at the top of the bowl. Another device (see Figure 17.14B) is a centerboard feeder. Basically, this consists of a hopper for the random dumping of parts. A blade with a slot in its upper edge is periodically pushed through the parts and catches some in its slot. When the blade is in its highest position, it allows the parts to run down a chute that feeds to the assembly below.

Individual Machine Automation

Automatic screw machines, multispindle automatic lathes, and gear cutting machines are examples of individual machine mechanization. Unless these machines have automatic stock loading and unloading, they are classified as semiautomatic. Magazine-type feeders

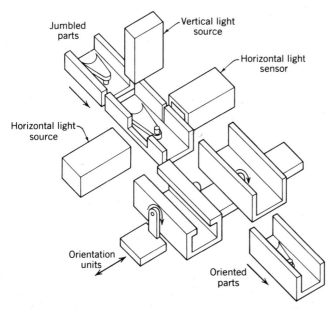

Figure 17.13
Use of light sources and computer choices for part orientation.

are now used to supply lathes with bar stock, gear blanks to the gear cutting machines, and other operations, to make them fully automatic.

The razor blade assembly is an example of automation. Figure 17.15 shows a spider that has undergone prior subassembly. Of the seven components in the complete assembly, three—the spider and two half-caps—make up a half-cap subassembly. These are put together on an in-line subassembly machine at a rate of 60 per minute. Spiders are fed into a chute, vibrating feeders supply half-caps to each side of the spider, half-caps are

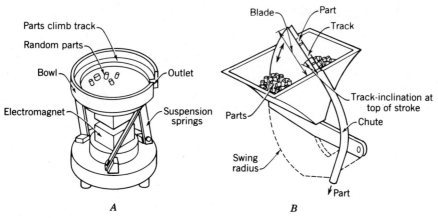

Figure 17.14
Bowl feeders. *A*, Vibratory. *B*, Centerboard.

426 MANUFACTURING SYSTEMS

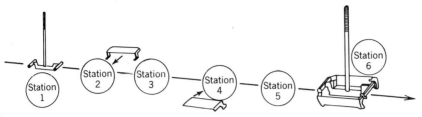

Figure 17.15
Sequence of half-cap assembly. 1, Load spiders onto track, escapes into fixture; 2 and 4, hopper feed of half-cap into spider; 3 and 5, snap half-cap onto spider; and 6, unload half-cap subassembly into magazine trays for final assembly.

sprung into place, and subassemblies are chuted into trays that serve as oriented storage magazines for the next machine assembly.

Complete razors (Figure 17.16) are assembled on the second machine at a rate of 30 razors per minute. This machine uses rotary indexing with work fixtures located on the perimeter. Vibratory bowl feeders feed guard extensions, outer tubes, inner tubes, and screws. The spider subassembly was previously completed and is now joined to the other units. This machine is an example of a *specific product machine automation*. This automation equipment is custom designed for a particular product.

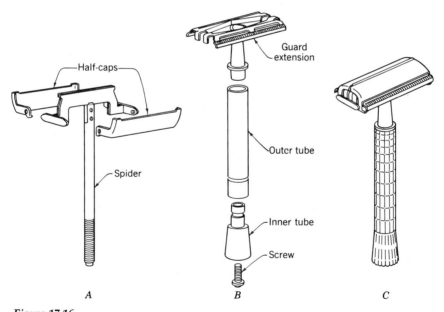

Figure 17.16
Two assembly machines complete this Gillette safety razor. *A*, Half-cap subassembly. *B*, Final assembly. *C*, Final product.

Adaptable Programming Assembly

Batch assembly is different from high-volume assembly as described by the razor example. In that case parts and subassemblies were limited and a large quantity was necessary. But outside of consumer products, high volume is not always characteristics. *Batch production* is more the norm. Batch production covers about 75% of all production. Batch production is labor intensive and not easily automated. The batch sizes can range from several to hundreds of the same product.

Figure 17.17 is an example of a demonstration project of *adaptable programming assembly*. This project, engineered by Westinghouse and Unimation, uses fractional horsepower motors, a product Westinghouse sells by the millions and manufactures in batches of a few hundred. These motors come in eight classes and 450 styles. A representative batch has a quantity of 600. An assembly line averages 13 style changeovers per day. Assuming an hour per changeover, there is as much time spent in setup as in cycle production. This is an application for adaptable programming assembly. Figure 17.17 is an assembly line to build end bell subassemblies, rotors, and stators.

The assembly process begins at the top center of the picture and preceeds counter-clockwise. Vision is used with the robots, and transmitted infrared light is the main source. One of the first stations determines the bell end orientation, then grips it, rotates, and places it on a special pallet that preserves the orientation through the entire assembly.

Figure 17.17
Adaptable programmable assembly system (APAS) facility.

Image-processing algorithms aid the interface between the camera and computer, which necessitates preprocessing hardware to increase speed. Some vision systems are binary, that is, black and white only, without shades of gray. Variations in the reflectivity of the painted surfaces preclude the use of reflected light. Instead the choice of infrared light source minimizes problems resulting from ambient light. The vision algorithm provides information about end bell characteristics such as total dark area, number, location, and size of holes, and compares that to the library of parts. The computer instructs the robot manipulator to accept or reject and dispose in a reject bin if necessary.

Other robots are *pick and place;* they add a felt wick and a plastic cage. Lubricants are added later to the parts. A gripper design is a three-stage device using different holding techniques for various pieces it might hold.

The system also incorporates "hard" automation, such as automatic screwdrivers fed by vibratory screw feeders. In this experimental system the automatic screw driving was a problem area because of misalignment, wrong screws, bad threads, and bad holes. If the station had been converted to robots, these problems could have been detected by sensors. Economic considerations are a choice in the system engineering of any automation.

Assembly Lines and Materials Moving

In assembling a simple or a complex product, a first step is breaking down the assembly into smaller steps. This facilitates material handling by assuring that parts are supplied in proper place and sequence. Materials handling is achieved by many methods. Figure 17.18 shows a bay crane, overhead track crane, magnetic chuck on a movable arm, motorized truck, and other simple conveyances. Installing automatic or semiautomatic handling equipment between machines already on line is successful and permits easy introduction of automation into existing production systems.

Assembly means fitting together individual parts to produce a product. The parts may first be combined into miniassemblies, then subassemblies, and finally, perhaps, final assembly. Figure 17.19 is a sketch of a highly flexible assembly line in which the parts are fitted manually. This line is a strong candidate for automation, FMS, or cellular assembly if economic conditions warrant. Robots may also assist the assembly process (Figure 17.20). Whether either of these choices is economical depends on many factors.

Figure 17.18
Material-handling methods.

Figure 17.19
Labor-intensive conveyor assembly.

In all instances assembly may be performed *in line;* that is, along a conveyor on which parts move or on pallets that assure accurate positioning. Alternatively, the assembly may be of the rotary kind with a carousel carrying the unit from station to station.

Towline or Wire Guidance

In this system workpieces are attached to pallet fixtures or platforms that are carried on carts towed by a chain located beneath the floor. The pallet fixture is designed so that it may be conveniently moved and clamped at successive machines in manufacturing cells. The advantage of this method is that the part is accurately located in the pallet, and it is correctly positioned for each machining or assembly operation.

With the ***wire guidance*** system, carts can move along a path determined by wire embedded in the floor. A cart picks up a finished palletized workpiece from the machining center and delivers it to an unload station elsewhere in the system.

Roller Conveyor

A conveyor consisting of rotating rollers may be used throughout the factory. The conveyor can transport palletized workpieces or parts that are moving at constant speed between the manufacturing cells. When a workpiece approaches the required cell, it can be picked

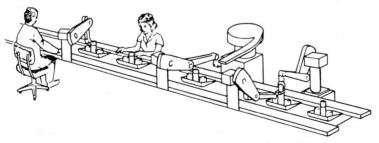

Figure 17.20
Robot-assisted placement of parts alongside manual assembly.

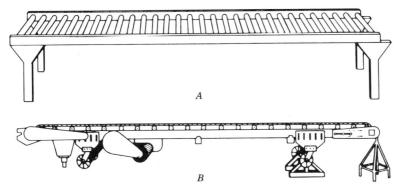

Figure 17.21
A, Roller conveyor. B, Belt, power-driven, and portable conveyor.

up by the robot or routed to the cell via a cross-roller conveyor. The rollers can be powered either by a chain drive or by a moving belt that provides for the rotation of the rollers by friction. Figure 17.21 shows a gravity roller conveyor and a belt, power-driven, portable conveyor.

Belt Conveyor

In this materials-moving system, either a steel belt or a chain driven by pulleys transports the parts. This system can operate by three different methods. In *continuous transfer* the workpieces are moving continuously and either the processing is performed during the motion or the cell's robot picks up the workpiece when it approaches the cell.

Synchronous transfer is mainly used in automatic assembly lines. The assembly stations are located with the same distance between them, and the parts to be assembled are positioned at equal distance along the conveyor. In each station a few parts are assembled by a robot or automatic device with fixed motions. The conveyor is of an indexing type; namely, it moves a short distance and stops when the product is in the station, and subsequently the assembly takes place simultaneously in all stations. This method can be applied where station cycle times are almost equal.

Power and free material handling allows each workpiece to move independently to the next manufacturing cell for processing.

ROBOTS

The Robot Institute of America defines the *industrial robot* as "a reprogrammable multifunctional manipulator designed to move materials, parts, tools, or other specialized devices through variable programmed motions for the performance of a variety of tasks." It is understood that an industrial robot must include an end effector, factory work, and stand-alone operation.

Applications

Robots are used in light manufacturing, casting and foundry, automotive, electrical, heavy manufacturing, and aerospace industries. Labor-intensive operations where end effectors and sensing equipment are necessary to replace the human worker constitute a growing field of application. In the sense of operations for manufacturing processes, robots are used for palletizing, spraying, grinding, welding, deburring, searching, machine load and unload, packaging, assembly, tool carrying, and line tracking. In the forging and foundry area alone robots are used for upsetting, die forging, press forging, heat treating, loading and unloading ovens and furnaces, and flame cutting.

The system, insofar as the robot is concerned, can be restricted to the robot cell for this discussion. The considerations for the robot are compatibility of the operation to the robot, selection, axes of motion, load capacity, power source, end effector, and operating temperatures. Additionally, these concepts are important: robot work zones, tooling fixtures and jigs, material conveyances, control console, programming methods, control software, system monitoring devices, and human factors. A robot welding cell is shown by Figure 17.22 that includes these considerations.

Robot Components

Robots consist of several major components, namely, the manipulator, the controller, and the power supply. The structure of a robot manipulator in general is composed of a *main frame* and a *wrist* at its end. The **manipulator** is a series of mechanical linkages and joints capable of motions in various directions to perform work. The manipulator simulates the movement of the human arm, wrist, hand, and fingers. The main frame is called the *arm,* and the most distal group of joints affecting the motion of the end effector

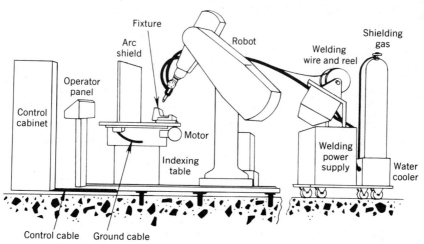

Figure 17.22
Welding cell using robot.

is called the *wrist*. The end effector can be a welding head, spray gun, machining tool, or a gripper containing on–off jaws, depending on the specific application of the robot. These devices mounted at the end of the robot that perform the work are known as the *robot tools*.

An industrial robot is an all-purpose machine equipped with computer memory capable of rotation and of replacing human labor by automatic performance of movements. Examples include the following:

1. **Manual Manipulator**. This type of robot is worked by an operator.

2. **Fixed-sequence robot**. This manipulator performs repetitive operations according to a predetermined pattern. The set of information cannot be easily changed.

3. **Variable-sequence robot.** This manipulator repetitively performs operations according to a predetermined set of rules whose set of information can be easily changed.

4. *Playback robot*. From memory, this manipulator can produce operations originally executed under human control. A human operator initially operates the robot to establish the rules. All the information relevant to the operations, sequence, conditions, and positions is loaded into memory. This information is recalled (or played back when required—hence, the term "playback" robot). The operations are automatically executed from memory. As an example, consider an arc-welding robot that is roughly programmed to weld automobile body sections together. At the time of the production it will be fine-tuned by operators who may hand guide it through the motions that are necessary for the compound surfaces of the body. These robots are sometimes called "teachable."

5. *NC robots*. Using media such as a punched tape and computer, this type of manipulator can perform a given task according to the sequence, conditions, and position as commanded via numerical data.

6. *Intelligent robot*. This robot uses vision, touch, or both to notice changes in the work environment or condition. Using its own decision-making ability, the robot proceeds with its operation.

Controller

The robot's *controller* or "brain" stores data and directs the movement of the manipulator. Controllers can be simple or complex, but a typical controller permits storage and execution of more than one program. The controller first initiates and terminates motions of the manipulator in a desired sequence and at specific points. Second, it stores position and sequences data in memory; third, it interfaces with the manufacturing operation. *Feedback* is often a part of the controller. The controllers range from simple step sequencers through pneumatic logic circuits, electrical and electronic sequencers, microprocessors, and minicomputers. The controller may be installed physically in the manipulator or have a special cabinet.

Robots may be further classified as nonservo or servo-controlled devices. **Nonservo robots** are often referred to as endpoint, *pick-and-place, bang-bang,* or *limited-sequence* robots. Nonservomechanism robots have directional control valves that are either fully opened or closed, thus limiting program and positioning capacity. Their arms travel only at one speed and can stop only at the endpoint of their axes. Significant features include relatively high speed, good repeatability to within 0.25 mm, limited flexibility in terms of program capacity, simple operation and programming, low maintenance, and comparatively low cost. These nonservo robots are hydraulic or pneumatically powered. Nonservo machines are used on high speed and precise tasks.

A typical operation for a *nonservo* robot is as follows: After the program starts, the sequencer/controller initiates signals to control valves on the manipulator's actuators. The valves open, admitting air or oil to the actuators and the members begin to move. Once physically limited by end stops, the valves remain open. Limit switches signal the end of the travel to the controller, which commands the control valves to close. The sequencer then indexes to the next step and the controller again signals the output. This may be an external device such as a gripper.

Servo controlled robots have one or more servomechanisms or motors that allow the arm and gripper to change direction in midair without having to trip a mechanical switch. They can very speed at any point in the work envelope. The available features of a nonservo robot include the following: ability to move heavy loads in a controlled fashion, maximum flexibility to program the manipulator to any position within the limits of travel, storage and execution of more than one program from memory via externally generated signals, and subrouting and branching. The end-of-arm positioning has a positioning accuracy of 1.5 mm and a repeatability of ±1.5 mm. Figure 17.23 is a six-axis industrial robot using electrohydraulic servocontrol systems. Payloads of 100 lb are possible.

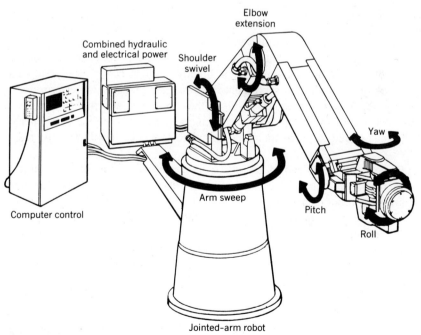

Figure 17.23
Jointed-arm robot.

The welding cell in Figure 17.22 may be remotely controlled and located at any attitude by computer control with solid-state electronics and use of combined hydraulic and electric power.

The operating sequence of a servo-controlled robot begins with the start of program execution. The controller addresses the memory location of the first command position and also reads the actual position of the various axes as measured by the position feedback. The two sets of data are compared and their differences, or "error," are amplified and transmitted as command signals to servo valves or motors for the actuators of each axis. This control is discussed in Chapter 16. The servo valves, operating at constant pressure, control flow to the manipulator's actuators. As the actuators move the manipulator's axis, feedback devices such as encoders, potentiometers, resolvers, and tachometers send position or velocity measurements, for example, back to the controller. These feedback signals are compared to the desired position, and new error signals are created, amplified, and transmitted to the servo valves.

The process continues until the error signal is effectually small enough to be ignored; then the error signal is null or zero, and feedback is stopped. The servo valves stop flow to the actuators, and the manipulators come to rest at the desired position. The controller then addresses the next memory location and responds to the data given by that address. This may be another positioning sequence for the manipulator or a signal to an external device. The entire program is repeated sequentially until the program is completed.

Robot Paths

Servo-controlled robots are further classified as *point to point* or *continuous path*. In many respects this classification is similar to NC (Chapter 16). Point to point is used for tool and part handling. The path through which the various members of the manipulator move when traveling from point to point is not directly programmed. Instead, endpoints are indicated. In point to point the servo-controlled robot is "taught" to perform its job one point at a time. A human operator positions the robot hand at a particular point in space and instructs the robot to store that position in memory. This procedure is repeated until the robot has stored in its memory the complete sequence it will be expected to perform.

With continuous-path servo-controlled robots, feedback and positioning requirements are more important. To "teach" a continuous-path robot its task, a human operator physically moves the robot manipulator through whatever series of motions it is expected to perform. These "learned" motions are stored in the controller for later playback.

Power to the manipulator actuator is by electric, hydraulic, or pneumatic means. A robot with hydraulic power usually consists of a motor-driven pump, filter, reservoir, and heat exchanger. Remote air compressors provide the power for "air logic" robots, and this compressor may serve other requirements.

A programming method is the technique for teaching a robot specific motion paths required to perform a job. *Manual* refers to off-line method for programming either a mechanical, pneumatic, or electrical memory. It involves setting the limit switches on each axis by presetting cams on a rotating step drum, connecting air logic tubing, or prewiring appropriate connections. *Lead-through* consists of maneuvering the robot arm from one path point to the next by means of a control console. ***Walk-through programming*** involves physically guiding the robot arm through the desired motions.

Coordinate Systems

Robots can be structured in four coordinate systems: Cartesian, cylindrical, jointed spherical, and spherical. These axes of motion, sometimes called *degrees of freedom,* refer to the separate motions a robot has in its manipulator, wrist, and base.

Robots having *Cartesian coordinate motions* travel in right-angle lines to each other as shown by Figure 17.24A. There are no radial motions. The profile of the robot's work envelope is rectangular in shape. Also called "rectangular robots," these robots tend to have greater accuracy and repeatability than other types, especially for heavier loads.

A *coordinate robot* (Figure 17.24B) has a horizontal shaft that goes in and out and rides up and down on a vertical shaft that rotates about the shaft. These two members can rotate as a unit on a base.

The *jointed spherical coordinate* robot, also called "jointed arm," can perform similar actions to a human's shoulder, arm, and elbow arrangement (Figure 17.24C). The work envelope approximates a portion of a sphere.

A spherical coordinate robot (Figure 17.24D) moves so that the work envelope forms the outline of a sphere. A robot with a cylindrical coordinate system moves so that the

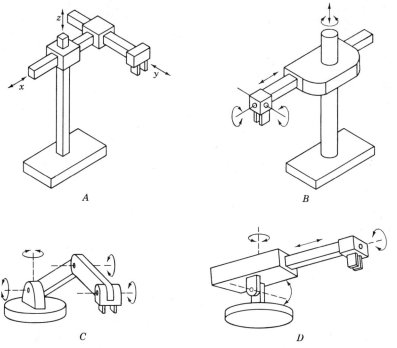

Figure 17.24
Coordinate systems. *A,* Cartesian. *B,* Cylindrical. *C,* Jointed arm. *D,* Spherical.

work forms the outline of a cylinder. The spherical coordinate robot, sometimes called "polar," has a configuration similar to a tank turret. The arm moves in and out and is raised and lowered through an arc while rotating about the base.

Robots have differing work envelopes. The envelope is defined as the area in space that the robot can touch with the mounting plate on the end of its arm. See Figure 17.25, in which the top and elevation views are given. Figure 17.25A shows the work envelope for the jointed arm; Figure 17.25B provides the diagram for the spherical coordinate robot.

The wrist, which is the unit mounted on the end of the robot's arm, has a tool or gripper installed in it. The wrist has three basic motions. *Yaw* refers to rotation in a

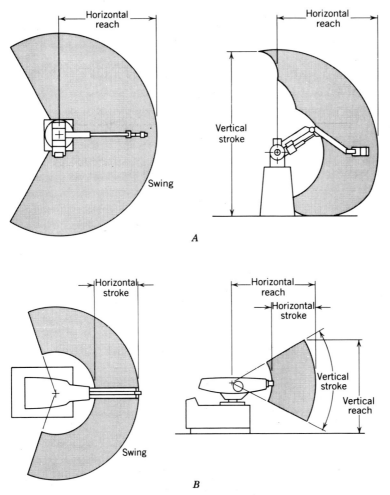

Figure 17.25
Work envelopes. *A*, For jointed arm. *B*, For spherical-coordinate robot.

horizontal plane through the arm. *Pitch* refers to the rotation in a vertical plane through the arm, and *roll* is rotation in a plane perpendicular to the end of the arm.

These six basic motions, or degrees of freedom, provide the capacity to move the end effector through a sequence as shown by Figure 17.26. Not all robots are equipped with six degrees, as it may not be economically wise.

End Effectors

This is analogous to the human hand and is sometimes called the *gripper or end-of-arm tooling*. Various factors determine the end effector use. For example, type of power, floor layout and work envelope size, work environment, part configuration, characteristic of the robot (payload, accuracy, reach, construction), part fixturing, cycle time, and maintenance. Cost, of course, is an important consideration. Figure 17.27 illustrates selection of end effectors.

Figure 17.27A is a clamp or crimp. Stud welding is possible with the tool shown in Figure 17.27B, and spare studs are fed down a tube. Torches for welding or heating are possibilities (Figure 17.27C). A ladle for the hot and dirty job of pouring molten metals is another application (Figure 17.27D). Spot-welding guns (Figure 17.27E), pneumatic nut runners, drills, impact wrenches, and tool changing (Figure 17.27F) are favorite tools.

Robotic Sensors

The robot may take on more humanlike senses and capabilities to perform a task. These senses and capabilities include *vision and hand–eye coordination, touch, and hearing*. The types of sensors used in robotics fall into the following three categories: vision, tactile, and voice sensors.

Applications for sensors are several. They may be used in support and direction for

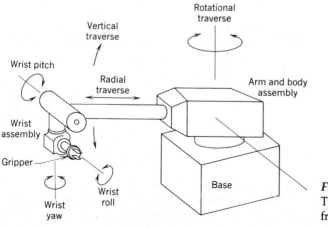

Figure 17.26
Typical six degrees of freedom in robot motion.

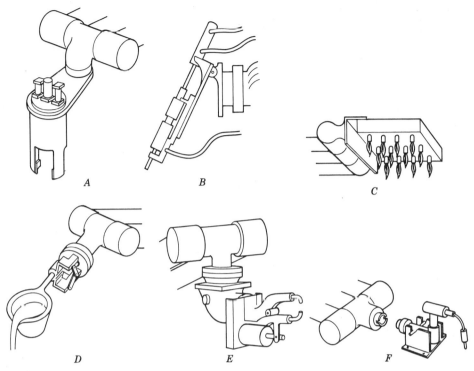

Figure 17.27
End effectors. A, Clamping, crimping, or nut running. B, Stud welding. C, Heating with flame torch. D, Pouring molten metal. E, Spot welding. F, Tool changing.

robots and in confirming assembly operations, inspection, simple fixturing requirements, and providing continuous production monitoring.

Vision Sensors

Robot *vision* uses a video camera, light source, and a computer programmed to process image data. The camera is mounted either on the robot or in a fixed position above the robot, so that its field of vision includes the robot's work volume. The computer software enables the vision system to sense the presence of an object and its position and orientations. Vision capability enables the robot to retrieve parts that are randomly oriented on a conveyor and to recognize particular parts that are intermixed with other objects. It will also perform visual inspection tasks and perform assembly operations that require alignment. Computerized vision systems will be an important technology in future automated factories.

A simple sensor may be a *photoelectric switch*. Using light rays, it determines if a characteristic of an object is present or absent. Door opening is a popular application. Another more complicated application will locate a hole and allow a fastening head to position on the hole.

Machine vision is a noncontact optical technique used for a variety of automatic tasks of stationary or moving objects in two or three dimensions. It uses a high-resolution video camera, laser or incandescent light source, microcomputers, and image-processing hardware and software to determine defined product features. The process of machine vision requires that as an image is received from the part or product, it must be converted to a useable form for processing by the computer. It is compared to a nominal part or standard set of information, and for inspection work a decision is made about the part regarding its acceptability.

A camera without photographic film is required. The camera uses electronic circuitry to determine the image. The circuitry is composed of an array of light-sensitive devices referred to as *pixels* or picture elements. Each pixel produces a signal proportional to the light striking it during an image capture cycle. Cameras are either "line" or "area." Pixels are arranged in a one-dimensional array for a line camera, but the area camera is two-dimensional (see Figure 17.28A).

Both cameras may be equipped with different lensing arrangements to provide for a field of view and spatial relationships; this is especially true for inspection tasks (see Figure 17.29B). Resolution for line cameras varies from 64 to 4096 pixels. Area cameras vary from 64 by 64 to 240 by 320 pixels.

Lighting for vision is important. Either lasers, fluorescent or incandescent lighting, or a combination of these are arranged to create spots, shadows, narrow lines, crosshairs, parallel lines, off-angle projections, or front or back illuminations to light the product. Figure 17.28C shows the arrangements for lighting. These arrangements cause high contrast to make the information of interest appear as a dark or light image as compared to the rest of the information in the camera's field of vision.

In the visualization of parts (e.g., stampings), there are four types of systems: range, contour, surface, and feature. The range system measures the distance that a part surface

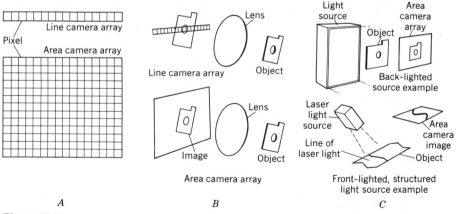

Figure 17.28
A, Electronic circuitry of vision is composed of pixels of light-sensitive devices.
B, Field of view and lensing. C, Light sources create spots, shadows, and contrast light and dark regions.

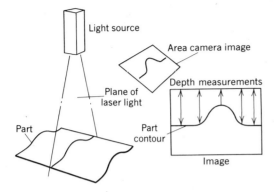

Figure 17.29
Coordinates for hundreds of two-dimensional points on the pixel array that are representative of the surface contour.

is from the face of the camera or datum surface. This measurement is to an accuracy of 0.1 mm. The method is a triangulation in which a spot of low-power laser is projected at the surface at a known angle. If the spot's image strikes the camera single-dimension array to the left or right of the normal position, the part is closer or further from the camera. The range sensor method is used to measure surface position, cavity depths, and hole depths, or to detect part features. An area camera used triangulation methods to define contour (e.g., Figure 17.29).

Tactile Sensors

Tactile sensors provide the robot with the capability to respond to contact forces between itself and other objects within its work volume. Figure 17.30 is a robot gripper with tactile sensitivity for assembly work and movement of parts.

The human touch feels various sensations such as shape, pressure, temperature, and texture. Immediately it is apparent if the object is hard, round, or pointed. The material is often identifiable by touch. In CAM, touch perception is more important. Robot

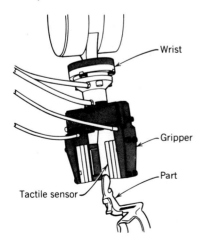

Figure 17.30
Robot gripper with tactile sensitivity.

operations for painting, welding, drilling, and basic part handling usually do not require touch perception. Future automatic manipulators will require tactile sensing. For example, Figure 17.31 shows a tactile sensor picking up the outline of a wrench end.

Diverse and complicated assembly will require tactile sensing, not simply touch, which means more than knowing that touch has occurred. A comprehensive ability to grope and identify shape, surface features, texture, force, and slippage are important future requirements. Tactile sensing or *taction* means continuous variable measurement of forces in a large surface area. *Touch* usually means simple contact or force at a single point. Diverse approaches are being considered; among them are the following.

1. *Strain gages*. These transducers, previously discussed in Chapter 11, do not have creep problems, fatigue, or hysteresis problems that other methods such as elastomers have. Strain gages are a proven technology. When the end-of-arm tool is grabbing a part, if the sensor is too soft there may be too much deflection. For the most part, strain gages are used as force monitors rather than tactile sensors. Some gages measure forces to 0.1% accuracy and distinguish between locations that are 0.001 in. (0.03 mm) from each other. Strain gage transducers monitor the reactions at the supports to a force acting anywhere on the end-of-arm platform. An electrical circuit such as a wheatstone bridge arrangement is necessary with strain gages.

2. *Silicon sensors*. Silicon chip microstructure has been applied to tactile sensors. The silicon sensors are three-dimensional mechanical devices etched in silicon using specialized integrated processing techniques. The silicon sensors may have nine elements mounted in a closely packed 3 by 3 array where each element is 2 by 2 mm. Each individual sensor is a three-dimensional etching of silicon oxide. The sensor measures only perpendicular forces, as it does not measure slip or shear forces. The sensor is shaped like a box, and the forces transmitted are transduced at the silicon diaphragm. A protective layer of hard plastic 250 μm thick covers the chip, and an elastomeric cover rests on that. Notice the elements of Figure 17.32. An electronic computer and special software are used to interpret the signals.

3. *Optical fiber-based sensor*. Fiber optic technology is displayed as a 16 by 16 array roughly 38 mm square. There are 256 individual sensing points with approximately 2.5-

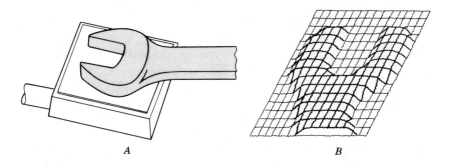

Figure 17.31
A, Tactile sensor. B, Outline displayed by sensor of wrench end.

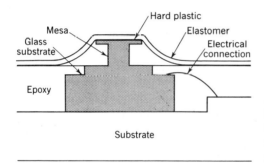

Figure 17.32
Single-element tactile sensor made by micromachining technology allowing mesa, diaphragm, and connections to be made from a silicon wafer.

mm spacing. Normal forces at the surface result in deflection of the light being coupled into the 256 receiving fibers yielding a corresponding reduction in the electrical output of the light detectors. The contact surface can be a flexible material. Using the signal and special software packages and processing the information, the shape, location, size, or weight can be found in real time. Processing software is necessary along with the sensors.

4. **Conductive elastomer.** This material measures the change of resistance based on the surface area contact. The outer layer is a conductive surface that seals the sensor from the environment. Beneath this surface are conductive V-shaped rows that change resistance when depressed. These rows have *piezoelectric properties*. A disadvantage to piezoelectric materials is that response to pressure is transient.

Voice Sensors

Voice programming can be defined as the oral communication of commands to the robot or other machine. Voice and sound recognition permit the system to recognize the symptoms of equipment malfunction such as air leaks, abnormal vibrations, or broken drill bits. Talking robots will warn factory personnel of machine breakdowns, rejected parts, and even low inventory levels.

Multiarmed Assembly Robot

Robots will continue to develop in the decades ahead. One future robot, nicknamed the "troikabot," can be expected to perform a variety of complicated assembly tasks at a single work station. Figure 17.33 is a CAM-generated solid model showing the work cell. This model has not been completely realized, but research and development are continuing. For example, the plans provide for three interacting six-degree-of-freedom arms to perform three-dimensional assembly tasks without fixtures. These assembly tasks would include parts mating, insertion, and use of gravity and friction forces. The robots use vision, electrical grippers, and a sensor mitt that is acquired and used by the arms.

The goal of the troikabot is a practical work cell extending robots to assembly processes. The robot recognizes that the *"factory of the future"* will be closely integrated such that

Figure 17.33
CAM-generated solid model depicting the troikabot assembly work cell.

order taking, CAD, CAM, inventory control, assembly, inspection, packaging, and delivery will be automatic. Second, the factory of the future will have the flexibility to manufacture a variety of products and be capable of changing products from time to time without major changes in equipment or arrangement. Finally, the factory will have the ability to plan and perform complex assembly tasks. The troikabot is a research and development effort to anticipate a comprehensive work cell composed of robots.

The *troikabot* is a highly integrated work cell that has three interactive robot manipulators, vision, multiple interactive sensors, and advanced electromechanical components such as sensor wrists, electronic grippers, and end-effector mitts. The grippers acquire these mitts and thereby extend their ability. The troikabot is programmed off-line and uses a hierarchical *real-time control* system. Three arms were chosen to permit free-space assembly with simple grippers. One robot holds the subassembly of the components while a second robot acquires a second component and inserts it into the subassembly. A third robot is required to clamp or regrip the subassembly so that the other robot can release its grippers without the danger of the subassembly falling apart.

ADAPTIVE CONTROL

Adaptive control is described as automatic on-line adjustment of the machine, process, or assembly operating variables to meet the objective of cost minimization or maximizing production output. Adaptive control will sense operating variables such as cutting forces, torque, temperature, tool wear rate, and surface finish. It responds by adjusting the revolutions per minute, feed, or depth of cut. The adaptive control may optimize an economic function subject to process constraints. In machining, the variables may be optimized within a prescribed region bounded by the manufacturing limits such as maximum torque or horsepower of the machine. An adaptation strategy varies the operating variables in real time as the machining progresses.

Part programmers predetermine speeds or feed rates, and often conservative values are selected for the operating variables that consequently slow down the system's production. Adaptive control promises to improve this operation.

A dedicated computer is needed for adaptive control. The adaptive control is basically a feedback system in which the operating variables automatically adjust themselves to the actual conditions of the process. Adaptive control is done in *real time*, that is, as the event is occurring. The computer converts this information into commands, which modify the speed or feed, for example, to optimize the cutting operation. Optimization may involve maximizing material removal rate between tool changes. If the forces increase excessively during machining because of a wearing tool, adaptive control reduces the speed or feed or both, to lower the forces to acceptable levels. High cutting forces may cause tool chipping or friction, excessive tool point bending, and loss of dimensional control and tolerances. A washing machine may be said to be adaptive if the water temperature cools down and the control system activates a heating element to reheat the water. On the contrary, a clothes washer is nonadaptive if after the out-of-balance signal, the washer will not redistribute the weight of the clothes in the spinning tank.

If the depth of cut or width of cut increases, a corresponding increase in force, torque, and power consumption is found. The system senses this increase and automatically reduces the feed to avoid tool breakage or to maintain cutting efficiency.

Adaptive control is possible if the *quantitative process* model between the dependent and independent variables are known and stored in computer memory. If the temperature in a machining operation is excessive, the computer has instructions to change the speed and increase the coolant flow to reduce the temperature. If the operation is to bend a metal sheet in a V-die, the data must be stored in the computer as to how springback varies with punch travel, material type, grade, and hardness.

Forces, torque, and power are effective as measured variables for *on-line monitoring* of surface finish and tool wear rate. Residual stresses developed in a machined surface are difficult to measure. Adaptive control has been applied in machining, grinding, sheet forming, assembly, and welding operations. Adaptive control can be linked with computer numerical control (CNC).

GROUP TECHNOLOGY

Group technology (GT) is a management philosophy based on the recognition that similarities exist in design and manufacture of discrete parts. In "family of parts manufacturing," GT achieves advantages on the basis of these similarities. Similar parts are arranged into part families. For example, a plant producing many part numbers of shafts, chassis, and castings will group the parts into families of their physical features, such as shafts. Each family possesses similar design and manufacturing characteristics. Efficiencies result from reduced setup times, lower in-process inventories, better scheduling, streamlined material flow, improved quality, improved tool control, and the use of standardized process plans. In some plants where GT has been implemented, the production equipment is arranged into machine groups or cells to facilitate work flow and parts handling. In product design there are also advantages obtained by grouping parts into families. A parts classification and coding system is required in a design retrieval system. GT is a prerequisite for computer-integrated manufacturing.

Parts classification and coding are concerned with identifying the similarities among parts and relating these similarities to a coding or a number system. Part similarities are of two types: *design attributes* (such as geometric shape and size) and *manufacturing attributes* (the sequence of processing steps required to make the part). It should be noted that GT is not a science with precise formulas, but rather is a tool to be developed in each plant.

Coding can be used in computer-aided process planning (CAPP). CAPP involves the computer-generation of an operations sheet or route sheet to manufacture the part. Discussion of the operations sheet is deferred to Chapter 25.

Part Families

A moderate-size plant may have many thousands of drawing numbers. The drawing number does not allow easy identification of similar parts unless that is called out on the drawing. The advantage of the part numbers, which are collected on a bill of materials for a product, is for the product listing. A *part family* is a collection of parts that are similar either because of geometric shape and size or because similar processing steps are required in their manufacture. The parts within a family are different, but are close enough to merit identification as members of the part family.

Figure 17.34 shows a collection of similar shafts. The parts, used as drill ends for mining drills, have many similar details, materials, and dimensions. Grouping these parts sequentially for manufacturing reduces setup, cycle time, and inventory, and improves shop efficiency.

The obstacle in changing over to GT from a traditional production shop is the problem of grouping parts into families. Methods to solve this problem are visual inspection, production flow analysis, and parts classification and coding systems.

Grouping parts into families involves an examination of the individual design and manufacturing characteristics of each part. The attributes of the part are uniquely identified by means of an alphanumeric code number. This classification and coding may be carried out on the entire list of active parts of the firm, or a sampling process may be used to establish the part families.

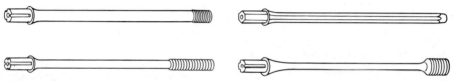

Figure 17.34
Sample of grouped parts.

QUESTIONS AND PROBLEMS

1. Give an explanation of the following terms:

   Job lot industries
   Manual assembly
   Work station
   Process efficiency
   CAM
   Partial automation
   Cell
   S-shaped layout
   Pallet transfer
   Mechanization
   Fixtures
   Automation
   Modular automation units
   Bowl feeders
   Pick-and-place robots
   Wire guidance
   Synchronous transfer
   Manipulator
   Variable-sequence robot
   Nonservo robot
   Walk-through programming
   Pixels
   Conductive elastomer
   Factory of the future
   Troikabot
   Real-time control
   Part families

2. You are required to perform an assembly process for less than 500 units. Would you choose a magazine loading system or a robot arm loading system?

3. In the general scheme of a computerized visual feeding system, what steps follow separation and holding of a single product component?

4. Provide an economic classification for discrete part production.

5. What is the essential purpose of the manufacturing system?

6. List the manufacturing system you might adopt for the production of a consumer item where production is: (a) 2 million units; (b) 200,000 units; (c) 20,000 units; (d) 2000 units; or (e) 200 units.

7. What separates moderate production from mass production?

8. Are industrial products normally job lot or moderate production?

9. What are the advantages and disadvantages of individual work stations?

10. If you were designing automation equipment, would you stress primary or secondary operations? Can automation be conducted without secondary operations?

11. State the classification of cost for the following items: machine operator, raw material ending up in the consumer product, robot, technician, machine operator, and engineer.

12. An automation cycle consists of seven operations using 17, 11, 14, 18, 21, 16, and 23 s. Indexing time between the stations is 14 s. If there are no means of providing for inventory between the stations, what is the process efficiency? Which operation is limiting in this in-line sequence?

13. An automation system is suggested as a replacement for a manual system. The first cost for the automatic equipment is $5 million, and annual net saving is anticipated to be $725,000. What is the payback?

14. The total budget for the automation system is $5 million. What amounts can be expected for robot equipment? For engineering and peripheral equipment?

15. What products are customarily produced with continuous-flow production processes?

16. Describe various layouts for cell production.

17. List various ways of moving the following materials: thousands of electronic wafers, several printed circuit boards, cast-iron automobile engines, turbine shroud, fractional horsepower motors, and light bulbs.

18. How does mechanization differ from automation?

19. List the steps for mechanizing the production of a wire paper clip with the following specifications: three U-bends, 0.037-in. wire, prebent length 3.675 in. and ductile cold-rolled AISI 1008 steel.

20. Material-handling equipment is expected to cost $80,000 installed. Annual net saving is $18,000. What is the payback?

21. Develop a list of practices for preventive maintenance on automation equipment producing razors.

22. Automation equipment produces bulletlike containers, and the four steps require 14 s each. There are five handling steps each 8 s in duration. What is the process efficiency?

23. Describe various methods to orient parts.

24. What are the possible ways for parts to be disoriented?

25. Sketch an optical means of assuring the orientation of a toothbrush before it is packaged.

26. What is the purpose of image-processing algorithms in vision orientation systems?

27. Describe the types of vision systems.

28. For vision application, it is important to have bright surfaces, contrast, ambient light, or no outside light?

29. What are the difficulties in assembly of products?

30. Define a robot and list several industrial applications.

31. What are a manipulator, wrist, controller, end effector, and control system for a robot?

32. Contrast the coordinate systems used for robots.

33. List the typical operating sequence for a nonservo and a servo-type robot.

34. What function does an error signal have in control of robots?

35. Write a paragraph description for handling parts such as printed circuit boards, automobile wheel hubs, Christmas tree ornaments, fruit, survey stakes, and clothing. Consider tactile sensors and their role in the handling. Assume a large volume of product to be handled.

36. Write a newspaper article of 250 words describing the troikabot. Indicate its functions, purpose, and role in production.

37. Describe various methods of assigning group technology code numbers for parts that are washerlike, except they differ in thicknesses and both inside and outside diameters. Additionally, compensate for material. Use a four-digit number, then a five-digit number.

CASE STUDY

THE ROUND PLATE COMPANY

The Round Plate Company specializes in round plates. This company has standardized its product into two families, shown in Figure 17.35 A and B. These plates are used for many applications such as bearings plates, post rests, flanges, bases, and any engineered design that starts with round bar stock in 12- and 10-in. OD and ends with these general shapes. Operations are limited to turning, drilling, tapping, and reaming, step milling and grinding.

Mr. Sander Friedman, owner of the shop, tells any potential customer, "I can win any bid starting with standard bar stock in two sizes and concluding with no more features than shown on the sketches." He has been talking to you about investing in his company.

A registered and legal financial prospectus about Round Plate suggests that the capitalization using shareholder's investment will result in a modern factory of the future designed for vertical market penetration in the round plate business. The prospectus gives various details

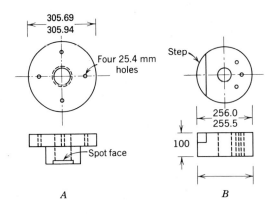

Figure 17.35
Maximum dimensions and general part family configuration for Round Plate Company's product.

about the product and factory. A summary is given here.

A sketch is given of the product, shown by Figure 17.35. Dimensions on the sketches are maximum, and other dimensions may be SI and are scaled to approximate missing dimensions. These dimensions are useful as limitations for standard pallets, fixtures, grippers, and robots required in manipulation and loading of machines. Bolt circles are important to the company's operation.

Round Plate's operation is a garage-style cell with CNC machines, robotic operation for loading, and night time supervision and inspection. According to the prospectus, Round Plate will employ the president, maintenance operator, toolroom grinder, computer programmer, manufacturing engineer, robot maintainer and teacher, accountant and bill collector, and finally, a materials handler for shipping, raw stock receiving, and chip and waste removal.

"We will operate the night shift without anybody," boasts Friedman. "Even the lights are out," he says. "This is a good time to buy shares." He looks at you expectantly.

A pause and then you reply. "Well, everything sounds pretty good, and I do like your idea. But I don't buy into businesses without first sleeping on it. I'll see you tomorrow and we will go over some details."

That night you begin to ponder the offer. You mutter, "What are the critical requirements?" Continue your questioning attitude and develop a series of questions for Mr. Friedman. Develop your comments along these general groupings: (1) systems engineering; (2) equipment and ancillary support in tooling, material movement, controls, and vision; (3) part family and group technology; and (4) marketing strategy and business factors.

Prepare these questions in a report.

CHAPTER 18
METAL CUTTING

In manufacturing products it is important that the processes be efficient and capable of producing parts of acceptable quality. After metals have been refined they are changed by a primary process to shapes and sizes suitable for fabrication. Final products are often obtained by machining shapes such as bar stock or plate to size. It is important that metal cutting principles be well understood to have economical application. Metal cutting principles are used in turning, planing, milling, and drilling operations as well as other processes performed by machine tools. Parts are produced by removing metal in the form of small chips. The cutting tool that removes these chips is the focus of many important principles.

METAL CUTTING THEORY

The simplest form of cutting tool is the single-point tool such as that found in a lathe cutoff operation or planer or shaper work. Multiple-point cutting tools are made up of two or more single-point tools arranged together as a unit. The milling cutter and broaching tool are examples of multiple-point cutters. But the discussion in this section deals mostly with *orthogonal tools* in which the cutting edge is perpendicular to the direction of the cut and there is no lateral flow of metal. Nor is there curvature in these idealized forms, and all parts of the chip have the same velocity. This form of cutting is illustrated in Figure 18.1.

We limit discussion to orthogonal or two-dimensional cutting rather than the more common oblique or three-dimensional cutting. A sketch of oblique cutting is given later.

In analyzing the cutting process it is assumed that the chip is severed from the workpiece by a shearing action across the plane as shown in Figure 18.1, although other theories exist on chip formation. Because the deformed chip is in compression against the face of the tool, a frictional force is developed. The work of making the chip must overcome both the shearing force and the frictional force.

In the *orthogonal cutting model* the tool can be considered stationary and the workpiece moving. An opposite motion pattern does not change the concept. The state of stress before and after the shear plane is a complicated plastic flow of metal. The shear plane ϕ is determined by the *rake angle* α of the tool and by the *friction* between the chip and the tool face.

The forces acting on the chip are shown in Figure 18.2 and are listed as follows:

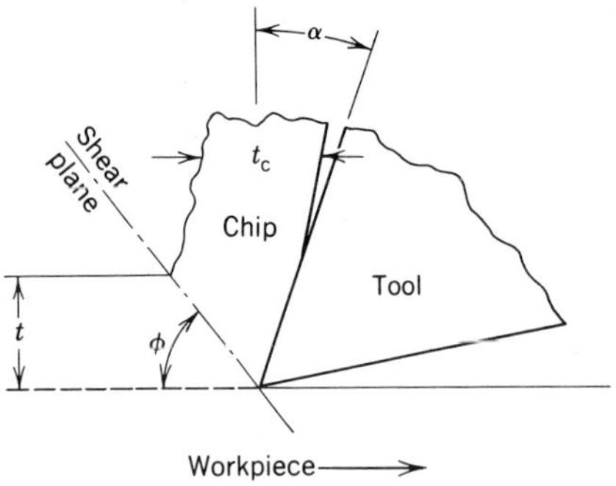

450   METAL CUTTING

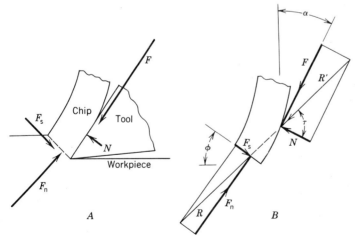

Figure 18.1
Orthogonal cutting tool model.

F_s = Resistance to *shear* of metal in forming the chip. This force acts along the line

F_n = Force *normal* to the shear plane; the resistance offered by the workpiece

N = Force acting on the chip *normal* to the cutting face of the tool; provided by the tool

F = Frictional resistance of the tool acting on the chip; acts against the chip as it moves along the face of the tool

Figure 18.2B is a *free-body diagram* showing the forces acting on the chip. Forces F_s and F_n are replaced by their resultant R and forces F and N by their resultant R'. This reduces the two combined forces to act as one on the chip. Although there are external couples on the chip that in actuality tend to curl it, they are neglected. Using a principle of mechanics with equilibrium existing when a body is acted on by two forces, they are equal in magnitude, opposite in direction, and, in the absence of a couple, colinear. The components of the force of the workpiece on the chip are the shear force F_s and the

Figure 18.2
A, Forces on chip. B, Two force triangles on free-body diagram.

normal compressive force F_n. Conversely, the forces of the chip on the workpiece are $-F_s$ and $-F_n$.

The shearing force and the angle of the shear plane are affected by the *frictional force* of the chip against the tool face. The frictional force in turn depends on a number of factors including the smoothness and keenness of the tool, whether or not a coolant is used, materials in the tool and workpiece, cutting speed, and the shape of the tool. A large frictional force results in a thick chip having a low shear angle, whereas the reverse is true if the frictional force is low. The efficiency with which metal is removed is higher when the friction force can be minimized.

The two force triangles of Figure 18.2B can be superimposed by placing the two equal forces R and R' together. The angle between F_s and F_n is a right angle, and together with F and N a circle may be shown in Figure 18.3. From this convention two more forces are drawn: the horizontal *cutting* force (F_c) and the **vertical** or *tangential* force (F_t). These two forces can be found during machining with a force **dynamometer**. F_c is the horizontal cutting force of the tool on the workpiece, and F_t is the force in the vertical direction to hold the tool against the work. The rake angle α can be determined by measurements of the tool in its holder. Even ϕ, the *shear angle*, can be approximated from photomicrographs. Another method results from measurement of the thickness of the chip (t_c) and the depth of cut or by measurement of the length of chip.

Various quantities can be determined using the force diagram. For instance, the **coefficient of friction** μ can be found in terms of the forces or

$$F = F_t \cos \alpha + F_c \sin \alpha$$

and

$$N = F_c \cos \alpha - F_t \sin \alpha$$

Therefore

$$\mu = \tan \tau = \frac{F}{N}$$

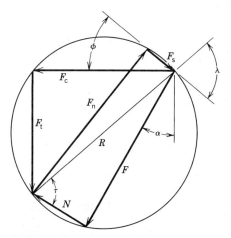

Figure 18.3
Relationship between forces in orthogonal metal cutting.

or

$$\mu = \frac{F_t + F_c \tan \alpha}{F_c - F_t \sin \alpha}$$

and

$$\tau = \text{Friction angle}$$

Customarily the chip is thicker than the depth of cut or (as in Figure 18.1), $t_c > t$. The ratio of the depth of cut t and the chip thickness t_c is called the *chip thickness ratio* r, and may be expressed as

$$r = \frac{t}{t_c} = \frac{\sin \phi}{\cos(\phi - \alpha)}$$

and solving for $\tan \phi$,

$$\tan \phi = \frac{r \cos \alpha}{1 - r \sin \alpha}$$

The ratio r can be found by measurements of the depth of cut and chip thickness by a micrometer. The back of the chip is rough, making an accurate value for the chip thickness unlikely. An alternate way is by using lengths rather than thicknesses. This assumes that the density does not change during cutting, and the volume of the chip is equal to the volume of the metal removed. We have already assumed that there is no lateral flow of the chip during cutting; this is especially true in orthogonal cutting. This correspondence gives equal volume before and after cutting or

$$tL = t_c L_c$$

where

L_c = length of chip
L = Corresponding length of material removed from workpiece

and

$$r = \frac{t}{t_c} = \frac{L_c}{L}$$

The velocity of the chip as it moves along the face of the tool is less than the cutting velocity. This holds because the chip is thicker than the depth of cut. Velocities can be found and the velocity relationships are shown in Figure 18.4.

In this figure the velocity of the workpiece is given as V_c. The chip moves along the cutting face of the tool with a velocity relative to the tool equal to V_f. The relative velocity of two bodies, the chip and the tool, is equal to the *vector difference* between their velocities relative to the reference body or

$$V_c = V_s - V_f$$

The velocity of the chip sliding along the cutting face of the tool is therefore

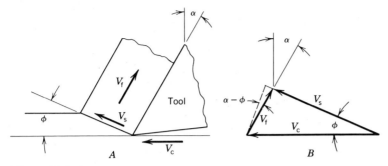

Figure 18.4
A, Tool and chip velocity relationship. B, Velocity polygon.

$$V_f = \frac{V_c \sin \phi}{\cos(\phi - \alpha)}$$

or

$$V_f = rV_c$$

From Figure 18.4 the velocity of the chip sliding along the shear plane is

$$V_s = \frac{V_c \cos \alpha}{\cos(\phi - \alpha)}$$

Consider the following example of the equations. A shaper is machining a steel block where the length of cut is 3 in. The length of the chip is measured after annealing and straightening as $L_c = 2.4$ in. The tool is known to have a rake angle $\alpha = 40°$. Other known conditions are $V_c = 35$ ft/min; depth of cut $t = 0.008$ in.; horizontal cutting force $F_c = 275$ lb; and vertical force $F_t = 95$ lb.

The coefficient of friction is found as

$$\mu = \frac{F_t - F_c \tan \alpha}{F_c - F_t \tan \alpha} = \frac{95 + 275 \times 0.839}{275 - 95 \times 0.839} = 1.7$$

The length of cut is the block length, or 3 in., and the chip ratio r is

$$r = \frac{L_c}{L} = \frac{2.4}{3} = 0.8$$

The thickness of the chip now becomes

$$t_c = \frac{t}{r} = \frac{0.008}{0.8} = 0.010 \text{ in. } (0.25 \text{ mm})$$

The shear plane angle ϕ is found as

$$\tan \phi = \frac{r \cos \alpha}{1 - r \sin \alpha} = \frac{0.8 \times 0.766}{1 - 0.8 \times 0.643} = 1.26$$

or
$$\phi = 52°$$

The velocity of the chip along the tool face is found as
$$V_f = \frac{V_c \sin \phi}{\cos(\phi - \alpha)} = \frac{35 \times 0.642}{0.978} = 23 \text{ ft/min } (0.12 \text{ m/s})$$

The velocity of shear along the shear plane is
$$V_s = \frac{V_c \cos \alpha}{\cos(\phi - \alpha)} = \frac{35 \times 0.766}{0.978} = 27.4 \text{ ft/min } (0.14 \text{ m/s})$$

Measurement of forces acting on the tool can be made with a *dynamometer*. Electronic load dynamometers are frequently used to measure these forces. As it is impossible to measure cutting forces at the point of the tool and workpiece, the reactions are measured away from the cutting point. Transducers and a platform are combined to measure one, two, or three forces and torques. A tool or workpiece is mounted on a platform. Figure 18.5 shows the workpiece mounted on a platform where transducers measure drilling feed, force, and torque. It is more convenient to mount the stationary workpiece on a platform than to attempt to measure reactions on the rotating drill.

Transducers measure deformations using a change of inductance, capacitance, or resistance (strain gages), as was discussed in Chapter 11. The transducers used in the load

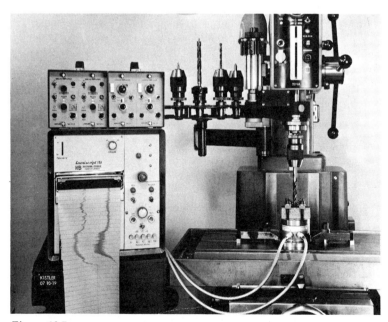

Figure 18.5
Two-channel dynamometer measuring drill thrust and moment. Dynamometer transducers mounted in platform.

cell of Figure 18.6 are *piezoelectric*. A piezoelectric force measuring principle differs in that as a force acts on a *quartz element*, a *proportional electric charge* appears on the loaded surface as shown by Figure 18.6. The piezoelectric properties of quartz are such that the crystals are sensitive either to pressure or shear in one axis. In this way components of cutting force or torque are measured independently.

Typical forces acting on an oblique cutting tool that might be measured by a dynamometer are indicated by Figure 18.7; they are cutting, thrust, and radial forces. Figure 18.8 indicates the approximate distribution of the forces. In most oblique tool cutting operations the cutting force is the most significant.

The forces on a cutting tool for a given material depend on a number of considerations.

1. Tool forces are not changed significantly by a change in cutting speed.
2. The greater the feed of the tool, the larger the forces.
3. The greater the depth of cut, the larger the forces.
4. Cutting force increases with chip size.
5. Thrust force is decreased if the **cutting tool nose radius** is made larger or if the side cutting edge angle is increased.
6. Cutting force is reduced as, the back rake angle is increased about 1% per degree.
7. Using a coolant reduces the forces on a tool slightly but greatly increases tool life.

The power required by machining can be found by different methods. A practical method is the use of a wattmeter or ammeter at the drive motor. The wattmeter is considered more accurate when the motor is fully loaded. The horsepower at the cutter can be found by subtracting the idle horsepower from cutting horsepower.

Horsepower can be calculated from measurement of the forces by a dynamometer and using F_c. This gives the horsepower at the spindle.

$$HP_s = \frac{F_c \times V_c}{33,000}$$

where

F_c = Cutting force, lb (N)

V_c = Cutting speed, ft/min (m/s)

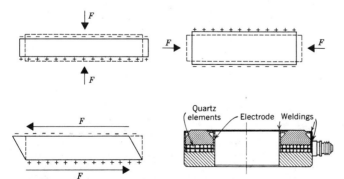

Figure 18.6 Diagram showing longitudinal, transverse, and shear effect on quartz element and construction of dynamometer load transducer.

456   METAL CUTTING

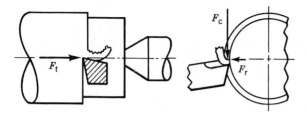

Figure 18.7
Forces acting on end of lathe tool: longitudinal thrust, F_t; tangential cutting, F_c; and radial, F_r.

Sometimes this formula is seen with two additional terms that provide for the two additional force vectors. Their contribution to the horsepower requirements are slight because the force times distance for these two terms is minor.

Let

$$HP_m = \text{Horsepower at motor}$$
$$= \frac{HP_s}{E}$$

where

$$E = \text{Efficiency of spindle drive, \%}$$

Suppose that information indicates that $F_c = 275$ lb, $V_c = 35$ ft/min, and $E = 65\%$. Then $HP_s = 0.29$ at the spindle and $HP_m = 0.45$ at the motor. The *metal removal rate* can be determined by the following:

$$Q = 12 \times t \times f_r \times V_c$$

where

$$Q = \text{Metal removal rate, in.}^3/\text{min (mm}^3/\text{min)}$$
$$t = \text{Depth of cut, in. (mm)}$$
$$f_r = \text{Feed, in. per revolution (ipr) (mm/r)}$$
$$V_c = \text{Cutting speed, ft/min (m/s)}$$

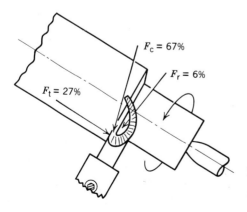

Figure 18.8
Distribution of forces on a single-point cutting tool.

Sometimes it is important to understand that *unit power* equations can be calculated, given the information.

$$P = \frac{HP_s}{Q}$$

where

P = Unit power, hp/in.^{3}/min (W/mm^3/s)

METAL CUTTING TOOLS

Geometry

To understand the cutting action of a single-point tool as applied to a lathe, refer to Figure 18.9. The tool has been ground to a wedge shape, the included angle being called the *lip* or *cutting angle*. The **side relief angle** between the side of the tool and the work is to prevent the tool from rubbing. The angle is small, usually 6° to 8° for most materials. The *side rake angle* varies with the lip angle, which in turn depends on the type of material being machined. If the cutting tool is supported in a horizontal position, the *back rake angle* is obtained by grinding. However, most toolholders are designed to hold the tool in an approximate position for the correct back rake. *End clearance* is necessary to prevent a rubbing action on the flank of the tool. The angles shown in Figure 18.9 are for a cutting tool mounted horizontally and normal to the workpiece. The effective angles can be changed by adjustment to the toolholder without changing the angles ground on the tool.

In grinding tools it should be noted that the lip or cutting angle varies with the kind of material being cut. The cutting angle must be keen enough to cut well with a minimum of power consumption, yet the edge must be sufficiently strong to withstand the tool forces involved and to carry away the heat generated. A compromise is necessary. In

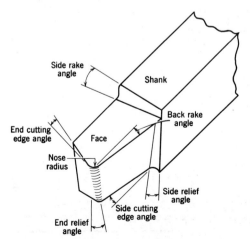

Figure 18.9
Nomenclature for a right-hand cutting tool.

general it is based on the hardness of the workpiece. Hard materials require a cutting edge of great strength with a capacity for carrying away heat. Soft materials permit the use of smaller cutting angles, around 22° for wood tools. Soft and ductile metals such as copper and aluminum require larger angles ranging up to 47°, whereas brittle materials, having chips that crumble or break easily, require still larger angles. An interesting variation in tool angles is that recommended for brass and duralumin. These materials work best with practically zero rakes, the cutting action being a scraping one. Because of the high ductility, the tool will dig in and tear the metal if a small cutting angle is used. Research has indicated approximate tool angles and cutting speeds for a number of materials. Table 18.1 shows recommended values for high-speed steel cutting tools.

In addition to the solid single-point tool, a carbide tip may be brazed on or inserted in a toolholder. Many styles of toolholders and inserts are available. The inserts are disposable and vary in geometry from triangular, square, circular, and diamond to other special shapers. Figure 18.10 is an exploded view of the components of a single-point tool using disposable carbide tool inserts. The insert in this figure is triangular. More heat is generated with carbide tooling, so an adequate coolant supply must be provided.

Tool Material

Present-day production practices make severe demands on machine tools. To accommodate the many conditions imposed on them, a wide variety of tool materials has been developed. The best material to use for a certain job is the one that will produce the machined part at the lowest cost. Desirable properties for any tool material include the ability to resist softening at high temperature, a low coefficient of friction, good abrasive resisting qualities, and sufficient toughness to resist fracture.

It is not uncommmon to see several types of tool materials operating on a given workpiece at one time. An example is in turning two different diameters wherein the

Table 18.1 **Tool Angles and Cutting Speeds for High-Speed Tools**

Material being cut	Side relief angle,°	Side rake angle,°	Back rake angle,°	End-clearance angle,°	Cutting speed	
					ft/min	m/s
Mild steel 1020	12	14	16	8	100	0.5
Medium-carbon steel 1035	10	14	16	8	70	0.4
High-carbon steel 1090	10	12	8	8	50	0.3
Screw stock, 1112	12	22	16	8	150	0.8
Cast iron	10	12	5	8	50	0.3
Aluminum	12	15	35	8	450	2.3
Brass	10	0	0	8	250	1.3
Monel metal	15	14	8	12	120	0.6
Plastics	12	0	0	8	120	0.6
Fiber	15	0	0	12	80	0.4

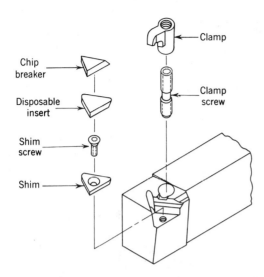

Figure 18.10
Exploded view of a tool with disposable, triangular, carbide cutting point insert.

cutting speed of the tool on a small diameter may be greatly different from the one simultaneously cutting a larger diameter.

The principal materials used in cutting tools are discussed in the following paragraphs.

High-Carbon Steel. For many years before the development of high-speed tool steels, carbon steels were used for all cutting tools, their carbon content ranging from 0.80% to 1.20%. These steels have good hardening ability and, with proper heat treatment, attain as great a hardness as any of the high-speed alloys. At maximum hardness the steel is brittle and, if some toughness is desired, it must be obtained at the expense of hardness. Depth-hardening ability (hardenability) is poor, limiting the use of this steel to tools of small size. Because these tools lose hardness at around 600°F (300°C), they are not suitable for high cutting speeds and heavy-duty work, their usefulness being confined to work on soft materials such as wood.

High-Speed Steel. *High-speed steels* (HSS) are high in alloy content, have excellent hardenability, and will retain a good cutting edge to temperatures of around 1200°F (650°C). The ability of a tool to resist softening at high temperatures is known as *red hardness* and is a most desirable quality. The first tool steel that would hold its cutting edge to almost a red heat was developed by Frederick W. Taylor and M. White in 1900. This was accomplished by adding 18% tungsten and 5.5% chromium to steel as the principal alloying elements. Present-day practice in manufacturing high-speed steels still uses these two elements in nearly the same percentage. Other common alloying elements are vanadium, molybdenum, and cobalt. Although there are numerous HSS compositions, they may all be grouped in the following three classes:

1. *18-4-1 high-speed steel.* This steel, containing 18% tungsten, 4% chromium, and 1% vanadium, is considered to be one of the best all-purpose tool steels.

2. *Molybdenum high-speed steel.* Many high-speed steels use molybdenum as the principal alloying element, because one part will replace two parts of tungsten. Molybdenum steels such as 6-6-4-2 containing 6% tungsten, 6% molybdenum, 4% chromium, and 2% vanadium have excellent toughness and cutting ability.

3. *Super high-speed steels.* Some high-speed steels have cobalt added in amounts ranging from 2% to 15%, because this element increases the cutting efficiency, especially at high temperatures. One analysis of this steel contains 20% tungsten, 4% chromium, 2% vanadium, and 12% cobalt. Because of the greater cost of this material, it is used principally for heavy cutting operations that impose high pressures and temperatures on the tool.

Cast Nonferrous Alloy. A number of nonferrous alloys containing principally chromium, cobalt, and tungsten with smaller percentages of one or more carbide-forming elements like tantalum, molybdenum, or boron, are excellent materials for cutting tools. These alloys are cast to shape, have high red hardness, and are able to maintain good cutting edges on tools at temperatures up to 1700°F (925°C). Compared with high-speed steels, they can be used at twice the cutting speed and still maintain the same feed. However, they are more brittle, do not respond to heat treatment, and can be machined only by grinding. Intricate tools can be formed by casting into ceramic or metal molds and finishing to shape by grinding. Their properties are largely determined by the degree of chill given the material in casting. The range of elements in these alloys is 12% to 25% tungsten, 40% to 50% cobalt, and 15% to 35% chromium. In addition to one or more carbide-forming elements, carbon is added in amounts of 1% to 4%. These alloys have good resistance to cratering and can resist shock loads much better than carbides. As a tool material they rank midway between high-speed steels and carbides for cutting efficiency.

Carbide. *Carbide cutting tool* inserts are made only by the powder metallurgy technique. The metal powders of tungsten carbide and cobalt are pressed to shape, semisintered to facilitate handling and forming to final shape, sintered in a hydrogen atmosphere furnace at 2800°F (1550°C), and finished by a grinding operation. Carbide tools containing only tungsten carbide and cobalt (~94% tungsten carbide and 6% cobalt) are suitable for machining cast iron and most other materials except steel. Steel cannot be satisfactorily machined by this composition because the chips stick or weld to the carbide surface and soon ruin the tool. To eliminate this difficulty, titanium and tantalum carbides are added, in addition to increasing the cobalt percentage. A typical analysis of a carbide suitable for steel machining is 82% tungsten carbide, 10% titanium carbide, and 8% cobalt. This composition has a low coefficient of friction and as a result has little tendency toward top wear or cratering. Because variation in composition alters the properties of carbide materials, several grades are available to accommodate the work to be done.

The red hardness of carbide tool materials is superior to all others, because it will maintain a cutting edge at temperatures over 2200°F (1200°C). In addition, it is the hardest manufactured material and has extremely high compressive strength. However, it is very brittle, has low resistance to shock, and must be very rigidly supported to prevent cracking.

Grinding is difficult and can be done only with silicon carbide or diamond wheels. Clearance angles should be held to a minimum. Carbide tools permit cutting speeds two to three times that of cast-alloy tools, but at such speeds that a much smaller feed must be used. From an economic point of view, carbide tools should always be used if possible. Machines using carbide tools must be rigidly built, have ample power, and have a range of feeds and speeds suitable to the material.

Micrograin carbide is a high-strength, high-hardness, fine-grain size tungsten carbide that is used when cutting speeds are too low for regular carbides but conventional tools will not hold up to wear. They are used increasingly for form and cutoff tools.

Carbide tools may be coated with a 0.002- to 0.003-in. (0.05- to 0.08-mm) bonded layer of titanium carbide, aluminum oxide, or titanium nitride to a tungsten carbide base. These coatings reduce the heat caused by the chip flowing over the tool and the effects of diffusion and adhesion. Sometimes they are only applied when cratering is most likely to occur. Aluminum oxide–coated tools will operate at almost twice the cutting speed of any other coating. Coated tools are not recommended for workpieces having a heavy scale or sand inclusions.

Diamond. Diamonds used as single-point tools for light cuts and high speeds must be rigidly supported because of their high hardness and brittleness. They are used either for hard materials difficult to cut with other tool materials or for light, high-speed cuts on softer materials where accuracy and surface finish are important. Diamonds are commonly used in machining plastics, hard rubber, pressed carbon, and aluminum with cutting speeds from 1000 to 5000 ft/min (5–25 m/s). Diamonds are also used for dressing grinding wheels, for small, wire-drawing dies, and in certain grinding and lapping operations.

Ceramic. Aluminum oxide powder, along with additives of titanium, magnesium, or chromium oxide, is mixed with a binder and processed into a cutting tool insert by powder metallurgy techniques. The insert is either clamped onto the toolholder or else bonded to it with an epoxy resin. The resulting material has an extremely high compressive strength but is quite brittle. Because of this the inserts must be given a 5° to 7° negative rake to strengthen their cutting edge and be well supported by the toolholder.

Silicon nitride (code-named S-8) tools are used for cast-iron machining applications. These *ceramic tools* have a tool life that was effective over 1500 cast-iron pieces, where coated tungsten carbide tools lasted through only 250 parts before dulling.

CHIP SHAPE AND FORMATION

Research has been done in studying the mechanics and geometry of chip formation, and the relationship of chip shape to such factors as tool life and surface finish. Tool chips have been classified into three types as shown in Figure 18.11. Type 1 (Figure 18.11A), a *discontinuous or segmental chip*, represents a condition in which the metal ahead of the cutting tool is fractured into fairly small pieces. This type of chip is obtained in machining most brittle materials such as cast iron and bronze. As these chips are produced,

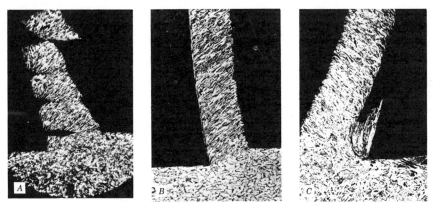

Figure 18.11
Basic chip types. *A*, Discontinuous. *B*, Continuous. *C*, Continuous with built-up edge.

the cutting edge smooths over the irregularities and a fairly good finish is obtained. Tool life is reasonably good, and failure usually occurs as a result of abrading action on the contact surface of the tool. Discontinuous chips can also be formed on some ductile materials if the coefficient of friction is high. However, such chips on ductile materials are an indication of poor cutting conditions.

An ideal type of chip from the standpoint of tool life and finish is the simple *continuous chip type 2* (Figure 18.11*B*), which is obtained in cutting ductile materials having a low coefficient of friction. In this case the metal is continuously deformed and slides up the face of the tool without being fractured. Chips of this type are obtained at high cutting speeds and are common when cutting is done with carbide tools. Because of their simplicity they can be analyzed easily from the standpoint of the forces involved. Continuous chips come off the bar stock as string from a ball and can be troublesome to handle and sometimes dangerous.

The type 3 chip (Figure 18.11*C*) is characteristic of those machined from ductile materials that have a fairly high coefficient of friction. As the tool starts the cut, some of the material, because of the high friction coefficient, builds up ahead of the cutting edge. Some of the workpiece may even weld onto the tool point, and is thus known as a *built-up edge or **BUE***. As the cutting proceeds, the chips flow over this edge and up along the face of the tool. Periodically a small amount of this BUE separates and leaves with the chip or is embedded in the turned surface. Because of this action, the surface smoothness is not so good as with the type 2 chip. The BUE remains fairly constant during cutting and has the effect of slightly altering the rake angle. However, as the cutting speed is increased, the size of the BUE decreases and the surface finish is improved. This phenomenon is also decreased by either reducing the chip thickness or increasing the rake angle, but on many of the ductile materials it cannot be eliminated entirely.

An analog of the stress patterns involved in metal cutting can be seen in Figure 18.12.

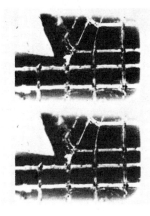

Figure 18.12
High-speed photography of beeswax being machined.

The photograph, which shows two motion picture frames of beeswax being cut at 50 ft/min (0.25 m/s), was taken at 7200 frames per second.[1] The lines were 0.050 in. (1.27 mm) apart. As the chip is formed the vertical lines are compressed together but remain parallel, thus indicating the action of the compressive forces causing the wax to expand in the direction perpendicular to the cutting direction. A shear angle of approximately 45° can be visualized.

Some investigators report that 97% of the work that goes into cutting is dissipated in the form of heat. Figure 18.13 shows the three *zones* in which the heat is generated. As the shear angle ϕ is increased, the percentage heat generated in the shear plane A will decrease because the plastic flow of the metal will take place over a shorter distance. The shear angle can be increased by applying a coolant and reducing the friction between the chip and the tool as well as by properly grinding the tool. Of all the cutting variables, cutting speed has the most effect on temperature. To increase the rate of metal removal, an increase in feed is much to be preferred over an increase in speed.

[1]S.N. Agrawal, R.D. Harris, and B.H. Amstead, "Basic Formation of Chips." Paper No. 560-WA-282, American Society of Mechanical Engineers, 1960.

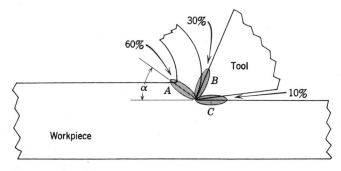

Figure 18.13
Approximate sources of heat in three zones.
A, Shear plane.
B, Friction plane.
C, Surface plane.

Chip Control

In high-speed production turning the control and disposal of chips is important to protect both the operator and the tools. Long curling chips snarl about the workpiece and the machine. Their sharp edges and high tensile strength make their removal from the work area difficult and hazardous, particularly when the machine is in operation. The *chip breaker* curls and highly stresses the chips so that they break into short lengths for easy removal from the machine. Various means can be provided to accomplish this as noted in Figure 18.14. Methods include the following:

1. Grinding on the face of the tool along the cutting edge a small flat to a depth of 0.015 to 0.030 in. (0.38–0.76 mm). This is known as a step-type chip breaker, and it may be either parallel with the edge or at a slight angle. The width varies according to the feed and depth of cut and may range from $1/16$ to $1/4$ in., (1.6–6.4 mm).

2. Grinding a small groove about $1/32$ in. (0.8 mm) behind the cutting edge to a depth of 0.010 to 0.020 in. (0.25–0.50 mm). The correct dimension for both the land distance and depth depends on the feed and should be increased slightly as the feed is increased.

3. Brazing or screwing a thin carbide-faced plate or clamp on the face of the tool. As the chip is formed it hits the edge of the plate and is curled to the extent that it breaks into short pieces.

4. Proper selection of tool angles controlling the direction of the curled chip. They force the chip into some obstruction and stress the chip to its breaking point.

Coolants

Improvement in cutting action may be accomplished by using solids, liquids, emulsions, or gases in the cutting process. In forming and cutting operations high temperatures develop as the result of friction and, unless temperatures and pressures are controlled, the metal surfaces tend to adhere to one another. Figure 18.13 indicates the principal locations of heat. A proper coolant can perform the following functions:

1. Reduce friction between chip, tool, and workpiece.
2. Reduce the temperature of the tool and work.
3. Wash away chips.

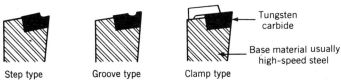

Step type   Groove type   Clamp type

Figure 18.14
Chip breakers used on single-point tools.

4. Improve surface finish.
5. Reduce the power required.
6. Increase tool life.
7. Reduce possible corrosion on both the work and machine.
8. Help prevent welding the chip to the tool.

A coolant should be nonobjectionable physiologically to the operator, harmless to the machine, and stable. It should also have good heat transfer characteristics, be nonvolatile and nonfoaming, provide lubrication, and have a high flash temperature.

Solids that improve cutting ability include certain elements in the workpiece such as graphite in gray cast iron or lead in steel. Liquids are principally in the form of water base or oil base solutions with certan additives in them to increase their effectiveness. Gases include water vapor, carbon dioxide, and compressed air. Most coolants are in liquid form because they may be directed on the tool at the proper place and are easily recirculated.

Chemical coolants are blends of chemical components dissolved in water. Their purpose is to cool, but they may be used for both cooling and lubricating. The chemical agents used are listed here.

1. Amines and nitrites for rust prevention.
2. Nitrates for nitrite stabilization.
3. Phosphates and borates for water softening.
4. Soaps and wetting agents for lubrication and to reduce surface tension.
5. Compounds of phosphorus, chlorine, and sulfur for chemical lubrication.
6. Chlorine for lubrication.
7. Glycols as blending agents and humectants.
8. Germicides to control bacterial growth.

The advantages of using a coolant result from cooling the tool and from reducing the friction, particularly between the chip and the tool. Because of the roughness of the chip as well as of the machined workpiece, coolant can be forced in small quantities to the cutting edge. The best coolant application is between the tool and the workpiece or, if possible, between the chip and the tool. Simple flooding of the cutting area is not so effective as directing the coolant to the *tool interface areas*. The vibration of the tool and workpiece help pump the coolant to the cutting edge. Capillary action and vaporization of the lubricant also help keep the cutting edge cool and lubricated.

Many kinds of coolants are used, depending on the kind of material being machined and the type of operation being performed. The following are some nonchemical coolants used for several of the common materials.

1. *Cast iron*. Compressed air, soluble oil, or worked dry. The use of compressed air necessitates an exhaust system to remove the dust.
2. *Aluminum*. Kerosene lubricant, soluble oil, or soda water. Soda water consists of water with a small percentage of some alkali that acts as a rust preventative.

3. *Malleable iron.* Dry or water-soluble oil lubricant. The latter coolant consists of a light mineral oil held in suspension by caustic soda, sulfurized oil, soap, and other ingredients that form an emulsion when mixed with water.
4. *Brass.* Worked dry, paraffin oil, or lard oil compounds.
5. *Steel.* Water-soluble oil, sulfurized oil, or mineral oil.
6. *Wrought iron.* Lard oil or water-soluble oil.

MACHINABILITY AND SURFACE FINISH

Machinability, or ease with which a given material can be cut, is greatly influenced by the kind and shape of cutting tool. It must be recognized, however, that machinability is a relative term and is expressed in such factors as the length of tool life, power required to make the cut, cost of removing a certain amount of material, or surface condition obtained. The most important of these factors so far as machining costs are concerned is tool life, and values of machinability given in handbooks are based on this factor only.

For many commercial applications the engineering design strength of the part is not so important as economical machining. For this reason materials are often selected because of the importance of tool life and machinability.

Plain-carbon steels have better machinability than alloy steels of the same hardness and carbon content. The addition of lead to steel, while teeming the ingots, adds to machinability although it makes the steel more costly, softer, and more ductile. Materials have been developed especially for certain applications. Automatic machine screw stock which is very easily machined is common, such as 12L14, a low-carbon leaded steel. The addition of a few hundredths of 1% of tellurium to steel will increase the machinability and cutting speed about 3.5 times, but the cost of the element is comparable to that of gold. The addition of moderate amounts of phosphorus as well as sulfur adds to machinability; phosphorus causes chips to be brittle and hence eliminates long, difficult chip formation.

Machinability tests must be conducted under standardized conditions if results are to be comparable. Such tests indicate the resistance of the material being cut, and the results are affected by its composition, hardness, grain size, microstructure, work-hardening characteristics, and size. Other influencing factors include type and rigidity of equipment, coolant properties, feed and depth of cut, and the kind of tool used. Tests conducted should simulate actual machining operations.

Two factors that affect the machinability of a metal are ductility and hardness. As hardness of a metal is increased, penetration by the tool is more difficult and machinability is decreased. The more ductile materials do not lend themselves to the formation of discontinuous chips and, in cases where continuous chips are formed, the increased ductility contributes to the speed at which a BUE occurs. Thus low ductility is an asset to good machinability.

Figure 18.15 is a idealized curve showing effects of these two characteristics. White case iron has high hardness and low ductility. When malleabilized it has higher ductility but a very large decrease in hardness; hence, it can be machined with greater ease. A

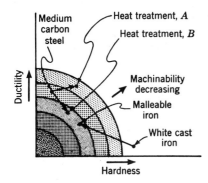

Figure 18.15
Relative effect of ductility and hardness on machinability.

medium-carbon steel may be heat treated in such a way that its machinability either increases or decreases. Unfortunately for most metals, hardness increases as ductility decreases, and the choice of a material or a heat treatment is a compromise between desired service conditions and machinability.

Good machinability does not necessarily mean a good surface finish or low forces or long tool life. Sometimes it refers to the costs associated with metal removal. Although these concepts are easy to state, a rigorous value consistent across many tool and part materials that is a *machinability rating* is difficult to find. As such it is an ideal feature that represents a variety of concepts that depend on the circumstances. Accelerated tests are unable accurately and consistently to provide reliable machinability ratings that are comparable with each other. However, there are four tests that are used to give broad machinability values.

1. A fixed-shape tool is set to cut at predetermined depth and feed. The cutting speed at which a tool can be run and still have a 60-min life is a measure of machinability.
2. The cutting tool wear rate is determined by inspection or by radioactive methods. Low wear rates indicate good workpiece machinability. This method is not suitable for carbide or ceramic tools.
3. A dynamometer is used to record the tool forces for a given set of cutting conditions. The material that can be turned at the highest speed before causing an arbitrary force on the dynamometer will have the best machinability.
4. Operate the tool until an unsatisfactory finish is apparent.

A good surface finish is affected by many variables in single- or multiple-point turning. The factors improving surface finish are light cuts, small feeds, high cutting speeds, cutting fluids, round-nose tools, and increased rake angles on well-ground tools.

TOOL LIFE

The life of a tool is an important factor in metal cutting, because considerable time is lost whenever a tool must be ground or replaced and reset. Cutting tools become dull as usage continues, and their effectiveness drops. At some point in time it is necessary to replace, index, or resharpen and reset the tool. ***Tool life*** is a measure of the length of time a tool will cut satisfactorily and machinability may be measured in a number of ways. Sometimes tool life is expressed as the mintues between changes of the cutting tool.

In general there are five basic types of wear that affect a cutting tool:

1. *Abrasion wear*. This type of wear is caused by small particles of the workpiece "rubbing" against the tool surface.

2. *Adhesion wear*. Plastic deformation and friction associated with high temperatures involved in the cutting process can cause a welding action on the surfaces on the tool and workpiece. The consequent stresses of the cutting process lead to fracture of the weld, causing degeneration of the cutting tool.

3. *Diffusion wear*. Diffusion wear is caused by a displacement of atoms in the metallic crystal of the cutting element from one lattice point to another. This results in a gradual deformation of the tool surface.

4. *Chemical and electrolytic wear*. A chemical reaction between the tool and workpiece in the presence of a cutting fluid is the cause of chemical wear. This electrolytic wear is the result of possible galvanic corrosion between the tool and workpiece.

5. *Oxidation wear*. At high temperatures oxidation of the carbide in the cutting tool decreases its strength and causes wear of the edge.

Wear is evident in two places on a tool (Figure 18.16). One is on the flank of the tool where a small *land* extending from the tip to some distance below is abraded away. On high-speed tools a failure is considered to have taken place if this land has worn 0.062 in. (1.58 mm), and for carbide tools when the wear land has reached 0.030 in. (0.76 mm). In Figure 18.16*B* the progressing amounts of wear are noted, and it is this length that is plotted on Figure 18.17*A*. Some velocities may not result in wear that is noticeable, for example, turning aluminum bar stock with carbide. Note velocity V_1 in Figure 18.17*A*. Wear also takes place on the face of the tool in the form of a small crater (Figure 18.16*A*)

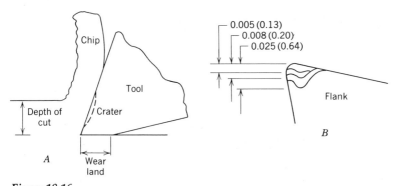

Figure 18.16
A, Wear of a cutting tool. *B*, Progressive flank wear measured during cutting tests.

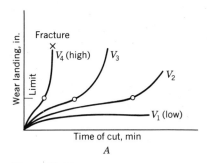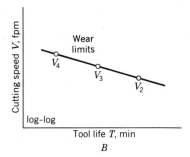

Figure 18.17
A, Several tool life curves for a constant velocity. B, Plotting time to reach wear land limit versus cutting velocity.

or depression behind the tip. This depression results from the abrading action of the chip as it passes over the tool face.

In 1906 Frederick W. Taylor reported the relationship between tool life and cutting speed as follows:

$$VT^n = C$$

where

V = Cutting speed, ft/min (m/s)
T = Tool life, min (s)
n = Exponent depending on cutting conditions
c = Constant equal to cutting speed for a tool life, min (60 s)

min (60 s)

Because tool life decreases as the cutting speed is increased, tool life curves are plotted as tool life in minutes against cutting speed in feet per minute (meters per second) or in cubic inches (cubic meters) of metal removed. In some cases the life is determined by surface finish measurements and in others by an increase in force on a dynamometer. Taylor tool life curves are shown as Figure 18.17B. One test might have a velocity V_4, and the time is noted that it takes to reach a specific limit of land wear, say, 0.030 in. for carbide tools. Similarly, another velocity different from V_4 is tested and the time it takes to reach a flank wear limit is plotted as in Figure 18.17A. The composite of these curves is Figure 18.17B, the time it takes to reach the specified standard amount of land wear such as 0.030 in. (0.08 mm) limit, for several velocities. This is the *tool life curve*.

With high velocities it is possible quickly or instantaneously to fracture the tool point, and severe damage or safety problems may result. Note V_4 in Figure 18.17A. The empirical coefficients are determined by statistical tests of cutting metal test logs under standard conditions and noting the time for flank wear to reach the specified limit against rotational velocity of the test log. These tests are usually undertaken with turning operations on lathes.

The exponent n depends on characteristics of the workpiece and tool material. Samples values are as in the accompanying tabulation.

	High-speed steel		Tungsten carbide	
Material	C	n	C	n
Stainless steel	170	0.08	400	0.16
Medium-carbon steel	190	0.11	150	0.20
Gray cast iron	75	0.14	130	0.25

When the cutting speed is plotted as a function of the tool life on log-log paper, a straight line results as shown in Figure 18.18. This is a tool life curve. The value of n can be determined by using the formula. Because the data of Figure 18.18 depend on a given set of conditions, results may not be applicable if the depth of cut or feed is changed. If either of these is increased, a reduction must be made in cutting speed to obtain the same tool life.

An intercept C is found at the intersection of the line and tool life of one minute. The slope is found using the equation

$$n = \frac{\log V_2 - \log V_1}{\log T_2 - \log T_1}$$

Note the locations of the points on the figure. Taking measuremens from the graph, $C = 190$ and $n = 0.1482$, the Taylor tool life equation for this curve is then $VT^{0.1482} = 190$. The equation can be used to find either velocity or tool life given the other quantity. For example, if a life of 50 min is desired, the velocity would be found as 106 ft/min. The HSS cutting edge would be resharpened or indexed in the case of a carbide insert after about 50 min of actual cutting time.

The correct choice of cutting velocity can enhance tool life, but at the same time, the

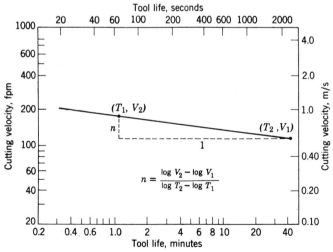

Figure 18.18
Effect of cutting speed on tool life for a high-speed steel tool.

tool should be used to its maximum capacity. Tool failures usually occur for the following reasons:

1. *Improper grinding of tool angles.* Cutting angles depend on the tool and workpiece. A brief listing of values is given in Table 18.1, but other values have been recorded in handbooks and in manufacturers' literature.

2. *Loss of tool hardness.* This is caused by excessive heat generated at the cutting edge. This situation is relieved by coolants or by reducing the cutting speed.

3. *Breaking or spalling of tool edge.* This may result by taking too heavy a cut or by too small a lip angle.

4. *Natural wear and abrasion.* Tools gradually become dull by abrasion. This process is **accelerated** by the development of a crater just behind the cutting edge. As the crater increases in size, the cutting edge becomes weaker and finally breaks off. This effect can be reduced by the proper selection of tool material.

5. *Fracture of tool by heavy load.* This condition is reduced if the cutting tool is rigidly supported with a minimum of overhang.

SURFACE INTEGRITY

A good surface finish is affected by many variables in single- or multiple-point machining. The factors improving surface finish are light cuts, small feeds, high cutting speeds, cutting fluids, round-nose tools, and increased rake angles on well-ground tools. Many of these features, while highly desirable, may shorten tool life, increase forces, and be uneconomical overall.

The outermost boundary of a body adjacent to the air is called a *surface*. When this surface is deformed by a sharp cutting edge, the term surface finish describes the boundary. When a rotating cylindrical body in contact with a tool has an existing surface removed, the substrate surface provides the new boundary. If the surface were magnified many times it would have the appearance of successive jagged mountain ranges. Despite the seemingly random natural appearance of a mountainous profile, there is mathematical form and geometry to the topography. This analogy extends to the microscopic texture of grooves turned by a tool on a lathe.

The importance of surface finish in present-day technology is recognized by the consumer and engineer. Excessive cost can be attributed to overfinish. On the other hand, surface finish contributes to the precision of the dimension. Engineering properties such as fatigue, hardness, and heat transfer are affected by surface finish.

Surface roughness is subject to measurement electronically and visually. Traditional methods of measurements were discussed in Chapter 11. Another device that provides a nondestructive surface examination is based on the **light section principle.** A diagrammatic sketch is given in Figure 18.19. According to this principle, an optical cut is made through the surface without touching or scratching, which are two problems with traditional methods. In operation an incandescent lamp, Q illuminates slit S. A razorlike beam projects through objective O_1 at a 45° angle to the workpiece surface. The band of light

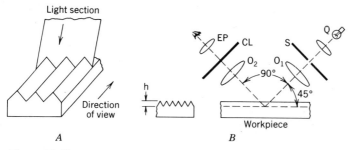

Figure 18.19
A, Light section upon workpiece. B, Light principle through lens system.

is observed through a microscope at the opposite 45° angle. The microscope objective O_2 has the same magnification as O_1. This fine band of light can observe the peaks and valleys, and a crossline reticle (CL) in the eyepiece (EP) can be shifted within the field of vision. Two measurements are possible: roughness height or the distance from the peak to the valley, and roughness width. The instrument can measure micron dimensions. With this approach it becomes possible to find surface finish effects that reasonably correspond to the tool point profile.

In the ideal surface we presume that the surface is assumed to be machined exactly as the geometric factors of the tool suggest. Chatter or BUE effects or effects of tool wear are ignored. In turning, the tool point places a screw thread helix upon the surface, but the penetration of the tool upon the surface is so slight that there is no noticeable impression of a screw thread. Note Figure 18.20 for the single-point effect on a rotating bar stock.

A model of the ideal surface is possible using the geometric analog. Two tool geometries are shown in Figure 18.21 because of their popularity in finishing operations. For the tool plan shown in Figure 18.21A the maximum surface roughness is given by

$$H_{max} = \frac{f}{\tan C_s + \cos C_e}$$

where

f = Feed distance, in. (mm)
C_s = Side angle, °
C_e = End angle, °

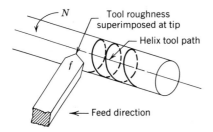

Figure 18.20
Helix effects caused by feeding a tool point along a bar stock.

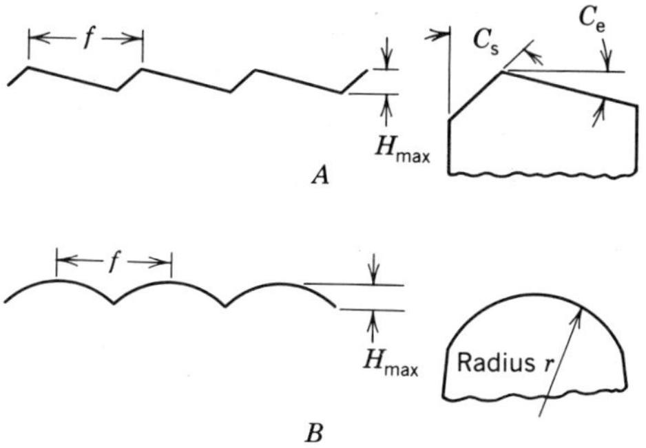

Figure 18.21
Surface geometry models.

For the tool plan shown in Figure 18.21B, surface geometry results in the following:

$$H_{max} = \frac{f^2}{8r}$$

where r is the radius of the tool.

H_{max} is not the same as the arithmetical average as shown by Figure 11.14, where the distance is measured by a midline-roughness datum. Thus those measures have to be multiplied by two to compare to H_{max}.

For example, assume a feed dimension of 0.008 in. and $C_s = C_e = 15°$ and the tool plan in Figure 18.21A with a maximum roughness of 0.002 in., or about 100 μin., which is not a fine finish. On the other hand, a round-nose tool, say with a $1/32$-in. radius, will result in a surface height maximum of 0.000256 in., or about eight times finer than the above tool. Now these values represent ideal figures, and in practice, the operator will determine steady-state roughness using standard methods. If roughness exceeds these values for a predetermined tool geometry, feed, tool, and workpiece material, the operation has exceeded machinability standards.

Metal cutting theory is often misinterpreted by experimental results, because the broad spectrum of materials, equipment, and complex nature of this process are such that uniform results between the possibilities make a practical theory hinge on the fortunate choice of many factors. The ramifications of the metal cutting business make it virtually impossible to evaluate even a few of the materials, tools, and equipment. An aggregate mathematical model conveying a complete theory is probably unobtainable in machinability.

CUTTING SPEEDS AND FEEDS

Cutting speed V_c is expressed in feet per minute (meter per second) and on a lathe is the surface speed or rate at which the work passes the cutter. In turning the work rotates, whereas for milling the tool rotates and the work is stationary. Cutting velocity may be expressed by the formula

$$V_c = \frac{\pi DN}{12}$$

where

$$D = \text{Diameter of work, in.}$$
$$N = \text{Rotary speed of workpiece, rpm}$$

The equivalent metric equation is given as

$$V_c = \frac{\pi DN}{60{,}000}$$

where

D is the Diameter of the work, mm.

The cutting speed in this expression is seldom unknown, because recommended cutting speeds for many tool and workpiece materials are found in textbooks and handbooks. These recommendations are based on tool life, surface finish, and machinability. Table 18.2 is very abbreviated and is used for problems given later. Most values of *cutting speeds and feeds* are given with a range, and our practice in this book will be to use the midpoint value. Note that feeds, inches per revolution (ipr), are given at the foot of the table and are used with cutting velocity.

Table 18.2 **Typical Cutting Speeds and Feeds**

	Cutting material			
	High-speed steel		Carbide	
Material being cut	Finish[a]	Rough[b]	Finish[a]	Rough[b]
Free-cutting steels, 1112, 1315	250–350 (1.3–1.8)	80–150 (0.4–0.8)	600–750 (3.0–3.8)	350–450 (1.8–2.3)
Carbon steels, 1010, 1025	225–300 (1.1–1.5)	75–125 (0.4–0.6)	550–700 (2.8–3.5)	300–400 (1.5–2.0)
Medium steels, 1030, 1050	200–275 (1.0–1.4)	70–120 (0.4–0.6)	450–600 (2.3–3.0)	250–350 (1.3–1.8)
Nickel steels, 2330	200–275 (1.0–1.4)	70–110 (0.4–0.6)	425–550 (2.1–2.8)	225–325 (1.1–1.6)
Chromium nickel, 3120, 5140	150–200 (0.8–1.0)	50–75 (0.3–0.4)	325–425 (1.7–2.1)	175–260 (0.9–1.3)
Soft gray cast iron	120–150 (0.6–0.8)	75–90 (0.4–0.5)	350–450 (1.8–2.3)	200–250 (1.0–1.3)
Brass, normal	275–350 (1.4–1.8)	150–225 (0.8–1.1)	600–700 (3.0–3.5)	400–500 (2.0–2.5)
Aluminum	225–350 (1.1–1.8)	100–150 (0.5–0.8)	450–700 (2.3–3.5)	200–300 (1.0–1.5)
Plastics	300–500 (1.5–2.5)	100–200 (0.5–1.0)	400–650 (2.0–3.3)	150–250 (0.8–1.3)

[a]Cut depth, 0.015 to 0.095 in. (0.38–2.39 mm). Feed, 0.005 to 0.015 ipr (0.13–0.38 mm/r).
[b]Cut depth, 0.187 to 0.375 in. (4.75–9.53 mm). Feed, 0.030 to 0.050 ipr (0.75–1.27 mm/r).

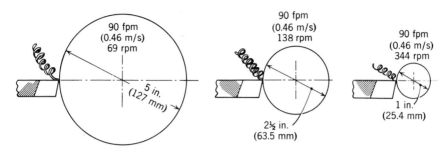

Figure 18.22
Relationship of revolutions per minute to surface velocity using $V_c = \dfrac{\pi DN}{12}$.

In lathe work the unknown factor is the rotary speed or the term N. If one refers to Figure 18.22, it will be noted that to maintain a recommended cutting speed of 90 ft/min it is necessary to increase the work revolutions materially as the diameter is decreased from 5 in. (127 mm) to 1 in. (25.4 mm).

The term feed refers to the rate at which a cutting tool or grinding wheel advances along or into the surface of the workpiece. For machines in which the work rotates, feed is expressed in inches (millimeters) per revolution. For machines in which the tool or work reciprocates, it is expressed in inches (millimeters) per stroke; and for stationary work and rotating tools it is expressed in inches (millimeters) per revolution of the tool. The following factors would necessitate a reduced feed for a given tool: greatly increased cutting speed, harder workpiece, more ductile workpiece, less coolant, dulling of the tool, or a reduced rigidity in the workpiece or machine. Values of feeds corresponding to the speeds given in Table 18.2 are enlarged upon in later chapters.

QUESTIONS AND PROBLEMS

1. Give an explanation of the following terms:
   Orthogonal chip
   Vertical cutting force
   Dynamometer
   Coefficient of friction
   Cutting tool nose radius
   Lip
   Side relief angle
   High-speed steel
   Carbide cutting tools
   Ceramic tools
   Discontinuous chip
   Continuous chip type 2
   BUE
   Heat zones
   Chip breaker
   Machinability
   Tool life
   Accelerated tool life
   Light section microscope
   Cutting speeds and feeds

2. What does the word orthogonal mean with regard to cutting tools?

3. An orthogonal-type cut is made of a pipe 1 $\frac{1}{2}$ in. in diameter. Only one rotation is turned before the pipe is stopped suddenly. The cutter rake angle = 35°; cutting speed = 35 ft/min; depth of cut = 0.015 in.; length of continuous chip for one rotation of the pipe = 2.4 in.; cutting force = 425 lb; and vertical force = 170 lb. Find the coefficient of friction, chip ratio, shear plane angle, and velocity of the chip.

4. List factors affecting the frictional force in single-point tools.

5. What are the reasons that account for the tangential force on a single-point cutting tool as used on a lathe being at least 10 times greater than the radially directed force?

6. Explain how tool forces are affected by changing cutting speed. What factors vary as a result?

7. How is the longitudinal force affected by an increase in side cutting edge angle?

8. Describe the ideal chip type. Why is it ideal?

9. What are the effects of a large frictional force on a cutting tool, and how can they be reduced?

10. What is a major disadvantage of long, continuous chips?

11. Sketch a left-hand cutting tool and label the angles.

12. Why are liquid coolants usually preferred?

13. A block 6 in. long is shaped. Data from a test give a rake angle = 10°; cutting speed = 70 ft/min; depth of cut = 0.025 in.; length of chip = 5.2 in.; horizontal cutting force = 950 lb; and vertical force = 315 lb. (a) Find the coefficient of friction, chip ratio, shear plane angle, and velocity of the chip. (b) Repeat for a 20° rake angle. (c) Repeat for a 30° rake angle.

14. An AISI 1045 steel block 6 in. long is orthogonally machined. All testing conditions are consistent except that the rake angle of the tool is changed. Constant data for the test is cutting speed = 75 ft/min; depth of cut = 0.010 in.; length of chip is substantially identical over the tests and was found to be 5.2 in. Other factors varied as in the accompanying tabulation.

Rake angle,°	Horizontal force, lb	Vertical force, lb
10	925	320
20	950	330
30	974	331
50	1075	340

Determine the coefficient of friction. Discuss your results and make conclusions about this test.

15. Data from a machinability test indicated $F_s = 750$ lb and $V_c = 375$ ft/min. Find the spindle horsepower. If spindle efficiency = 85%, what is the required motor power? Also in metric units.

16. A turning operation of a bar stock requires a feed rate of 0.018 ipr, depth of cut = 0.25 in., and a cutting speed of 275 ft/min. Determine the metal removal rate. Repeat for 0.015 ipr and 300 ft/min.

17. A boring operation indicated a cutting force of 225 lb and cutting speed of 180 ft/min. Find the spindle horsepower. The efficiency of the spindle = 92%. Find the motor horsepower. Convert to SI units.

18. For a depth of cut = 0.015 in., cutting velocity = 425 ft/min, and a feed of 0.064 ipr, find the metal removal rate. Also convert to SI units and make the calculations.

19. What are the advantages and limitations associated with using carbide tools?

20. The tool has a 60° included-angle sharp point and is used for making V threads. Find the roughness for a 0.020-in. feed. What is the approximate arithmetical average for this roughness using ordinary standards?

21. What is the effect of shear angle on the heat generated in cutting?

22. Why are chip breakers used on some tools and how are they constructed?

23. Set up on paper an experiment for measuring machinability using a dynamometer. Explain the data needed and why.

24. A gray cast-iron wheel is turned on a lathe. Find the cutting speed for 30 min of tool life with a tungsten carbide tool point. Repeat for 20 min. Repeat for 60 min.

25. What characteristics should an ideal coolant have?

26. How can the vibration of a tool and workpiece increase coolant flow?

27. The Taylor tool life equation is $VT^{0.1} = 172$. What is the expected average tool life for $V = 275$ ft/min? For $V = 180$ ft/min? Find the velocity for 18 min of life. For 25 min. For $V = 200$ ft/min.

28. How does ductility affect machinability?

29. What factors causes the tool to lose hardness?

30. If tool life needs increasing for economic reasons, what are the most significant steps you can take?

31. List advantages of the materials used for cutting tools.

32. Categorize tool materials.

33. How much faster in revolutions per minute must a lathe be driven if a 4-in. bar stock of aluminum is being finish turned, as compared to brass with high-speed steel? Rough turned with carbide? For a 6-in. diameter bar, compare between 1050-carbon steel and 5140 steel for rough turning with carbide.

34. Find the peak-to-valley roughness with a $3/64$-in. nose radius and a 0.010-in. feed. For a 0.005-in. feed. Repeat for a $3/64$ radius tool nose.

35. If 3 hp is being used to drive a lathe and 70% of this power is used for cutting, how much energy is dissipated in the form of heat? Assume 760 W equals 1 hp.

36. Using Figure 18.18, what is the tool life of the high-speed tool if it runs at 200 ft/min (1.0 m/s)?

37. Calculate the revolutions per minute that a lathe should be turned to cut the following; rough finish 4 in. (102 mm) diameter cast iron with high-speed steel; finish a 12-in. (300-mm) brass bar with high-speed steel; finish a $1/2$-in. (12.7-mm) diameter bronze bar with carbide.

38. Estimate the difference in machinability of SAE 1020 and SAE 1050 steel. Involve recommended cutting speeds.

39. If the feed on a lathe is set at 0.006 ipr (0.15 mm/r), what is the feed per minute for problem 37?

40. If a drill press operates at 120 rpm, how long would it take to drill a 2-in. (51-mm) deep hole if the feed is 0.007 ipr (1.78 mm/r)? Repeat for 150 rpm.

41. A stainless-steel material can be machined by either high-speed steel or tungsten carbide tools. What is the velocity for 30 min of cutting for these two materials? What percentage advantage will a 400-ft/min velocity provide with tungsten carbide for tool life?

42. Let $C_s = 15°$ and $C_e = 25°$ for a sharp tool point. Find H_{max} for a feed of 0.005 in. Repeat for $C_e = 15°$ and 0.008 in. feed.

43. A C-7 carbide tool was tested on AISI 8640 steel, 190 Bhn, and turned at several cutting velocities. Conditions were dry, depth of cut = 0.010 in., feed of 0.010 ipr, and a wear limit of 0.015 in. as selected by the engineer. Plot the tool life curve on log-f|log scales for the following (T, V) data: (60, 340), (44, 400), (30, 500), (11, 600), and (8, 700). From the graph find the values of n and C. What is the expected life for a velocity of 550 ft/min using the equation and the curve? Using the graph and equation, find the velocity for 25 min of life.

CASE STUDY

MACHINABILITY STUDY

Bob Williams, manufacturing engineer for Advanced Job Shop, is considering the prospect of a large new order for turning AISI 8140 steel bar stock and discussing the order with the Vice-President of Operations, Frank Essenburg.

"Frank, we are going to be making a lot of chips for this Acme job," Bob says, while looking at the purchase order. "It calls for turning diameters of at least 8 in., and we don't have much experience in that size and grade."

"This job doesn't have much profit in it, Bob," Frank comments while looking at the material specification. "We have to bid competitively to keep our shop labor working while we get through this slump. Perhaps there is something that you can do to make a better profit on the job, Bob. After all, we have that new lathe."

"Umm, with the quantity that we are to deliver, it might be useful to conduct a machinability test before releasing the order to the shop."

"What is a machinability test, Bob?"

Bob explains the function of the test, and at the end Frank remarks by saying: "Bob, it is important to AJS that we make money on this job, because of the risk involved. Do your testing and get back to me with the results and recommendations."

Bob reasons that a tool life test with various velocities will economize on the machining part of the job. He selects bar stocks identical to the new job, and instructs the lathe operator to run a test with the following conditions: depth of cut = 0.050 in. and a feed = 0.010 ipr. The lathe operator is instructed to vary the job with three different revolutions per minute values: $V_1 = 600$ ft/min, $V_2 = 525$ ft/min, and $V_3 = 480$ ft/min. The machinist is to examine the tool point for wear using a portable microscope especially made for measuring the length of wear land along the flank of the tool. From time to time, the machinist will stop the work, measure the wear, and note the time on a table that Bob constructs. The machinist provides Bob with the following:

V_1		V_2		V_3	
min	wear in.	min	wear in.	min	wear in.
3	0.005	6	0.007	5	0.004
6	0.011	11	0.013	9	0.005
9	0.022	13	0.018	13	0.010
10	0.040	15	0.025	19	0.017
11	Fracture	17	0.036	23	0.027
				25	0.035
				35	0.041

Bob puzzles over the data and begins to plot it similar to Figure 18.17A. He uses a graph to find the point for 0.030 in. of wear for each velocity. Then he uses log-log graph paper and constructs the tool life line and finds n and C from the drawn line. At that point, he ponders his results, and says: "Each part should require about 25 min of machining, and it is better to replace the tool after the part is finished than during a run." Help Bob out with his problem. What velocity should he consider for these parts? What suggestions can you provide to help the optimization for this work? Do all analysis and submit your recommendations for operation in a report.

CHAPTER 19
TURNING, DRILLING, BORING, AND MILLING MACHINE TOOLS

This chapter discusses the most common of production processes. Their importance to manufacturing cannot be understated, and in terms of the dollar effort nationally, fabrication work achieved by turning, drilling, boring, and milling ranks number one.

The oldest and most common machine tool is the lathe, which removes material by rotating the workpiece against a single-point cutter. Parts can be held between centers, attached to a face plate, supported in a jaw chuck, or held in a draw-in chuck or collet. Although this machine is particularly adapted to cylindrical work, it may also be used for other operations. Plane surfaces can be obtained by supporting the work in a face plate or chuck. Work held in this manner can be faced, centered, drilled, bored, or reamed. In addition, a lathe can be used for knurling, cutting threads, or turning tapers.

One of the simplest machine tools used in production and toolroom work is the drill press. *Drilling* produces a hole in an object by forcing a rotating drill against it. The same can be accomplished by holding the drill stationary and rotating the work, such as drilling on a lathe with the work held and rotated by a chuck. Although the drill press is essentially a single-purpose machine, a number of dissimilar operations are possible with other cutting tools on this machine tool.

Perhaps the Power Age was born through boring, as the boring mill made Watt's steam engine a reality by being able to machine round cylinders true throughout the cylinder's length. *Boring* is enlarging a hole that has already been drilled or cored. Principally, it is an operation of truing a hole that has been drilled previously with a single-point lathe-type tool. For this operation on a drill press, a special holder for the boring tool is necessary.

A *milling machine* removes metal when the work is fed against a rotating cutter. Except for rotation the circular cutter has no other motion. The milling cutter has a series of cutting edges on its circumference, each acting as an individual cutter in the cycle of rotation. The work is held on a table that controls the feed against the cutter. In most machines there are three possible table movements—longitudinal, crosswise, and vertical—but in some the table may also possess a swivel or rotational movement.

The milling machine is considered the most versatile of all machine tools. Flat or formed surfaces may be machined with excellent finish and accuracy. Angles, slots, gear teeth, and recess cuts can be made with various cutters. Drills, reamers, and boring tools can be held in the arbor socket. Because table movements have micrometer adjustments, holes and other cuts can be dimensioned accurately. Most operations performed on shapers, drill presses, gear-cutting machines, and broaching machines can be done on the milling machine. Heavy cuts can be taken with little appreciable sacrifice in finish or accuracy. Cutters are efficient in their action and tool life is excellent. In most cases the work is completed in one pass of the table. These advantages plus the availability of a variety of cutters make the milling machine indispensable in the shop and toolroom.

LATHE GROUP

Engine Lathes

The principal parts of a lathe are labeled in Figure 19.1. Controls on the side of the headstock allow selection of any one of many speeds, which are arranged in logical geometric progression. A combination electric chuck and brake is provided for starting, stopping, or jogging the work.

The *tailstock* can be moved along the bed of the lathe to accommodate different lengths of stock. It is commonly provided with a hardened ball bearing that is center mounted and that may be moved in and out by wheel adjustment, and with set-over screws at its base for adjusting the alignment of the centers and for taper turning.

The *lead screw* is a long, carefully threaded shaft coated slightly below and parallel to the bedways extending from the headstock to the tailstock. It is geared to the headstock and its rotation may be reversed. It is fitted to the carriage assembly and may be engaged or released from the carriage during cutting operations. The lead screw is for cutting threads and should be disengaged when not in use to preserve its accuracy. Below the lead screw is a *feed rod* that transmits power from the quick-change box to drive the

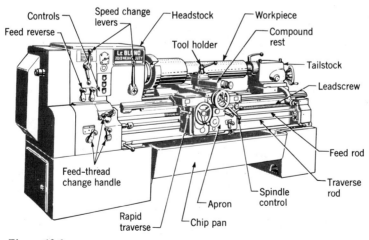

Figure 19.1
Heavy-duty engine lathe.

apron mechanism for cross and longitudinal power feed. Changing the speed of the lead screw or feed rod is done at the *quick-change gear box* located at the headstock end of the lathe.

The *carriage assembly* includes the compound rest, tool saddle, and apron. Because it supports and guides the cutting tool, it must be rigid and constructed with accuracy. Two-hand feeds are provided to guide the tool in a crosswise motion. The upper hand wheel or crank controls the motion of the compound rest, and because this rest is provided with a swiveling-adjustment protractor, it can be placed in various-angle positions for short-taper turning. A third hand wheel is used to move the carriage along the bed, usually to pull it back to starting position after the lead screw has carried it along the cut. The portion of the carriage that extends in front of the lathe is called an apron. It is a double-walled casting that contains the controls, gears, and other mechanisms for feeding the carriage and cross slide by hand or power. On the face of the apron are mounted the various wheels and levers.

Lathe size is expressed in terms of the diameter of the work it will swing. A 16-in. (400-mm) lathe has sufficient clearance over the bed rails to handle work 16 in. (400 mm) in diameter. A second dimension is necessary to define capacity in terms of workpiece length. Some manufacturers use maximum work length between the lathe centers, whereas others express it in terms of bed length. The diameter that can be turned between centers is somewhat less than the swing because of the allowance that must be made for the carriage.

There are varieties of lathes, and their design depends on the type of production or the nature of the workpiece. A *speed lathe* is the simplest of all lathes and consists of a bed, a headstock, a tailstock, and an adjustable slide for supporting the tool. Usually it is driven by a variable-speed motor internal to the headstock, although the drive may be a belt to a step-cone pulley. Because hand tools are used and the cuts are light, the lathe is driven at high speed. The work is held between centers or attached to a face plate on the headstock. The speed lathe is used in turning wood, centering metal cylinders, and metal spinning.

The *engine lathe* derives its name from the early lathes that obtained their power from engines rather than overhead belts. It differs from a speed lathe in that it has additional features for controlling the spindle speed and for supporting and controlling the feed of the fixed cutting tool. There are several variations in the design of the headstock through which the power is supplied to the machine. Light- or medium-duty lathes receive their power through a short belt from the motor or from a small cone pulley countershaft driven by the motor. The headstock is equipped with a four-step cone pulley that provides a range of four spindle speeds when connected directly from the motor countershaft. In addition, these lathes are equipped with *back gears* that, when connected with the cone pulley, provide four additional speeds. Figure 19.1 is a geared-head engine lathe. The spindle speeds are varied by a gear transmission, the different speeds being obtained by setting speed levers in the headstock. These lathes are usually driven by a constant-speed motor mounted on the lathe, but in a few cases variable-speed motors are used. A geared-head lathe has the advantage of a positive drive and has a greater number of spindle speeds available than are found on a step cone-driven lathe.

A *bench lathe* is a small lathe that is mounted on a workbench. It has the same features as engine lathes but differs from these lathes only in size and mounting. It is adapted to small work, having a maximum swing capacity of 10 in. (250 mm) at the face plate.

A *toolroom engine lathe* is equipped with all the accessories for accurate tool work, being an individually driven geared-head lathe with a considerable range in spindle speeds. It is equipped with center steady rest, quick-change gears, lead screw, feed rod, taper attachment, thread dial, chuck, indicator, draw-in collet attachment, and a pump for a coolant. Toolroom lathes are carefully tested for accuracy and, as the name implies, are especially adapted for making small tools, test gages, dies, and other precision parts.

Turret Lathes

Turret lathes are a major departure from engine or basic lathes. These machines possess special features that adapt them to production. The "skill of the worker" has been built into these machines, making it possible for inexperienced operators to reproduce identical parts. In contrast to this, the engine lathe requires a skilled operator and takes more time to produce parts that are dimensionally the same. The principal characteristic of this group of lathes is that the tools for consecutive operations can be set up in readiness for use in the proper sequence. Although skill is required to set and adjust the tools properly, once they are correct less skill is required to operate them and many parts can be produced before adjustments are necessary.

Horizontal turret lathes are made in two general designs known as the ram and saddle. In appearance they are much alike, and both may be used for either bar or chucking work. The ***ram-type turret lathe*** shown in Figure 19.2A is so named because of the way the turret is mounted. The turret is placed on a slide or ram that moves back and forth on a saddle clamped to the lathe bed. This arrangement permits quick movement of the turret and is recommended for bar and light-duty chucking work. The saddle, though capable of adjustment, does not move during the operation of the turret. Ram-type machines do not require the built-in rigidity of chucking machines, because the bar tools can be made to support the work.

A *saddle-type lathe* (Figure 19.2B) is used for chucking work and has the turret mounted

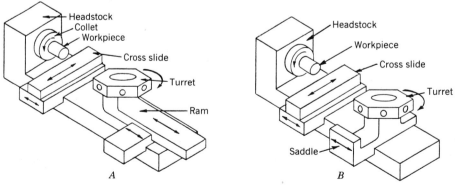

Figure 19.2
Horizontal turret lathes. *A*, Ram type. *B*, Saddle type.

directly on a saddle that moves back and forth with the turret. Because chucking tools overhang and are unconnected with the work through some sort of support, greater strain on both work and tool support results. Chucking tools must have rigidity. The stroke is longer, which is an advantage in long turning and boring cuts, and saddle mounts assist the rigidity.

Turret lathes are constructed similarly to engine lathes. The headstock in most cases is geared, with provision for 6 to 16 spindle speeds and double these with a two-speed motor. The spindle speeds as well as forward and reverse movement are controlled by levers extending from the head. The drive motor is sometimes located in or on the motor leg below the headstock and connected to the geared-head sheave by V-belts, or a flange-mounted motor may be directly connected to the drive shaft in the head.

The cross-slide unit on which the tools are mounted for facing, forming, and cutting off is somewhat different in construction from the tool post and carriage arrangement used on lathes. It is made up of four principal parts: cross slide, square turret, carriage, and apron. These parts are discernible in the various turret lathe illustrations. Some of the cross slides are supported entirely on the front and lower front ways, permitting more swing clearance. This arrangement is frequently utilized on saddle-type machines that are to be used for large-diameter chucking jobs. The other arrangement for mounting has the cross slide riding on both upper bedways and is further supported by a lower way. This is used on machines engaged in bar work and in other processes where a large swing clearance is not necessary. An advantage of this type is the added tool post in the rear, frequently used in cutting-off operations.

On top of the cross slide is mounted a square turret capable of holding four tools. If several different tools are required, they are set up in sequence and can be quickly indexed and locked in correct working position. To allow cuts to be duplicated, the slide is provided with positive stops or feed trips. Likewise, the longitudinal position of the entire assembly may be controlled by positive stops on the left side of the apron. Cuts may be taken with square-turret tools *simultaneously* with tools mounted on the hexagonal turret.

The outstanding feature is the *turret* in place of the tailstock. This turret, mounted on either the sliding ram or the saddle, carries the tools for the various operations. The tools are mounted in proper sequence on the various faces of the turret, so that as it indexes around between operations, the proper tools are brought into position. For each tool there is a stop screw that controls the distance the tool will feed. When this distance is reached, an automatic trip lever stops further movement of the tool by disengaging the drive clutch.

The difference between the engine and turret lathes is that the turret lathe is adapted to quantity production work, whereas the engine lathe is primarily used for miscellaneous jobbing, toolroom or single-operation work. The features of a turret lathe that make it a quantity production machine are these:

1. Tools may be set up in the turret in the proper sequence for the operation.
2. Each station is provided with a feed stop or feed trip so that each cut of a tool is the same as its previous cuts.
3. *Multiple cuts* can be taken from the same station at the same time, such as two or more turning or boring cuts.

4. *Combined cuts* can be made; that is, tools on the cross slide can be used at the same time that tools on the turret are cutting.
5. Rigidity in holding work and tools is built into the machine to permit multiple and combined cuts.
6. Turret lathes may be fitted with attachments for taper turning, thread chasing, and duplicating, and can be tape controlled.

Tape-Controlled Turret Lathe

A heavy-duty two-axis turret lathe with numerical control (NC) is shown in Figure 16.1; it is designed especially for heavy-duty production. The control provides automatic functioning of spindle speed, slide movement, feeds, turret indexing, and other auxiliary functions. The slant bed, inclined rearward from the vertical, provides maximum rigidity and operator accessibility to the work area. This machine can be set up quickly for small-lot jobs, normally changing only jaw chucks, control tape, and possibly one or two cutters.

Vertical Turret Lathe

A vertical turret lathe resembles a vertical boring mill, but it has the characteristic turret arrangement for holding the tools. It consists of a rotating chuck or table in the horizontal position with the turret mounted above on a cross rail. In addition, there is at least one side head provided with a square turret for holding tools. All tools mounted on the turret or side head have their respective stops set so that the length of cuts can be the same in successive machining cycles. It is, in effect, the same as a turret lathe standing on the headstock end, and it has all the features necessary for the production of duplicate parts. This machine was developed to facilitate mounting, holding, and machining of large-diameter heavy parts. Only chucking work is done on this kind of machine.

A vertical turret lathe shown in Figure 19.3 is provided with two cutter heads: the swiveling main turret head and the side head. Another side head is possible. The turret

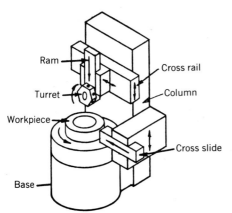

Figure 19.3
Vertical turret lathe.

and side heads function in the same manner as the hexagonal and square turrets on a horizontal lathe. To provide for angle cuts, both the ram and turret heads may be swiveled 30° right or left of center. The side head has rapid traverse and feed independent of the turret and, without interference, provides for simultaneous machining adjacent to operations performed by the turret. The ram provides another tool station on the machine that can be operated separately or in conjunction with the other two.

The machine can be provided with a control that permits automatic operation of each head including rate and direction of feed, change in spindle feed, indexing of turret, starting, and stopping. Once a cycle of operations is preset and tools are properly adjusted, the operator need only load, unload, and start the machine. Production rate is increased over those manually operated machines, because they operate almost continuously and make changes from one operation to another without hesitation or fatigue. By reducing the handling time and making the cycle automatic, an operator can attend more than one machine.

Automatic Vertical Multistation Lathe

Machines as illustrated in Figure 19.4 are designed for high production and are usually provided with either five or seven work stations and a loading position. In some machines two spindles are provided at each station, doubling the capacity for small-diameter work. The work to be machined is mounted in chucks, the larger machine having chucking capacities up to 18 in. (460 mm) in diameter. Both plain and universal heads may be used, the latter providing for tool feed in any direction. All varieties of machining operations can be performed, including milling, drilling, threading, tapping, reaming, and boring. The advantage of this machine is that all operations can be done simultaneously and in proper sequence. At each station except the loading one, an operation is performed that leads to the completion of the part when it has been indexed through all the working stations. The work cannot be indexed to the succeeding positions until the operation requiring the longest time is complete. Each time, when all operations are complete, the work spindles stop, the tools are retracted, and the worktable holding the spindles is indexed to the next position.

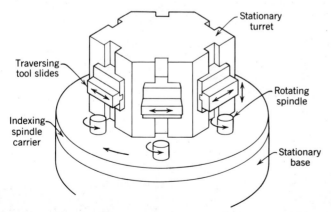

Figure 19.4
Multispindle vertical chucking lathe.

Automatic Lathes

Lathes that have their tools automatically fed to the work and withdrawn after the cycle is complete are known as *automatic lathes*. Because most lathes of this type require that the operator place the part to be machined in the lathe and remove it after the work is complete, they are perhaps incorrectly called automatic lathes. Lathes that are *fully automatic* are provided with a magazine feed so that a number of parts can be machined, one after the other, with little attention from the operator. Machines in this group differ principally in the manner of feeding the tools to the work. Most machines, especially those holding the work between centers, have front and rear tool slides. Others, adapted for chucking jobs, have an end tool slide located in the same position as the turret on the turret lathe. These machines may also have the two side tool slides.

Duplicating Lathes

Duplicating or ***tracer lathes*** reproduce a number of parts from either a master form or a sample of the workpiece. This machine is shown in Figure 19.5. Most any standard lathe can be modified for tracer work, or special automatic tracer lathes are available. Reproduction is from a *template,* either round or flat, which is generally mounted at the rear of the lathe. It is engaged by a stylus that is actuated by air, hydraulic, or electric means. On this type of lathe many kinds of cuts can be made using only a single-point tool.

Other models are usually furnished with a point-to-point NC system having a direct-reading decimal dial input. The tracer unit is an electromechanical system composed of three parts: an electric amplifier, a controllable mechanical power amplifier for positioning the cross slide, and a stylus that engages the master template. Another feature of the lathe is the automatically controlled two-position indexing tool block mounted above the workpiece. Once set, this machine will produce pieces without attention except for loading and unloading.

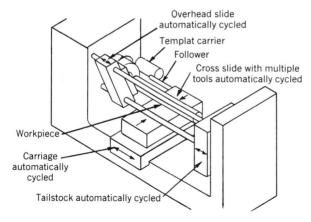

Figure 19.5
Duplicating lathe.

Automatic Screw Machine

The *automatic screw machine* was invented by Christopher N. Spencer of the Billings and Spencer Company in 1873. The principal feature of the invention provides a controlling movement for the turret so that tools can be fed into the work at desired speeds, withdrawn, and indexed to the next position. This was accomplished by a cylindrical or drum cam located beneath the turret. Another feature, also cam controlled, was a mechanism for clamping the work in the collet, releasing it at the end of the cycle, and then feeding unworked bar stock against the stop. These features are still used in about the same way as originally worked out.

The first machines of this type were used mainly for manufacturing bolts and screws. Because it can produce parts one after the other with little attention from the operator, it is called automatic. Most automatic screw machines not only feed in an entire bar of stock, but also are provided with a magazine so that bars can be fed through the machine automatically.

Automatic screw machines may be classified according to the turret or the number of spindles. Multispindle machines, however, are not usually spoken of as screw machines,

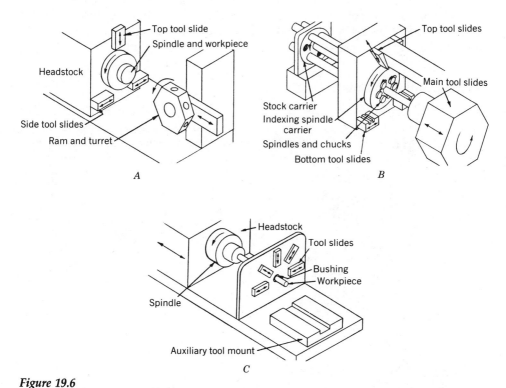

Figure 19.6
A, Single-spindle automatic screw machine. B, Multispindle automatic screw machine. C, Swiss automatic screw machine.

but rather as *multispindle automatics*. The work that the two machines do is the same, although there is a difference in the design and production capacity.

Figure 19.6A illustrates on automatic screw machine designed for bar work of small diameter. This machine has a cross slide capable of carrying tools both front and rear and a turret mounted in a vertical position on a slide with longitudinal movement. The tools used in the machine are mounted around the turret in a vertical plane in line with the spindle. Usual machining operations such as turning, drilling, boring, and threading can be done on these machines. The bar stock used, whether round, square, hexagonal, or of some special shape, is determined by the cross section preferred in the finished product. Collets that are sized for these commercial shapes are available.

Figures 19.7 and 19.6C are the end view and schematic of a *Swiss-type screw machine* developed for precision turning of small parts. The single-point tools used on this machine are placed radially around the carbide-lined guide bushing through which the stock is advanced during machining operations. Most diameter turning is done by the two horizontal tool slides, while the other three are used principally for knurling, chamfering, cutting off, and recessing. The stock is held by a rotating collet in the headstock behind the tools, and all longitudinal feeds are accomplished by a cam that moves the headstock forward as a unit. This forward motion advances the stock through the guide bushing and to the single-point tools, which are controlled and positioned by cams. By coordinating their movement with the forward movement of the stock, small diameters on slender parts can be held to tolerances ranging from 0.0002 to 0.0005 in. (0.005–0.013 mm).

Multispindle Automatic

Multispindle automatic machines are the fastest type of production machines for bar work. They are fully automatic in their operations and are made in a variety of models with two, four, five, six, or eight spindles. In these machines the steps of the operation

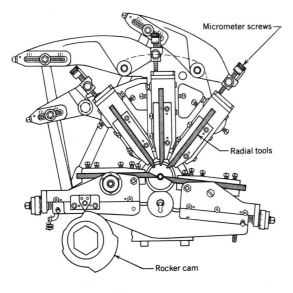

Figure 19.7
End view of Swiss-type screw machine showing rocker cam and tool control mechanism.

are divided so that a portion of it is performed at each of the several stations simultaneously, thereby shortening the time required to finish one part. One piece is completed each time the tools are withdrawn and the spindles indexed.

The construction of a multispindle automatic is shown in Figure 19.8. A schematic is shown in Figure 19.6B. The spindles carrying the bar stock are held and rotated in the stock reel. In front of the spindle is an end tool slide on which tools are placed in line with each of the spindles of the machine. The tool slide does not index or revolve with the spindle carrier, but moves forward and back to carry the end working tools to and from contact with the revolving bars of stock. The tooling section also includes one cross slide for each spindle position. Slides are independently operated and are used in connection with end slide tools for such operations as form turning, knurling, thread rolling, slotting, and cutting off. Tools for such operations as drilling and threading are mounted on the end tool slide.

Bars of stock are loaded into each spindle when it has been indexed to the first position. If automatic stock feeding is used, it is done in the lower spindle position at the rear of the machine. In operation the spindle carrier is indexed by steps to bring the bar of stock in each of the work spindles successively in line with the various tools held on the tool slides. All tools in the successive positions are at work on different bars at the same time. The time to complete one part is equal to the time of the longest operation plus the time necessary for withdrawing the tools and indexing to the next position. This time can frequently be reduced to a minimum by dividing the long cuts between two or more operations. Multispindle automatics are not limited to bar stock but may be provided with hydraulic or air-operated chucks for holding individual pieces.

A variety of parts can be produced by a multispindle automatic, the only limiting factor being the capacity of the machine. Long-run jobs are necessary to offset the high initial investment, maintenance, and tooling costs. Both single-spindle automatics and

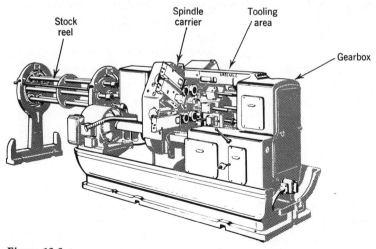

Figure 19.8
Multispindle automatic bar machine.

hand turret lathes have wide application, and in short- and medium-run work prove to be economical in operation. Each machine is good in its field, but care must be taken in making the initial selection.

DRILL PRESS GROUP

Drill Presses

Drill presses are very common, useful, and inexpensive relative to other machine tools. Drill presses are often classified according to the maximum diameter drill that they will hold, as in portable drilling units. The size of a sensitive or upright drilling machine is designated by the diameter of the largest workpiece that can be drilled. Thus a 24-in. (600-mm) machine has at least 12 in. (300 mm) clearance between the center line of the drill and the machine frame. The size of radial drilling machines is based on the length of the arm in feet (meters). Usual sizes are 4 ft (1.2 m), 6 ft (1.8 m), and 8 ft (2.4 m). In some cases the diameter of the column in inches (millimeters) is also used in expressing size.

Portable and Sensitive Drills

Portable drills are small compact drilling machines used principally for drilling operations that cannot be conveniently done on a regular drill press. The simplest of these is the hand-operated drill. Most portable drills equipped with small electric motors operate at fairly high speeds and accommodate drills up to $1/2$ in. (12.7 mm) in diameter. Similar drills using compressed air as a means of power are used where sparks from the motor may constitute a fire hazard.

The *sensitive drilling machine* is a small, high-speed machine of simple construction similar to the ordinary upright drill press. It consists of an upright, standard, horizontal table and a vertical spindle for holding and rotating the drill. Machines of this type are hand fed, usually by means of a rack and pinion drive on the sleeve holding the rotating spindle. These drills may be driven directly by a motor, belt, or a friction disk. The *friction disk drive,* which has considerable speed regulation, is not suitable for slow speeds and heavy cuts. Sensitive drill presses are suitable only for light work and are seldom capable of rotating drills over $5/8$ in. (16 mm) in diameter.

Upright Drills

Upright drills similar to sensitive drills have power feeding mechanisms for the rotating drills and are designed for heavier work. Figure 19.9A shows a 39-in. (1 m) machine with a box-type column. A ***box column*** machine is more rigid than a round-column machine and, consequently, is adapted to heavier work. These drilling machines can be used for tapping as well as for drilling.

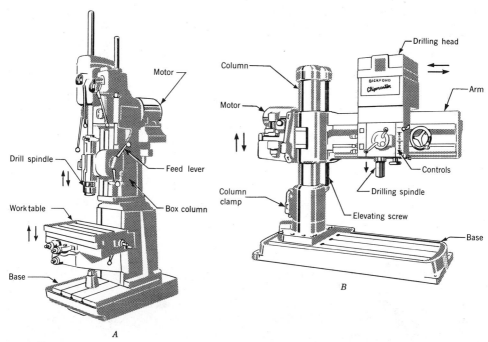

Figure 19.9
A, Thirty-nine inch (1 m) upright drill press. *B*, Radial drilling machine.

Radial Drilling Machine

The radial drilling machine is designed for large work when it is not feasible for the work to be moved. Such a machine (Figure 19.9*B*) consists of a vertical column supporting an arm that carries the drilling head. The arm may be swung around to any position over the work bed, and the drilling head has a radial adjustment along this arm. These adjustments permit the operator to locate the drill over any point on the work. Plain machines of this type will drill only in the vertical plane. On semiuniversal machines the head may be swiveled on the arm to drill holes at various angles in a vertical plane.

Gang Drilling Machine

When several drilling spindles are mounted on a single table, it is known as a ***gang drill***. Figure 19.10*A* shows a three-spindle schematic. This type is adapted to production work where several operations must be performed. The work is held in a jig that can be moved on the table from one spindle to the next. If several operations must be performed, such as drilling two different-sized holes and reaming them, four spindles are set up. With automatic feed control, two or more of these operations may be going on simultaneously, attended by only one operator. The arrangement is similar to operating several independent drill presses and is much more convenient because of its compactness.

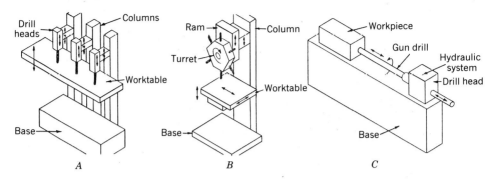

Figure 19.10
A, Gang drilling machine. B, Turret drilling machine. C, Gun drilling machine.

Turret Machines

A turret machine overcomes the floor space restriction caused by a gang drill press. A six-turret station NC drill press is shown in Figure 19.10B. The stations can be set up with a variety of tools. NC is also available. Two fixtures can be located on the worktable, thus permitting loading and unloading of one part while the other part is being machined. This reduces the machine cycle.

Multispindle Drilling Machines

Multispindle or cluster drilling machines, as shown in Figure 19.11, drill several holes simultaneously. The holes may not be the same diameter. They are production machines that will drill many parts with accuracy. Usually, a jig or fixture provided with hardened bushings is essential to guide the drills into the work.

A common design of this machine has a head assembly with a number of fixed upper spindles driven from pinions surrounding a central gear. Corresponding spindles are located below this gear and are connected to the upper ones by a tubular drive shaft and two universal joints. Three lower spindles carrying the drills can be adjusted over a wide area.

Multispindle drilling machines frequently use a table feed eliminating the movement of the heavy geared-head mechanism that rotates the drills. This may be done by rack and pinion drive, lead screw, or a rotating plate cam. The last method provides varying motions that give rapid approach, uniform feed, and quick return to starting position.

Transfer-Type Production Drilling Machines

Frequently designated as automated machines, they complete a series of machining operations at successive stations and transfer the work from one station to the next. They are in effect a production line of connected machines that are synchronized in their operation so that the workpiece, after being loaded at the first station, progresses auto-

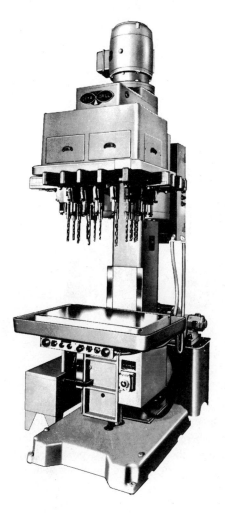

Figure 19.11
General-purpose, adjustable multispindle drilling and tapping machines.

matically through the various stations to its completion. The automatic machines are of the indexing-table or the in-line transfer types:

1. *Indexing table.* Parts requiring only a few operations are adapted to indexing-table machines, which are made with either vertical or horizontal spindles spaced around the periphery of the indexing table.

2. *Transfer type.* The **transfer machines** are provided with suitable handling or transfer means between stations. A simple and economical method of handling parts is to move them on a rail or conveyor between stations. When this is impractical because of the shape of the part, the part is clamped on to a holding fixture or pallet.

Figure 19.12 shows a 35-station *automatic transfer machine* that performs a variety of operations on transmission cases. Palletized workholding fixtures secure the trans-

494 TURNING, DRILLING, BORING, AND MILLING MACHINE TOOLS

Figure 19.12
Thirty-five station palletized transfer machine for transmission cases; 75 parts per hour production.

missions during all operations. Transfer machines range from comparatively small units having only two or three stations to long straight-line machines with over 100 stations. They are used primarily in the automobile industry where with full production schedules it is possible to offset their high initial cost by savings in labor. Products processed by these machines include cylinder blocks, cylinder heads, refrigeration compressor bodies, and similar parts.

Deep-Hole Drilling Machine

Several problems not encountered in ordinary drilling operations arise in drilling long holes in rifle barrels, long spindles, connecting rods, and certain oil well drilling equipment. As the hole length increases, it becomes more difficult to support the work and the drill. Rapid removal of chips from the drilling operation becomes necessary to ensure the operation and accuracy of the drill. Rotational speeds and feeds must be carefully determined, because there is greater possibility of deflection than when a drill of shorter length of used.

Deep-hole drilling machines have been developed to overcome these problems. These machines may be either horizontal or vertical. The work or the drill may revolve. Most machines are of horizontal construction, using a center-cut gun drill that has a single cutting edge with a straight flute running throughout its length (see Figure 19.10C). Oil under high pressure is forced to the cutting edge through a lengthwise hole in the drill. In gun drilling the feed must be light to avoid deflecting the drill.

BORING MACHINE TOOL GROUP
Boring Machines

Several machines have been developed that are specially adapted to boring work. The jig borer is constructed for precision work on jigs and fixtures. Similar in appearance to a drill press, it will drill and end mill in addition to bore. The vertical boring mill and the horizontal boring machine are adapted to large work. Although the operations that these machines perform can be done on lathes and other machines, their construction is justified by the ease and economy obtained in holding and machining the work.

Jig Boring Machine

Figure 19.13A is a schematic of a machine designed for locating and boring holes in jigs, fixtures, dies, gages, and other precision parts. *Jig-boring machines* resemble a vertical milling machine but are constructed with greater precision and are equipped with accurate measuring devices for controlling table movements. On typical machines positioning to ±0.0001 in. (±0.003 mm) can be dialed directly from a drawing. There are two sets of direct-reading dials, one for longitudinal and the other for transverse measurements. The operator sets the numbers on the dials to correspond with the dimensions on the drawing and, by pushing the button for each axis on the control panel, can automatically and accurately position the workpiece.

This machine is also designed to be operated by NC. Putting jobs on tape assures

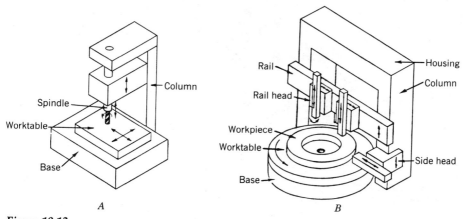

Figure 19.13
A, Jig-boring machine. B, Vertical boring mill.

accurate repetition, eliminates jigs and fixtures, and renders precision boring practical for for small-lot manufacturing.

Vertical Boring Mill

In the vertical boring mill the work rotates on a horizontal table in a fashion similar to the old potter's mill. The cutting tools are stationary except for feed movements, and are mounted on the adjustable-height crossrail. These tools are of the lathe and planer type and are adapted to horizontal facing work, vertical turning, and boring. This machine is sometimes called a *rotary planer,* and its cutting action on flat disks is identical with that of a planer. These machines, rated according to their table diameter, vary in size from 3 to 40 ft (0.9–12 m). A sketch is shown in Figure 19.13B.

Horizontal Boring Machine

The horizontal boring machine, nicknamed a "bar," differs from the vertical boring mill in that the work is stationary and the tool is revolved. It is adapted to the boring of horizontal holes, as can be seen in Figure 19.14. The horizontal spindle for holding the tool is supported in an assembly at one end that can be adjusted vertically within the

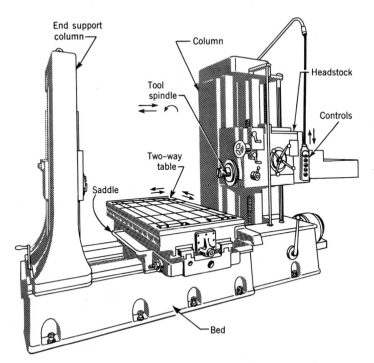

Figure 19.14
Table-type horizontal boring machine to perform boring, drilling, and milling operations.

limits of the machine. A worktable having longitudinal and crosswise movements is supported on ways on the bed of the machine. In some cases the table is capable of being swiveled to permit indexing the work and boring holes at desired angles. At the other end of the machine is an upright to support the outer end of a boring bar when boring through holes in large castings.

MILLING MACHINE GROUP
Milling Machines

Milling machines are made in a great variety of types and sizes. The drive may be either a cone pulley belt drive or an individual motor. The feed of the work may be manual, mechanical, electric, or hydraulic. There are a variety of table movements, as will be seen by the following milling machine descriptions.

Hand Milling Machine

The simplest milling machine is hand operated. It may have either the column and knee construction or the table mounted on a fixed bed. Machines operated by hand are used principally in production work for light and simple milling operations such as cutting grooves, short keyways, and slotting. These machines have a horizontal arbor for holding the cutter and a worktable that is usually provided with three movements. The work is fed to the rotating cutter either by manual movement of a lever or by a hand screw feed.

Plain Milling Machine

The plain milling machine is similar to the hand machine except that it is of much sturdier construction and is provided with a power feeding mechanism to control the table movements. Plain milling machines of the column and knee type have three motions: longitudinal, transverse, and vertical. Figure 19.15 shows a knee and column plain milling machine. Although it is a general-purpose machine, it is also used for quantity production work. Other models are available with universal or vertical head milling features. The machine uses stop dogs to control machine slide movements. Automatic table cycles are available. Hand wheels are able to control both the longitudinal and cross movements of the table, or power can be used for one move while the other is hand controlled. Cutters are mounted on a horizontal arbor rigidly supported by the overarm.

Vertical Milling Machines

A typical vertical machine, shown in Figure 19.16, is so called because of the vertical position of the cutter spindle. The table movements are the same as in plain machines. Ordinarily no movement is given to the cutter other than the usual rotational motion. However, the spindle head may be swiveled, which permits setting the spindle in a vertical plane at any angle from vertical to horizontal. This machine has a short axial

498 TURNING, DRILLING, BORING, AND MILLING MACHINE TOOLS

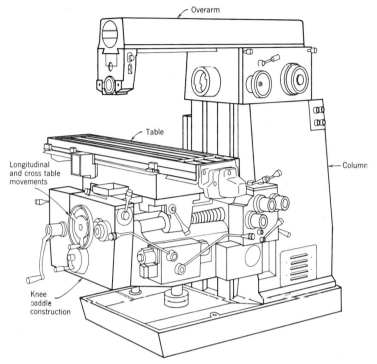

Figure 19.15
Knee and column milling machine.

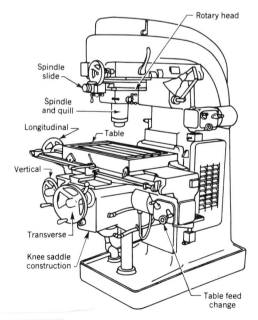

Figure 19.16
Rotary head vertical milling machine.

spindle travel to facilitate step milling. Some vertical milling machines are provided with rotary attachments or rotating worktables to permit milling circular grooves or continuous milling of small-production parts. End mill cutters are used.

Planer-Type Milling Machine

This milling machine receives its name from its resemblance to a planer. The work is carried on a long table having only a longitudinal movement and is fed against the rotating cutter at the proper speed. The variable table feeding movement and the rotating cutter are the principal features that distinguish this machine from a planer. Transverse and vertical movements are provided on the cutter spindle. These machines are designed for milling large work requiring heavy stock removal and for accurate duplication of contours and profiles. A hydraulically operated unit of this type is shown in Figure 19.17. Much work formerly done on a planer is now done on this machine.

Machining Centers

These NC machines are designed for small- to medium-lot production. The term *machining center* was unknown before NC, covered in Chapter 16. Refer to Figure 16.2 for an illustration. A machining center may refer to one or more NC machines that have multipurpose-capacity machining. It is incorrect to assume that these machines can do only milling. The machining center will mill, drill, bore, ream, tap, and contour, all in

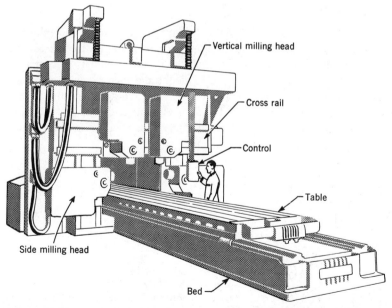

Figure 19.17
Planer-type milling machine.

a single setup. Depending on the machine, machining centers will start and stop the work, select and change tools, perform two- or three-dimensional contouring using linear or other interpolation techniques, feed in any of two or three axes approximately 0.1 to 99 in./min (40 μm/s to 40 mm/s), position any axis at a rapid traverse rate of 400 in./min (0.17 m/s), start or stop the spindle at programmed speed and direction of rotation, index the table to a programmed position, and turn the coolant on and off. Although these machines are versatile, the many features are optional depending on cost.

The machines, while costly, may replace several other machines. NC has made little improvement on the actual machining. Its savings contribution depends on auxiliary and supporting functions. To maximize all practical savings from this versatile manufacturing machine tool, inasmuch as possible the parts should be completely machined with one setup without transfer to several machines.

A tool changer automatically changes tools in 4 s or more, and cut-to-cut time is 8 to 10 s. A *tool-changer magazine* may have anywhere from 8 to 90 or more tools stored permanently or semipermanently. Accurate depth settings are possible by touching off tool points at work surfaces with a spindle hand wheel. This information is then recorded within the machine control unit. NC machining centers have switchable inch or metric programming.

The performance of these machines in cutting is 7 in.^{3}/min (1900 mm^3/s) for steel and 14 in.^{3}/min (3800 mm^3/s) for cast iron. The tolerances are boring, ±0.0002 in. (±0.005 mm); positioning, ±0.0001 in. (±0.003 mm); repeatability, ±0.0005 in. (0.013 mm); and depths to ±0.0001 in. (±0.003 mm).

Planetary Milling Machine

Planetary milling machines are used for milling both internal and external short threads and surfaces. The work is held stationary, and all movements necessary for the cutting are made by the milling cutters. At the start of a job the rotating cutter is in center or neutral position. It is first fed radially to the proper depth and then given a planetary motion either inside or around the work. The relation between the work and the cutters is illustrated by the line diagram in Figure 19.18. Typical applications of this machine include milling internal and external threads on all kinds of tapered surfaces, bearing surfaces, rear axle end holes, and shell and bomb ends.

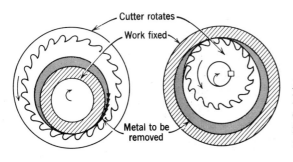

Figure 19.18
Planetary milling setup showing cutter action for both external and internal milling. Left, external milling. Right, internal milling.

MILLING MACHINE GROUP 501

Duplicating Machines

The production of large forming dies for automobile fenders, tops, and panels is an important use of duplicator machines. These machines reproduce a part from a model with no reduction or enlargement of size. A large-capacity machine is illustrated in Figure 19.19. This machine, known as a *copy milling* machine, has a true and mirror image-copying range of 177 by 98 in. (4.5 by 2.5 m) with a table size of 216 by 126 in. (5.5 by 3.2 m). The 7-in. (178-mm) diameter spindle has an infinitely adjustable speed from

Figure 19.19
A copy milling machine.

10 to 150 rpm. The models or templates used in this work are made of hard wood, plaster of paris, or other easily worked materials, because the only purpose they serve is to guide the tracer that controls the tool position.

QUESTIONS AND PROBLEMS

1. Give an explanation of the following terms:

   Tailstock
   Lathe size
   Turret lathe
   Ram-type turret lathe
   Duplicating lathe
   Multispindle automatic
   Sensitive drill press
   Friction disk drive
   Box column
   Gang drilling
   Transfer machines
   Deep-hole drilling machine
   Jig-boring machine
   Machining center
   Planetary milling

2. What functions does the carriage assembly in an engine lathe perform?

3. How are tools held for machining in an engine lathe? Also refer to earlier chapters.

4. Describe differences in horizontal turret lathes.

5. Distinguish between the cross slide on a turret lathe and that on an engine lathe.

6. What is the purpose of each of the following lathe parts: face plate, compound rest, lead screw, back gears, and tailstock?

7. Discuss the accuracy that can be expected in turning a part with a toolroom or production lathe.

8. Classify the differences in automatic screw machines that make small parts.

9. What problems will a lathe with a worn lead screw cause?

10. How can a hole be bored in a short piece of bar stock with threading on the OD taking place at the same time? On an engine lathe? On a turret lathe?

11. What is the purpose of the feed rod? If trip dogs are worn, what problems might occur?

12. How many "working" stations are there on an eight-station vertical chucking lathe?

13. List the kinematic sequence of movements for a Swiss screw machine.

14. A large casting is to be drilled. The supervisor suggests that a sensitive drill press be used. Comment on her suggestion.

15. A sheet metal part to be drilled has several diameters. What kind of drill press would you select, and why?

16. Select the appropriate type of drill press for the following parts: small 3-lb casting with one hole diameter; 25 units of a gear blank requiring drilling and reaming of the center hole; thick metal with a schedule of 10 different-sized holes; milling machine base made of cast iron; and stainless-steel tube 6 in. in diameter with 25 holes along a center line 72 in. long.

17. What type of machine would you recommend for making (a) tiny shafts for a watch; (b) 10,000 precision $5/8$-11 screws; 5 precision $5/8$-11 screws; threading 10,000 $1/2$-13 NC nipples?

18. Explain the terms multiple or single spindle when applied to stock feeders.

19. Describe the operation of a Swiss-type automatic screw machine.

20. A 16-in. lathe with a 4-ft. bed is to be employed in a shop making all SI-measured items. What is the size of this lathe in SI units?

21. What type of machine should be used for making $3/4$-10 by $1 1/2$ in. (9.5 by 38.1 mm) hexagonal head cap screws when the quantities are 10, 200, and 10,000?

22. List various machines for making holes in heavy castings when the quantity is several; when the quantity is 2500.

23. What is the difference between a drilling and a boring machine?

24. Radial drilling machines are used for what kind of parts?

25. Contrast the operation of a vertical boring machine with a horizontal boring machine. Give part examples for each type of machine.

26. Describe the operation of a tool changer. What machine has a tool changer?

27. What circumstances call for the use of a gun-drilling machine?

28. What type of milling machine do you select for the following parts: cutting teeth on a spur gear; milling a pocket in a casting; recessing a slot in bar stock; cutting a keyway in a round bar; and surfacing a curving face of a die?

29. How does a vertical milling machine differ from a vertical drill press? What kinds of parts would you run on each?

30. A planer miller has replaced much of the work handled by a planer. Substantiate the causes for this claim.

31. Fixed-bed milling machines are used for what kinds of work? Briefly describe their construction.

32. Indicate similarities between a turret lathe and a machining center. Also between a vertical boring mill and a jig-boring machine.

33. Describe the major parts of a machining center. Also refer to previous chapters.

CASE STUDY

AIRFRAME PART, P.N. 50532

Advanced Machine Shop has received an order for 275 units of part number 50532, which is used on a new commercial passenger jet. Part material is ASTM 4340 alloy steel, which is purchased as annealed bar stock in 4-in. diameters. Each bar is 12 ft long when received. The

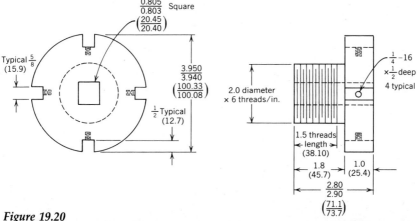

Figure 19.20
Case study.

drawing requires that the part test to Rockwell C45 hardness and have a surface finish no rougher than 64 μ in. arithmetic average. Other than the tolerances shown by the simplified sketch in Figure 19.20, unstated tolerances are commercial level or ±0.010 in. Other dimensions may be roughly scaled using the 1-in. dimension of the hub.

Jim Greene, the chief process planner, mentions to you, "Here is another job for that new jet. They want us to bid on this drawing. See what machines are necessary for the job and get back to me when you are finished. Oh yes, find out how many bars are necessary for the job. Don't forget to make allowances for the cutoff kerf and end facing stock."

As the new process planner, you have the job of selecting the possible machine tools that will transform the material from round 4-in. bar stock to finished dimension. Assume that the only machine tools available are those listed in this chapter. Form a rough operations sheet like the following:

Operation number	Machine tool description	Operation description

The columns are to be completed by listing successive operations 10, 20, . . ., choosing the machines, and writing operation descriptions. Follow that with sketches showing how each operation removes material. Use crosshatch conventions that correspond to the operations. Finally, determine the metal efficiency—that is, the ratio of the final metal in the parts to purchased bar stock. Assume that only 12-ft lengths of bar stock can be purchased.

CHAPTER 20
MACHINING OPERATIONS

In the production of fabricated parts machines, tools are important but insufficient by themselves. Chapter 19 introduced the turning, drilling, boring, and milling machine tools. In this chapter consideration is directed to the cutting tools, jigs, and fixtures, and the performance of these integrated production systems. It is the production system as a whole, the machine tool, cutter, attachments, and tooling, and its operation that is considered when one evaluates performance. If any part of the system is subpar, the system suffers. Cutting tools must run optimally, operators need to be trained and skilled, fixtures and attachments require design and construction, and the selection of the best machine by the process planner is important to economic success.

CUTTING TOOLS

In Chapter 18 cutting theory, machinability, tool life, materials for tools, and force and velocity patterns were discussed. Some general recommendations were given for single-point turning tools. The purpose of this section is to expand on *tools* (i.e., the point at which the material removal is ongoing) for turning, drilling, boring, and milling applications. Some of these tools can be used on several machine tool types; drills, for example, can be used on lathes, drill presses, and milling machines. Other tools are designed with the machine tool in mind, such as a shell end mill. Tools are either general or specific in purpose.

Turning Tools

Cutting tools are classified basically as single-point or multipoint cutters. For turning operations single-point cutters are used, and the theory and practice begins there. Three types of the wide variety of single-point tools available are shown in Figure 20.1. The solid high-speed steel insert is often ground by hand and placed in the toolholder as shown in Figure 20.1A. Back rake is usually included in the holder to simplify grinding. This tool (Figure 20.1A) is used mostly for laboratory and educational classes.

These high-speed steel cutter bits are often hand ground to form. See Figure 20.2 for various shapes. A lathe toolholder will have straight, left-hand, and right-hand shanks. *Left-hand shanks* permit machining operations close to the lathe chuck or face plate. *Right-hand shanks* are used for facing and machining operations that are close to the tailstock.

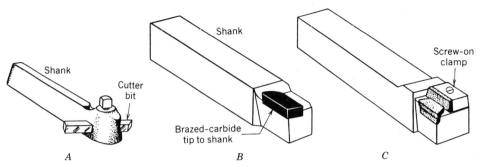

Figure 20.1
Toolholders. A, Solid high-speed steel insert bit. B, Brazed-carbide tip. C, Clamp-on type.

Carbide tools are costly and do not have as much resistance to shock as high-speed steel. Figure 20.1B demonstrates how the advantages of both materials are optimized. In this case a pocket is milled out of a high-carbon steel shank and a tungsten carbide tip is brazed in place. Brazed tip tools are used for general turning, drilling, reaming, and boring. As single-point turning tools they are adaptable when some job requires a special configuration. It may be necessary to grind the tip by diamond cutters. Tools with a **brazed-carbide tip** have brazing strains and have only one effective cutting corner.

Problems of brazed tips led to the disposable clamped-on insert shown in Figure 20.1C (see also Figure 18.10). When a cutting edge becomes dull the tool is unclamped and indexed to the next cutting edge. This saves time and avoids removing the cutter and regrinding, inserting, and adjusting to tolerance and dimension. Where negative rake is incorporated into the toolholder, both sides of the tool insert may be used. A **square**

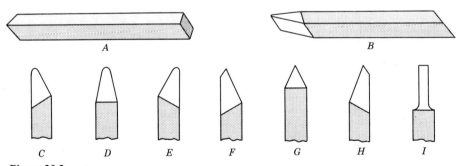

Figure 20.2
Lathe tool cutter bits. A, Cutter bit not ground. B, Cutter bit ground to form.
C, Left-hand turning tool. D, Round-nose turning tool. E, Right-hand turning tool.
F, Left-hand facing tool. G, Threading tool. H, Right-hand facing tool.
I, Cutoff tool.

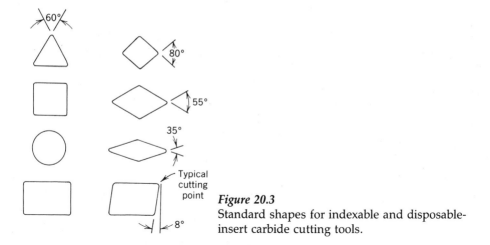

Figure 20.3
Standard shapes for indexable and disposable-insert carbide cutting tools.

insert would have eight cutting edges. These inserts are often referred to as *throwaways;* sometimes it is cheaper to throw them away after their use than regrinding. Common insert shapes are shown in Figure 20.3. Inserts can also be provided with built-in chip breakers.

Tool geometry and the nature of the angles were discussed in Chapter 18. Tool designations and symbols are shown in Figure 20.4. This right-hand tool travels from right to left. Positioning of the shank in the toolholder can affect these angles.

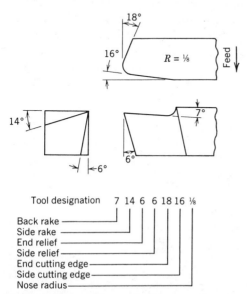

Figure 20.4
Cutting tool designations known as a tool signature.

Drills

A drill is a rotary-end cutting tool having one or more cutting edges and corresponding flutes that continue the length of the drill body. The *flutes,* which can be either straight or helical, provide passages for the chips and cutting fluid. Although most drills have two flutes, three or four may be used; the drill is then known as a core drill. Such drills are not used for starting a hole but for enlarging or finishing drilled and cored holes.

The most common drill is the *twist drill,* which has two flutes and two cutting edges. It is shown in Figure 20.5 with the various drill terms indicated. The drill can be provided with either a straight or a tapered shank. Tapered-shank drills are held and properly centered in the tapered socket of the drilling machine spindle. Drilling tools have a Morse taper of $5/8$ in./ft (5.209%), which is also standard for reamers and other similar tools. The tang at the end of the taper fits into a slot in the spindle socket to prevent slipping of the tapered surfaces. Straight-shank drills are generally held and properly centered in a drill chuck, but many are tanged and used with taper split sleeves. These drills, cheaper than those having a tapered shank, are used only for sizes up to $1/2$ in. (12.7 mm).

Several kinds of drills varying as to the number and angle of the flutes are shown in Figure 20.6. *Single-fluted drills* are used to originate holes and for deep-hole drilling. A two-fluted drill is the conventional type used for drilling holes. Either *interior* or external *oil channels* are used in production drilling. Three- and four-fluted drills are used principally for enlarging holes previously made. Both have greater productivity and improved finish than two-fluted drills. Other drills with various flute angles are available to give improved drilling to special materials and alloys.

Gun Drills

There are two kinds of straight-fluted *gun drills* used for deep-hole drilling (Figure 20.7). One, known as a trepanning drill (Figure 20.7A), has no dead center and leaves a solid core of metal. As the drill advances, the core acts as a continuous center guide at the point where the cutting is done. This prevents the drill from running to one side and makes it easier to maintain hole accuracy. The other type, known as a center-cut gun

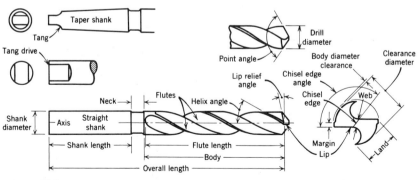

Figure 20.5
Standard twist drill and terms.

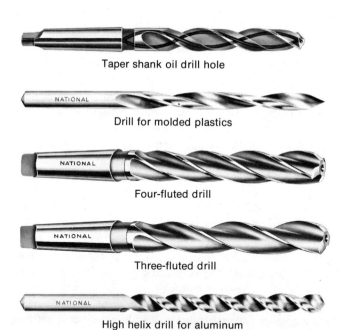

Taper shank oil drill hole

Drill for molded plastics

Four-fluted drill

Three-fluted drill

High helix drill for aluminum

Straight shank twist drill for mild steel

Figure 20.6
Types of drills.

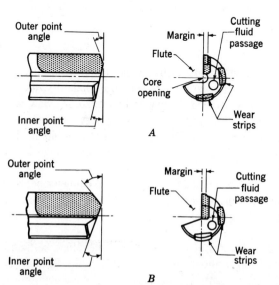

Figure 20.7
Straight-fluted gun drills. *A*, Trepanning drill. *B*, Center-cut gun drill.

drill (Figure 20.7B), has been the conventional drill for many years. It is still used for *deep-hole drilling* such as the drilling of blind holes where a core-type drill cannot be used. Both types are generally carbide tipped as shown in the figure.

Gun drills operate at much smaller feeds than conventional twist drills, but the cutting speeds are higher. Each type of drill has only a single cutting edge with a straight flute running throughout its length. Oil under considerable pressure is brought to the tip of the drill through the hole in the lip. The chips are carried out the hole along the flute of the drill as rapidly as they are formed. Greater accuracy and finish may be obtained in deep holes by the subsequent use of special reamers or broaching tools.

Gun drills and gun bores have a single-lip cutting action that is counteracted by bearing areas and lifting forces generated by coolant pressure. The supporting bearing arrangement forces the cutting edge to cut a true circular pattern and follow the direction of its own axis. Because of the single-point cutting action, the gun drill requires either a bushing or an accurate prebored start at the beginning of the hole. The tip of a gun drill has a (V-shaped cutting edge and is usually a brazed-carbide insert or a high-speed steel edge. Wear strips are located 180° and 90° from the cutting edge and consist of a wear-resistant material such as carbide.

Trepanning is used for large-diameter holes where an annular groove is cut axially through the workpiece, leaving a core of material called a slug. It is very efficient and cost effective, particularly if the slug is reusable.

Both gun drills and trepanning tools require a high-pressure stream of cutting fluid to lubricate and cool the cutting surface and to carry away the chips. Chips are removed by two methods, external or internal passages. External passages are usually a straight flute; fluid passes through the shank and out the top, carrying chips away through the flute. With an internal chip tool the chips pass through a chip mouth to the center of a tubular shank. Chip size must be carefully regulated to prevent clogging of the cutting tool. By using a heavy feed it is possible to install chip breakers to limit chip size. The internal chip tool is more torsionally rigid and cuts faster than the external design.

Special Drills

For drilling large holes in pipe or sheet metal, twist drills are not suitable because the drill tends to dig into the work or the hole is too large to be cut by a standard-size drill. Large holes are cut in thin metal by a *hole cutter* as shown in Figure 20.8A. Saw-type cutters of this design can be obtained in a wide range of sizes. For very large holes in thin metal a *fly cutter* is used. Such a cutter (Figure 20.8B) consists of tool bits held in a horizontal holder that can accommodate a range of diameters. Both cutters cut in the same path but one is set slightly below the other.

Spade drills (Figure 20.9) provide another method of making large-diameter holes in the $1^1/_2$ to 15 in. (38.1–380 mm) range. For holes over $3^1/_2$ in. (90 mm) they are the only drills provided as stock items. Materials used for spade drills either are of high-speed steel or are carbide tipped.

A drill designed for hardened steel operates at high speed and develops sufficient friction to anneal the steel and permit cutting without softening the drill point. Such drills,

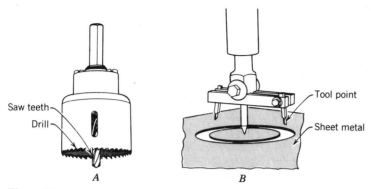

Figure 20.8
Cutters for holes in thin metal. *A,* Saw cutter. *B,* Fly cutter.

often carbide tipped, are made with point angles of 118° or greater. Some drills are made in combination with other tools such as the combination drill and tap or the drill and countersink. *Double-margin* drills and step drills are available to produce accurate holes as shown in Figure 20.10.

Reamers

Reaming is the enlarging of a machined hole to proper size with a smooth finish. A *reamer* is an accurate tool and is not designed to remove much metal. Although reaming can be done on a drill press, other machine tools are adapted to perform them. The material removed by a reamer depends on hole size and the material type. Whereas 0.015 in. (0.38 mm) is probably a good average, it can be as little as 0.005 in (0.13 mm) for a small reamer, or up to $1/32$ in. (0.8 mm) for a large one. When reaming strain-hardening alloys, enough stock—never less than 0.005 in. (0.13 mm)—must be removed to avoid rubbing on the work-hardened surface. Because of the small amount of stock removed by this process, reamed holes are round and have a smooth surface. The terms applying to reamers are shown in Figure 20.11.

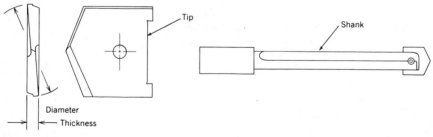

Figure 20.9
Spade drill and shank holder.

512 MACHINING OPERATIONS

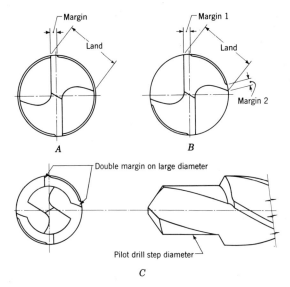

Figure 20.10
End views of conventional (*A*) and double-margin (*B,C*) drills showing construction differences. (*C*) Double-margin step drill. Pilot diameter not relieved, which improves piloting action.

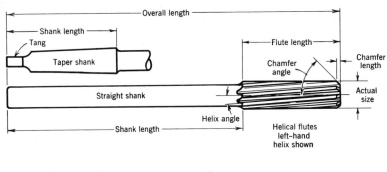

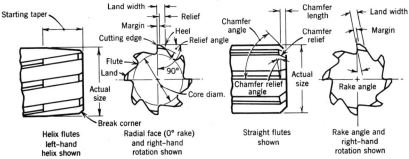

Figure 20.11
Standard reamer and terms.

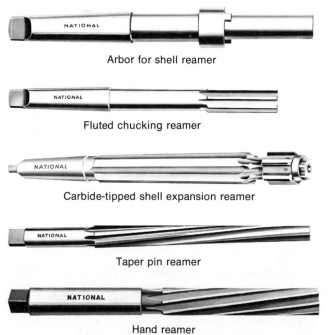

Figure 20.12
Types of reamers.

Illustrations of several reamers are shown in Figure 20.12. Hand reamers are finishing tools designed for the final sizing of holes. They are ground with a starting taper to provide easy entry of the reamer into the hole. This reamer, as well as most of the others, is made with both straight and spiral flutes.

Chucking reamers designed for use in machines are also made with both straight and spiral flutes. Spiral chucking reamers that have a free reaming action are used for materials difficult to ream. *Straight-fluted chucking reamers* are commonly used in turret lathes, drill presses, and lathes. Both reamers have a slight chamfer of 45° on the end. *Rose reamers,* which do all cutting on the beveled end, find some use in reaming cored holes. The main difference between the rose reamer and the fluted chucking reamer is that the former is relieved only on the chamfered end. *Shell reamers* consist of a shell-type end mounted on a tapered arbor. Slots in the reamer engage lugs on the arbor to obtain a positive drive. Shell reamers, used principally on turret lathes, are not recommended for removing excessive amounts of stock. *Taper pin reamers,* usually of small diameter, are quite long. They are made in both straight and spiral flute types, the latter being better for machine reaming. Taper reamers must be constructed as sturdily as possible, because the cutting edges are engaged in the cut through most of their length. *Expansion reamers* can be adjusted to compensate for wear or purposely to ream oversize holes. *Adjustable reamers* differ in that they can be manipulated to take care of a considerable range in size.

Taps and Threading Tools

A tap is a tool used to thread holes. It has a shank and a round body with several radially placed chasers. It may be thought of as a screw with teeth, and hardened to cut metals such as cast iron, steel, brass, and copper. Small taps are solid where large taps are solid or adjustable. A tap has two or more flutes that may be straight, spiral, or helical. They may be operated by hand or machine. Also, they are usually made of high-speed steel for rapid production and long wear. Because of their general importance for fastening, more material is devoted to their discussion. Additional material on threading may be found in Chapter 22.

When two parts are to be fastened together by a machine screw or cap screw, the hole in the untapped part is drilled larger than the outside diameter of the screw. The drill used for this operation is called a *clearance drill* because it provides a slight clearance between the part and the screw.

Boring Tools

Boring is enlarging holes previously drilled or bored. Drilled holes are frequently bored to eliminate any possible eccentricity and to enlarge the hole to a reaming size. Boring tools may also be used to finish holes to correct size, as is frequently done on large holes or on odd-sized holes for which no reamer is available.

Counterboring is enlarging one end of a drilled hole. The enlarged hole, which is concentric with the original one, is flat on the bottom. The tool is provided with a pilot pin that fits into the drilled hole to center the cutting edges. Counterboring is used principally to set bolt heads and nuts below the surface. To finish off a small surface around a drilled hole is known as *spot facing*. This is a customary practice on rough surfaces to provide smooth seats for bolt heads. If the top of a drilled hole is beveled to accommodate the conical seat of a flat-head screw, the operation is called *countersinking*.

Tools used in horizontal boring machines are mounted in either a heavy bar or a boring head, which in turn is connected to the main spindle of the machine. Most boring operations use a single-point cutter as shown in Figure 20.13, because they are simple

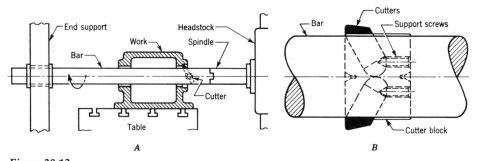

Figure 20.13
A, Straight boring on horizontal boring machine using line bar and end support.
B, Block-type boring cutter.

to set up and maintain. The bar serves to transmit power from the machine spindle to the cutter as well as to hold it rigidly during the cutting operation. The workpiece is normally stationary and the rotating cutter is fed through the hole. It is often necessary to provide additional support for the bar as shown in the figure. The bar must be long enough to reach the end support and also must provide the necessary longitudinal traverse for the machining operation.

For precision boring work on milling machines, jig borers, or drill presses, it is necessary to use a tool having micrometer adjustment. Such tools are held in a cutter head and rotate. Hence, any increase in hole size must be obtained by adjusting the tool radially from its center.

The most popular double-cutter arrangement is the **block type** shown in Figure 20.13B, which consists of two opposing cutters resting in grooves on the block. Screws are provided to lock the cutters in position as well as to adjust them. The entire assembly fits into a rectangular slot in the bar and is keyed in place. Cutters are ground while assembled in the block and are held in alignment by the center holes provided. The responsibility for tool accuracy and setup belongs to the toolroom personnel rather than the operator.

The boring tool commonly used in small machines such as lathes is a single-pointed tool, supported in a manner that permits its entry into a hole. This tool, shown in Figure 20.14A, is forged at the end and then ground to shape. It is supported in a separate holder that fits into a lathe tool post. For turret lathes, slightly different holders and forged tools similar to the one shown in Figure 20.14B are used. A modification of this tool is the boring bar shown in Figure 20.14C, which is designed to hold a small high-speed steel tool bit at the end. The bar supporting the tool is rigid and may be adjusted according to the hole length. Although the clearance, rake, and cutting angles of these tools should be similar to those recommended for lathe work, these angles cannot be used if the holes are small. Greater end clearance is necessary because of the curvature of the hole surface, and back rake is difficult to obtain because of the position of the tool. This may be seen

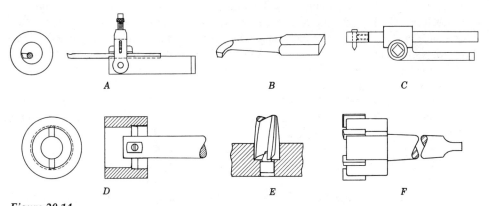

Figure 20.14
Types of boring tools. A, Light boring tool with bent shank. B, Forged boring tools. C, Heavy boring tool. D, Double-ended cutter or boring tool. E, Counterboring tool with pilot. F, Multiple-cutter boring tool.

by reference to the illustration showing the tool in working position. The side rake and side clearance angles have no restrictions. As the internal diameter is increased, properly shaped tools with correct angles can be used.

In production work, boring cutters with multiple cutting edges are widely used. These cutters, shown in Figure 20.14F, resemble shell reamers in appearance but are usually provided with inserted-tooth cutters that may be adjusted radially to compensate for wear and variations of diameter. Boring tools of this type have longer life than single-pointed tools and hence are more economical for production jobs. The counterboring tool shown in Figure 20.14E, provided with pilots to ensure concentric diameters, is designed to recess or enlarge one end of a hole.

Milling Cutters

The milling machine is versatile because of the large variety of cutters. These cutters are usually classified according to their general shape or by the way they are mounted, material used in teeth, or the method used in grinding the teeth.

1. *Arbor cutters.* These cutters have a hole in the center for mounting on an arbor.
2. *Shank cutters.* Cutters of this type have either a straight or tapered shank integral with the body of the cutter. These cutters are mounted in the spindle end.
3. *Face cutters.* These cutters are bolted or held on the end of short arbors and are generally used for milling plane surfaces.

A classification according to materials follows that of other cutting tools. Milling cutters are made of high-carbon steels, various high-speed steels, or those tipped with sintered carbide or certain cast nonferrous alloys. *High-carbon steel cutters* have a limited use because they dull quickly if high cutting speeds and feeds are used. Most general-purpose cutters are made of *high-speed steel,* which maintain a keen cutting edge at temperatures around 1000° to 1100°F (500°–600°C). Consequently, they may be used at cutting speeds 2 to 2.5 times those recommended for carbon steel cutters. *Cast nonferrous metals* (e.g., Stellite, cobalt, or Rexalloy) and carbide-tipped cutters have even greater resistance to heat and are especially adapted to heavy cuts and high cutting speeds. These materials either are used as inserts held in the body of the cutter or are brazed directly on the tips of the teeth. Cutting speeds of cast nonferrous and carbide cutters range from two to five times those recommended for high-speed steel.

Teeth in milling cutters are made in two general styles according to the method in sharpening. *Profile cutters* are sharpened by grinding a small land behind the cutting edge of the tooth. This also provides the necessary relief at the back of the cutting edge. *Formed cutters* are made with the relief (behind the cutting edge) of the same contour as the cutting edge. To sharpen these cutters the face is ground so as not to destroy the tooth contour.

The cutters most generally used, shown in Figure 20.15, are classified according to their general shape or the type of work they will do.

1. *Plain milling cutter.* A plain cutter is a disk-shaped cutter having teeth only on the circumference. The teeth may be either straight or helical if the width exceeds $5/8$ in.

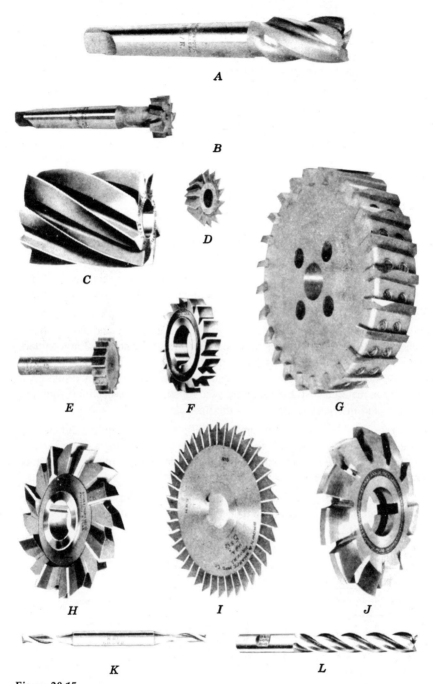

Figure 20.15
Types of milling cutters. A, Spiral-end mill. B, T-slot milling cutter. C, Plain cutter with helical teeth. D, Angle-milling cutter. E, Woodruff key-seat cutter. F, Plain milling cutter. G, Inserted-tooth cutter. H, Side-milling cutter. I, Metal-slitting saw cutter. J, Form-relieved cutter for gear teeth. K, Spiral double-end mill. L, Extra-long spiral-end mill.

(15.9 mm). Wide helical cutters used for heavy slabbing work may have notches in the teeth to break up the chips and facilitate their removal.

2. *Side milling cutter.* This cutter is similar to a plain cutter except that it has teeth on the side. Where two cutters operate together, each cutter is plain on one side and has teeth on the other. Side milling cutters may have straight, helical, or staggered teeth.

3. *Metal-slitting saw cutter.* This cutter resembles a plain or side cutter except that it is made very thin, usually $^3/_{16}$ in. (4.8 mm) or less. Plain cutters of this type are relieved by grinding the sides to afford clearance for the cutter.

4. *Angle-milling cutter.* All **angle-shaped cutters** come under this classification. They are made in both single- and double-angle cutters. The single-angle cutters have one conical surface, whereas the double-angle cutters have teeth on two conical surfaces. Angle cutters are used for cutting ratchet wheels, dovetails, flutes on milling cutters, and reamers.

5. *Form milling cutters.* The teeth on these cutters are given a special shape. They include convex and concave cutters, gear cutters, fluting cutters, corner-rounding cutters, and many others.

6. *End mill cutters.* These cutters have an integral shaft for driving and have teeth on both the periphery and the end. The flutes may be either straight or helical. Large cutters called shell mills have a separate cutting part held to a stub arbor as shown in Figure 20.16. Taking into account the cost of high-speed steel, this construction results in a considerable saving in material cost. End mills are used for surfacing projections, squaring ends, cutting slots, and in recess work such as die making.

7. *T-slot cutters.* Cutters of this type resemble small plain or side milling cutters that have an integral straight or tapered shaft for driving. They are used for milling T-slots. A special form is the *Woodruff* key seat cutter, which is made in standard sizes for cutting the round seats for Woodruff keys.

8. *Inserted-tooth cutter.* As cutters increase in size it is economical to insert the teeth made of expensive material into less expensive steel. Teeth in such cutters may be replaced when worn out or broken.

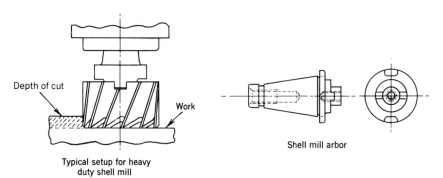

Figure 20.16
Shell mill and arbor.

Milling Cutter Teeth

A typical milling cutter with various angles and cutter nomenclature is shown in Figure 20.17. For most high-speed cutters positive radial rake angles of 10° to 15° are used. These values are satisfactory for most materials and represent a compromise between good shearing or cutting ability and strength. Milling cutters made for softer materials such as aluminum can be given much greater rake with improved cutting ability.

Usually only *saw-type* and *narrow, plain milling cutters* have straight teeth with zero axial rake. As cutters increase in width, a positive axial rake angle is used to increase cutting efficiency.

For milling with carbide-tipped cutters, negative rake angles (both radial and axial) are generally used. Improved tool life is obtained by the resultant increase in the lip angle; also, the tooth is better able to resist shock loads. Plain milling-type cutters with teeth on the periphery usually are given a negative rake of 5° to 10° when steel is being cut. Alloys and medium-carbon steels require greater negative rake than soft steels. Exceptions to the use of negative rake angles for carbide cutters are made when soft nonferrous metals are milled.

The clearance angle is the included angle between the land and a tangent to the cutter from the tip of the tooth. It is always positive and should be small so as not to weaken the cutting edge of the tooth. For most commercial cutters over 3 in. (75 mm) in diameter, the clearance angle is around 4° to 5°. Smaller diameter cutters have increased clearance angles to eliminate tendencies for the teeth to rub on the work. Clearance values also depend on the various work materials. Cast iron requires values of 4° to 7°, whereas soft materials such as magnesium, aluminum, and brass are cut efficiently with clearance angles of 10° to 12°. The width of the land should be kept small; usual values are $1/32$ to $1/16$ in. (0.8–1.6 mm). A secondary clearance is ground in back of the land to keep the width of the land within proper limits.

Much research on cutter form and size has proved that coarse teeth are more efficient for removing metal than fine teeth. A coarse-tooth cutter takes thicker chips and has freer cutting action and more clearance space for the chips. As a consequence, these cutters provide increased production and decreased power consumption for a given amount of

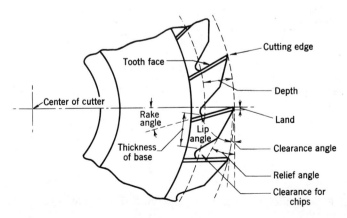

Figure 20.17 Milling cutter with nomenclature.

520 MACHINING OPERATIONS

metal removed. Also, fine-tooth cutters have a greater tendency to chatter than those with coarse teeth, but they are recommended for *saw cutters* used in the milling of thin material.

OPERATIONS

Operations can be performed in many diversified ways in turning, drilling, boring, and milling. These machines started the Industrial Revolution and continue to prove their great value. Operations make parts and products.

Lathe Operations

Operations on a lathe include turning, boring, facing, threading, and taper turning. For these operations a single-point cutter is fed along the revolving workpiece. Drilling and reaming require other types of cutters, but are also performed by the lathe group. A brief description of some of the turning operations follows.

Cylindrical Turning

The most common way to support work on a lathe is to mount it between centers as shown in Figure 20.18A. Holding work between centers allows the turning of the surface along its length.

When a flat surface is to be cut the operation is known as *facing*. The work is generally held on a face plate or in a chuck as illustrated in Figure 20.18B, but in some cases facing is also done with the workpiece between centers. Because the cut is at right angles to the axis of rotation, the carriage should be locked to the lathe bed to prevent axial movement.

Figure 20.19 shows various applications for right-hand and left-hand tools. Recall that the tools are ground with specific applications in mind.

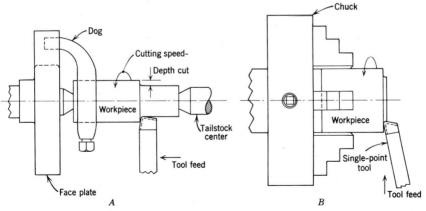

Figure 20.18
Lathe operations. *A*, Single-point tool in a turning operation. *B*, Facing cut.

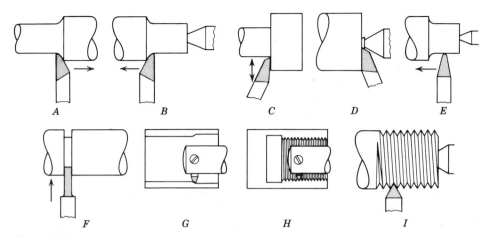

Figure 20.19
Lathe tools and applications. A, Left-hand turning tool. B, Round-nose turning tool. C, Right-hand turning tool. D, Left-hand facing tool. E, Threading tool. F, Right-hand facing tool. G, Cutoff tool. H, Boring tool. I, Inside-threading tool.

Taper Turning

Many parts and tools made in lathes have tapered surfaces, varying from the short steep tapers found on bevel gears and lathe center ends to the long gradual tapers found on lathe mandrels. The shanks of twist drills, end mills, reamers, arbors, and other tools are examples of taper work. Such tools, supported by taper shanks, are held in true position and are easily removed.

Several taper standards found in commercial practice are as follows:

1. *Morse taper.* Largely used for drill shanks, collets, and lathe centers. The taper is $5/8$ in./ft (5.208%).
2. *Brown and Sharpe taper.* Used principally in milling machine spindles, $1/2$ in./ft (4.166%).
3. *Jarno and Reed tapers.* Used by some manufacturers of lathe and small drilling equipment. Both systems have a taper of 0.6 in./ft (5.00%), but the diameters are different.
4. *Taper pins.* Used as fasteners. The taper is $1/4$ in./ft (2.083%).

Each of these standards is made in a variety of diameters and designated by a number. Accurate external tapers may be cut on a lathe in several ways:

1. Numerical control (NC) machines can cut tapers as a matter of course. Additionally, a straight chamfer, which is a type of taper, and radius operations are possible. See Chapter 16 for additional discussion of NC.

2. A taper-turning attachment on the lathe, illustrated in Figure 20.20A, is bolted onto the back of the lathe and has a guide bar that can be set at the desired angle or taper.

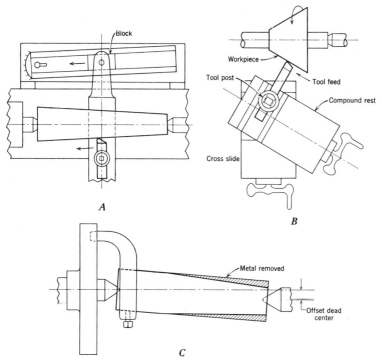

Figure 20.20
Taper turning. A, With attachment. B, With compound rest. C, Offsetting tailstock center.

As the carriage moves along the lathe bed, a slide over the bar causes the tool to move in or out according to the setting of the bar. Thus the taper setting of the bar is duplicated on the work.

3. The compound rest on the lathe carriage (Figure 20.20B) has a circular base and may be swiveled to any desired angle with the work. The tool is then fed into the work by hand. This method is especially adapted for short tapers.

4. Setting the tailstock center over (Figure 20.20C) is another method. If the tailstock is moved horizontally out of alignment $1/4$ in. (6.4 mm) and a cylinder 12 in. (305 mm) long is placed between centers, the taper will be $1/2$ in./ft (4.16%). However, a cylinder 6 in. (152 mm) long will have a taper of 1 in./ft (8.33%). Hence, the amount of taper obtained on a given piece depends on the length of the stock as well as on the amount the center is set over.

Internal tapers can be machined on a lathe by NC or by using the compound rest or the taper-turning attachment. Small holes for taper pins are first drilled and then reamed to size with a taper reamer.

Thread Cutting

While it is possible to cut all forms of threads, the engine lathe is usually selected when only a few threads are to be cut or when special forms are desired. The form of the thread is obtained by grinding the tool to the proper shape using a suitable gage or template. Figure 20.21 shows a cutter bit ground for cutting 60° V-threads and the gage that is used for checking the angle of the tool. This gage is known as a *center gage* because it is also used for gaging lathe centers. Special form cutters can also be used for cutting these threads. These cutters are previously shaped to the correct form and are sharpened by grinding only on the top face.

In setting up the tool for V-threads, there are two methods of feeding the tool. It may be fed straight into the work, the threads being formed by taking a series of light cuts as shown in Figure 20.21A. Cutting action occurs on both sides of the tool bit. Some back rake may be obtained, but it is impossible to provide any side rake on the cutting

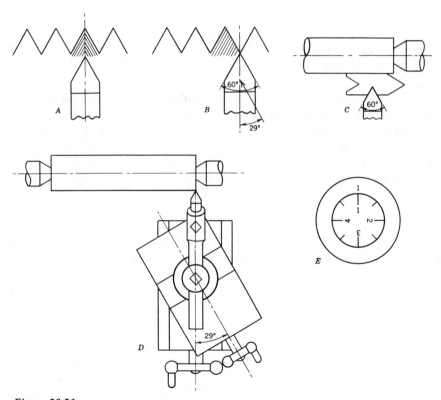

Figure 20.21
Method of setting a tool for thread cutting on a lathe. A, Straight feed. B, Feed at angle. C, Use of center gage for setting up threading tool. D, Method of setting up lathe for cutting V-thread. E, Threading dial.

tool. This method is satisfactory on materials such as cast iron or brass where little or no side rake is recommended. The second method, for cutting steel threads, feeds the tool in at an angle as shown in Figure 20.21B and D. The compound rest is turned to an angle of 29° and, by using the crossfeed on the compound rest, the tool is fed into the work so that all cutting is done on the left-hand side of the tool. The tool bit, being ground to an angle of 60°, allows 1° of the right-hand side of the tool to smooth off that side of the thread.

It is necessary that the tool be given a positive feed along the work at the proper rate to cut the desired number of threads per inch. This is accomplished by a train of gears located on the end of the lathe that drives the lead screw at the required speed with relation to the headstock spindle. This gearing may be changed to cut any desired pitch of screw. The lead screw in turn, engages the half-nuts on the apron of the lathe, providing positive drive for the tool.

After the engine lathe is set up, the crossfeed screw is set at some mark on the micrometer dial and a light cut is taken to check the pitch of the thread. At the end of each successive cut the tool is removed from the thread by backing off the crossfeed screw. This is necessary because any back play in the lead screw would prevent the tool from returning in its previous cut. The tool is returned to its original position, the crossfeed screw is set at the same reference mark, the tool is fed the desired amount for the next cut, and another cut is taken. These operations are repeated until the thread is cut to a proper depth.

Many engine lathes are equipped with a ***thread dial*** shown in Figure 20.21E. Close by the dial is a lever that is used to engage and disengage the lead screw with a matching set of half-nuts in the carriage. At the end of each cut the half-nuts are disengaged and then reengaged at the correct time, so that the tool always follows in the same cut. The indicator is connected to the lead screw by a small worm gear, and the face of the dial, which revolves, is numbered to indicate positions at which the half-nuts may be engaged.

Turret Lathe Operation

Once a turret lathe is properly tooled, an experienced machinist is not required for operation. However, skill is necessary in the selection, mounting, and adjustment of the tools. In small-lot production it is important that this work be done in as short a time as possible so as not to consume too much of the total production time, which consists of setup, work handling, machine handling, and cutting time. *Setup time* can be reduced by having the tools in condition and readily available. For short-run jobs a permanent setup of the usual bar tools on the turret is a means of reducing time. The tools selected are standard and, when permanently mounted, they may be quickly adjusted for various jobs. The *loading and unloading time,* which is the time consumed in mounting or removing the work, largely depends on the workholding devices used. For bar work this time is reduced to a minimum by using bar stock collets.

The time it takes to bring the respective tools into cutting position is a part of the *machine handling time*. This can be reduced by having the tools in proper position and sequence for convenient use and also by taking multiple or combined cuts whenever possible. The balance of the machine handling time is made up of the time necessary to change the speeds and feeds.

The *cutting time* for a given operation is controlled by the use of proper cutting tools, feeds, and speeds. However, additional time may often be saved by combining cuts as shown in Figure 20.22A. Combined cuts refer to the simultaneous use of both slide and turret tools. In bar work combined cuts are especially desirable, as additional support is given to the workpiece thereby reducing spring and chatter. In chucking work, internal operations such as drilling or boring may frequently be combined with turning or facing cuts from the square turret. Time also may be saved by taking multiple cuts—that is, having two or more tools mounted on one tool station. Figure 20.22B shows both boring and turning tools set up on one station of the turret.

For outside turning a single-cutter turner or box tool (Figure 20.22A) has been developed. As bar stock is supported only at the collet, additional support must be provided for heavy cuts to be taken. This is done by means of two rollers that contact the turned diameter of the stock and take up the thrust of the cutting tool.

To illustrate the method of tooling and sequence of operations for a given job, a basic hexagon turret setup for making necessary internal cuts on a threaded adapter is shown in Figure 20.23. This shows the details of the internal cuts required to machine the adapter. The various operations are given in the following:

1. The bar stock is advanced against the **combination stock stop and start drill** and clamped in the collet. The start drill is then advanced in the combination tool, and the end of the work is centered.
2. The hole through the solid stock is drilled to the required length.
3. The thread diameter is bored to correct size for the threads specified. A stub boring bar in a slide tool is used.
4. The drilled hole is reamed to size with the reamer supported in a floating holder.

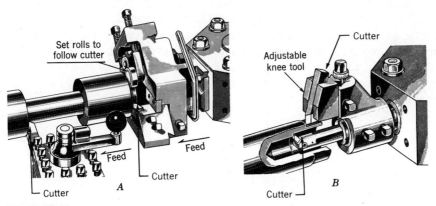

Figure 20.22
A, Combining cuts on bar work. B, Multiple cuts from hexagon turret.

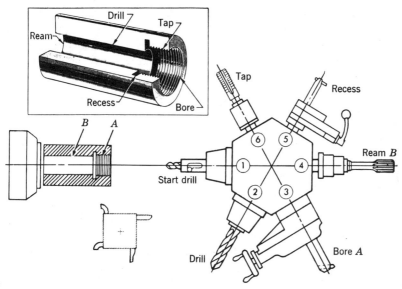

Figure 20.23
Hexagon turret setup illustrating the sequence of operations to handle required internal cuts on threaded adapter shown in insert.

5. A groove for thread clearance is recessed. For this operation a quick-acting slide tool is used with a recessing cutter mounted in a boring bar.
6. The thread is cut with a tap held in a clutch tap and die holder. This operation is followed by a cutting-off operation not shown in Figure 20.23. The cutoff would use the square turret on the carriage.

A few of the tools used in turret lathe work are illustrated in Figure 20.24. These tools are so designed that they may be quickly mounted in the turret and adjusted for use. In addition to the usual operations of drilling, boring, reaming, and internal threading shown in the figure, various other threading, centering, and turning operations are available. Internal threading is frequently done with collapsible taps to facilitate quick removal. For the same reason automatic die head chasers that open at the end of the thread are used for external threads.

Drilling

Production of holes with drills is seen in Figure 20.25. As hole making is important in manufacturing, the choice of the right tool depends on the application.

In evaluating drill performance, the material of the drill must not be overlooked. High-speed steel tools will accept about twice the cutting speed of carbon tool steel. For hard

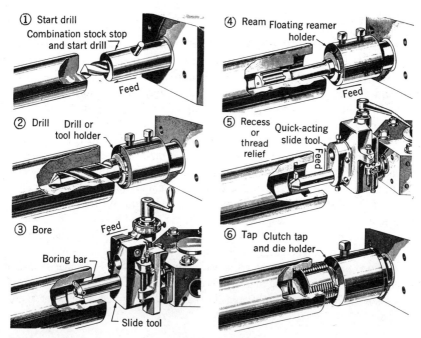

Figure 20.24
Setup for matching internal operations on threaded adapter.

and abrasive materials such as cast iron, drills tipped with tungsten carbide give excellent results, but for some hard steels and other materials these drills are not satisfactory. High-carbon, cobalt-bearing, superhigh-speed steels capable of drilling steels having a hardness of Rockwell C68 are used for drilling tough stainless steels and aerospace alloys. Also, many drills are given a thin, hard case surface treatment or are chrome plated to provide a hard wearing surface.

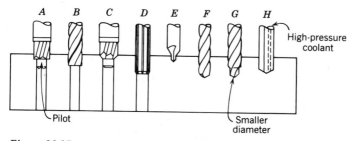

Figure 20.25
Operations for various drills. *A,* Counterboring. *B,* Core drilling. *C,* Countersinking. *D,* Reaming. *E,* Center drilling. *F,* Drilling. *G,* Step drilling. *H,* Gun drilling.

Drill Point Angle

To obtain good service from a drill it must be properly ground. The **point angle** should be correct for the material that is to be drilled. The usual point angle on most commercial drills is 118°, which is satisfactory for soft steel, brass, and most metals. For harder metals larger point angles, meaning the included angle, give better performance.

In Figure 20.26 two drills with point angles of 140° and 80° are shown. The thickness and width of the chips obtained from these drills are indicated by the letters T and W. Comparison of the two chips shows that the thickness T_1 for the 140° point angle is thicker than T_2 on the 80° point angle. Metal removed in the form of thick chips usually requires less energy per unit volume than when the same amount of metal is removed in the form of thin chips. In drilling hard and difficult to machine metals, the thicker chips allow some saving in power. It may also be noted that the width W_1 for the 140° point angle is less than W_2 for the smaller point angle. The larger width W_2, having a longer cutting edge, is useful in drilling materials creating some abrasive wear. The abrasive wear is distributed over a longer cutting edge and the cutting force per unit length is reduced. In addition, the corner angle for the 80° point drill (140°) is greater than that on the 140° point drill (110°), resulting in greater wear resistance to the drill at the corners. Materials such as soft cast iron and most plastics can best be drilled with point angles smaller than 118°.

Drill Helix Angle

Drill performance is affected by the *helix angle* of the flutes. Although this angle may vary from 0° to 45°, the usual standard for steel and most materials is 30°. The smaller this angle is made, the greater the torque necessary to operate at a given feed. As the angle is increased appreciably, the life of the cutting edge is reduced for some materrials. Drill efficiency is increased if the proper helix angle is used. For example, the angle for drilling copper, magnesium, and soft plastics should be approximately 35° to 45°, copper alloys 20° to 25°, hard plastics 17°, and soft to medium steel 24° to 32°. Tests show that there is a slight reduction in both torque and thrust as the helix angle

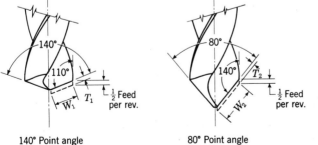

Figure 20.26 Point angle variation influences drill performance.

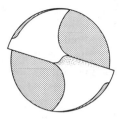 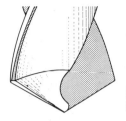

Figure 20.27
Spiral-point drill.

is increased, but it is of minor importance so far as the overall performance of the drill is concerned.

Drill Point

On most conventional drills there is a *chisel edge* at the end of the web that connects the two cutting lips (Figures 20.5 and 20.7). This chisel edge does not cut efficiently because of the large negative rake that exists not only at the center but all along the chisel edge. Actually, there should be a slight crown to the chisel edge that, if sufficient, will stabilize the drill when it is entering the work.

To improve drilling efficiency and reduce thrust, a self-centering drill point having a spiral edge has been developed (Figure 20.27) that has much better cutting action close to the drill axis. Although this drill is self-centering and has less negative rake, it is difficult to grind and requires a special grinder. An easier way to reduce the end thrust is by web thinning and point splitting, as shown in Figure 20.28. The split-point drill is used for drilling tough work-hardening steels and super alloys. Both of the drill point designs produce accurate holes with a minimum of oversize, and the thrust is much less than for chisel point drills.

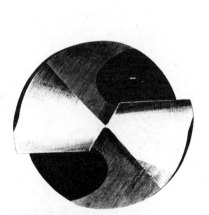 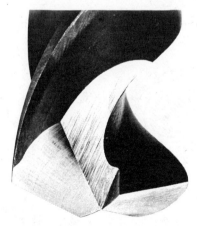

Figure 20.28
Split-point drill with thinned web to reduce end thrust.

Drill Cutting Fluids

To obtain best performance and long life for cutting edges, some cutting fluid should be used. The cutting fluid improves the cutting action between the drill and the work, facilitates removal of chips, and cools the work and tool. In production drilling, the matter of cooling is most important. To ensure long life, a cooling medium should be selected that will carry away the heat at the same rate that it is generated. A few metals with suggested coolants are given in the accompanying tabulation.

Metal	Coolant
Aluminum	Mineral lard, oil mix
Brass	Dry, mineral, lard oil mix
Cast iron	Dry, air jet
Malleable iron	Soluble oil
Soft steel	Soluble oil, sulfurized oil
Tool steel	Lard, soluble oil

Milling

There are two methods of feeding work to the cutter, as shown in Figure 20.29. Feeding the work against the cutter (Figure 20.29*A*) is usually recommended, because each tooth starts its cut in clean metal and does not have to break through possible surface scale. However, tests have proved that when the work is fed in the same direction as cutter rotation (Figure 20.29*B*), the cutting is more efficient, larger chips are removed, and there is less tendency for chatter. This method, called *climb* or down *milling,* is frequently used in production work where large cuts are to be taken and the surface of the work is free from scale.

Feed on milling machines is expressed in either of two ways. On some machines it

Figure 20.29
Methods of feeding work on milling machine. *A*, Conventional or up milling. *B*, Climb or down milling.

is expressed in thousandths of an inch per revolution (mm/r) of cutter. Such machines have feed changes ranging from 0.006 in. to as high as 0.300 in. (0.15–7.62 mm). The other way is to express the feed of the table in inches per minute (in./min) (mm/s), the usual range being from $1/2$ to 20 in./min (0.20–8.5 mm/s).

JIGS, FIXTURES, AND SECONDARY EQUIPMENT

Lathe Fixtures

Many workpieces machined on lathes can be held quickly with standard chucks and collets. However, a tool designer is sometimes required to design special fixtures for mounting irregularly shaped workpieces to the face plates of turret lathes. The advantage of special fixtures for turret or other production lathes are savings of operating time, uniformity of quality from improved centering and locating, and the possibility of heavier cuts because the workpiece is held more rigidly bolted to the face plate.

Chuck jaws are specialized to match the geometry of the part. Notice Figure 20.30*A* where a pointed-pin jaw holds in a tapered surface. Figure 20.30*B* shows a setup where the end of the workpiece is supported by a pilot insert. Without the pilot, the workpiece would deflect under the cutting loads. Long flanged workpieces can be given additional support with clamps, as shown by Figure 20.30*C*. The clamps may be simple screw type, or quick-acting features can be adopted.

Drill Jigs

Jigs are devices used in production drilling, tapping, boring, and reaming operations. They are not fixed to the machine but are normally hand held. Their function is to reduce the cost of the operation, increase production, assure high accuracy, and provide for interchangeability. A jig must hold the part and guide the cutting tools. Such a device is a good example of the transfer of skill from a mechanic to an accessory, thus the operation may be performed by an unskilled operator. This is illustrated in the drilling of four holes in a plate by using a plate or channel jig similar to the one shown in Figure 20.31. The jig is made with hardened-steel bushings that locate the positions of the four

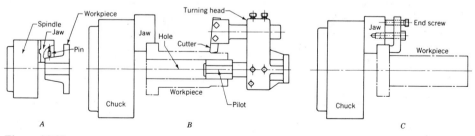

Figure 20.30
Fixtures. *A*, Pointed-pin jaw. *B*, Pilot to support work during turning. *C*, Holding clamps used with jaws.

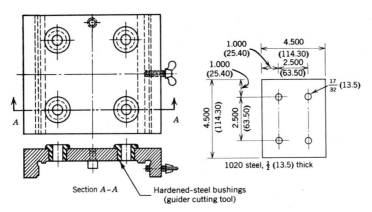

Figure 20.31
Plate-drilling jig.

holes accurately. Any number of plates may be clamped to this jig, and each part will be identical with the other.

Jigs perform the same function as a fixture but differ in appearance according to the shape and design of the part. Classification is based on their general appearance and construction. In Figure 20.32 is shown a ***box-type jig*** with open sides arranged for drilling two sides of a block. Figure 20.33 illustrates a table-type jig for drilling four holes in a flange. Other types in general use include templet, open, indexing, diameter, and universal.

A jig should be designed to provide quick and easy loading and unloading. Clamping devices must be positive, and the design should be such that there is no question about the proper location of the part in the jig. Clearance is usually provided under drill bushings to allow chips to escape without having to squeeze through the bushing. This is important if much metal is to be removed. Provision should be made for rapid cleaning of chips

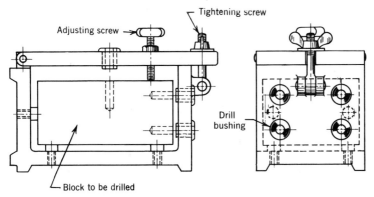

Figure 20.32
Box-type jig for drilling two sides of a block.

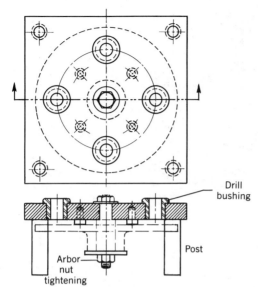

Figure 20.33
Table-type jig for drilling flange holes.

from the jig. Most jigs utilize standard parts such as drill bushing, thumb screws, jig bodies, and numerous other parts that can be ordered from catalogs. Jigs are not limited to drilling operations but are also used on tapping, counterboring, and reaming operations.

Fixtures

A fixture is a workholding and work-supporting device that is securely clamped or fixed to a machine. Fixtures, unlike jigs, do not guide the tool. Their primary function is to reduce the cost of the operation, increase production, and enable complex-shaped and often heavy parts to be machined by being held rigidly to a machine. Fixtures are used on lathes, turret lathes, milling machines, boring equipment, shapers, and planers.

Fixtures are made from gray cast iron or from steel plate by welding or bolting. They are bolted, clamped, or "set" with a low-melting alloy to the machine. A fixture has locating pins or machined blocks against which the workpiece is tightly held by clamping or bolting. To assure interchangeability, the locating devices are made from hardened steel. Many fixtures are massive because, like a machine frame, they may have to withstand large dynamic forces. Because all fixtures are between the workpiece and the machine, their rigidity and the rigidity of their attachment to the machine are paramount. Duplex fixtures have been built to allow for loading or unloading one side of a fixture while the machining operation is taking place on a part clamped to the other side.

Index Head

In addition to perishable tools like cutters or nonwearing *jigs and fixtures,* are a class of *attachments* that is interchangeable and serves a function to facilitate production. An *index or dividing head* is used to rotate the work through a certain number of degrees,

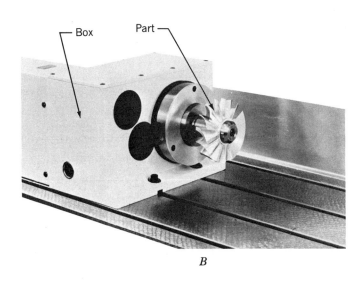

Figure 20.34
Index head. *A*, Mechanical index head for milling a spur gear. *B*, NC dividing head.

through a fraction of revolution, or while the table is feeding as when cutting helical gears. The head is supplied with a universal milling machine but may be used on other machines as well. In Figure 20.34A an index head and its footstock are shown mounted on the worktable of a machine. Because much of the work has to be supported between centers, both units are necessary.

The index head is designed principally for holding work between centers for machining grooves, gear teeth, and similar parts that have convoluted surfaces at a specified angular distance apart. Index heads are also controlled by NC, which does the job faster than geared index heads. Figure 20.34B is example of a NC dividing head.

PERFORMANCE

Once the turning, drilling, boring, and milling machines are operating and effectively making parts, attention turns to the performance of the operation. This usually means answering the question, "How much time does the operation require?" Additionally, how much metal is being removed or other performance questions are asked: If the performance is poor, then other methods or processes must be substituted in an effort to reduce the cost. Cost is a driving force in manufacturing.

Cutting Speeds

The amount of metal removal is a function of both the cutting speed and feed. The cutting speed, expressed in feet per minute (meters per second), is a measure of the peripheral speed as indicated by the following expression:

$$V_c = \frac{\pi DN}{12}$$

where

V_c = Peripheral velocity of surface, ft/min (m/s)
D = Diameter of cutter or work, in. (mm)
N = Revolutions per minute rpm

This equation was introduced in Chapter 16 and is applicable for turning, drilling, boring, and milling. In turning operations the diameter is the outermost dimension of the work being turned. For example, if a bar is being rough turned, the outside dimension is used. If the rough pass reduces the diameter by 1 in. (25.4 mm), the next calculation will use the smaller diameter. In facing or cutoff operations the outer diameter is used, even though the diameter is reducing during the cut. For boring the diameter is the internal dimension of the hole before being bored. In milling the diameter of the cutter serves as the diameter.

Usually the cutting speeds V_c are known from information based on machinability testing and experience. In this case the calculation will provide N, the rotary speed.

The tool or workpiece moves relative to the other member. For example, a turning tool mounted on the carriage moves along the bed, and its rate of velocity is expressed in inches per revolution (ipr). In milling, if the cutter is fixed (meaning that it is not traversing, although it is rotating), the table is moving or feeding at a velocity expressed in inches per revolution of the cutter. This relative velocity between the tool and workpiece leads to the following expression for turning-type operations.

$$t_m = \frac{L}{fN} = \frac{L\pi D}{12V_c f}$$

where

t_m = Machining time, min
L = Length of cut for metal cutting, in. (mm)
D = Diameter being cut, in. (mm)
f = Feed rate, ipr (mm/r)
N = Rotary cutting speed, $12V_c/\pi D$, rpm

The feed rate, also expressed in inches per minute, can be found as

$$f_m = fN$$

Whereas drill and threading feeds are expressed in inches per revolution (mm/r), they are sometimes converted to another expression for finding the duration of the operation as

$$t_m = Lf_{dt}$$

where f_{dt} is the drilling or tapping feed rate in minutes per inch.

Milling cutters are multitooth and the feed rate depends on the number of teeth in the cutter as well as the cutter diameter.

$$t_m = \frac{L\pi D_c}{12V_c n_t f_t}$$

where

n_t = Number of teeth on cutter
f_t = Feed per revolution per tooth

Length of Cut

The *length of cut* L is the distance for which the cutting tool or table is moving at a certain feed f velocity. This is much less than the "rapid-traverse" velocity, which may be from 100 to 500 in./min. The general relationship for this length is given as

$$L = L_s + L_a + L_d + L_{ot}$$

where

L_s = Safety length, in. (mm)
L_a = Approach length resulting from cutter geometry, in. (mm)
L_d = Design length of workpiece requiring machining
L_{ot} = Overtravel length resulting from cutter geometry, in. (mm)

L is the value used in finding the time or cost to machine. L_s (*safety length*) is necessary for any stock variations in length, because if the cutter is in rapid-traverse velocity and the trips that cancel the fast mode of table or cutter movement are set too short, it is possible for the cutter to bang into the workpiece, giving rise to possible damage and safety problems. Safety stock may vary from $1/64 \leq L_s \leq 1/2$ in. (12.7 mm).

Approach length depends on the cutter workpiece geometry. Approach is considered negligible for turning. For drilling, the drill point adds to the length of cut, and the trigonometric relationship gives the distance accounting for the 118° angle.

$$L_{ot} = \frac{D_c}{2 \tan 59} = 0.3 \, D_c$$

See the sketch of drill point approach in Figure 20.35A.

Milling also has an approach that depends on the diameter of the cutter and depth of cut, t. A roughing pass will always require an approach, but the overtravel may not be required and the cutter would stop at position 2 in Figure 20.35B. If the milling cutter is performing a finishing pass, it is customary to move the cutter off the workpiece to avoid rough tool marks at the end of the pass. Figure 20.35B illustrates a sidemill or slab milling cutter approach.

$$L_a = L_{ot} = \sqrt{\left(\frac{D_c}{2}\right)^2 - \left(\frac{D_c}{2} - t\right)^2} = \sqrt{t(D_c - t)}$$

where

D_c = Cutter diameter, in. (mm)
t = Depth of cut, in. (mm)

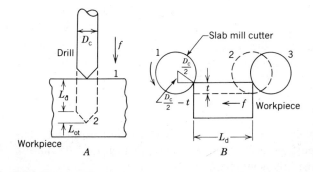

Figure 20.35 Approach and overtravel length. A, For drilling. B, For milling.

Table 20.1 **Machining Speeds and Feeds**

Material	Turning and facing (ft/min, ipr)			
	High-speed steel		Tungsten carbide	
	Rough	Finish	Rough	Finish
Stainless steel	150, 0.015	160, 0.007	350, 0.015	350, 0.007
Medium-carbon steel	190, 0.015	125, 0.007	325, 0.020	400, 0.007
Gray cast iron	145, 0.015	185, 0.007	500, 0.020	675, 0.010

Material	Plain milling Tungsten carbide		Slotting High-speed steel	
	Rough	Finish	Rough	Finish
Stainless steel	140, 0.006[b]	210, 0.005[b]	85, 0.002[b]	95, 0.0015[b]
Medium-carbon steel	170, 0.008	225, 0.006	85, 0.0025	95, 0.002
Gray cast iron	200, 0.012	250, 0.010	85, 0.004	95, 0.003

Drill diameter (in.)	Power feed drilling[a]		
	Stainless steel	Medium-carbon steel	Gray cast iron
$1/4$	0.55	0.20	0.20
$5/16$	0.65	0.23	0.23
$3/8$	0.65	0.25	0.25

Number of threads per inch	Center threading[a]	
	Steel	Stainless steel
32	0.18	0.33
20	0.15	0.30
16	0.19	0.33
10	0.32	0.48

[a]Times are given in minutes per inch.
[b]Feed per tooth.

Machining Cost

The foregoing calculations lead to finding machining cost or $C_o t_m$, where C_o is the productive hour cost for the machine and operator, in dollars per minute. Machining feeds and speeds that demonstrate these relationships are given in Table 20.1.

Drill Hole Size

Conventional, two-fluted drills will normally drill slightly oversize in most metals. The amount of hole oversize obtained from drills ranging from $1/8$ to 1 in. (3.2–25 mm) in diameter may be computed by the following relationships:

Average oversize = 0.002 + 0.005D (0.05 + 0.13D)
Maximum oversize = 0.005 + 0.005D (0.13 + 0.13D)
Minimun oversize = 0.001 + 0.003D (0.03 + 0.08D)

where D is the nominal drill diameter in inches (millimeters). These relationships apply not only to holes drilled in steel and cast iron but also to most nonferrous metals.

Rate of Metal Removal

The rate of metal removal is found for turning by

$$Q = 12 \times t \times f \times V_c$$

where

Q = Rate of metal removal, in.³/min (mm³/min)
t = Depth of cut, in.
f = Feed, ipr

For drilling, the rate of metal removal is given as

$$Q = \frac{D_c 2\pi f}{4}$$

where

D_c = Diameter of drill, in.
f = Feed rate, ipr

The rate of metal removal for milling is

$$Q = w \times t \times f$$

where w is the width of cut, in inches.

Horsepower

The horsepower required at the spindle can be found using the rate of metal removal for turning, drilling and milling by using

$$HP_s = Q \times P$$

where P is the unit horsepower, or horsepower per cubic inch per minute.
Let

HP_m = Horsepower at motor, hp (W)

$$= \frac{HP_s}{E}$$

where E is the percentage efficiency of the spindle drive. Sometimes it is important to understand that **unit power** equations can be calculated, given the information

Table 20.2 **Averge Unit Power Requirement**

Metal	Turning	Drilling	Milling
Stainless steel	1.5	1.3	1.6
Medium-carbon steel	1.4	1.2	1.4
Gray cast iron	1.8	1.4	1.8

$$P = \frac{HP_s}{Q}$$

where P is the unit power, hp/in.^{3}/min (W/mm^3). Some average unit power requirements are listed in Table 20.2.

Questions and Problems

1. Give an explanation of the following terms.

    Brazed tips
    Square insert
    Drill flutes
    Interior oil channels
    Gun drills
    Fly cutter
    Spade drill
    Chucking reamers
    Block-type boring cutter
    Arbor cutters
    Angle-milling cutters
    Morse taper
    Thread dial indicator
    Combination stock stop and start drill
    Drill point angle
    Climb milling
    Box jig
    Jigs and fixtures
    Attachments
    Length of cut
    Unit power

2. What is wrong with a lathe that turns slightly tapered surfaces instead of cylindrical surfaces? Assume that the small end is on the tailstock end.

3. Briefly describe the steps in cutting a V-thread on a lathe.

4. Explain the terms combined cuts and multiple cuts.

5. How does the box tool prevent bending of the bar stock when machining?

6. Describe the advantages and disadvantages of thin versus thick chips in drilling.

7. Sketch the coring, spot facing, counterboring, reaming, and countersinking operations.

8. Define a core drill and its applications.

9. When the same amount of metal is removed, does it require more or less energy with thick chips than with thin chips?

10. What effect does the helix angle have on drill performance?

11. What is the average, maximum and minimum oversize dimensions of a 13-mm drill in steel plates?

12. In drilling $^{11}/_{16}$-in. holes in a bearing cap, what variation in hole size is expected?

13. How is the coolant applied in deep-hole drilling?

14. What is the SI unit for the Morse taper?

15. List the types of operations that can be performed on a milling machine. On a lathe. On a drill press.

16. What are the differences between profile and formed tooth cutters?

17. Describe down milling. What are the advantages?

18. Compare the rake angles on milling and lathe cutters.

19. Sketch a plain milling cutter and indicate the rake angle, clearance angle, tooth face, land, and tooth depth.

20. Find the rate of metal removal for a 11.3-mm-long slot, 15 mm deep, and 5-mm-wide cut at a feed of 0.15 mm/s. If cutting takes place 40% of the time, how much metal is removed in 8 h?

21. A workpiece is 10 in. long and is to be milled to a depth of $1/8$ in. by a plain spiral cutter 6 in. in diameter. If the feed rate is 6 in./min, how long will it take to make the cut?

22. Stainless steel is to be rough and finish milled. What is the time to machine a flat of 10 by 20 in. for a cutter having a path width of 5 in.? The number of teeth for this plain carbide cutter is 12 and the diameter of the cutter is 4 in. How many revolutions per minute are suggested for the rough and finish passes? The rough pass removes 1 in., whereas the finish pass removes $1/16$ in.

23. An operator earns \$15/h; handling and metal cutting elements for this operation total 1.35 min. What is the cost for this operation?

24. The length of a machining pass is 20 in. and the part diameter is 4 in. OD. Velocity and feed for this material are 275 ft/min and 0.020 ipr. What is the time to machine?

25. Operator and variable machine expenses are \$60/h and handling time is 1.65 min. The length and diameter of a gray-iron casting are 8.5 by 8.6 in. Velocity is 300 ft/min, and feed is 0.020 ipr. Find the cost to handle and machine.

26. Stainless-steel material is to be rough and finish turned. Diameter and length is 4 by 30 in. Speeds and feeds for tungsten carbide tools are (350, 0.015) and (350, 0.007). Determine rough and finish time to machine. Find the rotary velocity of the bar stock. The rough pass removes $1/2$ in. on the diameter.

27. A surface 8 in. wide by 20 in. long is rough milled with a depth of cut of $1/4$ in. A 16-tooth cemented carbide cutter face mill 6 in. in diameter is used. The material is cast iron. What is the cutting length? Estimate the cutting time if $V = 120$ ft/min and tooth chip load $= 0.0012$ inches per tooth revolution. Find the cutter revolutions per minute. Repeat for $V = 150$ ft/min.

28. A stainless-steel part is tap drilled $5/16$ in. for a depth of 1 in. It is followed by a tap $3/16 - 16$ for a depth of $7/8$ in. Find the drilling length and drilling and tapping time. Repeat for steel.

29. The top of a square 250-mm block is to have a 65-mm slot machined on it. The end mill is 25 mm in diameter and only one pass is traced over the slot. Safety stock is 5 mm. Find the length of cut.

30. Find the length of cut for a 2-in. slab plain milling cutter removing $1/2$ in. thickness for a workpiece 10 in. in length. Safety stock is $1/16$ in. The cutter is to leave the stock free of milling marks. Repeat for $1/4$ in. thickness.

31. A bar of steel 2 in. (50 mm) in diameter is to be machined at 90 ft/min (0.5 m/s). What spindle speed should be used? Repeat for 125 ft/min. Repeat for 4 in.

32. Using carbide cutters, what should be the speeds for rough turning cast iron and stainless steel of the same diameter?

33. Assuming a 0.015-in. (0.38-mm) feed and 0.062-in. (1.57-mm) depth of cut in machining a 2-in. (50-mm) bar of SAE 1020 steel, what is the metal removal rate in cubic inches per minute (mm³/s) as a function of the cutting speed? What is the approximate horsepower at the spindle? If the spindle efficiency $= 92\%$, what motor horsepower is expected?

34. What would be the taper in inches per foot if a 1-m bar were turned with the tailstock set over 3 mm?

35. A 14-in. (360-mm) length of stock is to be

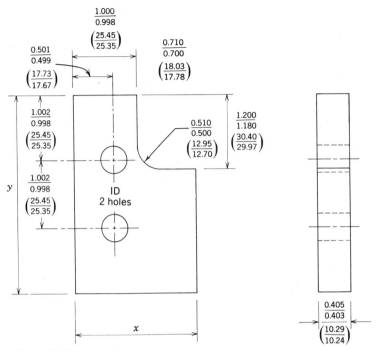

Figure 20.36
Spacer block.

tapered so that the diameter on one end is 3 in. (75 mm) and on the other end $2^7/_8$ in. (73 mm). How much should the tailstock be set over to cut this taper?

36. Using a carbide tool, what is the time to rough machine an aluminum shaft 3 in. (75 mm) in diameter and 24 in. (610 mm) long? The cutting speed is 600 ft/min (3 m/s), and a tool has a feed of 0.004 ipr (0.1 mm/r). Also in SI.

37. A rod $2^1/_2$ in. (63.5 mm) in diameter and turning at 120 rpm is to be cut off by a tool having a feed of 0.004 ipr (0.1 mm/r). What is the cutting time? Find the cutting time if the diameter is 4 in. and feed is 0.005 ipr.

38. A 4-in. OD tungsten carbide plain milling cutter having a 6-in. face and eight teeth is used for a $^3/_8$-in. depth cut. The drawing length is 11 in., safety stock $^1/_4$ in., and workstock material is medium-carbon steel. If the cutter is required to have overtravel, what is the length of cut? How much time will be used for a rough pass? If the finish depth of cut is $^1/_8$ in., how much time is necessary? The noncutting time for this operation is 2 min, and the machine and operator cost $45/h; what is the production cost for one part?

39. A spacer block, shown in Figure 20.36, is made of SAE 1020 cold-rolled steel and is required for a new product. The workpiece is cut to rough size and ground to the measurements given in the figure. A jig is necessary for drilling and reaming the two holes, and a third hole is subsequently milled to give the radius. A six-spindle gang drilling machine will be used. Sketch an inexpensive plate jig that will provide the required accuracy. For the jig the tolerances can be 20% of the workpiece tolerances. Use ground-position pins for location. Include a bill of materials list for the jig items. Two problems follow:

Problem	Dimension x	Dimension y	Dimension ID
A	$\frac{3.018}{3.006}$	$\frac{1.875}{1.865}$	$\frac{0.305}{0.304}$
	$\frac{76.65}{76.40}$	$\frac{47.63}{47.37}$	$\frac{7.75}{7.72}$
B	$\frac{4.268}{4.256}$	$\frac{3.125}{3.115}$	$\frac{0.2501}{0.2503}$
	$\frac{108.41}{108.10}$	$\frac{79.12}{79.375}$	$\frac{6.35}{6.36}$

CASE STUDY

NC MACHINED FORGING

Dick Crawford, manufacturing representative for Mestas Tooling, is retained as a consultant to improve the productivity of a 4140 steel housing forging being machined on an NC turret lathe (Figures 20.37 and 20.38). The first few pieces of a 6000-quantity lot have been run, and a cycle

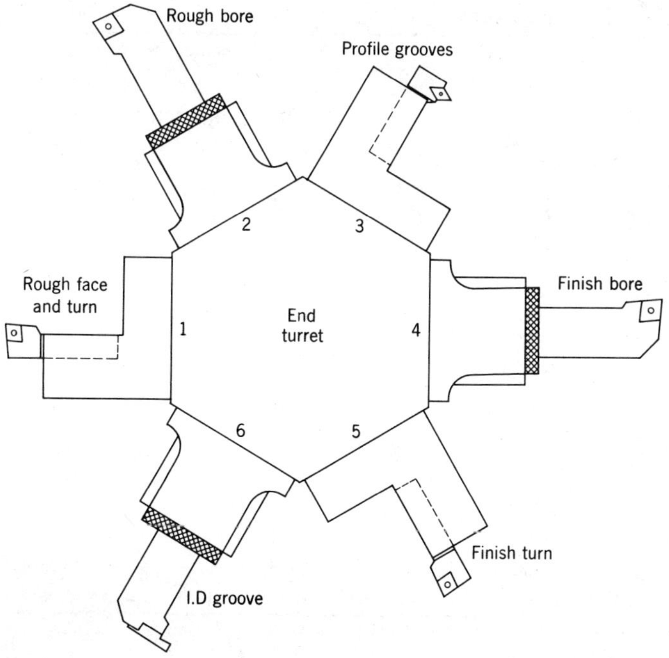

Figure 20.37
Turret tooling for machining 4140 steel forging.

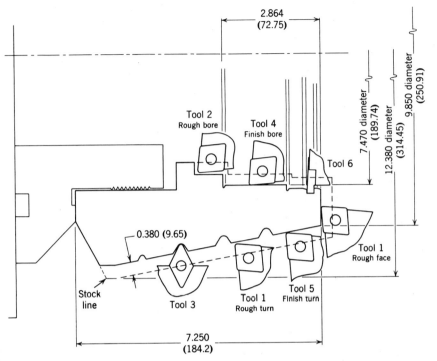

Figure 20.38
Position of tooling on workpiece.

Table 20.3 **Operational Data for Case Study**

Elements	ft/min (m/s)	ipr (mm/r)	Depth, in. (mm)	Diameter, in. (mm)	hp (kW)	Removal rate, in.³/min (mm³/s)	Cumulative time (min)
Rough face	360 (1.8)	0.030 (0.76)	0.250 (6.35)	9.85 (250.2)	29 (22)	33 (0.009)	0.45
Rough-turn taper	360 (1.8)	0.035 (0.89)	0.380 (9.65)	12.0 (350)	42 (31)	57.4 (0.016)	1.50
Rough bore	360 (1.8)	0.030 (0.76)	0.250/0.350 (6.35/8.89)	7.47 (189.7)	32 (24)	45.4 (0.012)	2.60
Profile grooves	390 (2.0)	0.010 (0.25)	0.250 (6.35)	Variable	Machined in four places		7.2
Finish bore	600 (3.0)	0.015 (0.38)	0.030 (0.76)	Variable	—	—	8.4
Finish face and taper	600 (3.0)	0.020 (0.51)	0.030 (0.76)	Variable	—	—	11.3
Groove ID	400 (2.0)	0.010 (0.25)	0.125 (3.18)	Variable	—	—	11.6

time of 24.3 min has been time studied. A total of 61 lb (27 kg) of metal is removed.

Dick observes that the first three elements are rough face, rough turn taper, and rough bore at 0.015 ipr (0.38 mm/r) and 0.190 in. (4.83 mm) depth. Dick suggests a change in the tool inserts, an increase of the feed rate to 0.035 ipr (0.89 mm/r), 360 ft/min (1.8 m/s), and combining two roughing passes to one for a 0.380-in. (9.65-mm) dimension. His recommendations are given by the operational data in Table 20.3, which is used for Figures 20.37 and 20.38.

To obtain these data, Dick retools the job and provides a turret and part layout. Dick says, "If you do it this way, there will be free-flowing, open-type chips that will reduce the feed force. Also, we use some of the turret stations for double duty."

Reconstruct the original forging section and show the progressive stock removal by cross-shatching. What is the saving resulting from Dick's recommendations if labor and the turret lathe cost $27/h? What is the saving for the first three elements if the original time is 6.6 min?

CHAPTER 21
SHAPING, PLANING, SAWING, AND BROACHING

In machines of the type discussed in this chapter, either the tool moves in a straight line across the workpiece or the workpiece moves linearly by the tool. There are some special variations of this movement, but they have minimal use. Although sawing might appear a totally different operation than shaping, which can utilize an orthogonal cutting tool, close examination of a magnified chip from a sawing operation resembles quite closely one that is generated on a shaper or planer. Likewise, the broaching operation produces a similar chip.

SHAPERS

A shaper is a machine with a reciprocating tool of the lathe type that takes a straight-line cut. By successive movement of the work across the path of this tool, a plane surface is generated. Perfection is not dependent on the accuracy of the tool as it is when a milling cutter is used for the same type of work. By employing special tools, attachments, and devices for holding the work, a shaper can also cut external and internal keyways, spiral grooves, gear racks, dovetails, and *T-slots*.

Classification of Shapers

According to general design, shapers can be classified as follows:

    A. Horizontal push cut
        1. Plain (production work)
        2. Universal (toolroom work)
    B. Horizontal draw cut
    C. Vertical
        1. Slotter
        2. Key seater
    D. Special purpose as for cutting gears

Power is applied to the machine by motor, either through gears or belt or by employment of a hydraulic system. The reciprocating drive of the tool can be arranged in several ways. Some older shapers were driven by gears or feed screws, but most shapers are now driven by an oscillating arm and crank mechanism.

Horizontal Push Cut Shapers

Construction. Figure 21.1 is a sketch of a plain horizontal shaper. Commonly used for production and general-purpose work, a horizontal shaper consisting of a base and frame that support a horizontal ram, is simple in construction. The *ram* that carries the tool is given a reciprocating motion equal to the length of the stroke desired. The *quick-return mechanism* driving the ram is designed so that the return stroke of the shaper is faster than the cutting stroke, which reduces the idle time of the machine. The toolhead at the end of the ram, which can be swiveled through an angle, is provided with means for feeding the tool into the work. A *clapper box* toolholder attached to the ram pivots at the upper end and flips up on the return stroke so as not to dig into the work.

The worktable is supported on a crossrail in front of the shaper. A lead screw in connection with the crossrail moves the work crosswise or vertically by either hand or power drive. A *universal shaper,* which has these same features, is provided with swiveling and tilting arrangements for machining at any angle. The swiveling adjustment acts about an axis that is parallel to the motion of the ram. The tilting feature is in the tabletop, which sets the table at an angle to the swiveling axis. Work is held to the worktable by bolting or fastening it in a vise or a fixture.

Shaper Drive. The shaper can be driven by a mechanical quick-return mechanism (Figure 21.2A) or by a hydraulic system (Figure 21.2B). The mechanical system consists of a rotating crank driven at a uniform speed connected to an oscillating arm by a sliding block that works in the center of the rather massive oscillating arm. The crank is contained in the large gear and can be varied by a screw mechanism. To change the position of the stroke, the clamp holding the connecting link to the ram screw is loosened and the ram positioner is turned. By turning the positioner screw the ram can be moved backward or forward to the correct cutting position as in Figure 21.2. The stroke length is varied by changing the length of the crank. The ratio of return to cutting speed is about 3 : 2.

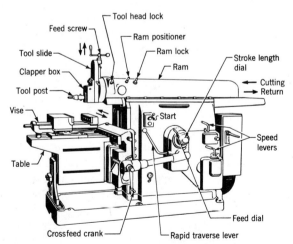

Figure 21.1
Plain horizontal shaper.

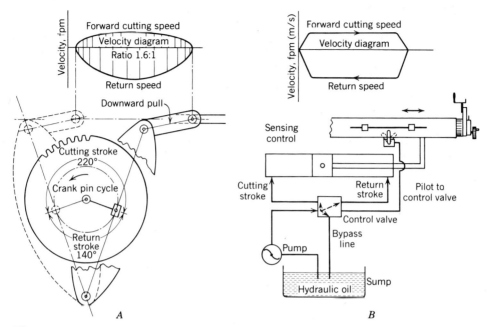

Figure 21.2
Shaper drives. A, Mechanical. B, Hydraulic.

Quick return in the hydraulic drive-type shaper is accomplished by increasing the flow of hydraulic oil during the return stroke. The hydraulic shaper is replacing mechanical shapers, because the cutting stroke has a more constant velocity and less vibration is induced in the hydraulic shaper. The cutting speed is generally shown on an *indicator* and requires no calculation. Both the cutting stroke length and its position relative to the work may be changed quickly without stopping the machine via handles at the side of the ram. Ram movement can be reversed instantly anywhere in either direction of travel. The hydraulic feed operates while the tool is clear of the work. Machine operation is quiet. The maximum ratio of return to cutting speed is about 2 : 1.

Cutting Speed. Cutting speed on horizontal shapers is defined as the average speed of the tool during the cutting stroke and depends primarily on the number of ram strokes per minute and the length of the stroke. If the stroke length is changed and the number of strokes per minute remains constant, the average cutting speed is changed. The ratio of cutting speed to return speed enters into the calculation, as it is necessary to determine what proportion of time the cutting tool is working. Thus with the ratio of cutting stroke to return stroke as 3 : 2, the cutter is working three-fifths of the time and the return stroke two-fifths of the time. The average cutting speed may be determined by the following formula:

$$V_c = \frac{2LN}{12C} = \frac{LN}{6C} = \text{ft/min (m/min)}$$

where

$$N = \text{Strokes per minute (spm)}$$
$$L = \text{Stroke length, in. (mm)}$$
$$C = \text{Cutting time ratio, } \frac{\text{cutting time}}{\text{total time}}$$

The number of strokes per minute for a desired cutting speed is then

$$N = \frac{V_c \times 6C}{L}$$

To determine the number of strokes required to complete a job and the total time required,

$$S = \frac{W}{f} = \text{Total number of strokes required}$$

$$T = \frac{S}{N} = \text{Total time, min}$$

where

$$W = \text{Width of work, in. (mm)}$$
$$f = \text{Feed, in. (mm)}$$

A general expression to determine the total time knowing the desired cutting speed and length of stroke is

$$T = \frac{SL}{V_c \times 6C} = \text{Total time, min}$$

Horizontal Draw Cut Shapers

This shaper is so named because the tool is pulled across the work by the ram instead of being pushed. Horizontal draw cut shapers are recommended for heavy cuts, being widely used for cutting large die blocks and machining large parts in railroad shops. During the cut the work is drawn against the adjustable back bearing or face of the column, thereby reducing the strains on the crossrails and saddle bearings. There is little vibration or chatter as a tensile stress is exerted in the ram during the cut.

Vertical Shapers

Vertical shapers or *slotters* (Figure 21.3) are used principally for internal cutting and planing at angles and for operations that require vertical cuts because of the position in which the work must be held. Applications are found on die work, metal molds, and metal patterns. The shaper ram operates vertically and has the usual quick-return feature like the horizontal-type machines. Work to be machined is supported on a round table having a rotary feed in addition to the usual table movements. The circular table feed permits the machining of curved surfaces, a process that is particularly desirable for many

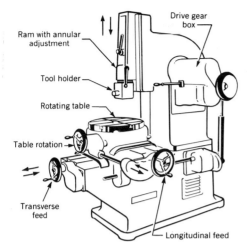

Figure 21.3
Vertical shaper.

irregular parts that cannot be turned on a lathe. Plane surfaces are cut by using either of the table crossfeeds. Another vertical shaper is the *key seater* and is designed for cutting keyways in gears, pulleys, cams, and similar parts.

PLANERS

A planer is a machine tool designed to remove metal by moving the work in a straight line against a single-edge tool. Similar to the work done on a shaper, a planer is adapted to much larger work. The cuts, which are mainly plane surfaces, can be horizontal, vertical, or at an angle. In addition to machining large work, the planer machines multiple small parts held in line on the platen. Planers are seldom used in production work as most plane surfaces are machined by milling, broaching, or grinding, but they are still employed for special purposes.

Classification of Planers

Planers may be classified in a number of ways, but according to general construction there are four types.

1. Double housing.   2. Open side.   3. Pit type.   4. Plate or edge.

Planer Drive

Hydraulic drives are used for planers. Uniform cutting speed is attained throughout the entire cutting stroke. The acceleration and deceleration of the table take place in so short a distance of travel that the time element need not be considered.

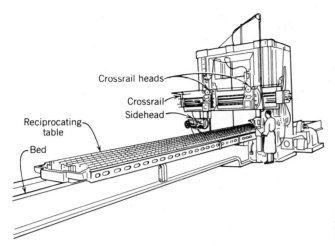

Figure 21.4
Double-housing planer.

Double-Housing Planers

This planer consists of a long heavy base on which the table or platen reciprocates. The upright housing near the center on the sides of the base supports the crossrail on which the tools are fed across the work. Figure 21.4 illustrates how the tools are supported both above and on the sides, and the manner in which they can be adjusted for angle cuts. They are fed manually or by power in either a vertical or a crosswise direction.

Open-Side Planers

This planer (Figure 21.5) has the housing on one side only. The open side permits machining the wide workpieces. Most planers have one flat and one double V-way, which allows for unequal bed and platen expansions. Adjustable dogs at the side of the bed control the stroke length of the platen. The accuracy of both the open-side and double-housing planer is determined by its rigidity and the manner in which the ways are machined.

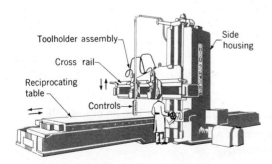

Figure 21.5
Open-side planer.

Figure 21.6
Pit-type planer.

Pit-Type Planers

A *pit-type planer* is massive in construction and differs from an ordinary planer in that the bed is stationary and the tool is moved over the work. Figure 21.6 shows a planer designed for work up to 14 ft (4 m) in width and 35 ft (11 m) in length. Two ram-type heads are mounted on the crossrail, and each is furnished with double clapper box toolholders for two-way planing. The two reversing housings that support the crossrail slide on ways and are screw driven from an enclosed worm drive at one end of the bed. All feeds are automatic and reversible and are designed to operate either at both ends of the planing stroke or at one end only.

Plate or Edge Planers

This special-type of planer was devised for machining the edges of heavy steel plates for pressure vessels and armor plate. The plate is clamped to a bed, and the carriage supporting the cutting tool is moved back and forth along the edge. A large screw drive is used for moving the carriage. Most edge planers use milling cutters instead of conventional planer tools for greater speed and accuracy.

Differences between Planers and Shapers

Although the planer and shaper are able to machine flat surfaces, there is not much overlapping in their fields of usefulness. They differ widely in construction and in the method of operation. When the two machines are compared the following differences may be seen:

1. The planer is especially adapted to large work; the shaper can do only small work.
2. On the planer the work is moved against a stationary tool; on the shaper the tool moves across the work, which is stationary.

3. On the planer the tool is fed into the work; on the shaper the work is usually fed across the tool.
4. The drive on the planer table is either by gears or by hydraulic means. The shaper ram can also be driven in this manner, but many times a quick-return link mechanism is used.
5. Most planers differ from shapers in that they approach more constant-velocity cutting speeds.

Tools and Workholding Devices

Tools used in shaper and planer work are the single-point type as used on a lathe but are heavier in construction. The holder should be designed to secure the bit near the centerline of the holder or the pivot point rather than at an angle as is customary with lathe toolholders. With the tip of the tool back, it tends less to dig into the metal and cause chatter.

Cutting tool shapes for common planer operations are usually tipped with high-speed steel, cast alloy, or carbide inserts. High-speed steel or cast alloys are commonly used in heavy roughing cuts and carbides for secondary roughing and finishing. Caution is necessary in using carbide tools on machines not equipped with an automatic lifting device for the tool on the return stroke. If the tool is permitted to rub the work, the cutting edge is likely to be chipped.

Cutting angles for tools depend on the tool used and the workpiece material. They are similar to angles used on other single-point tools, but the end clearance need not exceed 4°. Cutting speeds are affected by the rigidity of the machine, how the work is held, tool, material, and the number of tools in operation.

Worktables on planers and shapers are constructed with T-slots on their surfaces to hold and clamp parts that are to be machined. Most planner work is clamped directly to the table, and a wide variety of clamps, stop pins, and holding devices have been developed.

METAL SAWING MACHINES

An important first operation in manufacturing many parts is sawing materials and bar stock for subsequent machining operations. Although some machine tools can do cutting-off operations to a limited extent, special machines are necessary for mass production and work that requires cutting of shapes.

Hand sawing, used on simple jobs and in situations where the work cannot be taken to a power saw, is done with a thin flexible blade, usually 8 to 12 in. (200–300 mm) in length, held in a hacksaw frame that is provided with a suitable handgrip. The tooth spacing (pitch) will vary from 14 to 32 teeth per inch (0.6 to 1.3 teeth per millimeter). Although coarse-tooth saws allow more chip space, tooth spacing should be chosen on the basis of thickness and material being cut. An average pitch for handsaws is 18 teeth per inch (0.7 teeth per millimeter), but for thin materials and tubing a finer pitch is used.

Classification of Sawing Machines

Metal saws for power machines are made in *circular*, *straight*, or *continuous* shapes depending on the type of machine with which they are to be used. Various power-sawing machines are listed as follows:

A. Reciprocating saw
   1. Horizontal hacksaw
   2. Vertical sawing and filing
B. Circular saw
   1. Metal saw
   2. Steel friction disk
   3. Abrasive disk
C. Band saw
   1. Saw blade
   2. Friction blade
   3. Wire blade

Reciprocating Sawing Machines

The reciprocating power hacksaw, which may vary in design from light-duty, crank-driven saws to large, heavy-duty machines hydraulically driven, is simple in design and economical to operate. Machines vary in the method of feeding the saw into the work and the type of drive. They may also be designed for manual, semiautomatic, or fully automatic operation.

Methods of feeding can be classified as *positive* or *uniform-pressure* feeds. A positive feed has an exact depth of cut for each stroke, and the pressure on the blade will vary directly with the number of teeth in contact with the work. In cutting a round bar the pressure is light at the top of the bar and maximum at the center. A disadvantage of positive feed is that the saw is prevented from cutting fast at the start and the finish where contact is limited. With uniform-pressure feeds the pressure is constant regardless of the number of teeth in contact. This condition prevails in gravity or friction feeds. Here the depth of cut varies inversely with the number of teeth in contact, so that maximum pressure depends on the maximum load that a single tooth can stand. Some machines have incorporated both feed systems into their design. In all types the pressure is released on the return stroke to eliminate wear on the teeth.

The simplest type of feed is the gravity feed, in which the saw blade is forced into the work by the weight of the saw and frame. Uniform pressure is exerted on the work during the stroke, but some provision is usually made to control the depth of feed for a given stroke. Some machines have weights clamped on the frame for additional cutting pressure. Spring loading is another method for increasing cutting pressure. Positive-acting screw feeds with provision for overloads provide a means of obtaining a definite depth of cut for each cutting stroke. Hydraulic feeds afford excellent control of cutting pressures.

The simplest drive for the saw frame employs a crank rotating at a uniform speed. With this arrangement the cutting action takes place only 50% of the time because the time of the return stroke equals that of the cutting stroke. An improvement of this design provides a link mechanism that gives a quick-return action. Several link mechanisms are used including the Whitworth mechanism found on some shapers.

Figure 21.7 shows a reciprocating bar feed hacksaw equipped with automatic bar feed

METAL SAWING MACHINES   555

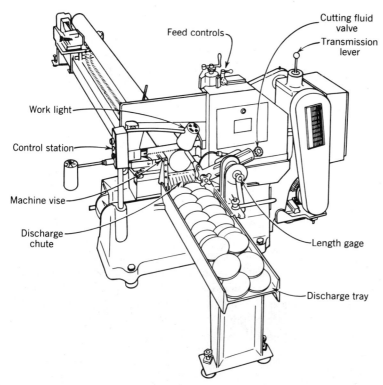

Figure 21.7
Automatic bar feed hacksaw machine.

and discharge track. Bars to be cut are loaded on a rolling dolly and vise and are moved forward by a chain arrangement. The usual cycle for automatic feed after the gage has been set is that a bar or bars move forward through an open vise, the vise is clamped, pieces are cut off by the saw, the saw is raised to the original position, the vise is opened, and so on until the length of bar has been cut up.

Hacksaw Blades. Power hacksaw blades are similar to those used for hand sawing. High-speed steel blades vary from 12 to 36 in. (300–900 mm) in length and are made in thicknesses from 0.050 to 0.125 in. (1.3–3.1 mm). The pitch is coarser than for hand sawing, ranging from $2^1/_2$ to 14 teeth per inch (0.1–0.6 teeth per millimeter). The tooth construction of most hacksaw blades is indicated in Figure 21.8A and B. The most common type is the *straight-tooth* design having zero rake. The undercut tooth, which resembles a milling cutter tooth, is used for the larger blades. For efficient cutting of ordinary steel and cast iron, a pitch as coarse as possible should be used to provide ample chip space between teeth. However, two or more teeth should always be in contact with the stock.

High-carbon and alloy steels require a medium-pitch blade, whereas thin metal, tubing, and brass require a fine pitch. To provide ample clearance for the blade while cutting, the teeth are set to cut a slot or *kerf* slightly wider than the thickness of the blade. This

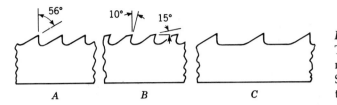

Figure 21.8
Tooth construction for metal saw blades. A, Straight tooth. B, Undercut tooth. C, Skip tooth.

is done by bending certain teeth slightly to the right or left, as may be seen in Figure 21.9. *Set* refers to the type of tooth construction on a saw. A *straight-tooth saw* has one tooth set to the right and the next tooth to the left. This type of saw is used for brass, copper, and plastic. On the *raker tooth saw* one straight tooth alternates with two teeth set in opposite directions. This tooth construction is used for most steel and iron cutting. A *wave set* consists of an alternate arrangement of several teeth set to the right and several teeth set to the left. This design is used in cutting tubes and light sheets of metal.

A lubricant is recommended for all power hacksaw cutting to lubricate the tool and to wash away the small chips accumulating between the teeth. Because there is little heat generated in most sawing operations, the problem is one of lubricating rather than cooling, and the cutting fluid should be chosen accordingly. Hacksaw machines cut between 40 and 160 spm, depending on the machinability of the metal. Surface speed is seldom specified, as it is not uniform throughout the stroke length.

Circular Sawing Machines

Machines using circular saws are commonly known as *cold sawing machines*. The saws are fairly large in diameter and operate at low rotational speeds. The cutting action is the same as that obtained with a milling cutter.

Figure 21.10 is a sketch of a cold sawing machine. This machine is hydraulically operated and saws round stock up to 10 in. (254 mm) in diameter with $\pm 1/64$ in. (± 0.4 mm) length tolerances. The saw is fed into the work, which is positively clamped by hydraulic, horizontal, and vertical vises. An automatic gripper-type feeder moves the stock when cutting to specified lengths.

Circular Metal Saws. Saws for rotating cutter machines are similar to the metal-slitting saws used with milling machines. However, metal-slitting saws are made only in diameters up to 8 in. (200 mm), which is not sufficient for large-size work. Solid blades with diameters up to 16 in. (400 mm) are used in circular sawing machines. Most large cutters

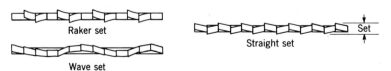

Figure 21.9
Types of set for metal saw blades.

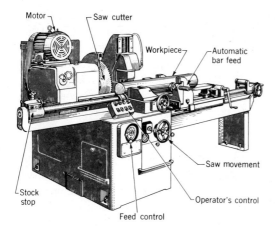

Figure 21.10
Cold sawing machine.

have either replaceable inserted teeth or segmental type blades (Figure 21.11). In the latter type the segments, each having about four teeth, are grooved to fit over a tongue on the disk and are riveted in place. Both inserted teeth and segmental type blades are economical from the standpoint of cutter material cost and have the additional advantage that worn teeth can be replaced.

The teeth are alternately ground so that one-half of them are 0.010 to 0.020 in. (0.25–0.50 mm) higher than the rest. The high teeth are for rough cutting and have a 45° chamfer on each side; the others are ground square across and are the finishing teeth for cleaning both corners. A clearance angle of around 7° is used for most steels and cast iron; if nonferrous metals are to be cut, this angle should be increased to 11°. Rake angles vary from 10° to 20°, smaller angles being for harder materials.

Cutting speeds of circular blades range from 25 to 80 ft/min (0.1–0.4 m/s) for ferrous metals. For nonferrous metals the cutting speed is from 200 to 4000 ft/min (1.0–20.3 m/s). The life of a saw is longer when the peripheral speed is not too high. The use of a lubricating fluid is recommended for all circular sawing work. Tolerances are $\pm {}^{1}/_{64}$ to $\pm {}^{1}/_{8}$ in. (± 0.4–± 1.5 mm) and surface finish is 250 to 1000 μin. (6350–25,400 nm).

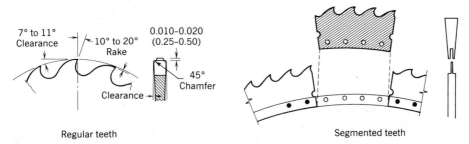

Figure 21.11
Tooth construction for circular saws.

Steel Friction Disks. Operating at high peripheral speeds, *steel friction disks* permit cutting through structural steel members and other steel sections. When the disk is rotating at rim speeds from 18,000 to 25,000 ft/min (90–125 m/s), the heat of friction quickly melts a path through the part being cut. About 30 s are required to cut through a 24-in. (600-mm) I-beam. Disks ranging in diameter from 24 to 60 in. (0.6–1.5 m) are available. They are usually furnished with small indentations on the circumference about $3/32$ in. (2.4 mm) deep. The disks are ground slightly hollow to provide side clearance. Water cooling is recommended.

Friction cutting is not limited by the hardness of the material. Stainless steel and high-carbon steel can be cut more easily than low-carbon steel. Cutting ability seems to depend on the structure of the metal and its melting characteristics rather than metal hardness. During cutting the tensile strength of the steel decreases quickly as the temperature increases. Steel is weakened to the extent that the friction disk pulls it away from the colder metal. The separation temperature is below the melting point of steel. Nonferrous metals cannot be cut satisfactorily by friction sawing, because these metals tend to adhere to the disk and do not separate readily as a result of the disk action.

Abrasive Disks. An abrasive-wheel machine suitable for either wet or dry cutting is shown in Figure 21.12. It will cut ferrous and nonferrous solids up to 2 in. (50 mm) in diameter or tubing up to $3^1/_2$ (90 mm). Resinoid-bonded wheels should be used at speeds around 16,000 ft/min (80 m/s) for dry cutting. High peripheral speed cuts more efficiently than low speed because the metal is heated rapidly and becomes soft for easy metal removal. For wet cutting rubber-bonded wheels operating around 8000 ft/min (40 m/s) are used. The surface speed is limited to this value to retain sufficient coolant on the wheel to prevent overheating. Cutting action depends entirely on the abrasive grains in the wheel and is unaffected by any metal softening. The finish and accuracy is 63 to 500 μin. (1600–12,000 nm) and $\pm 1/_{16}$ in. ($\pm$ 1.6 mm), which is better than steel friction blades.

Band Sawing Machines

Band Sawing. The sawing machines described thus far are designed for straight cuts and for cutting off. Figure 21.13 illustrates a cutting saw of the *band type*, which can

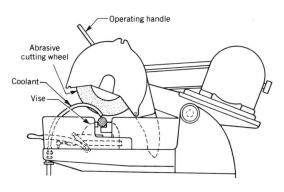

Figure 21.12
Abrasive-disk cutoff machine.

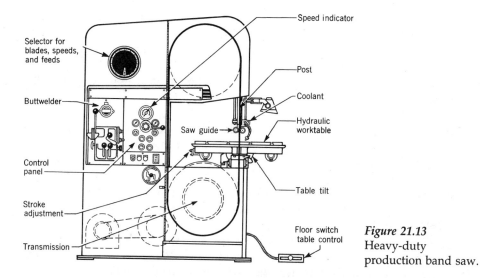

Figure 21.13
Heavy-duty production band saw.

be used for this work and additionally can cut irregular curves. This widens the application for the band saw, allowing work that formerly had to be done with other machine tools. Contour saving of dies, jigs, cams, templates, and other parts previously cut on other machine tools or by hand at greater expense may be done with band saws. Suitable and accurate continuous filing and polishing, both necessary operations in contour finishing, can be accomplished by band saws. Band sawing machines have variable speeds of 50 to 1500 ft/min (0.3–7.6 m/s) to accommodate most materials. Blade specifications vary as in hacksaws, and specific recommendations are available. Figure 21.14 shows the general relationships between material properties and cutting speed and feed. Specific recommendations accompany a saw. Computer control for band saws enables sawing intricate and complex parts including drawing dies and mold sections.

Band Friction Cutting. High-speed band sawing machines designed for friction cutting have a surface speed range of 3000 to 15,000 ft/min (15–75 m/s). Saws for these machines are selected carefully, as pitch (number of teeth per inch) varies from 10 for thick materials to 18 for thin materials (0.4–0.7 teeth per millimeter). Band friction cutting is limited to relatively thin ferrous metals and some thermoplastic materials.

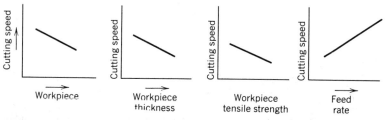

Figure 21.14
General relationships of cutting speed and feed for sawing and broaching.

Diamond Band Cutting. Diamond-impregnated band sawing cuts glass, carbide, ceramic, dies, and hard semiconductor materials. The speeds and feeds must be precisely controlled, and the workpiece is flooded with an appropriate cutting fluid. These saws are usually small in size because band cost is high and contour cutting often involves a radical change in blade direction. It is not uncommon for the blade to be round with the teeth covering the surface. This is referred to as a *wire blade*.

Band Filing and Polishing. When the machine is to be used for filing work, a file band replaces a saw band. The *file band* is constructed of files mounted on a flexible Swedish steel band. A snap joint is provided for quick fastening and unfastening for internal filing. A light to medium pressure is used on contour filing, and the filing speeds range from 50 to 200 ft/min (0.3–1.0 m/s). An advantage of filing is the continuous downward stroke. The absence of a backstroke greatly lengthens the life of the file and helps in holding the work onto the table.

Files used on this machine have the same materials, forms, and styles found on standard commercial files. *Single cut, double cut,* and *rasp cut* are the terms used in describing the cut of the file. Rasp cut differs from the other two in that the teeth are disconnected from each other, each tooth being made by a single punch. The coarseness of the teeth is described by the terms *rough, coarse, bastard, double cut,* and *smooth.* File cross sections are indicated by such terms as *flat, oval, half round,* and *mill*.

Band polishing calls for an endless band of emery cloth that is mounted in the same way as the band saws. At the point of work the cloth band is backed up by a rigid plate.

BROACHING

Broaching is the operation of removing metal by an elongated tool. Since each successive tooth removes metal, either each tooth must be larger than the preceding one or else each tooth is set higher than the previous one (see Figure 21.15). A part is completed in one stroke of the machine, and the last teeth on the cutting tool conform to the desired finished surface. In most machines the broach is moved past the work, but equally effective results are obtained if the tool is stationary and the work is moved. Many cuts, both external and internal, can be made on machines at a high rate of production. An accuracy of ±0.0005 in. (±0.013 mm) and a finish of 32 to 125 μ in. (800–3000 nm) are common. Ferrous materials with a hardness up to Rockwell C40 can be broached.

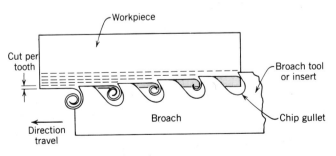

Figure 21.15 Broaching action.

A broaching machine consists of a workholding fixture, broaching tool, drive mechanism, and a supporting frame. Although the component parts are few, several variations in design are possible. A brief classification of broaching according to method of operation is as follows:

1. *Pull broaching.* The broaching tool is pulled through or across stationary work.
2. *Push broaching.* The broaching tool is pushed through or across stationary work.
3. *Surface broaching.* Either the work or the broaching tool moves across the other.
4. *Continuous broaching.* The work is moved continuously against stationary broaches. The path of movement may be either straight or circular.

Most broaching machines are horizontal or vertical in design. The choice of the design depends on the size of part, size of broach, quantity, and type of broaching. Vertical machines having the tool supported on a suitable slide are adapted for surface broaching. Both pull- and push-type internal broaching machines are constructed with this design. Horizontal machines pull the broach and are often used for internal and surface broaching of small- and medium-sized parts.

Most machines are hydraulically driven because of the large force requirements. Such a drive is smooth acting, economical, and adjustable for speed and length of stroke.

Broaching is a metal cutting operation that has been adopted for mass-production work because of the following features and advantages:

1. Both roughing and finishing cuts are completed with one pass of the tool.
2. Production rate is high because the actual cutting time is a matter of seconds. Rapid loading and unloading of fixtures minimize total production time and can be adapted to automatic production.
3. Either internal or external surfaces can be broached.
4. Any form that can be reproduced on a broaching tool can be machined.
5. Production tolerances are suitable to interchangeable manufacture.
6. Finishes comparable to milling work can be obtained. Burnishing shells incorporated as the final teeth on the broach improve the surface finish.

Broaching has the following limitations:

1. Cost of the tool is high, particularly for large or irregular-shaped broaches.
2. Short-run jobs are not advisable because of high tooling cost.
3. Parts to be broached must be rigidly supported and capable of withstanding the broaching forces.
4. Surface to be broached must be accessible.
5. Broaching is not recommended for removing a large amount of stock.

Broaching Machines

Vertical Single-Slide Surface Machines. The surface broaching operation being performed on the vertical machine shown in Figure 21.16 is simple and quick. Most machines

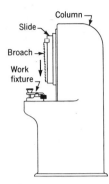

Figure 21.16
Vertical single-slide broaching machine.

of this type are provided with a receding table, so that the fixture can be loaded and unloaded while the broach is returning to its original position. The cycle is automatic and continuous except for the loading operation, which, if economics warrant, can be made automatic. A vertical double-slide machine differs from the single-slide machine in that it has two slides that operate opposite one another. The work is held on shuttle tables that move out during the unloading and loading operations while the ram returns to its starting position. While this is going on, the other ram is at work.

Vertical Push Broaching. An example of push broaching is shown in Figure 21.17A. A broach is finishing a round hole in a gear blank, which is more rapid than reaming or boring, and at the same time the hole can be held to accurate limits. Finished holes can be broached from holes previously drilled, punched, or cored. Push broaching requires comparatively short broaches of sufficient cross section to prevent column buckling resulting from the loads imposed during the operation. This type of broaching is also possible on a simple utility press.

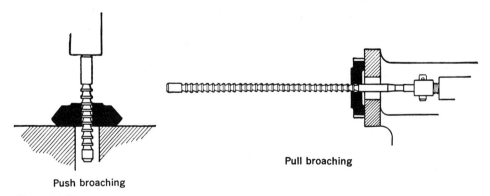

Figure 21.17
Round broaches for push- and pull-type machines. A, Push broaching. B, Pull broaching.

Vertical Pull-Down Broaching Machines. Vertical pull-down machines are adapted to internal broaching. The parts are placed in a fixture on the worktable, the pulling mechanism being in the base of the machine. Broaching tools are suspended above by an upper carriage. As the operation starts the broaches are lowered through the holes to be broached and are automatically engaged by the mechanism, which pulls them through the part. Upon removal of the work the tools rise, are engaged by the upper holders, and return to their starting position. Machines of this type have the advantage over pull-up machines in that the positioning of the part is easier and large parts are handled with less difficulty.

Vertical Pull-Up Broaching Machines. These machines are also adapted to internal broaching and are frequently preferred for small parts. In many machines there are four or more broaching tools. Although the general cycle of operation is similar to the pull-down machines, it is reversed. At the starting position the parts to be broached are placed over the shanks of the broaches then being held by the lower mechanism. As the broaches rise they engage the upper pulling mechanism, and the parts are then held against the lower side of the worktable. At the completion of the operation the parts fall and are deflected into a container. In this and most other broaching machines the operator has only to load the machine.

Horizontal Broaching Machines. Although horizontal broaching machines have surface broaching applications, they are generally used for internal broaching of large and medium-sized parts. A diagrammatic sketch of a horizontal broaching machine adapted for surface broaching is shown in Figure 21.18. Here the broach is pulled over the top surface of the workpiece held in the fixture. The hydraulic cylinder that pulls the slide and broach is housed in the right end of the machine. These machines operate at cutting speeds of 10 to 40 ft/min (0.05–0.2 m/s) and have return speeds around 100 ft/min (0.5 m/s). For broaching internal work the shank of the broach is manually threaded through the workpiece. Figure 21.17 also illustrates pull broaching on a horizontal machine.

Large, heavy-duty horizontal broaching machines are used in the high-speed production of cylinder blocks, intake manifolds, bearing cap clusters, and aircraft turbine disks. The cutting speed approaches 200 ft/min (1.0 m/s) and the stock removal is up to $1/4$ in. (6.4 mm) per stroke.

Broaching internal *keyways* is one of the oldest uses of the broaching process. A keyway broach and its adapter are shown in Figure 21.19. The adapter guides the broach and also assists in holding and locating the work. Broaching tools for this purpose are

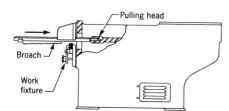

Figure 21.18
Horizontal broaching machine.

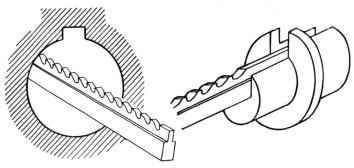

Figure 21.19
Keyway broaching.

extremely simple and can be obtained for general-purpose use. If multiple keyways or splines are to be cut, a single broach can be used with the work and indexed after each cut. This procedure is used only for large splines or in jobs where the quantity is small, because spline broaches can be obtained for any number of keys.

Internal gear broaches are similar to spline broaches except for the involute contours on the sides of the teeth. They may be made to cut any number of teeth and are used for broaching as small as 48 diametral pitch. This method of gear cutting is known as the form tooth process, and the accuracy of the teeth depends on the accuracy of the form cutter. External gear teeth may also be broached, but external gear cutting is usually limited to cutting teeth on sector gears where only a few teeth are involved.

An interesting development in horizontal broaching is cutting helical grooves or splines by pulling a broach through the part and at the same time either rotating the part or broaching tool according to the helix desired. This procedure has been adopted by many gun manufacturers in rifling small-caliber and light cannon gun barrels.

Rotary Broaching Machines. Rotary broaching consists of mounting the work in fixtures supported on a revolving table that moves past stationary broaches. These broaches are made in short sections so that they can be adjusted and sharpened easily. Rotary broaching machines limited to small parts are used for squaring distributor shafts, slotting, straddle milling, form milling, and facing small parts.

Continuous or Tunnel Broaching Machines. Continuous broaching machines are adapted only for surface broaching. This type of broaching machine consists of a frame and driving unit with several workholding fixtures mounted on an endless chain that carries the work in a straight line past the stationary broaches. A view of a continuous machine equipped with a motor-driven conveyor for removing work from the machine is shown in Figure 21.20. Loading is done by an operator who drops the parts in the fixtures as they pass the loading station. The work is automatically clamped before it passes into the fixture tunnel in which the broaches are held. After the fixtures pass the broach, they are automatically released by a cam; at the unloading position the work falls

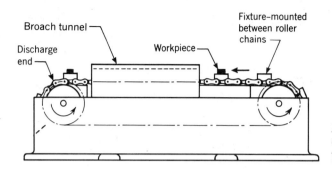

Figure 21.20
Continuous surface broaching machine.

out of the fixtures into the work chute. Production is high because the operator handles only the work in the loading position.

Some large *tunnel broaches* are designed for finish broaching the tops of engine cylinders and cylinder heads. The part is held in a fixture, which is pushed under the stationary broaches by a hydraulic ram. A wire cable then pulls the finished part onto an exit conveyor while the fixture returns to starting position. Six-cylinder engine blocks can be surfaced in this manner at the rate of 120 per hour.

Broaching Tools

Broaching tools differ from most other production tools in that they are usually adapted to a single operation. The feed of the tool must be predetermined, and once a broach is made the feed remains constant. Knowledge about the job, part material, and the machine are necessary before a broach can be made. In designing and constructing a broach the following information must be known:

1. Kind of material to be broached.
2. Size and shape of cut.
3. Quality of finish required.
4. Hardness of material.
5. Tolerance to be maintained.
6. Number of parts to be made.
7. Type of machine to be used.
8. Method of holding broach.
9. Pressure that the part can stand without breakage.

According to the method of operation, there are two kinds of broaches: pull or push broaches. Most internal broaching is done with pull broaches, because they can take longer cuts and can remove more stock than push broaches. Push broaches, which are necessarily short to avoid buckling under load, are used principally in sizing holes in heat-treated parts and for short-run jobs. They are also used in broaching blind holes.

The simplest broaches designed for flat surfaces can be made with either straight or angular teeth. Angular teeth produce a smoother cutting action. Because the entire length of such broaches can be supported on a slide, it is possible to produce them in short sections. Heavy-duty broaches frequently have inserted teeth to reduce the initial cost and facilitate replacements. Specially designed fixtures are often necessary to hold the workpiece, because cutting forces may fracture or deform the part if it is not properly supported.

Broaching Terms and Angles

Reference to Figure 21.21, showing a pull-type broach, illustrates some terms usually applied to broaches. Figure 21.22 shows an enlarged tooth form with terms and angles indicated. The top portion of a tooth is called the *land* and is ground to give a slight clearance. This angle, called *backoff* or *clearance angle*, is usually $1\frac{1}{2}°$ to $4°$ on the cutting teeth. Finish teeth have a smaller angle ranging from $0°$ to $1\frac{1}{2}°$. Regrinding on the lands of most broaching tools should be avoided because this changes the size of the broach. Sharpening is done by grinding the face or front edge of the teeth. The angle to which this surface is ground corresponds to the rake angle on a lathe tool and is called the *face angle, hook angle, undercut angle,* or *rake angle*. The rake angle varies according to the material being cut and in general increases as the ductility increases. Values of this angle range from $0°$ to $20°$, but for most steels a value of $12°$ to $15°$ is recommended. This angle has considerable effect on the force required to make the cut and the finish. A large angle might give excellent results, but from the standpoint of lengthening tool life a smaller angle is preferred. Frequently, the first cutting teeth are rugged in shape and have a small rake angle, while the finish teeth are given a large rake angle to improve the finish. Side rake angles of $10°$ to $30°$ are widely used on surface broaching to improve the finish of the cut.

For short broaches the pitch of a broach is

$$P = K \sqrt{L}$$

where

P = Pitch, in. (mm)
K = 0.35 for L, in.; 1.76 for L, mm
L = Length of cut, in. (mm)

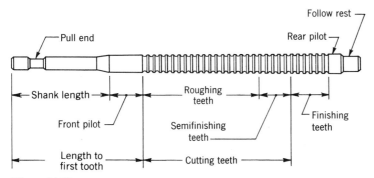

Figure 21.21
Internal pull-type broach.

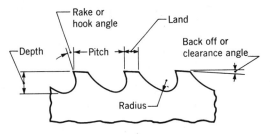

Figure 21.22
Tooth form on broach with principal terms and angles.

For small broaches the amount of material removed per tooth is approximately

$$D = \frac{0.10 \times (P - W)G}{L}$$

where

D = Depth of cut per tooth, in. (mm)
W = Length of land, in. (mm)
G = Gullet depth, in. (mm)

A longer broach will usually have at least two separate cutting areas, roughing and finishing. Each is treated separately. To find the minimum force to operate a broach, a number of factors must be known. With a surface broach the force may be calculated from

$$F = K_2 NDW$$

where

F = Minimum force, lb (kg)
K_2 = Constant
N = Number of broach teeth cutting at one time
W = Width of cut, in. (mm)

Approximations are $K_2 = 400$ for mild steels and cast iron, and 200 for nonferrous materials such as aluminum, zinc, and brass.

Burnishing

Sometimes holes are sized and the finish improved by burnishing. Burnishing is not a cutting operation, but consists of moving a very hard surface over the workpiece to remove irregularities and smear metal caused by prior machining. This can be done with a burnishing broach or a regular broach that has several burnishing shells following the finishing teeth. The amount of stock left for burnishing should not exceed 0.001 in. (0.03 mm), and for ordinary steel 0.0005 in. (0.013 mm) is sufficient. This amount should be distributed over three or four burnishing shells. The operation is one of cold working, which produces a hard, smooth surface.

QUESTIONS AND PROBLEMS

1. Give an explanation of the following terms:

   T-slot
   Ram
   Clapper box
   Slotter
   Pit planer
   Raker set
   Steel friction disk
   Double cut
   Keyway
   Tunnel broach

2. How is the feed obtained on a shaper? On a planer?

3. How is the length of stroke changed on oscillating arm shaper?

4. In a draw cut shaper the ram is in tension; in a push cut unit the ram is in compression. Which type can take the heavier cut and why?

5. How is return speed regulated on a hydraulic shaper?

6. How is the return speed regulated on a Whitworth drive-type shaper?

7. For what type of work are vertical shapers used?

8. What are the disadvantages of employing carbide tools on a shaper or planer?

9. What advantage does an open-side planer have over a double-housing planer?

10. Make a sketch indicating how parts are held by using T-slots.

11. For what type of job would a pit planer be employed?

12. In the conventional nonhydraulic shaper mechanism, describe why the speed of cutting is less than the return speed.

13. What type of tool would be necessary on a horizontal push cut shaper to cut a keyway in a gear?

14. How can a dial indicator be used on a shaper or planer to assure that a flat surface is perpendicular to the path of the ram?

15. If there is a heavy scale on the part to be machined, what type of cut or cuts will minimize tool wear and create the smoothest surface?

16. If the ratio of return to cutting speed is 3 : 2 on a shaper and the return speed is 200 ft/min (61 m/min), how long does it take to make a 10-in. (254-mm) cut? What is the cutting speed?

17. If the feed is 0.125 in. (3.2 mm) and the shaper makes 100 spm, how long does it take to surface a 7.5-in. (190-mm)-wide workpiece?

18. How does the feed rate affect problems 16 and 17?

19. How would the tool width affect the answer to problem 17?

20. Suppose the feed was 0.5 in. (13 mm) on a planer and the tool width was 0.25 in. (6.4 mm). What would the surface of the workpiece look like after machining?

21. If the feed on a shaper is expressed as $1/32$ in. (0.8 mm) per stroke, how long would it take to machine a square part $1\frac{1}{2}$ ft (460 mm) wide if the shaper makes 37 spm?

22. Sketch a quick-return mechanism that can be used for a reciprocating saw.

23. Describe how kerf is obtained for a saw cut. What is its purpose?

24. What factors should be considered in the selection of the pitch for a hacksaw blade?

25. What are the advantages and disadvantages of uniform feed on a hacksaw?

26. What kind of disks are used in abrasive cutoff machines? At what surface speeds are they operated?

27. Should a steel friction disk or an abrasive disk be used for (a) cutting a steel pipe; (b)

cutting a steel I-beam; (c) cutting aluminum pipe; (d) cutting very hard steel rod; (e) cutting glass?

28. Discuss the similarities between the method by which a tooth cuts on a saw and that of a single-point, orthogonal cutting tool.

29. In band saw work what determines the width of the saw, the number of teeth per inch (teeth per millimeter), and the set?

30. For what kind of material is the skip tooth band saw blade adapted?

31. Give an application for saw blades having sets such as (a) raker, (b) wave, and (c) straight.

32. What type of quick-return mechanism could be employed on a hydraulic broaching machine? Why is a quick-return mechanism an economical addition to a broaching machine?

33. What are the advantages and limitations of the broaching process?

34. Sketch a pull-type surface broach that could be used to make a dovetail slot in a part that had a rectangular slot milled in it. Label the principal parts.

35. How may rifling on a gun be made by using a broaching tool?

36. What advantage does a vertical pull-down machine have over a vertical pull-up machine?

37. What type of broaching machines do you recommend for broaching the following parts: keyway in gear, involute teeth on gear segment, top of engine cylinder, splines in gears, and cutting cap from connecting rod?

38. What is meant by burnishing and how is it done on a broaching machine?

39. How could the broach in problem 34 be modified to accommodate burnishing? What type of accuracy and surface finish might be expected?

40. Describe the quality of the surface finish of broached parts. How does it compare to the lathe, planer, and grinder?

41. For a shaft running at 900 rpm, find the diameter of an abrasive wheel for minimum peripheral speed.

42. A 14-in. (356-mm) diameter circular saw has 120 teeth, a feed rate of 0.003 in. (0.08 mm) per tooth, and a cutting speed of 60 ft/min (18.3 m/min). Determine the cutting time for sawing a 3.375-in. (85.7-mm) diameter steel bar.

43. If an abrasive wheel is to cut at 4000 ft/min, what must the revolutions per minute be if it is 260 mm in diameter?

44. A reciprocating hacksaw using a 16-in. (406-mm) high-speed steel blade with 12 teeth per inch (0.5 teeth per millimeter) is to cut a 3-in. (75-mm) steel bar. The machine operates at 80 spm with a uniform feed rate of 0.010 in. (0.25 mm) per stroke. If the tooth pressure will not exceed 8.5 lb (37.8 N) per tooth, what is the maximum pressure on the saw blade and the cutting time?

45. A rectangular piece of cold-rolled steel 2 in. (50.8 mm) thick is to be cut on a band saw having a 14 pitch (0.6 pitch per millimeter) blade and operating at 150 ft/min (0.76 m/s). If the average metal removed per tooth is 0.0001 in. (0.003 mm), how long will it take to cut through a width of 3 in. (75 mm)?

46. A horizontal broaching machine has a cutting speed of 24 ft/min (0.12 m/s), return speed of 36 ft/min (0.18 m/s), and a stroke of 38 in. (965 mm). If the starting and stopping time is 4 s and the loading time 10 s, find the machine output per hour for a production efficiency of 90%.

47. If the tooth length on a 16-in. (406-mm) hacksaw blade is 13.2 in. (335 mm), how many teeth are on the blade?

48. A short 15-in. (38.1-mm) broach is to be made to broach mild steel. What pitch should be used?

49. For the broach in problem 48, plot the material removed per tooth versus gullet depth for a land length of 0.125 in. (3.18 mm).

50. How much minimum force in pounds would be required to operate the broach described in problems 48 and 49 if the gullet depth was 0.375 in. (9.53 mm)?

51. What would be the total amount of material removed (i.e., the total depth of cut) for the conditions in problem 48 if gullet depth is 0.375 in. (9.53 mm) and the land length is 0.125 in. (3.18 mm)?

CASE STUDY

ACME BROACH COMPANY

The aluminum part shown in Figure 21.23 is to have a broached finish on the top surface. As the broach designer it is your job to choose the type and size of broaching machine, required (minimum force), and to specify the design of the broach so it can be built. The total amount of material to be removed is 0.031 in. (0.794 mm).

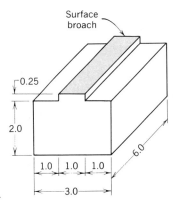

Figure 21.23
Case study.

CHAPTER 22
THREADS AND GEARS

A *screw thread* is a ridge of uniform section in the form of a helix on the surface of a cylinder. The terminology relating to screw threads is shown in Figure 22.1. The sequence designating a screw thread is the nominal size (fractional diameter or screw number), number of threads per inch, thread series symbol, and thread class. The nominal size is the basic major diameter.

The Unified (UN) system of screw threads has a designation system, as does the *ISO* or metric standard.

UNIFIED SYSTEM

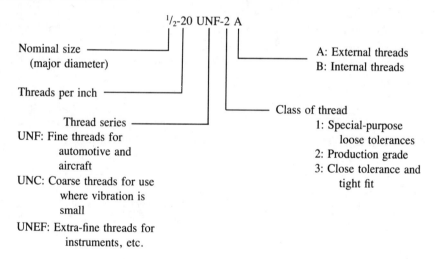

ISO SYSTEM

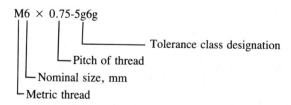

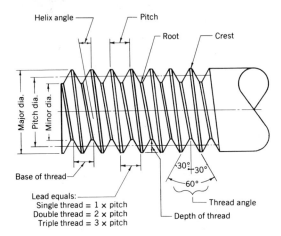

Figure 22.1
Screw thread terminology.

The tolerance class designation in the ISO system refers to the quality of thread; the first digit and letter determine the pitch diameter, a lower-case letter g being for external threads and an upper-case G for internal threads. Grade 6 is medium quality and 5 is slightly less. The second series, 6g, refers to the quality of the thread crest, 6 is again medium and a 9 would be highest. Again the g refers to an external thread. The letters may vary from e to h or E to H, where e or E is for a large allowance, and h or H is for a thread with little or no allowance.

If left-hand threads are designated, the term "LF" appears at the end of the thread specification.

Pitch is expressed by a fraction with 1 as the numerator and the number of threads per inch as the denominator. A screw having 16 single threads per inch has a pitch of $1/16$. It should be remembered that only on single-threaded screws does pitch equal the lead. In SI pitch is a distance between corresponding points on adjacent profiles.

Lead is the length a screw advances axially in one revolution. On a double-threaded screw the lead is twice the pitch; on a triple-threaded screw the lead is three times the pitch, and so on.

Screw threads are used as fasteners and to *transmit power* as illustrated in the screw jack. Threads also *transmit motion* when used in a lead screw on a lathe. Screw threads are employed in *measuring devices* such as micrometers.

TYPES OF SCREW THREADS

Screw threads have been standardized according to their cross-sectional form. The principal threads used worldwide are classified into two systems.

1. *Unified screw standards.* With slight modification, this threading system was formerly known as the American National Standard series. These threads are sometimes called inch threads as distinct from metric threads. (See Figure 22.2.)

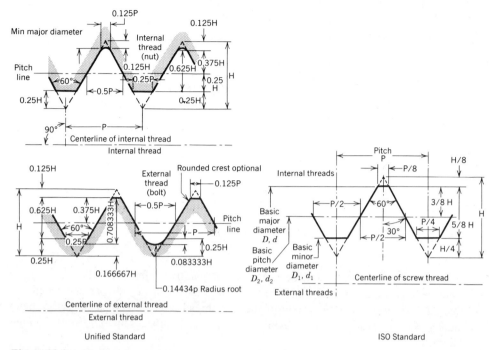

Figure 22.2
Profile comparison of Unified and ISO metric threads.

2. *ISO Standards*. The ISO system was promulgated by the International Organization for Standards and covers metric threads (See Figure 22.2.)

In the UN system there are a number of classifications, the principal ones being the fine, coarse, and extra-fine series. There are also eight other series of UN threads with constant pitches of 4, 6, 8, 12, 16, 20, 28, and 32 threads per inch. Whenever possible, selection should be made from the standard series (UN screw threads), with preference being given to the coarse- and fine-thread series. The coarse-thread series is generally used for the bulk production of screws, bolts, and nuts. The coarse-series threads (UNC) provide more resistance to internal thread stripping than the fine or extra-fine series. The fine-thread series (UNF) is used because of its high strength and in applications where vibrations occur. This series has less thread depth and a larger root diameter than the coarse-series threads. To prevent internal thread stripping a longer length of engagement is required for the fine series. The extra-fine series (UNEF) is used for equipment and threaded parts that require fine adjustment.

The ISO metric screw thread shown in Figure 22.2 has essentially the same basic profile as the UN screw thread basic form. Note the external and internal form is the same for ISO threads except when rounded roots are called for on external threads. The UN and ISO threads are not interchangeable. Tables of inch to metric conversions, such

Table 22.1 A Comparative Chart of UN Thread Series to ISO Metric Series[a]

UNC Threads	UNF Threads	ISO Metric threads
2-56 (2.18)[b]	2-64 (2.18)	M2 × 0.4 (2.00)
3-48 (2.51)	3-56 (2.51)	M2.5 × 0.45 (2.50)
4-40 (2.84)	4-48 (2.84)	M3 × 0.5 (3.00)
5-40 (3.18)	5-44 (3.18)	
6-32 (3.51)	6-40 (3.51)	
8-32 (4.17)	8-36 (4.17)	M4 × 0.7 (4.00)
10-24 (4.83)	10-32 (4.83)	M5 × 0.8 (5.00)
1/4-20 (6.35)	1/4-28 (6.35)	M6 × 1 (6.00)

[a]Threads are not mechanically interchangeable.
[b]Numbers in parentheses indicate major diameter in millimeters.

as given in Table 22.1, should be used for comparative reference only. The biggest difference between the UN and ISO series is the number of threads per unit length. In Table 22.1 several major diameters of the UN thread size are compared to the major diameter of the closest equivalent ISO metric thread size for sizes up to $1/4$ in. (6.35 mm) in diameter.

The basic profile for ISO and UN, as shown by Figure 22.1, is essentially the same. The principal differences are related to basic size, the magnitude and application of allowances and tolerances, and thread designations. The crest width of the nut is $1/4$ pitch, whereas that of the screw is $1/8$ pitch. The increased flat on the nut makes production easier and at the same time the nut is as serviceable. The shape of the thread crest and root is not mandatory; it can be either flat or rounded.

Square threads, shown in Figure 22.3, are suitable for transmitting power when there is a large thrust on one side of the thread. These threads cannot be cut with taps and dies and must be machined on a lathe. Another type similar to the square thread is known as a *buttress thread*. It has one side that slopes 45°; the other is perpendicular. Although this thread does transmit power, the thrust is only in one direction. Acme screw threads (Figure 22.3) have the advantage that wear may be compensated for by adjusting half-nuts in contact with the screw. They can be cut with taps and dies. *Worm threads*, similar

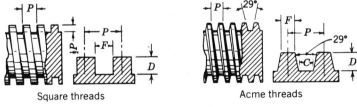

Square threads   Acme threads

Figure 22.3
Screw forms used for transmitting power.

to the Acme standard except that they have a greater depth, are used exclusively for worm gear drives.

Pipe threads have been standardized according to the American National Standard. To ensure tight joints the thread has a taper of $3/4$ in./ft (6.25%). The threads have the conventional V-shape except for the last four or five, which have flat crests and imperfect troughs. The usual method of cutting these threads is with suitable taps and dies, although they can be cut on a lathe by using the taper attachment.

METHODS OF MAKING THREADS

External threads may be produced by the following manufacturing processes:

1. Engine lathe.
2. Die and stock.
3. Automatic die head.
4. Milling machine.

5. Threading machine.
7. Die casting.
8. Grinding.

Internal threads may be produced by:

1. Engine lathe.
2. Tap and holder.
3. Automatic collapsible tap.

4. Milling machine.
5. Screw broach.

Taps and Dies

Taps are used for the production of internal threads. Figure 22.4 illustrates a *tap* with the parts of the tool labeled. The tool itself is a hardened piece of carbon or alloy steel resembling a bolt with flutes cut along the side to provide the cutting edges. For hand tapping these are furnished in sets of three for each size. In starting the thread the *taper tap* should be used, because it ensures straighter starting and more gradual cutting action on the threads. If it is a through hole no other tap is needed. For closed or blind holes

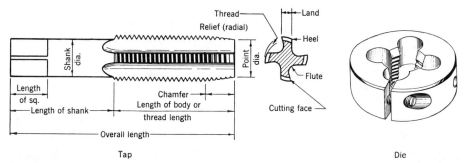

Figure 22.4
Tap and die.

with threads to the very bottom, the *taper, plug,* and *bottoming taps* should all be used in the order named. Other taps are available and are named according to the kind of thread they are to cut.

Where a hole is to be tapped, the hole that is drilled before the tapping operation must be of such a size as to provide the necessary metal for the threads. Such a hole is a *tap size hole*. For many years a 75% thread has been recommended, and most published tables show tap drill sizes that will yield a 75% thread. Tests have shown that a greatly reduced percentage of thread (a larger tap size hole) will give adequate strength to the fastening and greatly reduce the high torque required for tapping.

The length of a nut that will prevent stripping the threads when a bolt of the same material is pulled with a tensile stress is given by

$$t = 0.47d$$

where

t = Nut thickness, in. (mm)
d = Nominal bolt diameter, in. (mm)

Because nuts are usually made of slightly softer metal, $t = {}^7/_8 d$ is usually employed.

The most common method of cutting external threads is by the adjustable die shown in Figure 22.4. It can be made to cut either slightly undersize or oversize. When used for hand threading, the die is held in a die stock.

For successful operation of either taps or dies, consideration must be given to the nature of the material to be threaded. No tool can work successfully for all materials. The shape and angle of the cutting face also influence the performance. Another important factor is proper lubrication of the tool during the cutting operation; this ensures longer life of the cutting edges and results in smoother threads.

Taps and dies can also be used in machine cutting of threads. Because of the nature of the cutting operation they are held in a special holder, so designed that the tap or die can be withdrawn from the work without injury to the threads. This is frequently accomplished by reversing the rotation of the tool or work after the cut has been made. In some equipment this reversing action is faster than the cutting.

In small-production work on a turret lathe the tap is held by a special holder that prevents the tap from turning as the threads are cut. Near the end of the cut the turret holding the tool is stopped, and the tapholder continues to advance until it pulls away from a stop pin a sufficient distance to allow the tap to rotate with the work. The rotation of the work is then reversed and, when the tapholder is withdrawn, it is again engaged with the stop and held until the work is rotated from the tap. External threads can be cut with a die utilizing this same procedure, although most of the time such threads are cut with *self-operating dies*.

Thread Chasing

In production work *self-opening dies* and *collapsible taps* are used to eliminate backtracking of the tool and to save time. The tools have individual cutter dies known as **chasers** that are mounted in an appropriate holder and are adaptable to adjustment or replacement. With chasers more accurate work results, the cutters can be kept in proper adjustment, and there is no danger of damaging the cut thread as the tool is withdrawn. In some instances the tool is held stationary and the work revolves; in others the reverse procedure may be used. All precision screws require a lead screw feed to obtain accuracy.

Two types of *automatic die heads* are used. In one the cutters or chasers are mounted *tangentially*, as shown in Figure 22.5. In the other they are in a *radial* position. Radial cutters can be changed quickly; consequently, they are used for threading materials that are hard to cut. The die head commonly used on most turret lathes is of the stationary type. The work rotates and the chasers open automatically at the end of the cut so that they can be withdrawn from the work without damage. In threading machines the dies rotate and the work is fed to them, but otherwise the operation is the same.

Tapping Machines

Although tapping is done on drill presses equipped with some form of tapping attachment, most production tapping is done on specially constructed automatic machines. Nuts to be threaded are fed from an oscillating hopper to the working position; spindles are reversed at double the tapping speed; and nuts are discharged to containers.

A common type of tapping machine has a multispindle arrangement provided with taps having extra-long shanks. The tap is advanced through the nut by the lead screw and, upon completion of the threading, continues downward until the nut is released. The spindle then returns to its upper position with the tapped nut on its shank. When the shank has been filled with nuts, the tap is removed and the nuts are emptied.

Thread Milling

Accurate threads of large size, both external and internal, can be cut with standard hob-type cutters. For long external threads a threading machine similar in appearance to a lathe is used. Work is mounted either in a chuck or between centers, the milling attachment being at the rear of the machine. In cutting a long screw a single cutter is mounted in the plane of the thread angle and fed parallel to the axis of the threaded part as shown

Figure 22.5
Revolving tangent die head.

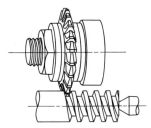

Figure 22.6
Single-thread milling cutter.

in Figure 22.6. The feed (f) in thread milling is expressed as the cutter advance per tooth, or inches per cutter tooth, by the following formula:

$$f = \frac{ds}{nN}$$

where

d = Nominal diameter of thread, in.
s = rpm of work
N = rpm of cutter
n = number of teeth in cutter

From this expression it is evident that the cutter load per tooth, which varies directly with the feed, can be changed by varying the cutter speed, work speed, or number of teeth in the cutter. This permits reducing the load on the cutter teeth so that deep threads can be cut in one pass.

For short external threads a series of single-thread cutters is placed side by side and made up as one cutter, having a width slightly more than that of the thread to be cut. The cutter is fed radially into the work to the proper depth and, while rotating a little over one revolution, completes the milling of the thread (Figure 22.6). Proper lead is obtained by a feed mechanism that moves the cutter axially while it is cutting.

Milling machines of the *planetary type* are also used for mass production of short internal or external threads. The milling head carrying the hob is revolved eccentrically about the rigidly held work, which is rotated simultaneously on its own axis. It is advanced by means of a lead screw for a sufficient distance to produce the thread.

Thread Rolling

Over 90% of bolts up to 2 in. (51 mm) in diameter have rolled threads. Threads can be rolled using any material that has sufficient ductility to withstand the forces of cold working without disintegration. An elongation of 12% or more in a 2-in. (51-mm) gage length is considered to be a rather good index of rollability. **Rollability** is the behavior of a metal during the rolling process. Because of the behavior of various types of metal during the formation of threads, rollability cannot be accurately determined by any single physical property of the metal. Certain factors that influence a material's resistance to plastic deformation are material hardness or lack of ductility, internal friction developed during plastic deformation, yield point of the material that must be exceeded in the

METHODS OF MAKING THREADS 579

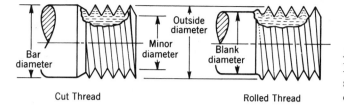

Figure 22.7
Illustrating stock material saving of rolled threads over cut threads.

process, and tendency of the material to work harden. Steels that work harden rapidly require greater pressures and the die life is reduced. However, work-hardened threads may eliminate subsequent hardening and grinding operations. Materials that have good rollability include the basic open-hearth steels, sulfurized steels up to 0.13% sulfur, 1300 series manganese steels, nickel steel, and many of the nonferrous alloys.

In thread rolling the metal on the cylindrical blank is cold forged under considerable pressure by the rolling action of the blank between either rotating cylindrical dies or reciprocating flat dies. The surface of the dies has the reverse form of the thread that is rolled. Rolling under pressure results in a plastic flow of the metal. The die penetrates to form the root of the thread and the displaced metal flows to form the crest. Less material is required for rolled threads over cut threads, as illustrated in Figure 22.7. This saving ranges from 16% to 25%. For example, the savings for a $1/2$-in., no. 13 thread screw size is 19%. Beginning stock diameter is approximately equal to the pitch diameter of the screw. In Figure 22.8 the blank diameters for both rolled and cut threads are indicated for several UNC threads.

Two methods are employed in rolling threads. In one the bolt is rolled between two flat dies, each being provided with parallel grooves cut to the size and shape of the thread. One die is held stationary while the other reciprocates and rolls the blank between the two dies. Figure 22.9 illustrates the method of rolling a screw between two soft boards

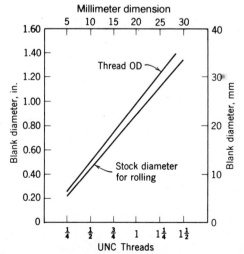

Figure 22.8
Blank diameter for rolled and cut threads.

580 THREADS AND GEARS

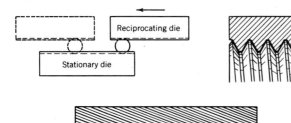

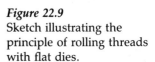

Figure 22.9
Sketch illustrating the principle of rolling threads with flat dies.

under pressure. Each board has impressed into its surface a series of angular, parallel lines. By reversing this illustration and starting with similar grooves in hardened steel, threads are rolled into a rod placed between them.

The other method, shown in Figure 22.10, employs either two or three grooved roller dies. In the two-die method the blank is placed on the work rest between two parallel, cylindrical rotating dies, and the right-hand die is fed into the blank until the correct size is reached, returning then to its starting position. The three-die machine utilizes cylindrical rotating dies mounted on parallel shafts driven synchronously at the desired speed. They advance radially into the blank by cam action, dwell for a short interval, and then withdraw.

Equipment is also available for producing large forms and work threads as shown in Figure 22.11. The $1\frac{1}{2}$-in. pitch worm gears are produced at the rate of 20 per hour. This machine, designed for forming operations requiring rolling forces of 50,000 to 200,000 lb (0.2–0.9 MN), is equipped with an air-operated workholding fixture. The fixture automatically retracts from the work area to unload the finished workpiece and insert a new blank. Because of the extreme depth of the teeth the workpiece is passed back and forth between the rolls until the required penetration is obtained. Other recent developments in the field of cold rolling include the production of small gears and splines.

Following are some of the advantages of the thread rolling process.

1. Improved tensile, shear, and fatigue strength properties.
2. Finer surface finish of 4 to 32 μin. (100–800 nm).
3. Close accuracy maintained.
4. Less material required.

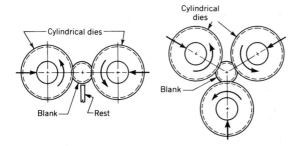

Figure 22.10
Thread rolling using either two or three cylindrical dies.

Figure 22.11
Thread rolling and forming machine.

5. Cheaper materials usable because of improvement of physical properties during rolling.
6. High production rate.
7. Wide variety of thread forms possible.

Limitations of the thread rolling process include the following:

1. Necessary to hold close blank tolerance.
2. Uneconomical for low quantities.
3. Can only roll external threads.
4. Cannot roll material having a hardness exceeding Rockwell C37.

Thread Grinding

Grinding is used as either a finishing or a forming operation on many screw threads where accuracy and smooth finish are required. This process is particularly applicable for threads that have been hardened.

Two types of wheels are used in thread grinding as illustrated in Figure 22.12. Shown in Figure 22.12A is a single wheel shaped to correct form that traverses the length of the screw. The wheel is rotated against the work, usually at speeds ranging from 750 to 10,000 ft/min (4–50 m/s), and at the same time traverses the length of the screw at a velocity determined by the pitch of the thread. With the feed ranging from $1^1/_2$ to 10 ft/min (0.008–0.05 m/s), the surface speed of the work is determined by the depth of grind and material.

Short threads may be ground by the ***plunge cut*** method, as shown in Figure 22.12B. The wheel is fed to the full thread depth before the workpiece is rotated. It then makes one revolution while traversing a distance equal to one pitch, thus completing the thread.

Most precision external threads are ground after heat treatment to eliminate possible distortion. Threads of 10 pitch or finer can be ground directly from solid hardened blanks. Coarse threads are previously roughed.

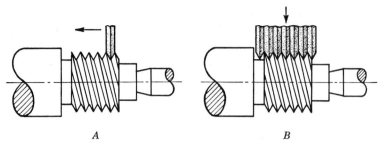

Figure 22.12
Methods used in thread grinding. *A*, Single-wheel traverse grinding. *B*, Plunge cut grinding.

GEARS

Gears transmit power and motion between moving parts. Positive transmission of power is accomplished by projections or teeth on the circumference of the gear. There is no slippage as with friction and belt drives, a feature most machinery requires, because exact speed ratios are essential. Friction drives are used in industry, where high speeds and light loads are required and where loads subjected to impact are transmitted.

When the teeth are built up on the circumference of two rolling disks in contact, recesses must be provided between the teeth to eliminate interference. The circumference upon which the teeth are developed is known as the *pitch circle*. It is an imaginary circle with the same diameter as a disk that would cause the same relative motion as the gear. All gear design calculations are based on the diameter of the pitch circle. A portion of a gear is shown in Figure 22.13.

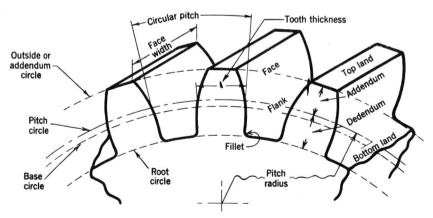

Figure 22.13
Nomenclature for involute spur gear.

Gear Nomenclature

The system of gearing used in the United States is known as the *involute system,* because the profile of a gear tooth is principally an involute curve. An ***involute*** is a curve generated on a circle, the normals of which are all tangent to this circle. The method of generating an involute is shown in Figure 22.14. Assume that a string having a pencil on its end is wrapped around a cylinder. The curve described by the pencil as the string is unwound is an involute, and the cylinder on which it is wound is known as the ***base circle.*** The portion of the gear tooth from the base circle at point a in the figure to the outside diameter at point c is an involute curve and is the portion of the tooth that contacts other teeth. From point b to point a the profile of the tooth is a radial line down to the small fillet at the root diameter. The location of the base circle on which the involute is described is inside the pitch circle and is dependent on the angle of thrust of the gear teeth. The relationship existing between the diameter of the pitch circle and base circle, D, is

$$D_b = D \cos \theta$$

where

D_b = Diameter of base circle

θ = Angle of thrust between gear teeth

The two common systems have their thrust angles or lines of action at $14^1/_2°$ and $20°$. Other angles are possible, but with larger angles the radial force component tending to force the gears apart becomes greater. If a common tangent is drawn to the pitch circles of two meshing gears, the line of action or angle of thrust is drawn at the proper angles ($14^1/_2°$) to this line. The base circles on which the involutes are drawn are tangent to the line of action.

Most gears transmitting power use the $20°$, full-depth, involute tooth form. These gears have the same tooth proportion as the $14^1/_2°$ full-depth involute but are stronger at their base because of greater thickness. The $20°$, fine-pitch involute gears are similar to the regular $20°$ involute and are made in sizes ranging from 20 to 200 diametral pitch. These gears are used primarily for transmitting motion rather than power. The $20°$ stub tooth gear has a smaller tooth depth than the $20°$ full-depth gear and is consequently stronger. Involute gears fulfill all the laws of gearing and have the advantage over some

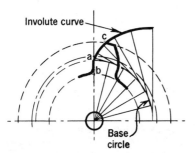

Figure 22.14
Method of generating an involute tooth surface.

Table 22.2 **American Gear Manufacturers Association Standard for Involute Gearing**

	20° Full depth	14¹/₂° Full depth	20° Fine pitch	20° Stub tooth
Addendum	$\dfrac{1}{P}$	$\dfrac{1}{P}$	$\dfrac{1}{P}$	$\dfrac{1}{P}$
Clearance	$\dfrac{0.250}{P}$	$\dfrac{0.157}{P}$	$\dfrac{0.2}{P} + 0.002$	$\dfrac{0.2}{P}$
Dedendum	$\dfrac{1.250}{P}$	$\dfrac{1.157}{P}$	$\dfrac{1.2}{P} + 0.002$	$\dfrac{1}{P}$
Outside diameter	$\dfrac{N+2}{P}$	$\dfrac{N+2}{P}$	$\dfrac{N+2}{P}$	$\dfrac{N+1.6}{P}$
Pitch diameter	$\dfrac{N}{P}$	$\dfrac{N}{P}$	$\dfrac{N}{P}$	$\dfrac{N}{P}$

other curves in that the contact action is unaffected by slight variation of gear center distances.

The nomenclature of a gear tooth is illustrated in Figure 22.13. The principal definitions and tooth parts for standard 14¹/₂° and 20° involute gears are discussed here.

The *addendum* of a tooth is the radial distance from the pitch circle to the outside diameter or addendum circle. Numerically, it is equal to 1 divided by the diametral pitch P.

The *dedendum* is the radial distance from the pitch circle to the root or dedendum circle. It is equal to the addendum plus the tooth clearance.

Tooth thickness is the thickness of the tooth measured on the pitch circle. For cut gears the tooth thickness and tooth space are equal. Cast gears are provided with some *backlash*, the difference between the tooth thickness and tooth space measured on the pitch circle.

The *face* of a gear tooth is that surface lying between the pitch circle and the addendum circle.

The *flank* of a gear tooth is that surface lying between the pitch circle and the root circle.

Clearance is a small distance provided so that the top of a meshing tooth will not touch the bottom land of the other gear as it passes the line of centers.

Table 22.2 gives the proportions of standard 14¹/₂° and 20° involute gears expressed in terms of diametral pitch P and number of teeth N.

Pitch of Gears[1]

The *circular pitch p* is the distance from a point on one tooth to the corresponding point on an adjacent tooth, and is measured on the pitch circle. Expressed as an equation,

---

[1]Metric gearing is based on the module (mod) instead of the diametral pitch P, as in the English system. The basic metric module formula is mod = D/N = amount of pitch diameter per tooth = millimeters per tooth measured on the pitch diameter. Also, mod = $1/P$ is expressed in millimeters. Also, mod P = 25.4.

$$p = \frac{\pi D}{N}$$

where

D = Diameter of pitch circle
N = Number of teeth

The *diametral pitch P,* often referred to as the pitch of a gear, is the ratio of the number of teeth to the pitch diameter. It may be expressed by the following equation:

$$P = \frac{N}{D}$$

Upon multiplying these two equations the following relationship between circular and diametral pitch results.

$$p \times P = \frac{\pi D}{N} \times \frac{N}{D} = \pi$$

Hence, knowing the value of either pitch we may obtain the other by dividing into π.

Gears and gear cutters are standardized according to diametral pitch. This pitch can be expressed in even figures or fractions. Circular pitch, being an actual distance, is expressed in inches and fractions of an inch. A 6-pitch gear (6 diametral pitch) is one that has 6 teeth per inch of pitch diameter. If the pitch diameter is 3 in., the number of teeth is 3×6 or 18. The outside diameter of the gear is equal to the pitch diameter plus twice the addendum distance or 3 in. $+ 2 \times \frac{1}{6}$, which is 3.333 in.

Any involute gear of a given diametral pitch will mesh properly with a gear of any other size of the same diametral pitch. However, in cutting gears of various diameters a slight difference in the cutter is necessary to allow for the change in curvature of the involute as the diameter increases. The extreme case would be a rack tooth, which would have a straight line as the theoretical tooth profile. For practical reasons the number of teeth in an involute gear should not be less than 12.

Gear Speeds

The speeds in rpms, s and S, of two meshing gears vary inversely with both the pitch diameters and the number of teeth. This may be expressed as follows:

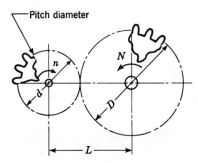

Figure 22.15
Nomenclature for meshing gear and pinion.

$$\frac{s}{S} = \frac{D}{d} = \frac{T}{t}$$

where D and d represent pitch diameters as indicated in Figure 22.15. T and t represent number of teeth on the gear and pinion.

$$\text{Center distance, } L = \frac{D + d}{2}$$

The speed ratio for a worm gear set depends on the number of teeth on the gear and the lead of the worm. For a single-threaded worm the ratio is

$$\frac{\text{rpm worm}}{\text{rpm gear}} = \frac{T}{1}$$

Kinds of Gears

The gears most commonly used are those that transmit power between two parallel shafts. Such gears having their tooth elements parallel to the rotating shafts are known as *spur gears,* the smaller of the two being known as a *pinion* (Figure 22.15). If the elements of the teeth are twisted or helical, as shown in Figure 22.16B, they are known as *helical gears.* These gears may be for connecting shafts that are at an angle in the same or different planes. Helical gears are smooth acting because there is always more than one tooth in contact. Some power is lost because of end thrust, and provision must be made to compensate for this thrust in the bearings. The *herringbone gear* is equivalent to two helical gears, one having a right-hand and the other a left-hand helix.

Usually, when two shafts are in the same plane but at an angle with one another, a

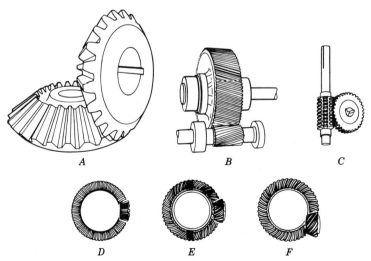

Figure 22.16
Special types of gears.

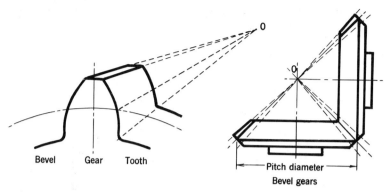

Figure 22.17
All elements of straight bevel gears converge at the cone apex of the gears.

bevel gear is used. Such a gear is similar in appearance to the frustum of a cone having all the elements of the teeth intersecting at a point, as shown in Figure 22.17. Bevel gears are made with either straight or spiral teeth. When the shafts are at right angles and the two bevel gears are the same size, they are known as *miter gears* (Figure 22.16A). ***Hypoid gears,*** an interesting modification of bevel gears shown in Figure 22.16F, have their shafts at right angles by they do not intersect as do the shafts for bevel gears. Correct teeth for these gears are difficult to construct, although a generating process has been developed that produces satisfactory teeth. *Zerol gears* (Figure 22.16D) have curved teeth but have a zero helical angle. They are produced on machines that cut spiral bevels and hypoids. *Worm gearing* is used where a large speed reduction is desired. The small driving gear is called a *worm* and the driven gear a *wheel*. The worm resembles a large screw and is set in close to the wheel circumference, the teeth of the wheel being curved to conform to the diameter of the worm. The shafts for such gears are at right angles but not in the same plane. These gears are similar to helical gears in their application, but differ considerably in appearance and method of manufacture. A worm gear set is shown in Figure 22.16C.

Rack gears, which are straight and have no curvature, represent a gear of infinite radius and are used in feeding mechanisms and for reciprocating drives. They may have either straight or helical teeth. If the rack is bent in the form of a circle, it becomes a bevel gear having a cone apex angle of 180° known as *crown gear*. The teeth all converge at the center of the disk and mesh properly with a bevel gear of the same pitch. A gear with internal teeth, known as an ***annular gear,*** can be cut to mesh with either a spur or a bevel gear, depending on whether the shafts are parallel or intersecting.

METHODS OF MAKING GEARS

Most gears are produced by some machining process. Accurate machine work is essential for high-speed, long-wearing, quiet-operating gears. Die and investment casting of gears has proved satisfactory, but the materials are limited to low-temperature-melting metals and alloys. Consequently, these gears do not have the wearing qualities of heat-treated steel gears. Stamping though reasonably accurate, can be used only in making thin gears from sheet metal.

Commercial methods employed in producing gears are summarized as follows:

A. Casting
   1. Sand casting
   2. Die casting
   3. Precision and investment casting
B. Stamping
C. Machining
   1. Formed-tooth process
      a. Form cutter in milling machine
      b. Form cutter in broaching machine
      c. Form cutter in shaper
   2. Template process
   3. Cutter generating process
      a. Cutter gear in shaper
      b. Hobbing
      c. Rotary cutter
      d. Reciprocating cutters simulating a rack
D. Powder metallurgy
E. Extruding
F. Rolling
G. Grinding
H. Plastic molding

Formed Tooth Process

A formed milling cutter, as shown in Figure 22.18, is commonly used for cutting a spur gear. Such a cutter used on a milling machine is formed according to the shape of the tooth space to be removed. Theoretically, there should be a different-shaped cutter for each size gear of a given pitch as there is a slight change in the curvature of the involute. However, one cutter can be used for several gears having different numbers of teeth without much sacrifice in their operation. Each pitch cutter is made in eight slightly varying shapes to compensate for this change. They vary from no. 1, which is used to

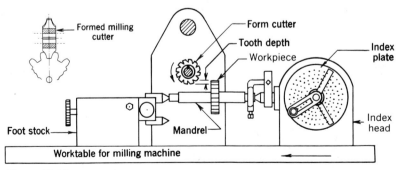

Figure 22.18
Setup for cutting a spur gear on a milling machine.

Table 22.3 **Standard Involute Cutters**

No. 1.	135 teeth to a rack
No. 2.	55 to 134 teeth
No. 3.	35 to 54 teeth
No. 4.	26 to 34 teeth
No. 5.	21 to 25 teeth
No. 6.	17 to 20 teeth
No. 7.	14 to 16 teeth
No. 8.	12 and 13 teeth

cut gears from 135 teeth to a rack, to no. 8, which cuts gears having 12 or 13 teeth. The eight standard involute cutters are listed in Table 22.3.

Setup of a milling machine to cut spur gears is illustrated in Figure 22.18. A discussion of this process is given in the chapter on milling. Formed milling is an accurate process for cutting spur, helical, and worm gears. Although sometimes used for bevel gears, the process is not accurate because of the gradual change in tooth thickness. When used for bevel gears at least two cuts are necessary for each tooth space. The usual practice is to take one center cut of proper depth and about equal to the space at the small end of the tooth. Two shaving cuts are then taken on each side of the tooth space to give the tooth its proper shape.

The formed-tooth principle may also be utilized in a broaching machine by making the broaching tool conform to the tooth space. Small internal gears can be completely cut in one pass by having a round broaching tool made with the same number of cutters as the gear has teeth. Broaching is limited to large-scale production because of the cost of cutters.

Template Gear Cutting Process

In the template process the form of the tooth is controlled by a template instead of by a formed tool. The tool itself is similar to a side-cutting shaper tool and is given a reciprocating motion in the process of cutting. The process is especially adapted to cutting large teeth, which would be difficult with a formed cutter, and also to cutting bevel gear teeth. Bevel gear teeth present a problem in cutting because the tooth thickness varies along its length, as may be seen in Figure 22.17. All elements of each tooth converge to point 0, the cone apex of the gear. In a bevel gear planer, diagrammatically shown in Figure 22.19, the frame carrying the reciprocating tool is guided at one end by a roller acting against a template, while the other end is pivoted at a fixed point corresponding to the cone apex of the gear being cut. Three sets of templates are necessary, one for the roughing cut and one for finishing each side of the tooth space. The gear blank is held stationary during the process and is moved only when indexed. This method of cutting produces an accurately formed tooth having the proper taper. Machines of this type are used only for planing teeth of very straight bevel gears. Most bevel gears are cut by various generating processes.

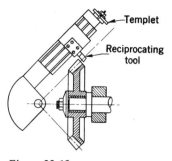

Figure 22.19
Schematic view of a bevel gear planer.

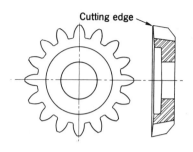

Figure 22.20
Gear shaper cutter.

Cutter Gear Generating Process

The cutter gear generating process for cutting involute gears is based on the fact that any two involute gears of the same pitch will mesh together. Hence, if one gear is made to act as a cutter and is given a reciprocating motion as in a shaper, it will cut into a gear blank and generate conjugate tooth forms. A gear shaper cutter of this description is shown in Figure 22.20, and Figure 22.21 shows how it is mounted in a Fellows gear shaper. In operation both the cutter and the blank rotate at the same pitch line velocity and, in addition, a reciprocating motion is given to the cutter. The rotary feed mechanism is so arranged that the cutter can be automatically fed to the desired depth while both cutter and work are rotating. The cutter feed takes place at the end of the stroke, at which time the work is withdrawn from the cutter by cam action. Figure 22.22 illustrates this process in cutting an internal gear. The operation of the machine is similar to the previous figure. In both, the generating action of the cutter and the blank is shown in Figure 22.23. The fine lines indicate metal removed by each cut in a given tooth space. Cutting action may occur on either the downstroke or the upstroke, whichever proves to be the best procedure. A shaper is shown in Figure 22.24.

The cutter gear method of generating gears is not limited to involute spur gears. By giving a spiral cutter a twisting motion on the cutting stroke, helical gears may be

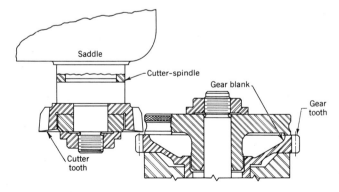

Figure 22.21
Sketch showing mounting of cutter and blank on a gear shaper.

METHODS OF MAKING GEARS 591

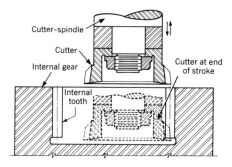

Figure 22.22
Sketch showing cutter setup for cutting an internal gear.

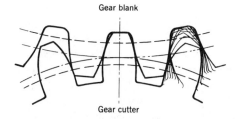

Figure 22.23
Generating action of Fellows gear shaper cutter.

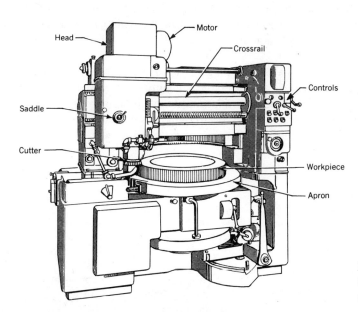

Figure 22.24
Gear shaper setup for cutting spur gears.

generated. Worm threads may be cut in a similar fashion. In addition, this process may be used to cut racks, sprocket wheels, gear-type clutches, cams, ratchet wheels, and other straight and curved forms.

An application of the cutter gear generator principle is the Sykes gear generating machine, known for its ability to cut continuous herringbone teeth. This machine employs two cutter gears mounted in a horizontal position as shown in Figure 22.25. In cutting herringbone gears the cutters are given a reciprocating motion, one cutting in one direction to the center of the gear blank and the other cutting to the same point when the motion is reversed. The cutters not only reciprocate but also are given a twisting motion according to the helix angle. Both the gear blank and cutters slowly revolve, generating the teeth in the same fashion as the Fellows shaper. Machines of this type are built in various sizes up to those capable of cutting gears 22 ft (7 m) in diameter.

Bevel Gear Generators

Straight bevel gears can be produced by two different types of generators: the *two-tool generator* with two reciprocating tools, and the *completing generator* with two multiblade rotating cutters. The principle involved is based on the fact that any bevel gear will mesh with a crown gear of the same pitch whose center coincides with the pitch cone apex of the gear. In Figure 22.26 the two cutting tools represent the sides of adjacent teeth of a crown gear. These tools are mounted on a cradle that rotates about the axis of the crown gear. At the same time the tools are given a reciprocating motion. The gear blank is also rotated about its axis at the rate it would have were it meshing with the crown gear. As the tools are simulating the respective positions taken by the crown gear, the correct form of tooth is cut. Both sides of a single tooth are cut on a single generating roll of the cradle; at the end of the generating roll the blank is withdrawn and indexed while the cradle returns to the starting position for the next cut. This cycle is repeated until all the teeth in the gear are cut. In general the tooth spaces are roughed out in a separate operation so that only a small amount of metal is removed by the reciprocating tools when finishing.

An advantage of this process is that a prior roughing cut is unnecessary, thus saving one handling of the blank. Cutter life is longer, gear quality is improved, and setup requires less time. Both processes will produce a localized tooth bearing in straight bevel gears. A slight crowning on the tooth surface localizes the tooth bearing in the center three-quarters and eliminates load concentrations on the ends of the teeth.

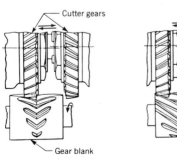
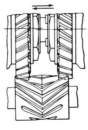

Figure 22.25
The cutters of Farrel–Sykes machines reciprocate, one cutting when the movement is in one direction and the other when the movement is reversed. Each ends its stroke at the center of the blank. These machines produce a herringbone gear.

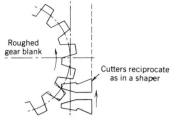

Figure 22.26
Cutting straight bevel gears with two reciprocating tools.

The method of cutting spiral bevel gears also uses the generating principle, but the cutter in this case is circular and rotates as a face milling cutter. The cutter is similar to Figure 22.27, which is shown cutting a hypoid pinion. The spiral teeth on gears cut by this process are curved on the arc of a circle, the radius being equal to the radius of the cutter. The blades of the cutter have straight cutting profiles to correspond with the tooth profile of a crown gear. The revolving cutters move through the same space as would be occupied by a crown gear tooth. As in the previous method the teeth are first rough cut before the true shape is generated. The rotating cutters may be designed to cut only one or both sides of the tooth space, the latter type of cutter having the advantage of more rapid production. Spiral bevel gears have an advantage over straight bevel gears in that the teeth engage with one another gradually, eliminating any shock, noise, or vibration in their operation. The hypoid gear can be cut in the machine just described.

Generating Gears with a Hob Cutter

Any involute gear of a given pitch will mesh with a rack of the same pitch. One form of cutting gears utilizes a rack as a cutter. If it is given a reciprocating motion similar to cutting on a Fellows shaper, involute teeth will be generated on the gear as it rotates

Figure 22.27
Close-up of a hypoid gear generating machine cutting a pinion.

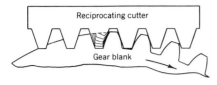

Figure 22.28
Rack-type cutter generating teeth for a spur gear.

intermittently in mesh with the rack cutter. This method is shown diagrammatically in Figure 22.28. Such machines require a long rack cutter to cut all the teeth on the circumference of a large gear, and for this reason they are little used.

The hobbing system of generating gears is somewhat similar to the principle just described. A rack is developed into a cylinder, the teeth forming threads and having a lead as in a large screw. Flutes are cut across the threads, forming rack-shaped cutting teeth. These cutting teeth are given relief and, if the job is viewed from one end, it looks the same as the ordinary form of gear cutter. This cutting tool, known as a *hob,* may be briefly described as a fluted steel worm. In Figure 22.29 is shown a hob in section and end view as it appears when cutting a gear.

Hobbing, then, may be defined as a generating process consisting of rotating and advancing a fluted steel worm cutter past a revolving blank. This cutting action is illustrated in Figure 22.30, where the teeth on a worm gear are being cut to full depth by a rotating hob. In this process all motions are rotary, there being no reciprocating or indexing movements. In the actual process of cutting, the gear and hob rotate together as in mesh. The speed ratio of the two depends on the number of teeth on the gear and on whether the hob is single threaded or multithreaded.

An elementary gear train for a hobbing machine is diagrammatically illustrated in Figure 22.31. The rotating hob in this figure is shown cutting a spur gear. Because the figure is oversimplified and the machine as shown can only cut gears having a specified number of teeth, change gears must be introduced in the drive to the worm shaft so that gears of any number of teeth can be cut. The mechanisms for feeding the hob or adjusting it at various angles are not shown. The hob cutting speed is controlled by change gears that vary the speed of the main drive shaft.

At the start of operation the gear blank is moved in toward the rotating hob until the proper depth is reached, the pitch line velocity of the gear being the same as the lead velocity of the hob. The action is the same as if the gear were meshing with a rack. As

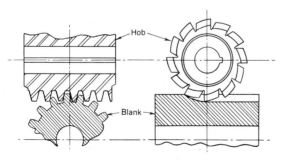

Figure 22.29
Cutting gear with a hob.

Figure 22.30
Gear hobbing machine.

soon as the depth is reached, the hob cutter is fed across the face of the gear until the teeth are complete, both gear and cutter rotating during the entire process.

Inasmuch as the hob teeth have a certain lead, the axis of the hob cannot be at right angles to the axis of the gear when cutting spur gears but must be moved an amount equal to the lead angle. For helical gears the hob must be moved around an additional angle equal to the helix angle of the gears. Worm gears may be cut with the axis of the hob at right angles to the gear and the hob fed tangentially as the gear rotates.

Gear hobs based on the rack principle will cut gears of any diameter and eliminate the need for a variety of hobs for gears having the same pitch but varying in diameter.

FINISHING OPERATIONS USED ON GEARS

The object of any finishing operation on a gear is to eliminate slight inaccuracies in the tooth profile, spacing, and concentricity so that the gears will have conjugate tooth forms and give quiet operation at high speeds. These inaccuracies are very small dimensionally, frequently not exceeding 0.0005 in. (0.013 mm), but even this amount is sufficient to increase wear and set up undesirable noises at high speeds.

To remedy these errors in gears that are not heat treated, such operations as *shaving* or *burnishing* are used. Burnishing is a cold working operation accomplished by rol-

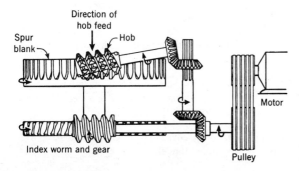

Figure 22.31
Hobbing machine gear train.

ling the gear in contact and under pressure with three hardened burnishing gears. Although the gears may be made accurate in tooth form, the disadvantage of this process is that the surface of the tooth is covered with amorphous or "smear" metal rather than metal having true crystalline structure, which is desirable from a long-life standpoint. More accurate results may be obtained by a shaving process, which removes only a few thousandths of an inch of metal. This process is strictly a cutting and not a cold-working process.

Rolling the gear in contact with a rack cutter and using a rotary cutter are two methods of shaving. Either will produce accurately formed teeth. Both external and internal spur and helical gears can be finished by this process.

Heat-treated gears can be finished either by *grinding* or by *lapping*. Grinding may be done by either the forming or the generating process. The disadvantage of gear grinding is that considerable time is consumed in the process. Also, the surfaces of the teeth have small scratches or ridges that increase both wear and noise. To eliminate the latter defect, ground gears are frequently lapped.

Gear lapping is accomplished by having the gear in contact with one or more cast-iron lap gears or true shape. The work is mounted between centers and is slowly driven by the rear lap. It in turn drives the front lap, and at the same time both laps are rapidly reciprocated across the gear face. Each lap has individual adjustment and pressure control. A fine abrasive is used with kerosene or a light oil to assist in the cutting action. The time consumed for average-sized gears is $1/2$ to 2 min per side of gear teeth. The results of lapping are demonstrated by longer wearing and quieter operating gears.

QUESTIONS AND PROBLEMS

1. Give an explanation of the following terms:

    ISO
    Lead
    Chaser
    Rollability
    Plunge cut
    Involute
    Base circle
    Hypoid gear
    Annular gear
    Shaving

2. On a triple-threaded screw as often found on a fountain pen, what is the lead?

3. What is the principal advantage of multi-threaded screws? What are the disadvantages?

4. Discuss the difficulty of comparing threads with the Unified and ISO standards.

5. Why is a pipe thread tapered?

6. Give the differences between a square thread, an Acme thread, and the buttress thread.

7. How could a collapsible tap be made?

8. What threading tool is used for internal threads in a blind hole?

9. How could you thread a piece of steel with a $3/4$-in., 10-thread die in order to have a $1/2$-in. lead?

10. What is the purpose of a bottoming tap? On what type of work would it be used?

11. How is the tap drill size for an internal thread determined?

12. Discuss the principal metallurgical and physical characteristics of a metal to be employed in thread rolling.

13. What advantage do cut threads have over rolled threads?

14. Describe the threads denoted by: $1/4$-20 UNC-3 BLF and M2.5 × 0.45-6G8F.

15. How should the following threads be made? (a) $1/2$ by $2\,1/2$ in. machine bolts; (b) lead screw on lathe; (c) square threads on jack; (d) internal threads in nuts; (e) Acme thread on end of rod.

16. Determine the pitch and depth of the following threads: (a) $1/2$ in.-13 UNC; (b) $3/4$ in.-10 UNC.

17. Determine the feed per tooth of a 24-tooth thread milling cutter turning at 50 rpm and the 2-in. workpiece at 10 rpm.

18. What percentage saving in material cost would be obtained in producing $5/8$ in.-11 UNC threads on 2-in. stud bolts if thread rolling were used in place of cutting dies?

19. Approximately how much stock would be wasted into chips if 10,000, $3/4$-16 UNC screws threaded 4 in. long were to be threaed by die as compared to rolling?

20. Determine the root diameter, pitch diameter, and lead angle for a $5/8$-11 UNC thread.

21. What is the minimum and most probably used nut thickness for the following diameter bolts: $1/4$ in., $1/2$ in., $3/4$ in., 1 in., 2.5 mm, 6 mm, 12 mm?

22. What is the function of the fillet between the base and root circle of a gear?

23. How are most bevel gears cut?

24. What is the difference in shaving and burnishing?

25. What is the principal difference between a $14 1/2°$ gear and a 20° gear?

26. A gear has a pitch diameter of 5 in.; what is the difference in the diameter of the base circles for a $14 1/2°$ and a 20° involute gear?

27. For a gear with a diametral pitch of 10, what is the difference in the addendum and total tooth depth for each of the types of gears listed in Table 22.2?

28. Sketch and show all diameters and dimensions of a $14 1/2°$ involute gear that has a pitch of 10 and 32 teeth. It is $1/2$ in. (12.7 mm) wide.

29. Sketch and show all diameters and dimensions of an 8-pitch, 20° stub gear that has 34 teeth.

30. Calculate the pitch of a gear that has a pitch diameter of 3.5 in. and 35 teeth.

31. A $14 1/2°$ involute gear has a thickness of $1 1/2$ in., an outside diameter of 11 in. (280 mm), and 60 teeth. Calculate all diameters and dimensions.

32. The diameter of the pitch circle of a $14 1/2°$ involute gear is 4.0 in. What is the diameter of the pitch circle in SI units? What is the diameter of the base circle? Cos $14 1/2°$ = 0.97.

33. The circular pitch of a gear is 0.314 in.; the pitch diameter is 5 in. What standard gear cutter should be used?

34. (a) In the gear train shown what is the pitch diameter of each gear if the circular pitch is 0.5283? (b) If shaft A turns at 1800 rpm what are the revolutions per minute of B?

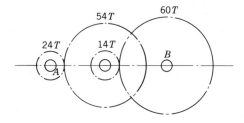

35. A spur pinion having 40 teeth is in mesh with a gear having 32 teeth. If a 10-pitch cutter is used in cutting the teeth, what should be the correct center distance between the two gears?

36. A 10-pitch, $14 1/2°$ involute spur gear has 64 teeth. Determine the base circle diameter and outside diameter of this gear.

37. Referring to the gear train in problem 34, what is the centerline distance between shafts A and B?

38. A 20° full-depth involute gear having 96 teeth and an outside diameter of 8.25 in. is to be cut with a single-tooth form cutter. Determine the pitch diameter, diametral pitch, tooth clearance, and addendum.

39. A standard involute $14\frac{1}{2}°$ gear has a pitch diameter of 5 in. and is cut with a 10-pitch cutter. How many teeth will the gear have and what is the total tooth depth?

CASE STUDY
LOTUS GEAR WORKS

Lotus Gear Works does specialty cutting of spur gears as a job shop. Customers come to Lotus with "down" problems on equipment and expect quick turnaround. Most customers know very little about gears and do not give specifications. All gears are cut on milling machines equipped with dividing heads.

A customer seeks Lotus' advice in that she has a gear in the gear box of a large construction crane that has worn teeth. The company manufacturing the crane has gone out of business. The customer says, "I can only see the gear teeth as they go by; I don't know the gear's size, but I need another one just like it. The gear it meshes with has an OD of 15 in., 133 teeth, a width of 3 in., and has a $14\frac{1}{2}$ stamped on it. The centerlines of the two shafts holding the two gears are 22.07 in. apart. That's all I know. Can you make me a gear?"

Because Lotus has had jobs like this before, you are asked to design the gear. What are the dimensions you decide on? Depth of cut? Number of teeth? What cutter should be used? What can be said about the speed of the two meshing gears? Turns of the index handle between cuts? Your design should show a sketch of the gear and all calculations.

CHAPTER 23
GRINDING AND ABRASIVE MACHINES

To grind means to abrade, to wear away by friction, or to sharpen. In manufacturing it refers to the removal of metal by a rotating abrasive wheel. Wheel action is similar to a milling cutter. The cutting wheel is composed of many small grains bonded together, each one acting as a miniature cutting point.

Because heat, small chips, and loose grit are released during grinding, coolants are almost always directed to the interface between the grinding medium and the workpiece. Coolants clean the abrasive wheel or belt in order that it will be free cutting. Coolants are usually water based, with soluble oil additives that prevent rust and add to the cooling effectiveness. Synthetic coolants are also available.

GRINDING

The grinding process has the following advantages:

1. It is the only way, exclusive of lasers and special welding techniques, to machine materials that have a hardness of greater than Rockwell C50.
2. Varying amounts of material may be removed in the process. Parts having high hardness are usually machined or formed in the annealed state, hardened, and then ground.
3. Because of the many small cutting edges inherent in the wheel, grinding produces fine finishes. Surface roughnesses of 16 to 90 μin. (0.4–2200 μm) are commonplace.
4. Grinding can finish work to accurate dimensions quickly. Because in general only a small amount of stock is removed, the grinding machines require close wheel control. It is possible to grind work to ± 0.0002 in. (± 0.005 mm) tolerance.
5. Grinding pressure is minimal. This characteristic permits magnetic chucks for holding the work in many grinding operations.

ABRASIVE MACHINING

Abrasive machining is a primary stock removal process. It is not a finishing operation like conventional grinding or lapping. Not limited to bonded-wheel operations, it includes coated-belt and free abrasive processes. Coated abrasive belts consist of the abrasive

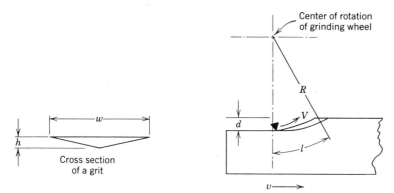

Figure 23.1
Grit and wheel geometry for a grinding wheel.

grain, backing, and bond. Consider the grain as the tool, the backing as the toolholder, and the bond as the agent clamping the tool to the holder. Abrasive machines, powered by 300-hp (200-kW) motors, remove metal from forgings, castings, and various stock shapes. The process provides fast metal removal, good surface finish, close size control, little need for holding fixtures, and adaptation to automation. Although up to $1/2$ in. (12.7 mm) of metal can be removed in a matter of seconds, there is the problem with heat dissipation generated in abrasive machining.

Each grit removes a chip. The depth of cut is zero, as the abrasive grit initially touches the workpiece. As the workpiece moves and the wheel revolves, the depth of cut increases to a maximum as shown in Figure 23.1. A more complex situation occurs in grinding cylindrical pieces. Some grits remove a maximum-size chip and others practically none. Figure 23.1 indicates the prevailing assumptions for a simple explanation of the mathematics associated with grinding. The grit height h is much smaller than its width w, after the wheel is dressed, although initially the grit would have a random slope. Thus,

$$r = \frac{w}{h}$$

where

r = Ratio of width to height of a grit
w = Width of a grit, in. (mm)
h = Height of grit, in. (mm)

A sharp grit will cut a chip, but a dull grit will be pulled from the wheel and washed away. From Figure 23.1 it can be shown

$$l = \sqrt{2Rd} = \sqrt{Dd}$$

where

l = Length of chip, in. (mm)
R = Radius of grinding wheel, in. (mm)

$$= \tfrac{1}{2}D \text{ (diameter), in. (mm)}$$

$$d = \text{Depth of cut, in. (mm)}$$

If the grinding wheel moves over a given length, the total volume of metal removed is expressed as

$$Q = 12v \times W \times d$$

where

Q = Quantity of material removed, in.³/min (mm³/min)

v = Velocity of workpiece past the wheel or velocity of the centerline of the wheel relative to the workpiece, ft/min (m/min)

W = Width of wheel, in. (mm)

The value of r is usually taken to be from 10 to 20,[1] and the average volume per chip is

$$q = \tfrac{1}{4}w \times h \times l$$

where q is the volume of an average individual chip, in.³ (mm³).

The number of chips per unit of time (n) is

$$n = VWc$$

where

V = Peripheral wheel speed, in./s (mm/s)

c = Number of active grits per in.² (mm²)

Combining the above equations, the mean **chip thickness** is found to be approximately

$$t = \sqrt{\frac{4v}{rcV}\left(\frac{d}{D}\right)^{1/2}}$$

where t is the chip thickness, in. (mm).

These equations[2] lead to several conclusions about grinding. Because the horsepower or energy required is proportional to the volume of chips removed, the effect of almost any variable can be estimated.

Although these equations are based on rough assumptions, it is important to notice that smoother finishes are obtained as t decreases. The effect of the variables on surface finish is readily discernible.

GRINDING AND ABRASIVE MACHINES

Grinding machines are designed principally to finish parts having cylindrical, flat, or internal surfaces. The surface largely determines the grinding machine; thus a machine

---

[1] Nathan H. Cook, *Manufacturing Analyses,* Readings Mass., Addison-Wesley, 1966.
[2] SI equations are not consistent dimensionally.

grinding cylindrical surfaces is called a cylindrical grinder. Machines designed for special functions such as tool grinding or cutting off are designated according to their operation.

A classification of grinding machines according to the surface generated or work done is as follows:

A. Cylindrical grinder
   1. Work between centers
   2. Centerless
   3. Tool post
   4. Crankshaft and other special applications
B. Internal grinder
   1. Work rotated in chuck
   2. Work rotated and held by rolls
   3. Work stationary
C. Surface grinder
   1. Planer type (reciprocating table)
      a. Horizontal spindle
      b. Vertical spindle
   2. Rotating table
      a. Horizontal spindle
      b. Vertical spindle
   3. Disk
   4. Loose grit
   5. Flap wheel
   6. Wire sawing
D. Universal
   1. Cylindrical work
   2. Thread form work
   3. Gear form work
   4. Oscillating
E. Tool grinder
F. Special grinding machines
   1. Swinging frame, snagging
   2. Cutting off, sawing
   3. Portable, offhand grinding
   4. Flexible shaft, general purpose
   5. Profiling, contouring
G. Surface preparing
H. Abrasive grinding belt, single multihead
I. Mass media
   1. Barrel tumbling
   2. Vibratory

Cylindrical Grinder

This machine is used primarily for grinding cylindrical surfaces, although tapered and simple formed surfaces may also be ground. They may be further classified according to the method of supporting the work. Diagrams illustrating the essential difference in supporting the work between centers and centerless grinding are shown in Figure 23.2. In the centerless type the work is supported by the **work rest,** the regulating wheel, and the grinding wheel itself. Both types use plain grinding wheels with the grinding face as the outside diameter.

A hydraulic, center-type cylindrical grinding machine is illustrated in Figure 23.3. Three movements are incorporated.

1. Rapid rotation of the grinding wheel at the proper grinding speed, usually 5500 to 6500 ft/min (28–33 m/s).

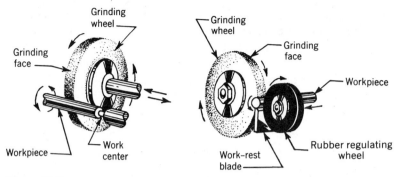

Figure 23.2
Methods of supporting work between centers and centerless type of cylindrical grinding.

2. Slow rotation of the work against the grinding wheel at a velocity to give best performance. This varies from 60 to 100 ft/min (0.30–0.51 m/s) in grinding steel cylinders.
3. Horizontal traverse of the work back and forth along the grinding wheel to grind the entire surface of a long piece or plunge grinding with a wheel wide enough to cover the desired surface.

The work should be traversed nearly the entire width of the wheel during each revolution. In finishing the traverse may be reduced to one-half the width of the wheel and the depth of cut reduced.

The depth of cut is controlled by feeding the wheel into the work. Roughing cuts around 0.002 in. (0.05 mm) per pass may be made, but for finishing this should be reduced to about 0.0002 in. (0.005 mm) per pass or less. In selecting the amount of

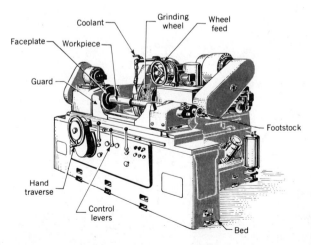

Figure 23.3
A 10 × 36 in. (250 × 915 mm) plain cylindrical grinder.

604 GRINDING AND ABRASIVE MACHINES

infeed, consideration is given to the size and rigidity of the work, surface finish, and the decision of whether or not to use a coolant.

Where the face of the wheel is wider than the part to be ground, it is not necessary to traverse the work. This is known as "plunge cut" grinding. The grinding speed of the wheel is stated in terms of surface feet per minute; that is,

$$V_c = \pi D_c \times N$$

where

V_c = Cutting or grinding speed, ft/min (m/min)

D_c = Diameter of grinding wheel, ft (m)

N = Revolutions of the wheel per minute, rpm

The *tool post grinder* is used for miscellaneous and small grinding work on a lathe. The grinder is held on the tool post and fed across the work, the regular longitudinal or compound rest feed being used. A common application of this grinder is the truing up of lathe centers.

Centerless grinders are designed so that they support and feed the work by using two wheels and a work rest as illustrated in Figures 23.2 and 23.4. The large wheel is the grinding wheel and the smaller one the pressure or regulating wheel. The *regulating wheel* is a rubber-bonded abrasive having the frictional characteristics to rotate the work at its own rotational speed. The speed of this wheel, which may be controlled, varies from 50 to 200 ft/min (0.25–1.02 m/s). Both wheels are rotated in the same direction. The *rest* assists in supporting the work while it is being ground, being extended on both sides to direct the work travel to and from the wheels.

The axial movement of the work past the grinding wheel is obtained by tilting the wheel at a slight angle from horizontal. An angular adjustment of 0° to 10° is provided in the machine for this purpose. The actual feed can be calculated by this formula.

$$F = \pi dN \sin \alpha$$

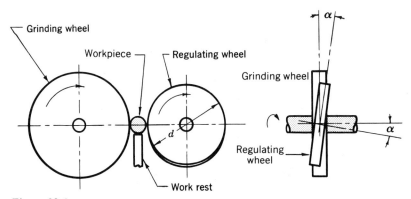

Figure 23.4
Principle of centerless grinding.

where

F = Feed, in./min (mm/min)
N = rpms
d = Diameter of regulating wheel, in. (mm)
α = Angle inclination of regulating wheel

Centerless grinding may be applied to cylindrical parts of one diameter. In production work on such parts as piston pins, a magazine feed is arranged and the parts may go through several machines in series before the part is finished. Each grinder removes from 0.0005 to 0.002 in. (0.013–0.05 mm) stock.

Where parts are not uniformly of the same diameter or where they require form grinding as does a ball bearing (Figure 23.5), an *infeed* centerless grinder is used. The method of operation corresponds to the plunge cut form of grinding. The length of the section to be ground is limited to the width of the grinding wheel. The part is placed on the work rest and is moved against the grinding wheel by the regulating wheel. Upon completion the gap between the wheels is increased either manually or automatically, and the work is ejected from between the wheels.

A third type of centerless grinding called *end feed* has been devised for short-taper work. Both wheels are dressed to their correct taper and the work is automatically fed in from one side to a fixed stop. The advantages of centerless grinding are as follows:

1. Chucking or mounting the work on mandrels or other holding devices is avoided.
2. The work is rigidly supported and there is no chatter or deflection of the work.
3. The process is rapid and especially adapted for production work.
4. Accuracy is easily controlled.
5. As a true floating condition exists during the grinding process, less grinding stock is required.
6. Machine operators rather than highly skilled machinists are used.

Disadvantages include the following:

1. Work with flats and keyways cannot be ground.
2. In hollow work there is no assurance that the outside diameter will be concentric with the inside diameter.
3. Work having several diameters is not easily handled.

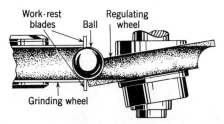

Figure 23.5
Centerless grinding of ball bearings.

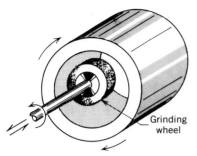

Figure 23.6
Sizing to dimension by internal grinding.

Internal Grinders

The work done on an internal grinder is diagrammatically shown in Figure 23.6. Tapered holes or those having more than one diameter may be accurately finished in this manner. There are several types of internal grinders.

1. The wheel is rotated in a fixed position while the work is slowly rotated and traversed back and forth.
2. The wheel is rotated and at the same time reciprocated back and forth through the length of the hole. The work is rotated slowly but otherwise has no movement.
3. The work remains stationary and the rotating wheel spindle is given an eccentric motion according to the diameter of hole to be ground. This type of grinder is frequently called the planetary type and issued for work that is difficult to rotate. In actual construction the wheel spindle is adjusted eccentrically in a larger one that rotates about a fixed axis. The wheel spindle is driven at high speed and at the same time rotates about the axis of the large spindle.
4. Work is rotated on the outside diameter by driving rolls, thus grinding the bore concentric with the outside diameter. This arrangement lends itself to production work because loading is simplified and magazine feed may be used.

A sketch of a centerless internal grinder is shown in Figure 23.7. Three rolls—regulating, supporting, and pressure—support and drive the work. Centerless grinders of this type can be arranged for automatic loading and unloading by swinging the pressure

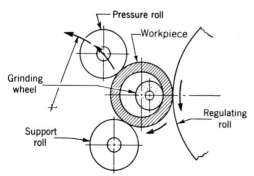

Figure 23.7
Centerless internal grinding.

roller out of the way at the end of the cycle. Advantages of internal centerless grinding include elimination of workholding fixtures and ability of the machine to grind both straight and tapered holes.

Because internal grinding wheels are small in diameter, the spindle speed is much higher than for cylindrical grinding in order to attain surface speeds up to 6000 ft/min (30.5 m/s). Most toolroom grinding is done dry, but common practice on production work is to grind steel wet and bronze, brass, and cast iron dry. Metal allowance for internal grinding depends on the size of the hole to be ground; shop practice suggests an allowance around 0.010 in. (0.25 mm).

Surface Grinding

Grinding flat or plane surfaces is known as surface grinding. Two general types of machines have been developed for this purpose: those of the planer type with a reciprocating table and those having a rotating worktable. Each machine has the possible variation of a horizontal or vertical positioned grinding wheel spindle. The four possibilities of construction are illustrated in Figure 23.8.

A line diagram of a surface grinder with the principal parts labeled is shown in Figure 23.9. This machine is provided with hydraulic control of table movements and wheel crossfeed. Straight or recessed wheels (types 1, 5, and 7 in Figure 23.16) grinding on the outside face or circumference are used. Machines of this type are adapted to reconditioning dies, grinding machine tool ways, and other long surfaces.

Reciprocating table grinders can have a vertical-spindle design, the grinding being done by a segmental or ring-shaped wheel. These machines grind gear faces, thrust washers, cylinder head surfaces, and other flat-surfaced parts.

A high-powered vertical rotary surface grinder set up for grinding a large part is shown

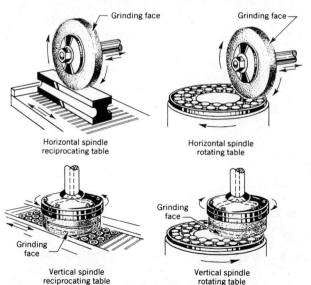

Figure 23.8
Types of surface grinding machines.

608 GRINDING AND ABRASIVE MACHINES

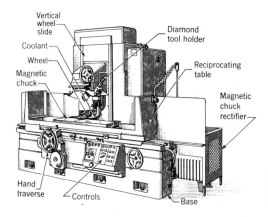

Figure 23.9
Horizontal spindle, reciprocating table surface grinder.

in Figure 23.10. Similar machines can remove 600 lb of metal per hour (270 kg/h). The spindle of the machine can be tilted while grinding to reduce the area of wheel in contact with the work, which produces deeper penetration, less heating, and better utilization of spindle horsepower. The spindle is returned to a perpendicular position, presenting a flat wheel surface to the workpiece for finish grinding. Flatness accuracy of 0.0005 in. (0.013 mm) over the full diameter is obtainable.

Tool and Cutter Grinders

In grinding tools by hand (known as *offhand grinding*), a bench or pedestal type of grinder is used. The tool is hand held and moved across the face of the wheel continually to avoid excessive grinding in one spot. This type of grinding is used to a large extent for single-point tools. Quality results depend on the skill of the operator.

For sharpening miscellaneous cutters a universal-type grinder is used. It is equipped with a universal head, vise, headstock and tailstock, and attachments for holding tools

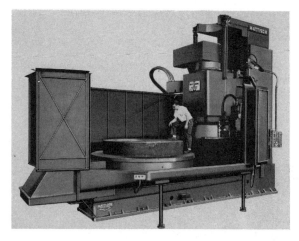

Figure 23.10
High-powered vertical rotary surface grinder for rapid metal removal.

Figure 23.11
Precision jig grinder.

and cutters. Though essentially designed for cutter sharpening, it can also be used for cylindrical, taper, internal, and surface grinding. Accuracy is of paramount importance in toolroom work, particularly when grinding form tool cutters and special shapes. Some grinders have optical magnifications of $10\times$, $20\times$, and $50\times$ for monitoring progress and accuracy.

A *jig grinder* having a selection of grinding speeds from 6700 to 175,000 rpm can be used for prototype, toolroom, or production work. Notice Figure 23.11. Accuracies of ±0.0001 in. (±0.003 mm) are obtainable. This machine is available with numerical control of table and saddle for production work. It has the appearance of a jig borer but operates at speeds too high for drilling and boring.

Surface Finishing

Honing. *Honing* is a low-velocity abrading process. Because material removal is accomplished at lower cutting speeds than in grinding, heat and pressure are minimized, resulting in excellent size, surface finish, and metallurgical control of the surface. The cutting action is obtained from abrasive sticks (aluminum oxide and silicon carbide) mounted on a metal mandrel as in Figure 23.12. Distortion is minimized because the work floats and is not clamped or chucked. For small-diameter bores a one-piece mandrel having a U-shaped cross section is used. There are two integral shoes and a narrow honing stone that provide a three-line (unevenly spaced) contact with the work circle. The abrasive stone is mounted on a wedge-actuated holder. The work is given a slow reciprocating motion as the mandrel rotates, thus generating a straight and round hole. Parts honed for finish remove only 0.001 in. (0.03 mm) or less; however, certain inaccuracies can be corrected in amounts up to 0.020 in. (0.51 mm). Coolants are essential

610   GRINDING AND ABRASIVE MACHINES

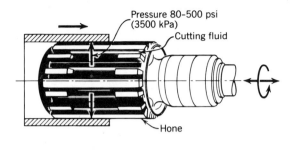

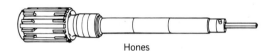
Hones

Figure 23.12
Honing using a mandrel with honing sticks on an expanding core.

to flush away small chips and to keep temperatures uniform. Sulfurized mineral base or lard oil mixed with kerosene is generally used.

Honing gives a smooth finish with a characteristic crosshatch appearance. The depth of hone marks can be controlled by variations in pressure, speed, and type of abrasive.

Lapping. Lapping produces geometrically true surfaces, corrects minor surface imperfections, improves dimensional accuracy, or provides a very close fit between two contract surfaces. Although it is a material-removing operation, it is not economical for that purpose. The amount of material removed is usually less than 0.001 in. (0.03 mm).

Lapping is used on flat, cylindrical, spherical, or specially formed surfaces in contact with a *lap*. The two have relative motion with one another in such a way that fresh contacts between a grit and the part are constantly being made. Loose abrasive carried in some vehicle such as oil, grease, or water is used between the lap and work to do the necessary abrading. Sometimes the abrasive is in the form of a bonded wheel and the lapping operation is similar to centerless and vertical-spindle surface grinding. Metal laps must be softer than the work and for machine lapping are usually made of close-grained gray iron. Other materials like steel, copper, lead, and wood are used where cast iron is not suitable. By having the lap softer than the work, the abrasive particles (usually boron carbide, silicon carbide, aluminum oxide in fine-screened sizes, or flour) become embedded in the lap and cause the greater wear to occur on the hard surface. In lapping carbide tools and jewels, diamond particles permanently embedded in copper laps are most successful.

Vertical lapping machines are used for both flat and cylindrical lapping. These machines have two laps: a lower one that supports the work and rotates at relatively slow speeds, and a stationary upper lap. The upper lap floats on the work and supplies pressure for the abrading action. Cylindrical work is loosely held and guided in a plate-type holder so that it travels on an off-radial axis, the work propelling the holder from motion received from the lower lap. A similar holder is used in lapping flat surfaces where the holder propels the work. Either it is provided with drive pins that impart a rotary and gyratory motion, or it is given a planetary motion. In either method the work contacts the entire surface of the lap in an ever-changing path. Commercial accuracy can be held to 0.000024 in. (0.00061 mm) and to even closer limits if needed. Products commonly finished by this process include gages, piston pins, valves, gears, roller bearings, thrust washers, and optical parts.

Superfinishing. All machining operations as well as the usual grinding processes leave a surface coated with fragmented, noncrystalline, or *smear metal* that, though easily removed by sliding contact, results in excessive wear, increased clearances, noisy operation, and lubrication difficulties. *Superfinishing* is a surface-improving process that removes this undesirable fragmentation metal, leaving a base of solid crystalline metal. It is somewhat similar to honing in that both processes use an abrasive stone, but it differs in the motions given to the stone. This process, which is a finishing process and not a dimensional one, can be superimposed on other finishing operations.

In cylindrical superfinishing (Figure 23.13A), a bonded-form abrasive stone, having a width about two-thirds of the diameter of the part to be finished and the same length, is operated at low speed and pressure. The motion given to the stone is oscillating with an amplitude of $1/16$ to $1/4$ in. (1.6–6.4 mm) at about 450 cycles per minute (cpm) (7.5 Hz). The stone pressure is 3 to 40 psi (21–275 kPa). If the part is of greater length than the stone, an additional longitudinal movement of either stone or work is necessary. The work is rotated at a speed of about 50 ft/min (0.25 m/s) and during the operation is flooded with a light oil that carries away the minute particles abraded from the surface by the short, oscillating stone strokes. The stone action is similar to a scrubbing movement and removes all excess and fragmented metal on the surface.

Superfinishing flat surfaces is illustrated in Figure 23.13B. A rotating cup-shaped abrasive stone is used with the work resting on a circular table carried by a rotating spindle. An additional oscillating movement can be given to the stone, but because both it and the work are rotating, this action is not so important in developing a continually changing path of the abrasive particles. Superfinishing spherical surfaces is similar to

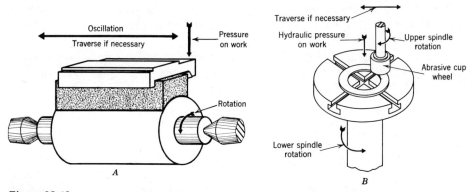

Figure 23.13
Motions between abrasive stone and work. *A*, Cylindrical superfinishing. *B*, Flat superfinishing.

that used for flat surfaces except that a formed-cup spindle is at an angle to the work spindle and no oscillating motion can be used.

Abrasive-Belt Grinding

This method is used for stock removal and surface preparation. Sometimes it is termed *high-energy grinding*. It is performed using a tensioned abrasive belt over precision pulleys at speeds between 250 and 6000 ft/min (1.27–30.48 m/s). Tension is maintained by springs or a hydraulic ram as the belt heats and stretches. The surface speed of a belt is approximately the same as the drive roller that pulls the belt. There may be one or more idler rollers of the same or a different size. The liner belt speed is

$$V = \pi KDN$$

where

V = Belt velocity, ft/min (m/min)
D = Diameter of belt roller, ft (m)
N = rpm of drive motor
K = Constant for belt slippage, usually 0.85 to 0.96

Heat is caused by belt slippage on the drive roller and by the abrasive cutting. Air and liquid coolants may be employed. On a 50-grit belt 8 in. (203 mm) wide and 96 in. (2438 mm) long there are about 1 million abrasive grains. Each grain passes the workpiece about 500 times a minute. If only 1 grit in every 1000 came in contact and removed a chip, an enormous number of chips and hence prodigious quantities of metal are capable of being removed. With heavy-duty machines this process can be almost as economical as a milling process for removing metal.

Figure 23.14 shows a belt grinding machine. Major areas of application include preparing flats, tubing and extrusion, and finishing partially fabricated stampings, forgings, and castings. Some belt grinding machines use wet belts and impervious plastic-bonded cloths. Surface finish is comparable to light milling and turning, and there is minimal work hardening and warpage resulting from heat generation. In some machines employing up to 150 hp (112 kW), the table rotates and the individual parts may have oscillatory motion as well.

On these machines a 24-in (600-mm) or wider abrasive belt can attain stock removal rates of 30 in.^{3}/min-in.^{2} (13 mm^3/s-mm^2) of contact on cast iron. Stock removal depth of 0.100 to 0.250 in. (2.54–6.35 mm) is possible.

Mass Media Finishing

Barrel Finishing. *Barrel finishing* or *tumbling* is a controlled method of processing parts to remove burrs, scale, flash, and oxides as well as to improve surface finish. Widely used as a finishing operation for many parts, it attains a uniformity of surface finish not possible by hand finishing. It is generally the most economical method of cleaning and

GRINDING AND ABRASIVE MACHINES   613

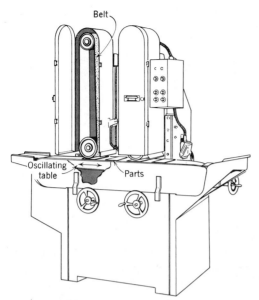

Figure 23.14
Dual-head, belt grinding machine.

surface conditioning large quantities of small parts. Materials that may be barrel finished include all metals, glass, plastics, and rubber.

Parts to be finished are placed in a rotating barrel or vibrating unit (Figure 23.15), with an abrasive medium such as water or oil and usually some chemical compound or detergent to assist in the operation. As the barrel rotates slowly, the upper layer of the work is given a sliding movement toward the lower side of the barrel, causing the abrading or polishing action to occur. A speed of 15 rpm will produce light action, whereas a speed of 30 rpm produces rapid abrasion. The time employed varies from a few minutes to 10 h. The same results also may be accomplished in a vibrating unit in which the entire contents of the container are in constant motion.

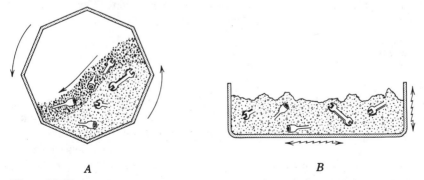

Figure 23.15
Two methods of barrel finishing. *A*, Slide action. *B*, Vibratory finishing machining.

Machines for tumbling can be tub shaped where batches of parts are processed. Also, mass media finishing can be continuous instead of batch type. Parts enter the conveyor tub of a vibratory finishing machine and eventually move along an open oscillating chute to a separating screen where the abrasive medium drops away. The tumbling abrasives are automatically returned to the starting point. These systems can be integrated with automatic handling equipment.

Abrasive-Media Flow Deburring Machines

These machines deburr internal edges or surfaces by a controlled-fore flow of an abrasive-laden semisolid grinding medium. The deburring machines hold the workpiece and tooling in a position relative to the medium. Fixtures can hold the workpiece in a position and contain, direct, or restrict the medium flow to areas of the workpiece where abrasion is desired. In application the number of holes or surface area does not necessarily influence the cycle time. Abrasion occurs in areas where medium flow is restricted.

Abrasive-media flow machines will deburr a hidden or secondary burr and improve internal surface finish These machines remove burrs that are inconvenient (or impossible) to reach by manual deburring methods. The deburring also will result in a smoother and brighter finish.

The flow of the medium is three-dimensional. The medium is forced up through a fixture and the workpiece into a top cylinder. After a predetermined volume of medium or time has been achieved, the medium returns from the top cylinder back through the workpiece and fixture to the bottom cylinder. Different media may be selected for different surface finish and burr removal. The medium cylinder diameter and pressure can vary depending on the machine. Hydraulic force is also used to clamp the workpiece and to give the desired medium pressure. Depending on the size and shape of the part, the fixtures may hold one or several workpieces.

Miscellaneous Finishing Operations

Wire Brushing. Rotating *brushes* with wire bristles are used to clean castings and to remove scratches, scale, sharp edges, and other surface imperfections. Tampico brushes with an abrasive compound may be used if the material is not too hard. Little metal is removed by brushing, and a satinlike finish appears on the surface. Usually a buffing operation is necessary if high polish is required.

Polishing. Cloth wheels or belts coated with abrasive particles are used for polishing operations. Though not considered precision metal-removing processes, such operations can remove sufficient metal to blend scratches and other minor surface imperfections. Both wheels and belts are flexible and will conform to irregular and rounded areas. Wide belts are used to polish plates, sheets, and other large metal parts. The amount of metal removed and the surface finish are controlled by the characteristics of the material being polished, belt speed, pressure, and grit size.

Polishing wheels are produced in disks of cotton cloth, canvas, leather, felt, or similar materials glued or sewed together to provide the required face width. Metal side plates may be used for stiffening. These wheels are coated with glue or a cold cement and immediately rolled in a trough containing abrasive grains. After drying a second application of glue and abrasive may be given. When the wheel is dry the hard, abrasive-coated surface is cracked by striking diagonally across the wheel face with an iron bar. This action breaks up the layer of glue and abrasive into small areas that provide flexibility and good polishing action to the wheel. Aluminum oxide and silicon carbide of various grit sizes are used as abrasives. Usually a part is passed over several wheels of decreasing grit size before the final polish is obtained.

Buffing. Buffing is a final operation to improve the polish of a metal and to bring out maximum luster. The wheels are similar to polishing wheels and are generally made of cotton, hemp cloth, flannel, linen, or sheepskin. They are charged with a fine abrasive such as rouge, tripoli, or amorphous silica. Buffing is frequently performed before plating.

ABRASIVES

Abrasives for grinding, honing, lapping, and superfinishing are bonded to a tool suitable for specific processes. They are hard materials that have been processed to cut or wear away softer materials. A classification of the common abrasive materials used for wheels and special shapes follows.

- A. Natural
    1. Sandstone or solid quartz
    2. Emery, 50% to 60% crystalline Al_2O_3 plus iron oxide
    3. Corundum, 75% to 90% crystalline Al_2O_3 plus iron oxide
    4. Diamonds
    5. Garnet

- B. Manufactured
    1. Silicon carbide, SiC
    2. Aluminum oxide, Al_2O_3
    3. Boron carbide, B_4C
    4. Zirconium oxide, ZrO_2

For many years it was necessary to rely on natural abrasives in manufacturing grinding wheels. *Sandstone* wheels, though available, are rarely used. Although they are cut from high-grade quartz or sandstone, they do not wear evenly because of variations in the natural bond.

Corundum and *emery* have long been used for grinding purposes. Both are composed of crystalline aluminum oxide in combination with iron oxide and other impurities. Like sandstone, these minerals lack a uniform bond and are seldom used in production work.

Diamond wheels made with a resinoid bond are useful in sharpening cemented carbide tools. In spite of high initial cost they have proved economical because of their rapid cutting ability, slow wear, and free cutting action. Very little heat is generated with their use, which is an added advantage.

Silicon carbide was discovered during an attempt to manufacture precious gems in an electric furnace. The hardness of his material according to Mohs' scale is slightly over 9.5, which approaches the hardness of a diamond. Raw materials consisting of silica sand, petroleum coke, sawdust, and salt are heated in a furnace to around 4200°F (2300°C) and held there for a considerable period of time. The product consists of a mass of crystals surrounded by partially unconverted raw material. After cooling the material is broken up, graded, and crushed to grain size. Silicon carbide crystals are very sharp and extremely hard. Their use as an abrasive is limited because of brittleness.

Aluminum oxide is made from the claylike mineral *bauxite,* which is the main source of aluminum. Aluminum oxide is slightly softer than silicon carbide, but it is much tougher. Most manufactured wheels are made of aluminum oxide.

MANUFACTURE OF GRINDING WHEELS

The process of making grinding wheels is similar for all materials, sizes, and shapes. The procedure is as follows:

1. The material is reduced to small size by being run through roll and jaw crushers. Fines are removed by passing the material over screens between crushing operations.
2. The material is passed through magnetic separators to remove iron compounds.
3. Dust and foreign material is removed by a washing process.
4. The grains are graded by being passed over vibrating standard screens. A standard 30-mesh screen has 30 meshes per inch or 900 openings per square inch (0.595 mm nominal sieve opening). The no. 30 size material passes through a no. 30 screen and is retained on the next finer size, which in this case is no. 36.
5. Grains are mixed with bonding material, molded or cut to proper shape, and heated. The heating or burning procedure depends on the bonding agent.
6. The wheels are bushed, trued, tested, and given a final inspection.

BONDING PROCESSES

The bonding process is performed to join the abrasives into a usable form. Because grinding operations are accomplished at high speeds and the wheels are dense, high bonding strength is necessary for structure and safety. The principal processes are described as follows.

1. *Vitrified process.* The abrasive grains are mixed with claylike ingredients that are changed to a glasslike material upon being "burned" at high temperature. The addition of water is required and the wheels are shaped in metal molds under a hydraulic press. Wheels produced by these methods are dense and accurately shaped. The time for burning

varies with the wheel size, being anywhere from 1 to 14 days. The process is similar to firing tile or pottery.

Vitrified wheels are porous, strong, and unaffected by water, acids, oils, and climatic or temperature conditions. About 75% of all wheels are made by this process. Recommended wheel peripheral speeds vary from 1500 ft/min (7.62 m/s) when grinding titanium to a maximum safe velocity of 12,000 ft/min (60.96 m/s) for specially reinforced grinding wheels. The maximum standard speed generally used is 6500 ft/min (33.02 m/s).

2. *Silicate process*. In this process sodium silicate is mixed with the abrasive grains and the mixture is tamped in metal molds. After drying several hours the wheels are baked at 500°F (260°C) from 1 to 3 days.

Silicate wheels are milder acting than those made by other processes and wear more rapidly. They are suitable for grinding edges of tools where the heat must be kept to a minimum. This process is also recommended for large wheels because they tend not to crack or warp in the baking process.

3. *Shellac process*. The abrasive grains are first coated with shellac by being mixed in a steam-heated mixer. The material is then placed in heated steel molds and rolled or pressed. Finally, the wheels are baked a few hours at a temperature around 300°F (150°C). This bond is adapted to thin wheels as it is strong and elastic.

4. *Rubber process*. Pure rubber with sulfur as a vulcanizing agent is mixed with abrasive by feeding the material between heated mixing rolls. After it is rolled to thickness the wheels are cut out with dies and vulcanized under pressure. Very thin wheels can be made by this process. Wheels having this bond are used for high-speed grinding (9000–16,000 ft/min; 45–81 m/s) because they afford rapid removal of the stock. They are used as snagging wheels in foundries and also as cutting-off wheels.

5. *Bakelite or resinoid process*. The abrasive grains in this process are mixed with a thermosetting synthetic resin powder and a liquid solvent, then molded and baked. This bond is very hard and strong; wheels made by this process can be operated at speeds of 9500 to 16,000 ft/min (48–81 m/s). They are used for general-purpose grinding and in foundries and billet shops for snagging purposes because of their ability to remove metal rapidly.

GRINDING WHEEL SELECTION

Selection of a grinding wheel for a definite purpose is important. Many factors complicate the choice among a variety of wheels. The most important ones are listed here.

1. *Size and shape of wheel*. Standard shapes are shown in Figure 23.16. Different wheel faces are available.

2. *Kind of abrasive*. The choice of silicon carbide or aluminum oxide largely depends on the physical properties of the material to be ground. Silicon carbide wheels are recommended for materials of low tensile strength such as cast iron, brass, stone, rubber,

618 GRINDING AND ABRASIVE MACHINES

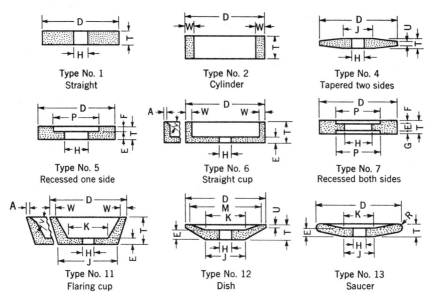

Figure 23.16
Standard grinding wheel shapes.

leather, and cemented carbides. Aluminum oxide wheels are best used on materials of high tensile strength like hardened steel, high-speed steel, alloy steel, and malleable iron.

3. *Grain size of abrasive particles.* In general, coarse wheels are used for fast removal of materials. Fine-grained wheels are used where finish is important. Coarse wheels may be used for soft materials, but generally a fine grain should be used for hard and brittle materials.

4. *Grade or strength of bond.* The grade depends on the kind and hardness of the bonding material. If the bond is very strong and capable of holding the abrasive grains against the force tending to pry them loose, it is said to be hard. If only a small force is needed to release the grains, the wheel is said to be soft. Hard wheels are recommended for soft materials, and soft wheels for hard materials.

5. *Structure of grain spacing.* The structure refers to the number of cutting edges per unit area of wheel face as well as to the number and size of void spaces between grains. Soft, ductile materials require a wide spacing. A fine finish requires a wheel with a close spacing of the abrasive particles.

6. *Kind of bond material.* The vitrified bond is most commonly used. Where thin wheels are required or high operating speed or high finish is necessary, other bonds are more advantageous.

A standard system of marking grinding wheels, adopted by the American National Standards Institute, is shown in the accompanying chart.

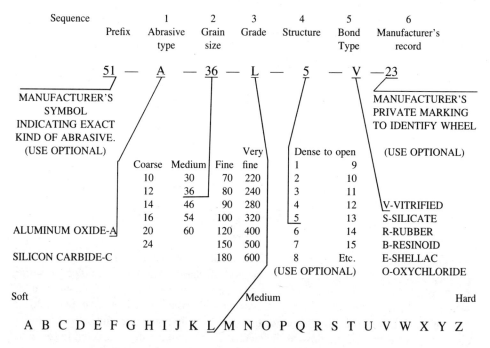

STANDARD MARKING SYSTEM CHART

COATED ABRASIVES

When abrasive particles are glued to paper or other flexible backings, they are known as *coated abrasives.* Common products include "sandpaper," abrasive disks, and belts like those used on belt grinding machines. Any of the abrasives used in wheel manufacture may be applied in this way. The most common type is "sandpaper," which frequently identifies this entire group. The abrasive in sandpaper is a flint quartz that is mined in large lumps and then crushed to size and graded. Another important natural abrasive is garnet, a red mineral. Of the several kinds of garnet known, the one called almandite is the best for abrasive coatings. It is much harder and sharper than flint and, when broken down, breaks into crystals with many cutting edges. Other natural abrasives are emery and corundum. The two manufactured abrasives that have wide application are silicon carbide and aluminum oxide.

Types of backing include paper, cloth, fiber, paper cloth, and cloth fiber. Each backing has certain characteristics that make it beneficial for a particular application.

An important phase in the manufacture of coated abrasives is the application of the abrasive particles on the backing material. The abrasive grains must be securely held to give the best cutting action for the work to be done. For severe service, closed coating where the grains completely cover the surface of the backing is recommended. When

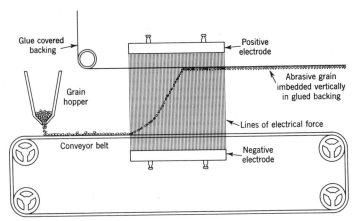

Figure 23.17
Electrocoating method for applying abrasive grains to glue-coated backing.

increased flexibility is required and there is no tendency for the particles to become loaded or clogged, an open coating is used. A method of coating known as the ***electrocoating*** process is illustrated in Figure 23.17. This method uses the principle that opposite-charged particles attract one another. The abrasive particles pass on a conveyor belt into an electrostatic field. As the particles enter the field they first stand on end, aligning themselves in the direction of the flow of the electric force, and are then attracted to the glue-covered backing, which is traveling in the same direction as the other belt. The abrasive grains are imbedded on end and are equally spaced on the backing.

Most coated abrasive machines utilize either belts or disks. On disk grinders a metal disk forms the backing for the coated abrasive. The belt or abrasive belt machines pass over a contact wheel and an idler, and the grinding or polishing is done against the contact wheel. When the grinding is done between the wheel and the idler, the belt is supported by a platen. Contact supporting wheels are made of steel, rubber, phenolic resin, or cloth depending on the type of work to be done. Hard contact wheels provide faster stock removal rate and soft wheels a better finish.

Not all coated abrasives are graded in the same manner. Flint paper and emery cloth have their own systems, whereas garnet and manufactured abrasives are graded by another system. Table 23.1 will be of assistance in the selection of the proper grades to use.

MASS MEDIA ABRASIVES

An important factor in the successful operation of barrel finishing is selection of the abrasive. Manufactured aluminum oxide and silicon carbide media in various sizes and shapes have proved to be satisfactory because of their cutting ability. Selection of the abrasive depends on the size and shape of the part to be processed, workpiece material, and the finish desired.

Abrasives may be obtained in the same range of grit sizes as is available for wheels.

Table 23.1 **Table Comparative Grit Sized—Approximate Comparison of Grit Numbers**[a]

	Durite Metalite	Adalox Garnet	Flint	Emery cloth	Emery polishing paper
Extra fine	600				4/0
					3/0
	500				2/0
	400	400(10/0)			0
	360				
	320	320(9/0)			$1/2$
Very fine	280	280(8/0)			1
	240	240(7/0)			1-G
	220	220(6/0)	Extra fine		2
Fine	180	180(5/0)		Fine	3
	150	150(4/0)			
	120	120(3/0)	Fine		
Medium	100	100(2/0)		Medium	
	80	80(0)	Medium		
	60	60($1/2$)		Coarse	
Coarse	50	50(1)	Coarse		
	40	40($1 1/2$)			Very coarse
Very coarse	36	36(2)	Extra coarse		
	30	30($2 1/2$)			
	24	24(3)			
Extra coarse	20	20($3 1/2$)			
	16	16(4)			
	12				

[a]Table prepared by Behr-Manning Division of Norton Company. Metalite and Adalox are made from aluminum oxide manufactured abrasive. Durite is made from silicon carbide manufactured abrasive. Garnet, flint, and emery are natural abrasives.

Figure 23.18
Miscellaneous sizes and shapes of ceramic-based tumbling nuggets.

When plastic parts are to be processed, rubber-lined barrels are recommended. In finishing rubber articles, parts are quickly frozen by adding a CO_2 refrigerant so that the flash breaks off easily. The fatigue strength of the steel parts is raised with barrel finishing.

The *tumbling nugget* sizes vary from $3/32$ to 2 in. (2.4–50 mm) and come in a variety of shapes, as Figure 23.18 shows. The shapes are classed as random or preformed. The shapes of the preforms are triangular, rodlike, or spherical. Random shapes are preferred where part geometry avoids lodging of the nugget and where the vigor of the tumbling action provides reasonable abrasive life.

QUESTIONS AND PROBLEMS

1. Give an explanation of the following terms:

   Chip thickness
   Work rest
   Tool post grinder
   Hone
   High-energy grinding
   Barrel finishing
   Coated abrasive
   Electrocoating
   Tumbling nugget

2. Discuss the approximate grinding wheel surface speeds.

3. Plot a curve of surface speed for a 12-in. (305-mm) diameter grinding wheel versus grinding wheel revolutions per minute. Speeds range between 1000 and 2,500 surface ft/min.

4. If the maximum safe speed of a grinding wheel is 8000 ft/min, make a table showing the maximum safe revolutions per minute for grinding wheels from 4 in. to 24 in. in 1-in. increments.

5. Discuss the principal advantages of grinding as a machining operation.

6. If a part is to be hardened to Rockwell C54, would the grinding operation normally be done before or after heat treatment? Why?

7. Approximately what grinding wheel speeds are used for (a) offhand grinding; (b) centerless grinding; (c) abrasive-belt grinding; (d) grinding between centers; and (e) surface grinding?

8. What inaccuracies can be eliminated by honing? Does honing remove material?

9. What is Mohs' hardness and how does it relate to manufactured abrasives?

10. Make a table showing the various bonding materials for grinding wheels and show their comparisons for speed, elasticity, approximate cost, and principal use.

11. What can be said in general with regard to the surface speed of a grinding wheel as compared to the surface speed of the workpiece of a cylindrical grinder, a centerless grinder, and surface grinders?

12. Select a suitable grinding wheel for each of the following applications, indicating wheel type, wheel size, bonding material, kind of abrasive, and wheel speed: sharpening high-speed steel milling cutter; grinding flat hardened steel; cutting off thin-walled aluminum tubing; finishing flat plates of soft steel; snagging iron casting; cutting glass bar stock.

13. How is the "through speed" or feed regulated on a centerless grinder? What are the ways "through speed" can be doubled? Examine the equations.

14. Make a plot showing the feed on a centerless grinder versus regulating wheel revolutions per minute for a 5° angle of inclination. Use increments of 100 rpm to a value of 2000. Regulating wheel equals 6-in. OD.

15. Make a plot showing the feed on a centerless grinder versus angle of inclination for a 8-in. OD regulating wheel revolving at 700 rpm.

16. If on a surface grinder using a 10-in. (254-mm) wheel a depth of cut of 0.009 in. (0.229 mm) is being ground, what is the average length of chip generated?

17. Refer to problem 16. For a wheel with 1-in. and a width of a grit equal to .01 in. find the volume of an average chip. The number of active grits equals 150/in.² Wheel speed equals 75.0 ft/min.

18. Refer to problems 16 and 17. Find the average chip thickness.

19. Plot a curve showing the effect of workpiece velocity on chip thickness.

20. Assume a grinder having a 20-in. OD × 2-in. wide wheel. Depth of cut is .002 in. and relative velocity is 500 fpm. A grit is .005 in. wide and its height is 0.008 in. The number of grits equals 750/in.² Find the average chip thickness.

21. What are the principal limitations when considering a centerless grinder for grinding shafts?

22. What type of grinding machine should be used for grinding or finishing the following parts: piston pins, carbide tools, flat washers, ball bearings, inside and outside of tapered roller bearing housings, form gear cutters, and large rolls?

23. What is a soft grinding wheel? What kind of work is it used for?

24. In lapping, what would be the consequence of having a situation in which the workpiece was softer than the lap?

25. How does the process of mass media tumbling compare to grinding? What kinds of work do you perform with tumbling?

26. What type of process is most adaptable to deburring?

27. For an abrasive-belt grinder that has a 9-in. (229-mm) driving wheel, specify the revolutions per minute at which it should be driven if a belt speed of 15 m/s is desired and K is taken as 0.90.

28. Explain the operations that should be made on a forged open-end wrench after broaching to prepare it for chrome plating.

29. What are the limitations of natural abrasives?

30. What are the characteristics of the following grinding wheels marked:

    32-C-100-P-11-B

    47-A-24-E-14-R

31. If the regulating wheel of a centerless grinder is 12 in. (300 mm) in diameter, is tilted to an angle of 5°, and is rotating at 30 rpm, what is the feed in inches per minute?

32. In problem 31, what is the feed in millimeters per second?

33. How long would it take to remove $1/16$ in. (1.6 mm) from the top surface of a case-iron plate $1/2 \times 4 \times 12$ in. (12.7 × 101 × 305 mm) on a coated-abrasive machine?

34. To grind off 0.012 in. (0.30 mm) on the diameter of a cylinder 14 in. (356 mm) long, what depth of cuts and at what approximate feed should a wheel 1 in. (25.4 mm) in width be used?

35. What type of abrading process gives the surface a high degree of finish with a cross-hatch appearance?

CASE STUDY

KARLTON GRINDING COMPANY

The Karlton Grinding Company is a large, full-service grinding jobbing shop. The part, a shaft connector shown in Figure 23.19, is of hardened steel and has been previously machined and heat treated. It is slightly oversized. All the surfaces except the ends and the angle must be ground

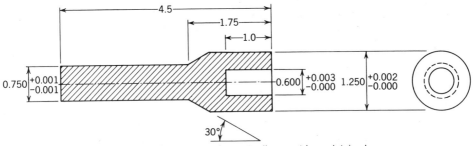

Figure 23.19
Case study.

on 20,000 pieces. Karlton needs work for its centerless grinding machines, but can they be used on this part? Choose what grinding machines will be used on this part and prepare an operations plan similar to that shown.

It is recognized that when a part such as this reaches the shop floor, 95% of the time it will be simply sitting in a bin or on a flat. Less than 5% of the time will it actually be worked on. Although your estimate of times may be quite inaccurate, they will give an approximation of cost. Figure machine time and inspection time at $40/h; shop helper time at $20/h. The shop helper moves the parts from one machine to another. What is your estimate of Karlton's cost for this job? What would be the advantage of a continuous process for such a part is a car manufacturing company adopts the part for mass production?

OPERATIONS PLAN

Part name _____

Date raw
material received _____

Date sent to shop _____

Number received _____

Number sent _____

Operation number	Machine	Estimated time/pc	Date started	Date finished	% Passing inspection

CHAPTER 24
SPECIAL PROCESSES AND ELECTRONIC FABRICATION

Of the many processes that exist, some are useful and popular, whereas others are found only rarely. The needs of manufacturing have led to many special processes, most of which are of recent origin. Several manufacturing processes that do not fall under categories of traditional methods will be considered in this chapter. These processes are sometimes called the *nontraditional* methods. Such manufacturing processes may be used to cut hard materials that cannot be cut easily, may employ methods to give a special effect or finish, or are so new that their impact on manufacturing processes is unclear.

The processes also involve manufacturing electronic products that are highly specialized. The manufacture of chips, resistors, printed circuit boards, and their assembly is an important opportunity for manufacturing students.

SPECIAL MACHINING PROCESSES

The shaping of parts made from carbides and other metals difficult to machine has for many years been limited to diamond wheel grinding. Because of the expense of diamond wheels and the time required for grinding, effort and research have been directed toward the development of more economical methods. There are no less than four fundamental types of machining energy being used: mechanical, chemical, electrochemical, and thermoelectric. These can be subdivided into a number of special processes, each having some special use or advantage. This chapter will describe the applications and economies to be gained. All processes remove metal from either soft or hard metals, but the rate of metal removal in soft metals is slower than by conventional machining methods.

Ultrasonic Machining

Ultrasonic machining, a vibratory process, was designed to machine hard, brittle materials. As shown in Figure 24.1A, it removes material by abrasive grains that are carried in a liquid between the tool and the work and that bombard the work surface at high velocity. This action gradually chips away minute particles of material in a pattern controlled by the tool shape and contour. A *transducer* causes an attached tool to oscillate

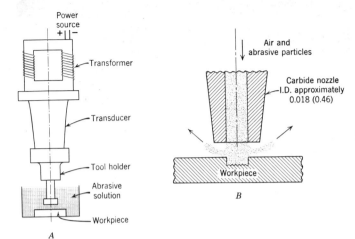

Figure 24.1 Ultrasonic and abrasive machining. A, Schematic diagram of the ultrasonic process. B, Abrasive jet cutting with aluminum oxide particle, 15 to 40 μm.

linearly at a frequency of 20,000 to 30,000 Hz and at an amplitude of 0.0005 to 0.004 in. (0.013–0.10 mm). The tool motion is produced by being part of a sound wave energy transmission line that causes the tool material to change its normal length by contraction and expansion. The toolholder is threaded to the transducer and oscillates linearly at ultrasonic frequencies, thus driving the grit particles into the workpiece. The cutting particles boron carbide, silicon carbide, and aluminum oxide are of a 280-mesh size or finer, depending on the accuracy and the finish desired.

The metal removal rate is slow or about 0.022 in.^{3}/min in tungsten carbide. Major factors influencing material removal rates, surface roughness, and accuracy are amplitude and frequency of the tool oscillation, impact forces, tool material, abrasive, and content of the slurry.

The tools are made of brass or soft steel and must match the surface to be machined. Tolerances of 0.002 in. (0.05 mm) can be maintained with 280 grit, or by using finer grit a tolerance of 0.0005 in. (0.013 mm) can be held. Operations that can be performed include drilling, tapping, coining, and the making of openings in all types of dies. Ultrasonic machining is used principally for machining materials such as carbides, tool steels, ceramics, glass, gem stones, and synthetic crystals. Advantages of this process include the absence of thermal stresses, low tooling costs, and the use of semiskilled workers for precision work. Holes with a curved axis, nonround holes, or holes of any shape for which a tool can be made are candidates.

Abrasive-Jet Machining

Abrasive-jet machining is a mechanical process for cutting, deburring, and cleaning hard, brittle materials. Cutting rates are generally slow, and focusing the stream of particles can be a problem in tooling and motion control. The cutting action is cool because the fluid tends to cool the surface while carrying the abrasive. It is similar to sandblasting but uses much finer abrasives with particle size and velocity under close control. The cutting action is shown schematically in Figure 24.1B. Air or CO_2 is the carrying medium for the abrasive particles that impinge on the workpiece at velocities about 500 to 1000 ft/s (150–300 m/s). Aluminum oxide or silicon carbide powders are used for cutting, whereas softer powders such as dolomite or sodium bicarbonate are used for cleaning, etching, and polishing. Powders are not recycled because of possible contamination, which is apt to clog the system.

Abrasive-jet machining cuts fragile materials without damage. Other uses include frosting glass, removing oxides from metal surfaces, deburring, etching patterns, drilling and cutting thin sections of metal, and cutting and shaping crystalline materials. It is not suitable for cutting soft materials, because abrasive particles tend to become embedded. Compared with conventional processes, its material removal rate is slow. For glass it is 0.001 in.^{3}/min (0.273 mm^3/s).

Water Jet Machining

Water jet machining or fluid jet machining is a process that utilizes a high-velocity stream of water as the cutting agent. Jet nozzles are approximately 0.0015 to 0.0067 in. (0.038–0.17 mm) in diameter and operate at velocities of 2000 to 3000 ft/s (600 to 900 m/s). Pressures range from 30,000 to 55,000 psi (2100–3900 kg/cm^2). Higher pressures reduce seal life from about 800 to 150 h. At these velocities the jet can cut through wood, plastics, textiles, and in some cases ceramics, steel, and titanium. A limitation of this process is the lack of suitable pumping equipment. The abrasive may be garnet or even silica, which reduces costs. Applications include aerospace beveling of edges for subsequent welding or cutting off sprues and gates by foundries.

Electrical Discharge Machining

Electrical discharge machining (EDM) is a process that can remove metal with good dimensional control from any soft or hard metal. It cannot be used for machining glass, ceramics, or other nonconducting materials. The machining action is caused by the formation of an electrical spark between an *electrode* shaped to the required contour and the workpiece. The conventional EDM machine cuts metals by electrical discharge or "spark erosion" between the metal to be cut (−ve charge) and an electrode (+ve charge). This cutting takes place in a nonconductive fluid known as a *dielectric*.

Because the cutting tool has no contact with the workpiece, it can be made of a soft, easily worked material such as brass. The tool works in conjunction with a fluid such as mineral oil or kerosene, which is fed to the work under pressure. The function of the coolant is to serve as a dielectric, to wash away particles of eroded metal from the workpiece or tool, and to maintain a uniform resistance to flow of current. The tank is filled with the dielectric fluid, and the workpiece and electrode end are submerged. An electrode is chosen depending on the shape of the cut needed. It is positioned on the top of the workpiece leaving a small gap between.

After connecting the electrode to a +ve charge, spark erosion takes place and it is a "miniature thunderstorm" between the two metals. Flashes of "lightning" in rapid succession occur. Each one produces a tiny crater in the surface of the two metals. *Metal evaporation* occurs where the flash strikes.

Equal amounts of material are *not* removed from both plates. By an appropriate choice of materials, electrode of copper, workpiece of steel, and a skillful selection of the opening and closing times of the automatic switch, more material is removed from the steel than from the copper.

During the process the dielectric constantly flows through the tank, requiring filtration. Also, erosion creates heat, so the dielectric has to be cooled. The capacity of work of the conventional EDM machine is measured by the rate of material removal, in cubic inches per minute.

Figure 24.2A is a diagram of a simple arrangement for electrical discharge machining. A condenser parallel with the electrode and workpiece receives a charge of direct current through a resistor. As the condenser is energized, its potential rises rapidly to a value sufficient to overcome the dielectric fluid between the electrode and work. The gap distance is *servocontrolled* so as to maintain a fairly constant potential in order to bring about an electrical breakdown of the dielectric fluid between the electrode and work. This distance is only a few thousandths of an inch (~0.05 mm), so hand control is difficult to maintain. Regardless of the electrode tool area, sparking occurs at the point where the gap is the smallest. The current density at this point is high and of sufficient force to erode small particles from the workpiece. These small particles of metal are vaporized or melted by the spark, cooled by the electrolyte, and flushed from the gap between the electrode and the workpiece. The rate of metal removal is not fast as compared with commercial machining and may vary from a small fraction of 1 in.^{3} to around 15 in.^{3}/h (70 mm^3/s). The best surface finishes are obtained with slow rates of metal removal. Most machines have a frequency selector that controls the number of sparks per second between the electrode and the workpiece. As the spark frequency increases, the surface finish is improved as a result of the reduction of energy per spark. Frequency of the sparks may range from 500 to 500,000 sparks per second.

The *electrode is the cutting tool*. Although it is not subject to much heat, it should have a high melting temperature and be a good electrical conductor. The wear rate of electrode materials varies with the material to be machined, but its selection is determined by the material cost and how it is made. Acceptable electrode materials include tungsten carbide, copper tungsten, graphite, copper, brass, and zinc alloys. The only requirement of the workpiece material is that it must be a good conductor of electricity.

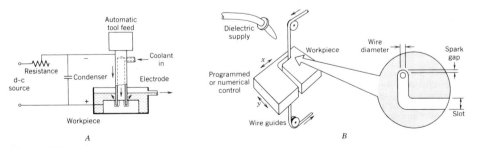

Figure 24.2
A, Diagram for electrical discharge machining. B, Traveling-wire electrical discharge machining.

SPECIAL MACHINING PROCESSES 629

Figure 24.3
Electrical discharge machine (EDM) with power unit and numerical control system, especially designed for the production of multiple-hole or complex cavity patterns.

Figure 24.3 shows an electrical discharge machine equipped with numerical control (NC) for the positioning with greater accuracy of the electrode from one hole to the next.

Electrical discharge machining is used for salvaging hardened parts, machining carbide stock, producing dies and metal molds for stamping, forging, and jewelry manufacture, as well as for making numerous parts from hard metals. A tap that is broken inside a hole is "salvaged" by EDM, as it is possible to erode away the tap without damaging the tapped hole. With this process close tolerances can be maintained and finishes of 8 to 10 μin. (200–250 nm) are possible.

Traveling-wire electrical discharge machining, a metal cutting process that removes metal with an electrical discharge, is suited for production of parts having extraordinary workpiece configurations, close tolerances, the need of high repeatability, and hard-to-work metals. Wire electrical discharge machining produces a wide array of parts such as gears, tools, dies, rotors, and turbine blades. It is appropriate for small to medium-size batch quantities. Actual machining times may vary from half an hour to 20 h. It uses the heat of an electrical conductor to vaporize material; thus, essentially no cutting forces are involved and parts can be machined with fragile, complex geometries. The sparks are generated one at a time in rapid succession (pulses) between the electrode (wire) and the workpiece. The sparks must have a medium in which to travel, so a flushing fluid (water) is used to separate the wire and workpiece. Hence the one requirement is that the workpiece must be electrically conductive. A vertically oriented wire is fed into the workpiece continuously traveling from a supply spool to a take-up spool so that it is constantly renewed. Note Figure 24.2B.

A power supply provides a voltage between wire and workpiece. By means of an adjustable setting one can determine the pulse amplitude and pulse duration, the on and off times (in microseconds). On time refers to metal removal; off time is the period during which the gap is swept clear of removed metal via flushing. So both the intensity of the spark and the time it flows determine the energy expended and consequently the amount of material removed per unit time.

Electrochemical Machining

The electrochemical process is based on the same principles used in electroplating, except that the workpiece is the anode and the tool is the cathode as indicated in Figure 24.4. It is in effect a *deplating operation*. The development of this process is a result of the successful machining of hard and tough materials as well as complicated configurations. Low microinch surface finishes are standard, tool wear is negligible, and the stock removal rate exceeds other nontraditional machining processes. An average rate of stock removal is 1 in.^{3}/min of metal (270 mm^3/s) for each 10,000 A of current.

In this process electrode accuracy is important because the surface finish of the electrode tool will be reproduced in the surface of the workpiece. Copper is frequently used as the electrode, but brass, graphite, and copper tungsten also find use. The tool must be electrically conductive, easy to machine, corrosion resistant, and capable of conducting the quantity of current needed. Although there is no standard electrolyte, sodium chloride is more generally used than others.

The accuracy of the product is greatly influenced by the accuracy of the electrode form and its surface finish. It is also affected by irregularities in electrolyte flow or current

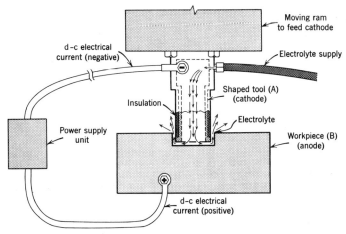

Figure 24.4
Diagram of electrochemical machining system.

flow. The electrolyte enters the gap between the electrolyte and the work at pressures ranging from 200 to 350 psi (1.4–2.4 MPa), whereas the flow rate may reach 150 gal/min (0.01 m^3/s) for high pressures. The current flow must be maintained at a constant density if a uniform gap is to be maintained. A temperature increase in the electrolyte improves the surface finish, but when temperature increases the metal removal rate is also accelerated, increasing the size of the gap. This changes the gap resistance, so that less current flows and the metal removal rate is lowered back to normal. When operating parameters are properly chosen, small variances are self-adjusting.

Electrochemical machining performs stress-free cutting of all metals, has high current efficiency, and can produce complex configurations difficult to obtain by conventional machining processes. Aside from cavity sinking as in die work, holes may be drilled, external surfaces shaped, circular pieces turned, and contour machining performed. In the steel industry these machines are used to extract test specimens from ingots, castings, and rolled shapes. In such processes all metal is removed by electrochemical decomposition.

In the several electrical machining processes described there are advantages not usually found in conventional machining.

1. Hard or soft *conductible materials* can be cut.
2. The cutting tool can be of soft material because it does not touch the workpiece.
3. There is no appreciable heating so no metallurgical changes or stresses caused by elevated temperatures occur.
4. Cutting forces are not involved.
5. Multiple operations can be carried out simultaneously.
6. Surface finishes can be maintained at from 5 to 10 μin. (125–250 nm).
7. Manual deburring is unnecessary.

Electrochemical Grinding. Electrochemical grinding, also known as *electrolytic grinding*, is similar to electrochemical machining but differs slightly in that the metal is removed by electrochemical decomposition plus some abrasive action, which accounts for only 10% of metal removal. Figure 24.5 illustrates the process.

In electrochemical grinding a metal disk with embedded abrasive particles serves as

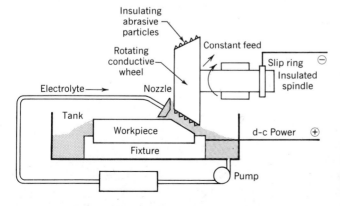

Figure 24.5
Schematic of electrochemical grinding machine.

the cathode. The workpiece is the anode and the electrolyte, used in much the same fashion as a coolant, completes the electrochemical circuit. Figure 24.6A shows an electrochemical grinder with the various parts labeled. In operation the abrasive particles maintain the proper spacing between the disk and the workpiece. Diamond abrasive is ordinarily used, but because the electrochemical process accounts for 90% of the cutting action, wheel wear is not great. Although electrochemical grinding is particularly adaptable to the sharpening of carbide tools and the grinding of chip breaker grooves, many other grinding applications are possible such as Figure 24.6B. Electrochemical grinding is cool; burrs are not generated and it produces a surface finish ranging from 8 to 12 μin. (200–300 nm). An average metal removal rate is 0.010 in^3/min (2.7 mm^3/s) per 100 A.

Laser Beam Machining. The term *laser* is an acronym for "light amplification by stimulated emission of radiation." Simply stated, it is a very strong monochromatic beam of light that is highly collimated and has a very small beam divergence. Laser beam machining, a **thermoelectric process,** is largely accomplished by material evaporation, although some material is removed in a liquid state. Figure 24.7 is a pictorial view of a laser head. A relatively weak light flash is amplified in the ruby because certain of the chromium ions in the ruby emit photons as the light beam bounces back and forth in it. This released energy from the ruby accelerates the intensity of the beam of light that leaves the rod and is focused on the workpiece. The ruby laser is most efficient when kept very cold, and liquid nitrogen at $-320°F$ ($-196°C$) serves this purpose. The light flash operates best when warm; hence, hot air is circulated over it. The vacuum chamber between the ruby and the flash acts as an insulator and enables the two temperature

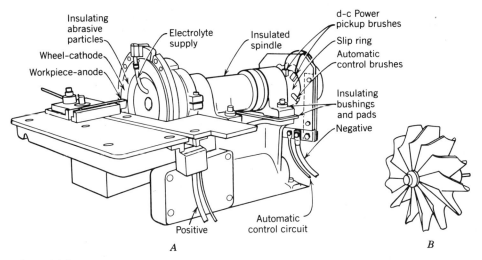

Figure 24.6
A, Setup for tool sharpening on electrochemical grinding machine. Short-circuiting between wheel and workpiece is prevented by abrasive particles projecting from wheel. A conductive electrolyte is used in the process. B, Curved form on Stellite turbine impeller ground by electrochemical grinder.

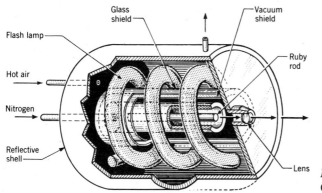

Figure 24.7
Optical laser.

climates to be maintained. The lamp operates from 1 flash every 3 min to 12 flashes per minute. The laser energy is applied to the workpiece in less than 0.002 s.

In addition to ruby lasers, other types are gaseous-state lasers that utilize CO_2 and other gases, liquid-state lasers, and semiconductor lasers. All four involve the use of a well-defined beam of light. Because the metal removal rate is very small, they are used for such jobs as drilling microscopic holes in carbides or diamond wire-drawing dies and for removing metal in balancing high-speed rotating machinery.

Lasers will machine through transparent materials and can vaporize any known material. They have small heat-affected zones and work easily with nonmetallic hard materials. Major limitations to the process are the high cost of the equipment, low operating efficiency, difficulty in controlling accuracy, and its use primarily for small parts.

One important application of lasers is in the area of welding. Carbon dioxide (gas) lasers are used in welding, but they are limited to metal thickness of less than 0.02 in. (0.5 mm). Pulsed ruby lasers have also been used successfully for welding thicker materials. (Welding is discussed in Chapter 8.) Another application is in the area of metal cutting, where the carbon dioxide laser is most widely used. This laser, operating continuously, can cut any material if the beam is focused and a jet of gas is used to concentrate the beam. Figure 24.8 shows a ***gas-assisted laser*** cutting head cutting a steel plate.

Electron Beam Machining. Like laser beam machining, *electron beam machining* is a thermoelectric process. It is similar because of the high temperatures and high thermal energy densities that can be achieved. In electron beam machining, heat is generated by high-speed electrons impinging on the workpiece, the beam being converted into thermal energy. At the point where the energy of the electrons is focused, it is transformed into sufficient thermal energy to vaporize the material locally. The process is generally carried out in a vacuum. The principles involved in this process are discussed in Chapter 8 in connection with electron beam welding. Figure 24.9 illustrates the arrangement of the gun and beam directors, which focus the electrons magnetically.

Although the metal removal rate is approximately 0.0001 in.^{3}/min (0.002 cm^3/min) the tool is accurate and is especially adapted for micromachining. Beams can be concentrated on spots as small as 0.0005 in. (0.00127 mm) in diameter. There is no significant

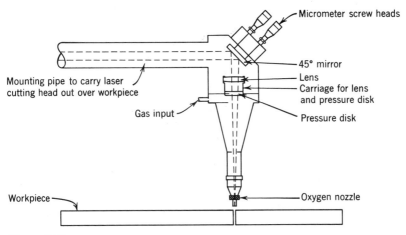

Figure 24.8
Gas-assisted laser cutting head.

heat-affected zone or pressure on the workpiece, and extremely close tolerances can be maintained. Limitations of the process include high equipment cost, the need for skilled operators, and a vacuum chamber that restricts the workpiece size. The process results in X-ray emission, which requires that the work area be shielded to absorb radiation. The process is used for drilling holes as small as 0.002 in. (0.05 mm) in any known material, cutting slots and shaping small parts for the semiconductor industry, and machining sapphire jewel bearings.

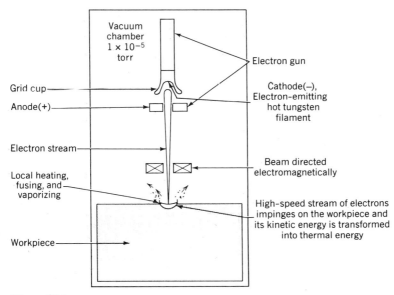

Figure 24.9
Electron beam machining.

Elevated-Temperature Machining

The shear strength of the metal is reduced when the workpiece is heated, and plastic deformation ahead of the cutting tool is accomplished with less power. The chips tend to be continuous and the temperature at the *tool–chip interface* does not increase in proportion to the temperature of the workpiece. Tool life is increased and cutting speeds may be doubled. Unfortunately, the expense of heating the workpiece is usually prohibitive. Heating may be accomplished by passing a current through the work, electric arc, acetylene flame, induction heating, or a radiofrequency heating apparatus. Because the greatest efficiency comes from heating the shear zone just ahead of the cutting tool, radiofrequency resistance heating is the most economical for magnetic materials, whereas the tungsten inert-gas torch can be used for heating nonmagnetic materials.

Figure 24.10A shows a radiofrequency heating apparatus that has been designed to concentrate heat in the area of the shear zone. Radiofrequency current tends to flow in the path of least impedance; hence, the return conductor is designed and placed in such a manner that the lowest impedance, and consequently the highest temperature, is focused ahead of the tool and in the shear zone.

The process is limited because the effects of elevated temperatures on the workpiece may cause a metallurgical change or distortion of the part, and heating the workpiece is expensive.

Plasma arc cutting torches employ a restricted tungsten arc through which gas is made to flow. A plasma is a gas that has been heated to a sufficiently high temperature to become partially ionized. Plasma cutting refers to metal cutting of the type accomplished by oxy-fuel torches. The operation is normally used for cutoff or rough shaping of plates or bars. These plasmas develop temperatures that approach 60,000°F (33,000°C). Such a torch can be used to replace rough machining operations such as turning and planing. Observe Figure 24.10B, which shows the plasma working on a turned diameter. Although

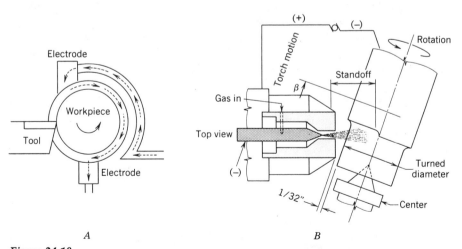

Figure 24.10
A, Elevated temperature machining using radiofrequency heating. B, Plasma arc machining using torch.

it is effective in cutting all metals regardless of metal hardness, there is a resulting rough surface and possible surface damage resulting from oxidation and overheating. Metal removal rates of 7 in.^{3}/min (115 cm^3/min) have been reported.

Cold-Temperature Machining

At the opposite extreme from elevated-temperature machining is the method of keeping metals at very low temperatures or *cold-temperature machining,* which reduces tool tip temperatures. Ways to accomplish this include surrounding the area with a cold mist about $-117°F$ ($-83°C$), using a dry-ice coolant, or deep freezing the part itself.

Temperatures at the tool–work interface may sometimes exceed 2160°F (1180°C) when machining tough high-alloy material. Problems arising from this heat include galling, seizing, work hardening, low-temperature oxidation, and short tool life. Conventional practice reduces the surface speed to prolong tool life, but the resulting reduced output has spurred the search for a better way. With parts mist-cooled, tool life is improved for some alloys from 100% to 300% over conventional machining.

CHEMICAL ENERGY

Chemical Milling. Chemical milling or *chem milling* is a controlled etching process in which metal is removed to produce complex patterns, lightweight parts, tapered-thickness sheets, and integrally stiffened structures. It is the adaptation of an old process to one that can successfully remove metal as in a machining process. The development of this process started in the aircraft industry where many complicated machining problems exist in the removal of excess metal to reduce weight. Although most chem milling has been done with aluminum alloys, any metal for which an etching solution is available can be processed in this manner. Observe Figure 24.11.

The process, which is relatively simple, consists first of thoroughly cleaning the sheet or part to be etched. It is then prepared for the etching process by masking those areas not to be affected with a chemically resistant coating. If the entire area is to be reduced, this is unnecessary. The part is then submerged in a *hot alkaline solution,* where metal in the unprotected area is eroded. The amount of metal removed depends mainly on the

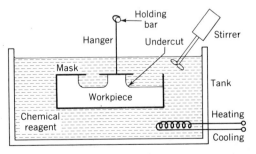

Figure 24.11
Chemical milling.

amount of time the part is in the hot solution. Finally, the part is neutralized and rinsed, and the masking material is removed.

The success of chemical milling can be controlled by suitable masking employing organic coatings that are capable of resisting the action of hot alkaline solutions. Masking is applied by either a dip or an airless spray technique, and two or four coats are required depending on the materials. Coatings are then air-cured or baked to increase the etchant resistance of the mask. Patterns or templates outlining the area where metal is to be removed are placed on the mask and marked along the edges with a knife. After the mask is scribed it is removed from the area where metal removal is to take place. Adhesive tapes applied to surfaces are excellent for chem milling panels to several thicknesses. Tapes are removed progressively after the etching begins if several panel thicknesses are required. Electroplating copper upon the area to be masked may also be used, but it is somewhat expensive. Photosensitive coating masking is frequently used for complex designs like that shown in Figure 24.12.

Where a uniform weight reduction is necessary, no masking is required. Similarly, sheet metal formed parts, which require a heavy sheet thickness for the operations involved, may be uniformly reduced in weight without masking. Another instance in which no masking is required is in the tapering of sheets (skins) for airplane use. Here the sheet to be tapered is gradually immersed or withdrawn from the solution. The operation of chem milling is uniform on all exposed areas; neither are internal stresses developed nor is there any change in the metal structure. Tests indicate that the physical properties of the metal are not impaired by this process if proper control and etchants are used. However, bend and fatigue strength can be reduced somewhat by improper etching, the result of which is rough surface and consequent notch effects. Normally, a surface roughness of 50 to 60 μin. (1300–1500 nm) is obtained, which is comparable to surfaces on many die castings and ground parts.

One unusual problem involved in chemical milling is that of undercutting. As the chemical dissolves the bottom of a hole, it also attacks the side of the hole underneath the resist. The rate at which this takes place is called the ***etch factor,*** expressed as the ratio of the side penetration over the total depth of the cut (Figure 24.13). The etch factor may vary from less than $1/3$ to more than 2, depending on the material and depth of penetration. It must be included in the design.

In comparison with machine milling, the following advantages are claimed for chemical milling:

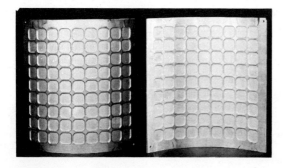

Figure 24.12
Wafflelike panels, an example of chemical milling.

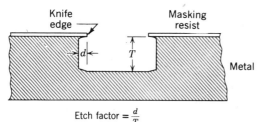

Etch factor = $\frac{d}{T}$

Figure 24.13
Undercutting in chemical milling.

1. Material can be removed uniformly from all surfaces exposed to the etching solution.
2. Material can be removed after parts are formed to shape.
3. Sheets and structural members can be uniformly tapered.
4. Highly skilled operators are not required.
5. Close tolerance can be maintained, and surface finish is good.
6. Operating costs are generally less than in machine milling. Also, equipment cost is less.

This process is limited in its applications for the following reasons:

1. Aluminum is the only metal being chem milled on a commercial scale.
2. On parts that can be machined while lying flat, conventional machining is more economical and accuracies greater than with chem milling.
3. Surface roughness is 50 μin. (1300 nm) higher.
4. The depth of cut is limited when masking is used; gases collect under the mask and cause uneven etching.
5. Masking techniques for certain conditions are expensive.
6. Gas generated in the process must be carried away.

Chemical Blanking. This form of chemical material removal is used to produce thin metallic parts by chemical action such as shown in Figure 24.14. A chemically resistant image of the part is first placed on the sheet, which is then exposed to chemical action by immersion or spraying.

Chemical metal removal does not require an electric current to carry the metal away in an *electrolyte*. It is known as *electroless etching*. In this type of etching the metal is converted chemically to a metallic salt, which is carried away as the etchant is replaced. In general there are two procedures for chemical removal. In one there is no attempt to control the location of the metal removed and the entire workpiece is affected. In the other, certain areas are covered with a protective coating to resist metal removal. The latter is generally spoken of as a selective metal removal process.

Chemical milling is a *selective metal-reducing* process for the purpose of weight reduction. Other selective processes include decorative etching, printed-circuit etching, and chemical piercing and blanking. The sequence of operations in the selective processes

Figure 24.14
Examples of thin parts produced by chemical blanking.

is essentially the same except for ***chem blanking,*** where a photographic-resistant image is placed on both sides of the metal blank.

The first step in chemical blanking is to prepare an accurate image of the part to be made. The metal should be chemically cleaned to eliminate all dirt, grease, and oxides. After cleaning the metal is immersed in a tank containing the photographic resist and then hung up to drain and dry. The metal coated with this photographic resist material, when exposed to ultraviolet light, will polymerize and remain on the panel. When developed, this *polymerized layer* acts as a barrier to the etching solution. For blanking, both sides of the metal panel must be exposed simultaneously so that the metal is removed on both surfaces. After printing, the panel is developed in a spray to remove the coating except in the areas of the workpiece that have been converted into etch-resistant images. Figure 24.15 shows the steps for blank preparation. In the etching machine shown in Figure 24.16, both sides of the metal are sprayed and then washed and dried. The photoresist may or may not be removed from the parts.

In producing ***printed circuit boards*** a copper cladding from 0.35 to 1.4 mm is etched away from an epoxy paper or epoxy fiberglass base material. Chemical resist applied either photographically or with silk screens protects the copper cladding where electrical connections in the circuit are desired.

The advantages of chemical blanking are as following:

1. Extremely thin metal can be worked without distortion. Most blanking is under $1/16$ in. (1.6 mm) in thickness.

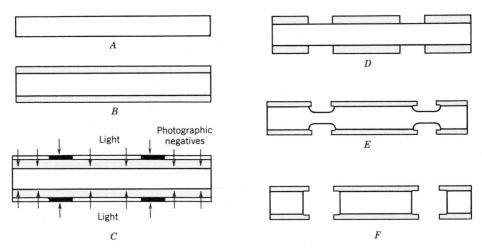

Figure 24.15
Steps in the photographic resist process. A, Clean metal. B, Metal coated with photoresist on both sides. C, Photographic negative and exposure. D, Resist developed and hardened and partially removed. E, Partially etched. F, Fully etched and separated.

2. No burrs are left on the edges.
3. Hard and brittle materials can be protected.
4. Setup and tooling costs are low.
5. Design change costs are low.

Limitations of this process are as follows:

Figure 24.16
Horizontal conveyorized spray etcher used in chemical blanking.

1. Skilled operators are required.
2. Etchant vapors are quite corrosive.
3. Maximum metal thickness is small.
4. Good photographic facilities are necessary.

Chemical Engraving. Chemical engraving is used to produce such parts as nameplates and other parts that customarily are produced on a pantograph engraving machine. It is similar to chemical blanking, except that the lettering or design is on one side only. Figures or letters may be either depressed or raised. If the letters are depressed they can be filled with paint.

This process can be used on most metals, including hard-to-work metals such as stainless steel. Fine detail is possible and the process is less expensive than other methods generally used on flat work.

ELECTROFORMING

Electroforming is one of the special processes for forming metals. Parts are produced by *electrolytic deposition* of metal upon a conductive removable mold or matrix. The mold establishes the size and surface smoothness of the finished product. Metal is supplied to the conductive mold from an *electrolytic solution,* in which a bar of pure metal acts as an anode for the plating current. The process differs from plating in that a solid shell is produced that is later separated from the form upon which it was deposited.

Electroforming is particularly valuable for fabricating thin-walled parts requiring a high order of accuracy, internal surface finish, and complicated internal forms that are difficult to core or machine. It may also be used to advantage in producing a small number of parts that would otherwise require expensive tooling.

Electroforming Process

The first step in production is to fabricate a *negative image* of the part. This is known as a matrix, mold, or pattern, which may be either permanent or expendable. Permanent molds can be used if there is sufficient draft to withdraw them without damage to the formed part—for example, in producing metal fountain pen caps, trumpet bells, and circular to rectangular transition-area waveguides. Such molds are generally machined from metal and are economical when many parts are to be made.

When it is impossible to use permanent patterns, *expendable* ones that are either chemically soluble or have a low melting temperature can be used. Soluble metals have the advantage of good internal finish and close tolerance. Also, they may often be made cheaply by die casting or plastic molding. Principal materials include aluminum, zinc alloys, and plastics. Low-melting materials like wax and lead–tin–bismuth alloys can be molded at low cost but are easily scratched. Both the fusible and the soluble molds have their principal use in complex internal forms that would be difficult or impossible to make by other processes.

Because some of the materials used for forms are nonconducting, they must first be coated with a metallic film. This can be done in a variety of ways, including brushing, spraying, and chemical reduction. Wax molds can be coated with graphite. The conductivity of the film must not be too low, and good electrical contacts are important.

After the forms are prepared they can be placed in the electrolytic solution and processed. Figure 24.17 shows the interior of a tank for nickel electroplating solutions. The tank is equipped with an automatic device to control the solution level and temperature. For rapid deposition the solution should be agitated; this is usually done by air. When sufficient time has elapsed to build up the required thickness, the part is removed from the bath, rinsed, and stripped from the mandrel.

Materials Used

All metal that can be used for plating can also be electroformed. Copper, nickel, iron, silver, zinc, lead, tin, cadmium, gold, aluminum, and a few others fall into this category. Copper, nickel, iron, and silver, used extensively in electroforming, possess properties such as good reproducibility, resistance to corrosion, electrical conductivity, good bearing surface, and adequate strength, which are required in most products. A *dendritic structure,* aligned normally to the conductive form surface (Figure 24.18) is common to all electroformed metals. Although the physical properties depend on the characteristics of the metal used, they are also influenced by the rate of deposition, plating temperature, and other bath variables. Dense structures can be obtained and, for some metals, the properties are changed materially by heat treatment.

Metallurgically, parts made by electroforming do not differ materially from parts made

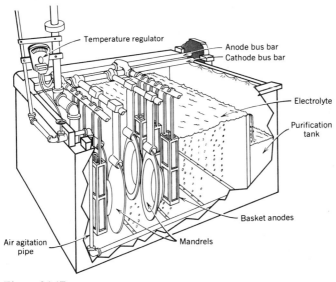

Figure 24.17
Tank interior for nickel electroplating solutions.

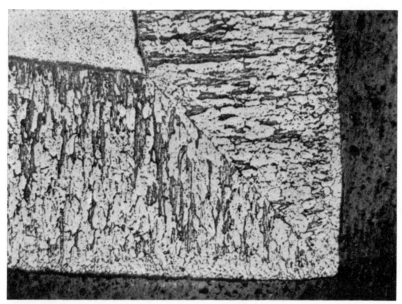

Figure 24.18
Cross section of copper deposit from acid sulfate bath. Sharp right angle shows weak plane formed by juncture of columnar crystals growing from sides of the angle. Magnification ×50.

by other processes. The importance of electroforming as a process is its ability to produce complex parts requiring intricate detail that are almost impossible to make by other processes.

Some advantages of electroforming are as follows:

1. Extreme dimensional accuracy can be held on surfaces next to the conducting form. Identical parts can be made with practically no dimensional variation.
2. Surface finishes of 8 μin. (200 nm) or less can be maintained.
3. Parts of extreme thinness can be made.
4. Laminated metals can be produced.
5. Extreme metal purity can be obtained.
6. Intricate internal or external surfaces difficult to form by other processes can be produced (see Figure 24.19).
7. Surfacing of parts to provide special physical or metallurgical properties is possible.

When compared with most normal processes, electroforming is not an economical method of fabrication, except for precision parts that are costly to produce by other processes. Limitations of electroforming are as follows:

1. Rate of production is slow.
2. Cost is high.

Figure 24.19
Intricate waveguide shape electroformed with nickel.

3. Accuracy of exterior surfaces cannot be controlled.
4. Process is confined to relatively thin products seldom exceeding $3/8$ in. (9.5 mm).
5. Selection of materials is limited.
6. Sharp internal angles should be avoided if design permits.

METAL SPRAYING

Several kinds of guns have been developed for spraying *molten metal* onto prepared base metals. The designs depend on the form of the metal when introduced to the gun, the material being sprayed, and the temperature that is required. With the melting-temperature range available, almost any metal, alloy, or ceramic material can be sprayed. Materials are supplied as rods, as wire, or in powder form.

Metallizing

Many metals in wire form are sprayed from a gun of the type shown in Figure 24.20. The wire is drawn through the gun and nozzle by a pair of rollers, melted by an oxyacetylene flame, and then blown by compressed air to the prepared surface. Although any metal that can be drawn into wire can be used in this gun, those most used are alloys such as steel, bronze, aluminum, and nickel.

Because the bond between the sprayed metal and the parent metal is entirely mechanical, it is important that the surface of the metal to be ***metallized*** be properly prepared before spraying. The usual method of cleaning and preparing the surface is by blasting with sharp silica sand or angular steel grit. Cylindrical objects may be prepared by rough-machining the surface. Either of these methods roughens the surface and provides the

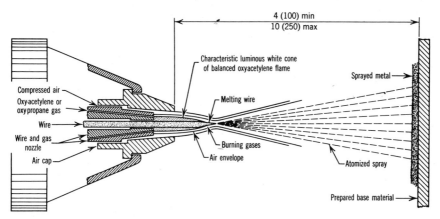

Figure 24.20
Nozzle cross section of metal spray gun.

necessary *interlocking surfaces* or keys to make the plastic metal adhere. The molten metal is blown with considerable force against the surface, causing it to flatten out and interlock with surface irregularities and adjacent metal particles. The sprayed metal itself provides a suitable surface for successive coatings and permits building up a layer of considerable thickness.

The change in the physical properties of metal applied in this manner is an increase in porosity and a corresponding decrease in the tensile strength of the material. The reason for this is that the bond is mechanical and not fusile as it is for welding. The compressive strength is high and there is some increase in hardness.

Metal Powder Spraying

A process, sometimes referred to as **thermospray,** utilizes metal and other materials in a powder form. These are placed in a small container on top of the gun (Figure 24.21)

Figure 24.21
Thermospray gun for spraying powdered metals and ceramics.

646 SPECIAL PROCESSES AND ELECTRONIC FABRICATION

and fed by gravity to the gas mixture. Here they are carried to the nozzle and melted almost instantly by an oxyacetylene or hydrogen gas flame. No compressed air is required, as there is sufficient force from the gas flame to carry the atomized metal at high speed to the surface being sprayed.

Powder materials developed for this process are designed to provide unusual surface properties such as a thermal barrier, resistance to oxidation or corrosion, high hardness, and extreme wearing ability. Materials that spray successfully include stainless steel, bronze, tungsten carbide, and various alloys.

Plasma Flame Spraying

A cross section of a plasma flame spray gun is shown in Figure 24.22. A gas (nitrogen, argon, or hydrogen) is passed through an electric arc, causing it to become ionized and raised to temperatures in excess of 30,000°F (17,000°C). Material to be sprayed is introduced in this gas stream, melted, and transported to the object being coated. The high-velocity gas stream, as it leaves the nozzle, is known as a *plasma jet,* a stream of ionized conducting gas.

The applications of *plasma flame spraying* are similar to those of powder metal spraying. Because of the extremely high temperatures possible, the *plasma torch* is

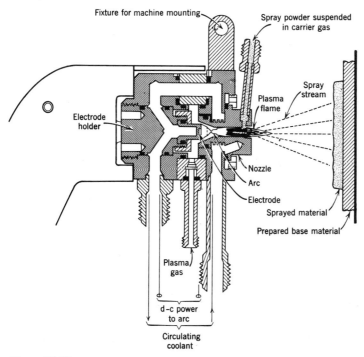

Figure 24.22
Plasma flame spray gun.

particularly useful in spraying high-temperature metals and refractory ceramics. A few of the materials that may be applied include tungsten, zirconium oxide, cobalt, chromium, and aluminum oxide.

The success of metal spraying is largely a result of the surface properties obtainable and how rapidly the metal can be applied. There is no distortion in the parts being surfaced, nor do internal stresses develop. Practically any metal can be applied to any other metal surface and even to other base surfaces such as wood and glass.

METALLIC COATINGS

With few exceptions, any marketable product of metal must be surface finished. Although the primary purpose of a coating or finish may be to improve the appearance and sales value of the item, coatings are used to give permanent resistance to destructive influences resulting from wear, electrolytic decomposition, and contact with weather or corrosive atmospheres.

Before metals can be coated, the surface must be properly prepared for good adhesion. Parts can be cleaned by various methods depending on material, size, surface peculiarities, and the coating to be applied. The basic methods are mechanical, such as blasting and tumbling, or chemical processes such as alkaline, acid, or organic agents and electrolytic cleaning.

In general the coating process is the application of a finite thickness of some material over the metal or is the transformation of the surface by chemical or electrical means to an oxide of the original metal. A discussion of some of the important processes follows.

Electroplating

Electroplating has long served as a means of applying decorative and protective coatings to metals. Most metals can be electroplated, but the common metals deposited in this way are nickel, chromium, cadmium, copper, silver, zinc, gold, and tin. In commercial plating the object to be plated is placed in a tank containing a suitable electrolyte. The *anode* consists of a plate of pure metal. The object to be plated is the *cathode*. The tank contains a solution of salts of the metal to be applied. A direct current having a density of 6 to 24 V is required for the plating operation. When the current is flowing, metal from the anode replenishes the electrolyte solution while ions of the dissolved metal are deposited on the workpiece in a solid state. The properties of the plated material and rate of deposition depend on current density, temperature of electrolyte, condition of surface, and properties of workpiece material.

Chrome Plating

For wear and abrasive resistance the outstanding metal for plating metallic surfaces is chromium. Coatings are seldom less than 0.002 in. (0.05 mm) thick and may be considerably more. Any measure of hardness or abrasive resistance is to some extent a function of the metal on which it is plated as well as of the chromium deposit itself.

The electrolytic process consists in passing an electric current from an anode to a cathode (the cathode being the object on which the metal is deposited) through a suitable chromium-carrying electrolytic solution in the presence of a catalyst. The catalyst does not enter into the electrochemical decomposition. A solution of chromic acid with a high degree of saturation is used as the *electrolyte*. The surfaces must be thoroughly polished and cleaned before operations start. Because the rate of deposition is fairly slow, the work must remain in the tanks several hours for heavy plating.

Chromium has proved satisfactory for wear-resisting parts because its extreme hardness exceeds most other commercial metals. According to the Brinell scale the hardness of plated chromium ranges from 500 to 900. This wide variation is a result not of the metal but of the methods and equipment used.

Galvanizing

Galvanizing is a zinc coating used extensively for protecting low-carbon steel from atmospheric deterioration. It offers a low-cost coating that has reasonable appearance and good wearing properties. An improved appearance known as the spangle effect can be produced by small additions of tin and aluminum. Zinc baths are usually maintained at about 850°F (450°C). Rolls, agitators, and metal brooms are used to remove the excess zinc from the product. Continuous and automatic processes are used for sheet and wire coating. Zinc coatings may also be applied by spraying molten zinc on steel, by sheradizing, which is the tumbling of the product in zinc dust at elevated temperatures, and by electroplating. The zinc dust must be rigidly controlled through the use of Hydro Sonic cleaners or scrubbers because of the environmental problems it creates.

Galvanized steel is familiar as seen in highway guard rails, light poles, transmission towers, and the 2-gal pail. A coating of 0.004 to 0.008 in. (0.10–0.20 mm) protects the steel from corrosive attack in atmospheres such as salt air. This protection continues even after small areas of base metal have been exposed, because the *galvanic action* of the coat sacrifices itself as zinc is more electrochemically active than steel. The comparison of tin and zinc coatings is shown in Figure 24.23.

Tin Coating

Tin coatings are often applied to sheet steel to be used for food containers, tin can manufacturers using approximately 90% of the tin produced. Although many tin coatings

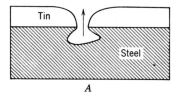
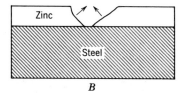

Figure 24.23
Sacrificial action of zinc continues to protect base metal after the area is exposed.

are now applied by *electrotinning,* a process where parts are immersed in an electrolyte and a current passed from the electrode to the work, the hot-dip method is still used considerably. Tin can be applied easily without affecting the base metal by dipping at temperatures of approximately 600°F (300°C). In most cases the tin coating is about 0.0001 in. (0.003 mm) thick as compared to only about 0.00003 in. (0.0008 mm) in electrotinned sheets. Porosity is greater in plated tin coatings, and when the containers are used for food, a lacquer seal is necessary.

Other Plating Metals

Copper, often used as an undercoat for subsequent nickel plating, provides good adhesion of the metal and improved appearance. It is seldom used alone as a plating except for stopoffs in selective carburizing and for electroforming heavy deposits. Nickel plating is most popular for protecting steel or brass from corrosion and for presenting a bright appearance. Lead has only limited commercial use, primarily as a protective coating against certain acids. For nonferrous articles used in food handling, silver plating is widely used.

Parkerizing

Parkerizing is a process for making a thin phosphate coating on steel to act as a base or primer for enamels and paints. In this process the steel is dipped in a 190°F (90°C) solution of manganese dihydrogen phosphate for about 45 min. Bluing is a process of dipping steel or iron in a 600°F (300°C) molten bath of nitrate of potash (saltpeter) for about 1 to 15 min. There are many salts that can be used to color brass and steel by dipping at elevated temperatures, but most of these have limited application and differing degrees of permanence.

Anodizing

Anodizing is an oxidation process developed for aluminum. An electrolyte of sulfuric, oxalic, or chromic acid is employed with the part to be anodized as the anode. Because the coating is produced entirely by oxidation and not by plating, the oxide coating is a permanent and integral part of the original base material. Although the coating is hard, it is porous, which is an advantage from a decorative point of view. The oxide coating enables organic coatings and dyes to be successfully applied to the surface of aluminum. Colored aluminum tumblers and pitchers are examples of this process. Magnesium is anodized in a somewhat similar manner.

Calorizing

Calorizing is a process designed to protect steel from oxidation at high temperatures. Aluminum is diffused into the metal surface at elevated temperatures, forming a film of aluminum oxide that protects the underlying metal from oxidation. The process is used for treating parts for furnaces, oil refineries, dryers, and kilns.

Hard Surfacing

Hard surfacing is the application to a wearing surface of some metal or treatment that renders the surface highly resistant to abrasion. Such processes vary in technique. Some apply a hard surface coating by fusion welding; in others no material is added and the surface metal is changed by heat treatment or by contact with other materials.

The several properties required of surfaces subjected to severe wearing conditions are hardness, abrasion resistance, and impact resistance. Hardness is easily determined by several known methods, and an accurate comparison of metals for this property can be obtained readily. Tests for wear or abrasion resistance have not been standardized, and it is difficult to obtain meaningful results. In general, wear testing must simulate the service conditions for each type of hard-facing material. The statement that "the wear resistance of a material is a function of the method by which it is measured" has been confirmed by both practical experience and research. All factors considered, hardness is probably the best criterion of wear resistance. Ability to withstand wear and abrasion usually increases with the hardness of the metal.

The classification of the various processes used for obtaining a hard surface does not point out that there is a great difference in the hardness that can be obtained, nor does it include heat-treating methods that are used to produce a hard surface or interior.

METHODS OF PRODUCING HARD SURFACES

   I. Heat treatment
      A. Carburizing—heating in contact with solids, liquids, or gases containing carbon
      B. Special-case-hardening processes—contact with gases or liquids containing cyanide, nitrogen, carbon
      C. Induction hardening—electric heating and rapid quenching
      D. Flame hardening—heating with torch and rapid quenching
  II. Metal spraying—with wire or powders
 III. Metal plating—electrolytic deposit of chromium, cobalt, and tungsten
  IV. Welding processes
      A. Employing high-carbon and steel alloys
      B. Employing nonferrous alloys—chiefly chromium, cobalt, and tungsten
   V. Employing sintered carbides as inserts or as screened sizes
  VI. Casting or spinning process—chiefly nickel boron

When thick coatings of hard materials are required, it is necessary to use some form of welding. The *hard-facing materials* used as electrodes or filler rods are classified roughly as "overlay" and "diamond substitute" types. The overlay materials include such metals as high-carbon steel, ferrous alloys of chromium and manganese, and numerous nonferrous alloys containing principally cobalt, manganese, and tungsten. The hardness of these materials varies considerably, ranging from Rockwell C40 to C70. The "diamond substitutes," materials that are among the hardest available, include tungsten, boron, tantalum, carbides, and chromium boride. On the Mohs' scale they fall between 8.5 and 9.5.

High-carbon welding rod with a carbon content ranging from 0.9% to 1.1% is the most economical hard-facing material to apply from the standpoint of initial cost. Such rods form a tough surface of moderate hardness ranging from Rockwell C30 to C45. Increased hardness and wear resistance can be obtained by alloying steel with such elements as nickel, manganese, molybdenum, and chromium. The limit of hardness for such coatings is around Rockwell C55 and, because many of the alloys result in austenitic deposits, their hardness can be increased by cold working. Corrosion resistance of most of these materials is good, as well as resistance to impact, and no heat treatment is required after application.

In the nonferrous group are included all rods that are made up of elements other than iron, although small percentages of iron may be present. The principal elements in this group are tungsten, chromium, molybdenum, and cobalt. The average room temperature hardness of this group is about the same as that of the ferrous alloy group. A high percentage of this hardness is retained while the rods are at red heat, which adds greatly to their wear-resisting power. In severe abrasive work considerable heat is developed by friction that acts on the minute areas of particles in contact. The effect of this heat is to soften the metal on these areas and cause them to wear away. If the metal in contact can retain a hardness at a relatively high temperature, it has a much greater resistance to wear than metals that do not have this property. In such cases the initial hardness is not a true criterion of the wear-resisting ability of the metal.

Both the electric arc and oxyacetylene processes can be used in applying this material, the latter being preferred. Better control of the deposit is obtained and there is less dilution of the rod with the parent metal. There is also minimum loss of the expensive rod material by *volatilization and spattering*.

The so-called diamond substitutes constitute the hardest materials that are available for hard surfacing. These materials, generally spoken of as *cemented carbides,* include tungsten carbide, tantalum carbide, titanium carbide, boron carbide, and chromium boride, or a combination of these and other carbides with a suitable cementing agent. In tungsten carbide, which is one of the most common of the group, the usual analysis by percentage is tungsten 81.4%, cobalt 12.7%, carbon 5.3%, and iron 0.6%. The cobalt serves as a binder and adds to the ductility of the carbide. It may vary from 5% to 13%. Tantalum carbide is 87% TaC with 13% of some binder. Usually the binder is either a combination of molybdenum and iron or one of tungsten carbide and cobalt. Boron carbide contains about 78.2% boron and 21% carbon with a trace of silicon and iron. It is usually known by the symbol B_4C. Many similar carbide materials, the compositions of which are not generally known, are manufactured under special trade names.

Carbide material cannot be applied like other hard-surfacing materials because of its high melting temperature, and thus it is furnished either in the form of small inserts or in screen sizes. Inserts can be applied by a brazing or sweating-on process or placed in melted or puddled metal and then surrounded by metal from a steel or hard-surfacing welding rod. Screened sizes of crushed carbide particles can be applied conveniently by putting the particles in steel tubes. The steel sheath melts like an ordinary welding rod and fuses to the metal. The carbide particles do not melt but are distributed through the molten metal and are held fast when the metal cools. Screen sizes can also be applied by mixing the particles with a suitable binder and casting them into rods. These rods can be used like other hard-surfacing welding rods. Carbide materials in powder form can also be sprayed with a plasma flame gun.

All these materials have hardnesses approaching that of a diamond, and on the Mohs' scale they range from 9 to 9.5. This hardness is maintained to a large extent at a red heat. Because of their extreme hardness and brittleness, diamond substitutes do not have a high strength rating and are not suitable where shock and impact conditions exist. This difficulty is partially eliminated by the elements being properly supported with a tough binding material. Another characteristic of these surfacing materials is that they do not respond to heat treatment or cold working and retain their initial hardness under all conditions. They are not suitable for casting, although a few hard materials, principally boron alloys with an iron base, can be processed in this manner.

ELECTRONIC FABRICATION

By definition an *integrated circuit* is a group of inseparably connected circuit elements fabricated in place within a substrate. A *substrate* is a waferlike piece of insulation material that may serve as a physical support or base and thermal sink for a printed pattern of circuitry. An integrated circuit is basically a single functional block that contains many individual devices (transistors, resistors, capacitors, etc.), which is known as a chip. Most often the substrate or base of the chip is made of *silicon*.

The structure of the integrated circuit is complex in the topography of its surface and in its internal composition. Each element of this device has an intricate *three-dimensional* architecture that must be reproduced identically in every circuit. The structure is composed of layers, each of which is a detailed pattern. Some of the layers lie within the silicon wafer, and others are stacked on top. The manufacturing process consists in forming this sequence of layers precisely in accord with the plan of the circuit designer.

A large-scale integrated circuit contains tens of thousands of elements, yet each element is so small that the complete circuit is typically less than a quarter of an inch on a side. But production of these circuits is to fabricate them many at a time on a larger **silicon wafer** 3 or 4 in. in diameter. A wafer is passed through many stages where a complete microelectronic circuit is composed on this substrate, and these circuits are separated into individual dice or chips. These circuits or chips are "packaged" or fastened to a metal stamping. Fine wire leads are connected from the bonding pads to the electrodes of the package, and a plastic cover is molded around each die. The units are separated from the metal strip and later inserted into the printed circuit board individually. This is discussed later.

Fabrication of the Wafer

Raw *silicon* is first reduced from its oxide, which is the main component of raw sand. A series of chemical steps are taken to purify it until the purity level reaches 99.999999%.

A charge of pure silicon is placed in a crucible and brought to the melting point of silicon, 2588°F (1420°C). An inert gas prevents the addition of unwanted impurities at this point. But desired impurities known as *dopants* are added to the silicon to produce a specific type of conductivity characterized by either *positive (p type)* charged *carriers* or *negative (n type)* ones.

In the process of **wafer shaping,** a large single crystal is grown from the melt by inserting a perfect single crystal "seed" and slowly withdrawing it. Single crystals 3 to 4 in. in diameter and several feet long can be pulled from the melt. The uneven surface of a crystal as it is grown is ground to produce a cylinder of standard diameter, typically 3.94 in. (100 mm). The cylinder is cut into *wafers* with a thin high-speed diamond saw. The wafers are ground on both sides and then polished on one side. The final wafer is about 0.002 in. (0.5 mm) thick. These final steps are carried out in an absolutely *clean environment*. There can be no defects, polishing scratches, or chemical changes on the final surface.

Silicon has a dominant role as the material for microelectronic devices because its oxide, *silicon dioxide,* forms on the surface when heated in the presence of oxygen or water vapor. The film serves as an insulator. It can be used as a mask for the selective introduction of dopants. This layer of silicon dioxide (SiO_2), about 0.75 μm thick, is formed on the chip to prevent such materials as arsenic, antimony, and boron from diffusing into the silicon chip. The thickness of silicon dioxide is gassed onto the wafers as the wafers are loaded in boats or fixtures that are inserted into ovens at about 2000°F (1100°C). The silicon dioxide thickness of 0.004 in. (0.1 mm) will grow in 1 h at a temperature of 1920°F (1050°C) in an atmosphere of pure oxygen. Several hundred wafers can be oxidized simultaneously in the fixture.

It must be emphasized that the construction of the silicon *substrate* is a complex process. The silicon chip is not produced by machining or other metal working processes but is actually *grown* through a chemical process. It must also be noted that all of the *diffusion* and growing processes that are to be described are actual formings of resistors, transistors, capacitors, and so forth. The description of the fabrication of a semiconductor microelectronic circuit is a highly simplified version of an actual process.

Photolithography in the Fabrication of Microelectron Circuits

The electronic circuits of old have been replaced by chips of semiconductor material with all the electronic elements closely interconnected in tiny, complex circuits. One of the technologies responsible for this change is the process of *photolithography,* by which a microscopic pattern can be transferred from a *photomask* to a material layer in an actual circuit on a silicon wafer.

The most basic photolithographic process involves etching the pattern into the silicon dioxide using contact photolithography. Once a layer of silicon dioxide is produced on the silicon wafer, the wafer is covered by a material called **photoresist.** This is done by applying a drop of the photoresist to the wafer and then spinning the wafer rapidly until a thin film of material is affected by exposure to ultraviolet radiation. Note these steps on Figure 24.24.

The resists are applied to a thickness of about 2 μm, usually by spinning, and then

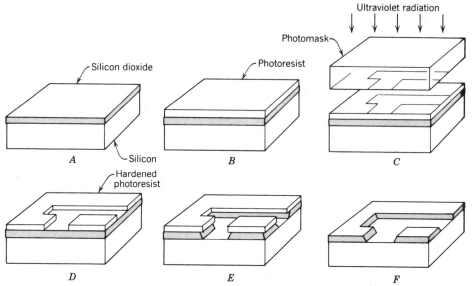

Figure 24.24
A, Oxidized layer on wafer of pure silicon. *B*, Coated with layer of light-sensitive material called photoresist. **C**, Exposed to ultraviolet light through photomask. *D*, Exposure renders the photoresist insoluble in a developer solution and a pattern of photoresist is left wherever mask is opaque. *E*, Wafer immersed in hydrofluoric acid solution, which attacks silicon dioxide only. *F*, Photoresist pattern removed by another chemical treatment.

prebaking. In the next operation where the chip and photographic plate are used together, cleanliness is extremely important. A vacuum printer is used to prevent contamination. The photographic plate then serves as a stencil. The wafer is exposed to a carbon arc light source for several minutes. After exposure the resist is developed and then washed in deionized water to *remove the unpolymerized material*. The remaining resist is then baked at 300°F (150°C) for 10 min. To remove the unprotected oxide layer, the chip is etched with hydrofluoric acid.

The mask contains the *opaque printed pattern*. The mask may either be brought into contact with the wafer or it may be held slightly above the wafer. The selection of the technique is based upon conditions of mask life and sufficient resolution. The structure is now flooded with ultraviolet radiation.

Under the opaque areas of the mask the *photoresist is unaffected*. However, in the transparent areas the photoresist becomes insoluble in the developer solution. Thus when the structure is washed in the developer solution, those areas under the opaque pattern of the mask are removed while the rest of the structure is unaffected.

Photoresists are of two types: negative resists, where polymerization is caused by exposure to ultraviolet light, and positive resists, where polymerization is degraded by exposure to light. A photosensitive resist is a lacquerlike material that, when exposed to light, is converted to a film that adds to the support and resists chemical action. Where it is not exposed to the light, it washes away in the developer. When the wafer is exposed to an etching solution of hydrofluoric acid, the areas of silicon dioxide exposed by the missing photoresist are removed.

The remaining photoresist is removed by another chemical solution, the desired pattern in the silicon dioxide is obtained, and the wafer is prepared again for the next masking process. By this method successive layers can be built up on the wafer.

The use of silicon as the base for these circuits is important to the photolithographic process. *Silicon dioxide can be etched away* in certain areas while leaving the silicon base unaffected, thus allowing the formation of the specified patterns on the wafer. Once all the patterns have been photolithographed onto the wafer and the circuits are completed, the elements are tested and then packaged. Throughout the entire fabrication of the circuit, extreme cleanliness must be maintained in the fabrication facility, because even minute dust particles can ruin a microelectronic circuit.

The process is repeated as many times as necessary, with the final step of the depositing of a thin layer of aluminum over the entire integrated circuit. The aluminum is selectively etched to leave the desired conductor pattern interconnecting the proper devices on the integrated circuit as well as to provide connecting pads that will later be wirebound to the external leads prior to packaging.

Photolithography is unquestionably one of the most important elements of microelectronic technology, allowing the fabrication of specified layers on a silicon wafer. However, photolithographic processes, like most processes in the field of microcircuit production, are continuously being changed and improved upon as circuit elements are designed smaller and more cost-effective production methods are sought. New technologies will undoubtedly allow the fabrication of even more complex microelectronic circuits in the future.

Layout of the Microcircuit

The microcircuit begins with conceptualization of the circuit design. Much of this work is done with the help of computers, which allow changes in the circuit design by just typing in the proper commands or by the use of light pens. When the design is completed, sets of plates called photomasks are produced from the circuit layout in the computer memory. Each photomask contains a pattern for a layer on the circuit. These patterns are transferred to the silicon wafer by the photolithographic process, thus building layer upon layer on the microelectronic circuit.

An early step in circuit fabrication is the *layout*. This phase is often called the **artwork**. To print the circuit, first a photographic mask must be made. The mask is a photographic reduction of a set of *cut-and-strip masters* typically 250 to 500 times larger than the actual size of the circuit. At each stage of this process, including the final stage when the entire circuit is completed, the layout is checked by means of detailed computer-drawn plots. Because the individual circuit elements can be as small as a few micrometers across, the checking plots must be magnified. Thus these plots are 500 times larger than the final size of the circuit. Eventually the photomask becomes a glass plate about 5 inches on a side (125 mm), and has a single circuit pattern *repeated* many times on its surface. This is often called a step and repeat. These plates are transferred to the wafer fabrication facility, where they will be used to produce the physical structure.

Another step in the construction of the mask is *photography*. High-precision camera equipment must be used to ensure a sharp, clear image on the reduction. Also, the camera and master must be clean and rigidly supported. A dust spot could render the final circuit nonoperational. A "clean room" where conditions can be controlled is necessary. Long exposures are often taken. Therefore, the camera must not be affected by any vibrations, which would prove disastrous.

The previous insights give a basic understanding of the construction of a silicon integrated circuit. The process is a complex one that deals with chemistry, physics, and engineering sciences. Consider some of the economics of the production processes up to this point:

1. *Batch processing*. One important attribute of silicon integrated devices is that numerous devices are processed as one unit up to the stage where leads are attached and they are packaged. This allows a high degree of process control and device similarity with relatively low part cost. A piece is one integrated chip at this point.

2. *Processing simplicity*. The number of processes that are involved in the fabrication of a silicon integrated device is small when compared to the total number of separate processes required to fabricate the components of the conventional equivalent circuit.

3. *Materials*. In the silicon integrated device a small number of different materials are used.

Component-Sequencing Machines

It has been estimated that more than 90% of commonly used printed-circuit board components can be *automatically* prepared for mounting on the board, sequenced, and inserted with special machines for this task.

The **electrical component sequencer** is a computer-controlled machine that selects electrical components, such as resistors, capacitors, diodes, and integrated circuits, and places them on a tape in a specific order. The tape is used by other automatic machines that place these components onto circuit boards at high rates of speed.

The sequencer has the capability of sequencing any series of standard components. The number of components in a certain sequence may be from 10 components to over a hundred. The lot or run may consist of any desired amount of these sequences. A "sequence" is one printed circuit board.

The operator initiates the computer program to operate the machine. The sequencer is fully automatic, requiring only periodic adjustments and corrections (replacing missing components, adjusting alignment, clearing scrap, etc.). Input reels have only one type of electrical component and their selection depends on the components for the printed circuit board. These input reels are previously sequenced. The output reel yields the arrangements of the components in successive order for the printed circuit boards.

The operating mechanisms, such as the heads that align and cut off component leads, are powered pneumatically. A machine may have the ability to stop feed automatically

Figure 24.25
Radial lead component sequencer-inserter machine.

when a component is missing from a sequence when indicated by an optical sensor. The computer also indicates which part is missing. Figure 24.25 is an example of a radial lead component sequencer with loading rates to 25,000 components per hour.

Insertion Machines

Component insertion machines automatically insert electrical components into printed circuit board holes. The components have two or more wires or leads, and they run axially or radially from the body of the component. The microelectronic components, with 4, 8, 16, and 32 leads, can also be inserted using these machines. Two-lead components are mounted in paper edges known as a tape, not to be confused with NC perforated tape.

Raw materials for the machine are a printed circuit board and a reel or tape that has the components. The components have been previously sequenced and placed on the tape. The roll of ordered components is mounted on the insertion machine and the components are fed onto the upper machine head.

Some components are axial shaped, such as resistors, and have two wires or leads extending from the ends of the resistor. The wire diameters are about 0.015 to 0.037 in. (0.94 mm) depending on the machine or design. The body length of the resistor may vary. The variable distance that is programmed is from center to center of the form leads.

The rolls have many *sequences*. One sequence may have sufficient numbers for the axial shape components as required by the printed circuit board. The machines are controlled by NC, which directs table positioning and machine head motion.

Printed circuit boards are loaded into a fixture mounted upon a movable table. The boards are located in the fixture using pins and are clamped. A variety of fixture-holding devices are possible with one or more boards or x–y fixtures.

The machine process is as follows. The table moves to a designated position over the printed circuit board as commanded by the NC control. The upper machine head cuts the component from the roll of tape and inserts it. A lower machine head raises and cuts the leads to length and bends them to either 90° or 45°. The heads retract and the table moves to a new position. The process of insertion continues.

Component-inserting machines in the circuit board industry also insert microchips that have been *packaged in a certain size case*—usually 0.3 in. or 0.6 in. wide (7.6–15 mm). The packages can have any number of leads but usually have 8, 16, or 32. The machine can be powered mechanically or hydraulically.

Setup of the machine requires that a computer program be loaded into memory. The computer controls the movement of the table in the x–y plane, movement of the package-loading head in the z direction, and the correct choice of package corresponding to its position on the circuit board.

Printed Circuit Board Stuffing

The word *stuffing* refers to manual insertion, clipping of the extra length of the lead, and folding over of the lead to hold the chip or electronic component against the underside of the printed circuit board. The work is the act of ***inserting the leads*** through the holes in the printed circuit board. This manual work may be done for low volume or occasional rework of printed circuit board assemblies. No machines are used in the process, but a few hand tools are available when needed.

The worker will start by adding each element to the printed circuit board one at a time. Every element is carefully checked for its identification number. After an element is in place on the board, it is then soldered to the board. One or more parts may be soldered to the board at the same time.

Once all of the parts are attached to the board, the worker will clean it by brushing it with alcohol. The board is then dried with an air hose. Next the printed circuit board is sent to the inspector, where it must pass an inspection. If any errors are found, they are fixed and it is reinspected. When the board has passed the inspection it is chemically sealed with hysol.

Wave Soldering Machines

A wave soldering machine applies a layer of molten solder to the underside of a printed circuit board. After solidification the solder electrically secures all of the inserted components into place. The machine consists of a conveyor that mechanically transports the circuit boards through the system. Machines vary but a typical process for the board is as follows: (1) flux applies a liquid layer to remove surface oxides and to give the filler metal the fluidity to wet the joint surfaces completely; (2) heating elements raise the temperature for proper solder adhesion; (3) air knives are used to remove excess flux; (4) a pass is made over the crest of the liquid solder wave at a set height; and finally (5) the washing cycle, which may vary from a series of washes, rinses, and air knives to a dishwasher for small lots.

Wave solder machines are able to solder different sizes of boards at variable rates of speed. Maximum and minimum rates are determined by the design of the machine and the application limitations. One or two operators may perform the loading and unloading.

Smaller machines used for smaller lot quantities make use of conveyor racks that can be adjusted to different board sizes. Larger machines have adjustable-width conveyors and are used for large lots; sometimes they run continuously for one or two shifts.

Figure 24.26 is a schematic of a unit that is intended for a large quantity of the same size boards. Two operators are required.

Single-Wire and Flat Cable Termination Machines

Wire terminal applicator machines apply strip terminals used in manufacturing electrical connections on cables, jumpers, and related assemblies. These terminals are found with the popular microcomputers, such as the RS232C standard. Most machines will accommodate a range of wire gauges and will specify a necessary crimp height for each size. Operation is simple, and specially trained operators are not required.

Growing industrial acceptance of *flat ribbon cable* has led to increased production. The demand has resulted from advantages of faster insulation stripping and mass termination on conductors offered by the connectors. In utilizing these machines, production

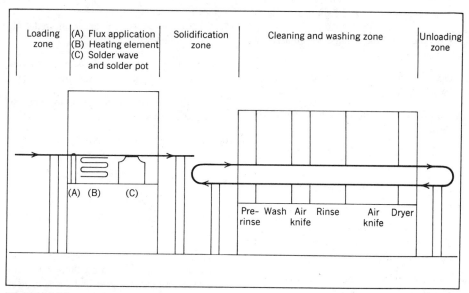

Figure 24.26
Automatic wave soldering machine, variable board sizes and speed rates.

QUESTIONS AND PROBLEMS

1. Give an explanation of the following terms:

   Ultrasonic machining
   Abrasive-jet machining
   Water jet machining
   Electrochemical machining
   Conductible materials
   Thermoelectric process
   Gas-assisted laser
   Plasma arc
   Cold-temperature machining
   Step and repeat
   Chem milling
   Etch factor
   Chem blanking
   Printed circuit manufacture
   Electrolytic solution
   Electroforming
   Metallizing
   Thermospray
   Plasma flame spraying
   Electroplating
   Anodizing
   Silicon wafer production
   Wafer shaping
   Photoresist
   Artwork
   Sequencing machines
   Lead attachment

2. What are the disadvantages of abrasive-jet machining?

3. Prepare a brief description of the electroforming process.

4. Name at least six products that are made by electroforming.

5. Describe the steps for producing microelectronic chips.

6. What are the advantages of the electroforming process?

7. How are printed circuit boards made? Consider only the etching and drilling of the boards, not their loading.

8. Describe the spark and where it occurs in electrical discharge machining.

9. Describe the process for making the wafer.

10. How does sheradizing differ from galvanizing?

11. In electrical discharge machining, why must the workpiece be an electrical conductor? Would the process work for machining ceramics?

12. What is the function of a dopant?

13. How are hard-surfacing materials such as carbides applied to oil well drilling tools?

14. Prepare the list of processes or steps for making a microelectronic chip.

15. What makes molten metal from a spray gun adhere to a revolving shaft?

16. How is photography used in the production of chips?

17. What is the safety problem connected with electron beam machining?

18. Contrast the functions of a component-sequencing machine and automatic stuffing.

19. Describe the ultrasonic machining process and state its advantages.

20. Describe in some detail how electroforming differs from the plating process.

21. Why is wave soldering superior to hand soldering printed circuit boards?

22. List the steps for wave soldering printed circuit boards.

23. Referring to Chapter 18, suggest how cold temperature applied at the appropriate places can affect the three principal zones of heat generation in single-point orthogonal cutting.

24. What is the difference between electrochemical machining and electrochemical grinding? Is metal removed in the same way in both cases?

25. Explain how material is removed in the electrochemical machining process.

26. What is the purpose of a chemical resistant?

27. Describe the chem milling process.

28. How does chemical blanking differ from chemical milling?

29. Can electroformed parts be plated? Why?

30. How does metal powder spraying differ from the method of applying sintered carbides?

31. Describe the procedure for chemically blanking a part.

32. Describe the process in which hollow steel tubes are filled with hard-surfacing material and then used as "welding rods."

33. Convert the following dimensions to micrometers and nanometers: 4 μin., 40 μin., 400 μin.

34. Convert 500 ft/s, 0.001 in.^{3}/min, and 1 in.^{3}/min to SI units.

35. In an electrochemical grinding process, 0.040 in. (1.02 mm) is to be removed. Calculate the thickness removed by abrasive action.

36. In problem 35, how long would it take to remove the 0.040 in. (1.02 mm) over a 1-in.^{2} (625 mm^2) area if the current were 200 A?

37. List applications for photoresists.

CASE STUDY

ELECTROCHEMICAL MACHINING

"Well, I assume all of you have read the sales order to increase the titanium compressor parts from two per day to eight per day."

Charles Kahng, president of Acme Tool and Die, a medium-size machine shop, was addressing Bill Wainwright, his production manager, and Lynne Eastwood, vice-president for manufacturing.

Bill Wainwright spoke up first. "I just don't see any way we can fill the orders. This part is a really tough machining job. It's about all we can do to produce the two per day we are making now."

Charles replied, "I know we'll need additional production capacity, so I asked Ms. Eastwood to look into the possibility of using electrochemical machining for this larger order. Lynne, what have you found out?"

"A 10,000-A, 12-V electrochemical machine can produce this part in about 45 min. Allowing another 15 min for replacing the finished part with a new blank, we can easily make one per hour or eight per day. The basic machine costs $35,000; we would also need an electrolyte system for about $65,000, and a power supply unit for about $40,000. Installation cost would be about $10,000. We would need to add 500 ft (150 m) of water line at $5/ft ($16.40/m), 400 ft (120 m) of 440-V, 550-A three-phase power line at $25/ft ($82/m), and 150 ft (45 m) of drainage line at $75/ft ($245/m). And, of course, before we can produce anything, we would have to make an electrochemical machining form of the part, which will cost an additional $10,000."

The president continued, "These are high initial costs, but once we get into production we should only have to spend about $25/h for labor, repair, and overhead and about another $5 per part for electrical power. Bill, how does this compare with the present system?"

"Well, we spend about $98 per part for labor, repair, and overhead right now. I don't think any other material or administrative costs would change if we went to the electrochemical machining process. What do you think, Lynne?"

"I think that's a reasonable assumption. Is there any other information you need, Charles?"

"No, I think that's everything. Thank you both for your help. I'll get back to you if I need more input."

What is the total initial investment for the ECM machine? What is the cost savings per part between the present process and ECM? Given this cost savings, how many parts will it take to recover the initial investment? Assuming 300 work days per year, how many years will it take to recover the investment?

CHAPTER 25
OPERATIONS PLANNING AND COST ESTIMATING

Many production processes are required to convert raw material into finished parts or products. A single manufacturing operation is insufficient, and several or many are required even for simple parts. A manufacturer who performs only one type of an operation is rare.

The purpose of this chapter is to describe the planning of the production process, which is customarily called operations planning. This plan is also known as an "op sheet," and it designates the sequence of the operations starting with raw material and concluding with the finished product. Once this planning stage is concluded, cost estimating is started. The purpose of a *cost estimate* is to find the cost of the manufacturing operations and ultimately the price. Cost estimating precedes the actual production, and thus consists of forecasting future costs.

Cost estimates will include the cost of labor, material, machine costs, overhead, and profit. The important objective of the manufacturing enterprise is to return a *profit* to the owners of the firm. Profit in turn is used to pay taxes, buy new equipment, and pay dividends to the company owners.

OPERATIONS PLANNING

The systematic determination of the methods to manufacture a product competitively and economically is called operations planning. It is the stage between design and production. The plan of manufacture considers functional requirements of the product, quantity, tools and equipment, and eventually the costs for manufacture. In a sense operations planning is a detailed specification and lists the operations, tools, and facilities.

Business Objectives

Operations planning is a responsibility of the manufacturing organization. A number of functional staff arrangements are possible. This process leads to the same output despite organizational differences. The following are business objectives for operations planning.

1. *New product manufacture.* A new design may not have been produced before or, alternatively, new manufacturing operations may be introduced for the product. Unless there is planning, the product introduction will be helter-skelter.

2. *Sales.* Opportunity for greater salability of an existing or new product can develop from different colors, materials, finish, or functional and nonfunctional features. The sales and marketing departments will provide advice on the direction to help manufacturing planning.

3. *Quantity.* Changes in quantity require different sequences, tools, and equipment. The planning differentiates for these fluctuations. If volume increases, the chance is for lower cost; in contrast, if volume decreases the cost should not increase out of reason. If quantity decreases too much, it is appropriate for the *op planner* to recommend that production may no longer be economical.

4. *Effective use of facilities.* Operations planning can often find alternate opportunities for the plant's production facilities to take up any slack that may develop. Seasonal products, which might be popular in the summer, need to be complemented by an alternative product for the winter season. For example, companies that produce sporting equipment may use the same facilities to produce tennis rackets and skis.

5. *Cost reduction.* Various opportunities become available if the company has an ongoing cost reduction effort. Suggestion plans, value analysis, redesign, and directed and systematic effort along these lines involve operations planning.

Production Analysis

Before operations, tools and equipment, and labor types can be identified, several policy questions need to be answered. The questions and answers are interrelated, so it may be necessary to backtrack several times in an iterative style. This stage is known as **production analysis**.

Design considerations are important. For example, a part may be cast or welded; or an eyelet may be rolled or stamped; or they may be purchased or made internally. These are common choices. These contrasts occur by the hundreds in a moderate-sized facility. But the eventual choice will have an important bearing on the production operations.

Material specifications have a significant effect on production. High- or low-grade carbon steels or alloy steels affect the production operation. There may be common characteristics among the various parts. Production might be organized on parts that are made from sheet or bar stock. By contrast material might be purchased on the basis that the company has machines that effectively make sheet metal products but are inefficient with bar stock.

Tooling may be a deciding factor in the analysis. Tooling can be of two types: *perishable or capital.* **Perishable tooling** is drill bits, small cutters, and the like. Capital tooling is jigs, fixtures, and ancillary support tooling that requires tool design and construction. Perishable tooling is low cost as compared to capital tooling. A company may make a process decision on the availability of jigs and fixtures and the equipment to use the tooling. Avoidance of new tooling cost is a popular policy because of its magnitude and the likelihood of the tools becoming obsolete.

METAL CUTTING ANALYSIS

Involved in a metal cutting operation cycle is load work (LW), advance tool (AT), machining, retract tool (RT), and unload work (UW). These elements are shown in Figure 25.1A and are labeled "one work cycle." After a number of parts are machined, tool maintenance is required, which includes removing the tool and replacing or regrinding the tool point and reinserting the tool ready for metal cutting. This is shown in Figure 25.1B. The mathematical model most frequently used in the study of machining estimating describes the cost of a single-point tool rough-turning operation.

Handling Time

Operation unit cost is composed of handling, machining, tool changing, and the tool cost. *Handling time* is the minutes to load and unload the workpiece from the machine. It can also include the time to advance and retract the tool from the cut and the occasional dimensional inspection of the part. It is independent of cutting speed and is a constant for a specified design and machine. Table 25.1 is an example of cycle handling data. Decimal minutes are adopted for cycle work rather than seconds or hours because it is more widely understood. We define

$$\text{handling cost} = C_o t_h$$

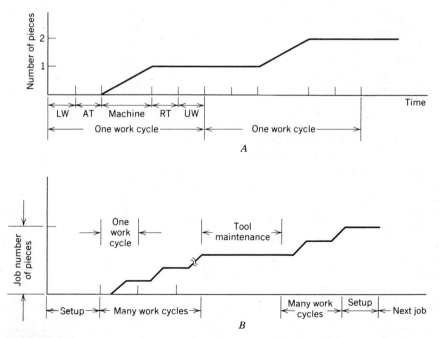

Figure 25.1
Operation items including setup, cycle, and tool maintenance.

Table 25.1 Handling Cycle Time Elements

Elements	Time (min)
All Machines	
Start and stop machine	0.08
Change speed of spindle	0.04
Engage spindle or feed	0.05
Air-clean part and fixture	0.06
Inspect dimension with micrometer	0.30
Brush chips	0.14
Turret Lathe	
Advance turret and feed stock	0.18
Turret advance and return	0.04
Turret return, index, and advance	0.08
Cross-slide advance and engage feed	0.07
Index square turret	0.04
Cross-slide advance, engage feed, and return	0.14
Place and remove oil guard	0.09
Milling Machine	
Pick up part, move, and place; remove and lay aside	
5 lb	0.13
10 lb	0.16
15 lb	0.19
Open and close vise	0.14
Seat with mallet	0.11
Wipe off parallels	0.26
Pry part out	0.06
Clamp, unclamp vise, $1/4$ turn	0.05
Quick-clamp collet	0.06
Clamp and unclamp hex nut	each 0.21
Numerically Controlled Turret Drill Press	
Pick up part, move, and place; pick up and lay aside	
To chuck:	
2.5 lb	0.10
5.0 lb	0.13
10.0 lb	0.16
Clamp and unclamp	
Vice, $1/4$ turn	0.05
Air cylinder	0.05
C-clamp	0.26
Thumb screw	0.06
Machine operation	
Change tool	0.06
Start control tape	0.02
Raise tool, position to next x–y location, advance tool to work	0.06/hole
Index turret	0.03/tool

where

$$C_o = \text{Direct labor wage, \$/min}$$

$$t_h = \text{Time for handling, min.}$$

Figure 25.2A is an example of handling cost plotted against cutting speed. C_o does not include overhead costs.

Machining time is the time that the tool is actually in the feed mode or cutting and removing chips. This has been discussed previously but it is again considered.

$$t_m = \frac{L}{fN} = \frac{L\pi D}{12Vf}$$

$$\text{Machining cost} = C_o t_m$$

where

t_m = Machining time, min

L = Length of cut for metal cutting, in. (mm)

D = Diameter, in. (mm)

V = Cutting speed, ft/min (mm/s)

f = Feed rate, ipr (mm/r)

N = Rotary cutting speed, rpm = $12V/\pi D$

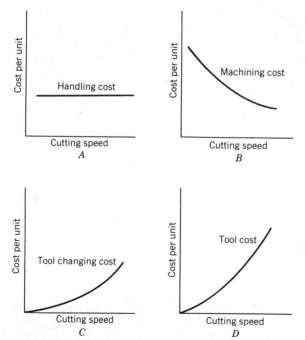

Figure 25.2
Graphic costs for four parts of machine turning economics.

The feed rate f is substantially less than rapid travel velocity of various machine elements. Each material has special turning and milling cutting speeds and feeds as determined by testing or experience. Values will be different for roughing or finish. A roughing pass will remove more material but does not satisfy dimensional and surface finish requirements. The material removal rate is in minutes per inch or per millimeter for drilling and tapping. Table 20.1 is an example of machining data. As cutting velocity increases, the unit cost decreases, and this is shown by Figure 25.2B.

The turning length, which is being machined in a lathe, is at least equal to the final drawing dimension and is usually greater because of additional stock for roughing or finishing. In a lathe facing element the diameter will vary from the center of the bar stock circle to the outside diameter (or the stock OD), and the facing length is at least equal to one-half of the diameter. The length of the cut includes safety length, approach, design length, and overtravel. This was discussed in Chapter 20.

The diameter D may be either the workpiece or the tool. When a lathe turning operation is visualized, the diameter is the largest unmachined bar stock dimension. In turning the periphery it is the maximum diameter resulting from the raw stock or previous element. For milling and drilling the diameter is the cutting tool diameter. A drill $1/2$ (12 mm) in diameter or a rotary milling cutter 1 in. (25 mm) in diameter are examples. The cutting speed velocity V has the dimensions of feet per minute (meters per second). Its value depends on many factors, and production time data will consider these effects. Table 20.1 shows a small sampling of speeds and feeds that will be used for the examples and problems.

Tool Life

Cutting tools become dull as they continue to machine. Once dull, they are either replaced by new tools or removed, reground, and reinserted in the toolholder. Empirical studies can relate tool life to cutting velocity for a specified tool and workpiece material. Two popular tool materials are high-speed steel (HSS) and tungsten carbide. Most studies of tool life are based on Taylor's tool life cutting speed equation. This was discussed in Chapter 18, but now is given more consideration.

$$VT^n = K$$

where

T = Average tool life, minutes per cutting edge

n, K = Empirical constants resulting from regression analysis and field studies, $0 < n \leq 1, K > 0$

The average tool life T can be found as

$$T = \frac{K^n}{V}$$

Table 25.2 provides a limited set of tool life data.

If a tool life equation is $VT^{0.16} = 400$, we can find either V or T given the other

Table 25.2 **Taylor Tool Life Parameters**

	High-speed steel		Tungsten carbide	
Material	K	n	K	n
Stainless steel	170	0.08	400	0.16
Medium-carbon steel	190	0.11	150	0.20
Gray cast iron	75	0.14	130	0.25

variable. If $V = 200$ ft/min, we can expect 76 min of machining before the tool must be indexed to a new corner.

Tool Changing Cost

The third cost is the tool changing cost per operation. Define it as

$$\text{Tool changing cost} = \frac{C_o t_c t_m}{T}$$

where t_c is the tool changing time in minutes.

The tool changing time t_c is the time to remove a worn-out tool, replace or index the tool, reset it for dimension and tolerance, and adjust for cutting. The time depends on whether the tool being changed is a disposable insert or a regrindable tool for which the tool must be removed and a new one reset. In lathe turning and milling there is the option of an indexable or regrindable tool. The drill is only reground. In Figure 25.2C we see the relationship of tool changing cost to cutting speed.

Define the following as

$$\text{Tool cost per operation} = \frac{C_t t_m}{T}$$

where C_t denotes the tool cost, dollars.

Tool Cost

Tool cost C_t depends on the tool being a disposable tungsten carbide insert or a regrindable tool for turning. For insert tooling, tool cost is a function of the insert price, and the number of cutting edges per insert. For regrindable tooling the tool cost is a function of original price, and total number of cutting edges. As the speed increases, the cost for the tool increases, as shown in Figure 25.2D. Table 25.3 provides tool costs and changing times.

The total cost per operation is composed of these four items. Machining cost is observed to decrease with increasing cutting speed while tool and tool changing costs increase. Handling costs are independent of cutting speed. Thus we can say that unit cost C_u is given as

$$C_u = \sum \left[C_o t_h + \frac{t_m}{T}(C_t + C_o t_c) + C_o t_m \right]$$

Table 25.3 Tool Changing or Indexing Time and Costs

Operation	Time and Costs
Time to index a turning type of carbide tool	2 min
Time to set a high-speed tool	4 min
Large milling tool replacement	10 min
Remove drill, regrind, and replace	3 min
Cost per tool cutting corner for turning, carbide	$3
Cost for high-speed steel tool point	$5
Cost per milling cutter, 6-in. carbide	$1500
Drill cost	$3

Upon substitution of t_m and T and after taking the derivative of this equation with respect to velocity and equating the derivative to zero, the minimum cost may be found as

$$V_{min} = \frac{K}{\left[\left(\frac{1}{n} - 1\right)\left(\frac{C_o t_c + C_t}{C_o}\right)\right]^n}$$

which gives the velocity for the unit cost of a rough-turning operation. In this development, we give no recognition to revenues that are produced by the machine. Consequently, V_{min} identifies the minimum velocity without revenue considerations.

Occasionally to avoid **bottleneck situations** there is a need to accelerate production at cutting speed greater than that recommended for minimum cost. In these expedited operations, we assume the tool cost to be negligible, or $C_t = 0$. If the costs in the basic model are not considered, the model gives the time to produce a workpiece, and we develop

$$T_u = t_h + t_m + \frac{t_c t_m}{T}$$

where T_u is minutes per unit. The production rate (units per minute) is the reciprocal of T_u. The equation that gives the cutting speed that corresponds to maximum production rate is

$$V_{max} = \frac{K}{\left[\left(\frac{1}{n} - 1\right) t_c\right]^n}$$

The tool life that corresponds to maximum production rate is given by

$$T_{max} = \left[\left(\frac{1}{n} - 1\right) t_c\right]$$

Example

Now consider an operation optimization problem of machining grade 430F stainless steel, 1.750-in. (44.45-mm) OD bar stock. The cutting length is 16.50 in. plus $1/32$ in. for approach, giving 16.35 in (419.8 mm). The turning operation will use a tungsten carbide, insertable and indexable eight-corner tool (about the size of a dime), which costs $3 per corner. Time to reset the tool is 2 min. Handling of the part is 0.16 min and the operator wage is $15.20/h. The Taylor tool life equation is $VT^{0.16} = 400$ (in U.S. customary units) for the tool and workpiece material. Feed of rough-turning element is 0.015 ipr (0.38 mm/r) for a depth of cut of 0.15 in. (3.8 mm) per side. These data are included in Tables 20.1, 25.1, 25.2, and 25.3. If the four items of the cost equation are plotted with several velocity values at the x variable, we have Figure 25.3, which shows the optimum at 200 ft/min (0.2 m/s).

Preparation of operations sheets is more than computing Figure 25.3. Often this analysis may not be possible, and operations sheets are created using other approaches. Use of prepared tables for *operation optimization* is a more common way to understand the total operation. But Figure 25.3 is a classic presentation, and it is important for a thorough understanding.

OPERATIONS SHEET PREPARATION

The operations sheet is fundamental to manufacturing planning. It is also called "route sheet," "traveler," or "planner." There are many styles, and each plant has its own form.

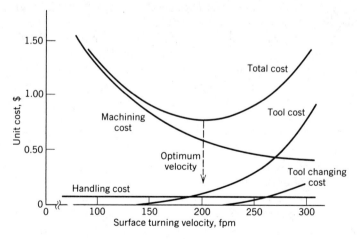

Figure 25.3
Study of the various effects of cutting speed upon handling, machining, tool, and tool changing cost to give optimum.

The purpose of the operations sheet is to select the machine, process, or bench that is necessary for converting raw material into product, provide a description of the operations and tools, and indicate the time required for the operation. The order of the operations is special, too, as this *sequence* indicates the various steps in the manufacturing conversion. The shop will use these instructions in making the part.

Information Required

Before operations planning can begin, one must have engineering drawings, marketing quantity, material specifications, a listing and specifications of the company's machines, processes, and benches. Job descriptions of the workers and their skills are necessary. The preparation of the ***operations sheet*** occurs at the same time as direct labor estimating. After the operations are listed, the "op sheet planner" will refer to company or published data that provide information for estimating the time to do the work designated by the operations sheet.

Each operations sheet has a title block indicating the material kind, part number, date, quantity, and operations planner. The part number and description are removed from the design and repeated on the title block. The engineer will indicate the lot quantity and material specification. Knowing the theoretical amount of material required by the design, the operations planner will add material to cover losses for scrap, waste, and shrinkage and multiply by the cost per pound (dollars per kilogram) rate of the material.

Sequence of the Operations

The sequence of the operation number and selection of the *machine, process,* or *bench* are made to manufacture the part. Consider Figure 25.4, illustrating a stainless-steel shaft that requires several operations. A horizontal turret lathe machine might be selected to face, turn, and cut off the part. The original material is a 12-ft bar of stainless steel, grade 430 Ferritic (430F). The operations sheet column headed "description of operations" is an abbreviated instruction to the shop that they will follow in making the part or subassembly. For operation 10 the instructions to the shop are listed on the form and correspond to Figure 25.5. Notice operations 20, 30, and 40. The vertical mill, horizontal mill, and a numerical control (NC) turret drill press will complete the selections for the equipment or stations.

The operations are numbered customarily as 10, 20, 30, and so on, the first time the operations are planned. Afterwards if an operation is found necessary between 20 and 30 it might be numbered as 25, and so on.

The entries for the description of operations are abbreviated, but with the drawing and knowledge of the machine, the operator is able to execute the requirements. Tools or gages can be listed. In Figure 25.5 the cut dimensions are listed as "face 0.015," meaning to remove 0.015 in. from the face of the bar stock. Notice Figure 25.6A, which shows the cutting tool facing the bar stock end. The other passes are indicated for operation 10, which corresponds to Figure 25.5. In a similar way the other dimensions are stated for the other machines. A collet fixture and a nesting vise are required for operations 20,

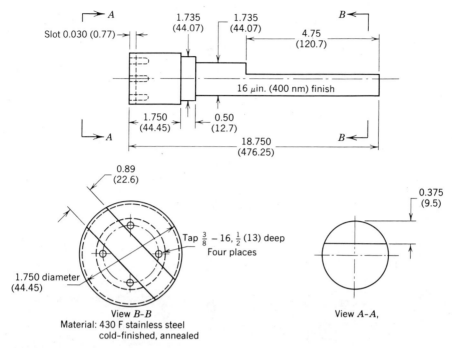

Figure 25.4
Typical part studied for cost estimating.

30, and 40. This will require that a tool designer later design tools to match the requirements of the op sheet.

Setup and Cycle

The columns in Figure 25.5 headed "setup hour," "cycle hour/100 units," and "unit estimate" are important. The engineer or planner will make selections of time to manufacture the part for the operations that are selected. Tables of estimating data or a computer data base will be available for this selection.

Setup includes work to prepare the machine, process, or bench for producing the parts or the cycle. Starting with the machine in a neutral condition, setup includes punch in and out, paperwork, obtaining tools, positioning unprocessed materials nearby, adjusting, and inspecting. It also includes return tooling, cleanup, and teardown of the machine ready for the next job. The setup does not include the time to make parts or perform the repetitive cycle. Setup estimating is necessary for job shops and companies whose parts or products have small- to moderate-quantity production. As production quantity increases, the effect of the setup value lessens its prorated unit importance, although its absolute value remains unchanged. Setup is measured in hours.

674   OPERATIONS PLANNING AND COST ESTIMATING

Part no. 4943806   Ordering quantity 1000
Part name Pinion   Lot requirement 200

Material 430F Stainless steel, 1.780 ± 0.003 in. cold-finished 12-ft bars = 1000 pieces
Unit material cost $22.47

Workstation	Operation no.	Description of operations (list tools and gauges)	Setup hour	Cycle hour/ 100 units	Unit estimate	Labor rate	Labor + overhead rate	Cost for labor + overhead
Turret lathe	10	Position Face 0.015 Turn rough 1.45 Turn rough 1.15 Turn finished 1.110 Turn 1.735 Cut off to length 18.750 (carbide tools)	3.2	10.067	0.117	18.35	1.70	3.65
Vertical mill	20	End mill 0.89 slot with 3/4 H.S.S. end mill (collet fixture)	1.8	7.850	0.088	19.65	1.85	3.20
Horizontal mill	30	Slab mill 4.75 × 3/8 (Nesting vise H.S.S. Tool)	1.3	1.500	0.022	19.65	1.80	0.78
NC turret drill press	40	Drill 5/8 holes – 4x Tap 3/8 - 16 (collet fixture)	0.66	5.245	0.056	17.40	2.15	2.10
					Unit material, labor, and overhead cost			32.20

Figure 25.5
Operation sheet for typical part, giving estimate.

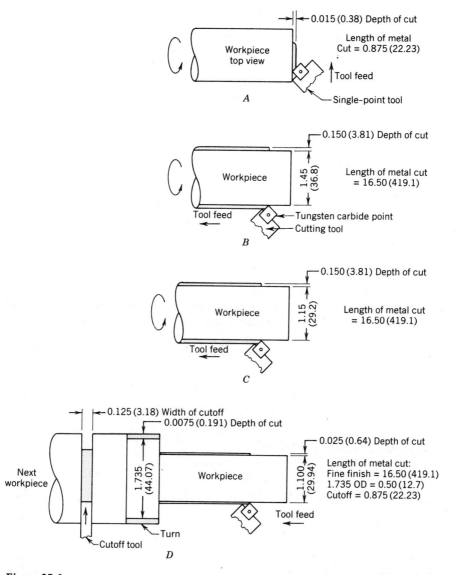

Figure 25.6
Machining cuts for operation 10 using turret lathe. A, Facing element. B, First pass, rough-turning element. C, Second pass, rough-turning element. D, Final turret lathe elements.

Cycle time *or run time* is the work needed to complete one unit after the setup work is concluded. It does not include any element involved in setup. Besides finding a value for the setup, the planner finds a unit estimate for the work from the listed elements, which have the dimension of minutes. These times include *allowances* in addition to the work time that take into account *personal* requirements, *fatigue* where work effort may be excessive because of job conditions and environment, and legitimate *delays* for operation-related interruptions. Because the allowances are included in the time for the described elements, and several or many operations, the allowed time is *fair*. The concept of fairness implies that a worker can perform the work throughout the day.

But the development of the cycle time is often done separately from the op sheet, and special forms or computer programs may do this work. When the work is separate, the answers are transferred to the op sheet and expressed in hours per 100 units. This dimension (hours per 100 units) is calculated from minutes per unit, which is the customary calculation for a cycle estimate.

The operation is broken down or detailed into elements that are described in estimating tables. These elements may be listed on a standardized company form, or marginal jottings on the operations sheet may suffice, or even scratch pad calculations may be followed. Computer-based estimating is possible. The purpose of the formal or informal elemental breakdown is identical: a *listing of elements* that will do the work is visualized and this listing is coordinated with the company data or manual.

$$T_u = \Sigma \text{ Operation elements from tables}$$

where T_u is the minutes per cycle.

The engineer will select the appropriate elements for the job from the tables. Inasmuch as the cycle elements are expressed in minutes, pieces per hour are found using

$$\text{Pieces per hour} = \frac{60}{T_u}$$

The operations sheet requires cycle hours per 100 units. This is found using

$$H_s = T_u \times \frac{100}{60}$$

where H_s is the cycle hours per 100 units.

The column headed "unit estimate" is a computation that is shown on the operations sheet, but the calculation is made using the setup, cycle hours per 100 units, and the lot quantity. This unit estimate quantity varies and is an important fact found on the operations sheet.

$$\text{Unit estimate} = (SU/N + H_s/\text{unit})$$

where SU is the setup hours for operation as recorded from data tables. The unit estimate includes a prorated share of the lot number N. H_s for the purposes of this equation is per unit.

Cost and Price from the Operations Sheet

The column headed "cost for labor and overhead" is found by multiplying the unit estimate by the "labor and rate," in dollars per hour, by the "labor plus overhead rate" for each operation. Overhead is a cost item that considers depreciation of machines, tool cost, space, power, heat, and other indirect costs. After multiplying by the labor and overhead rate, we then have the *machine hour cost* for that operation. When unit material cost is added along with machine hour costs for the operations, the full cost is found. From this point a variety of estimating and pricing practices become possible. A sales price is finally computed using the company's standard methods.

One popular pricing technique is called cost plus or *markup*. This sets price as proportional to cost or

$$P = C_t + R_m(C_t)$$

where

P = Unit price, $
C_t = Total cost of manufacturing, development, and sales, $
R_m = Markup rate on cost, decimal

The price of the pinion can be found. Assume that the company's markup rate is 18%, so the price = 1.18 (32.20) = \$38.00. The difference between price and cost is ***profit***, or \$5.80 for the pinion in Figure 25.4.

Metal Cutting Operations Sheet Problem

Suppose that an estimate is to be made for a stainless-steel shaft as shown by Figure 25.4. This part requires a turret lathe, drill press, milling machine, and various tools.

The estimator will examine special tables for each work station. For operation 10 the estimator will use Tables 25.1 and 25.4, and make selections of appropriate elements for doing the work. Table 25.5 will be formed for operation 10. The results of Table 25.5

Table 25.4 **Setup Elements Time**

Element	Time (h)
1. Punch in and out, study drawing	0.2
2. Turret lathe	
First tool	1.3
Each additional tool	0.3
Collet fixture	0.2
Chuck fixture	0.1
3. Milling machine	
Vise	1.1
Angle plate	1.4
Shoulder cut milling cutter	1.5
Slot cut milling cutter	1.6
Tight tolerance	0.5
4. Drill press	
Jig or fixture	0.14
Vise	0.05
NC turrets	
First turret	0.25
Additional turrets	each 0.07

Table 25.5 **Estimate for Operation 10**

Calculation of Machining Elements

Element	Dimension (in.)	Depth of cut (in.)	Length cut L_d (in.)	Safety stock (in.)	Length (in.)	Diameter (in.)	Velocity (ft/min)	Feed (ipr)	Time (min) t_m
Facing		0.015	0.875	1/32	0.906	1.750	350	0.015	0.08
Rough turn	1.45	0.15	16.5	1/32	16.53	1.750	350	0.015	1.44
Rough turn	1.15	0.15	16.5	1/32	16.53	1.45	350	0.015	1.20
Fine finish	1.110	0.025	16.5	1/32	16.53	1.15	350	0.007	2.03
Turn	1.735	0.0075	0.5	1/32	0.53	1.750	350	0.007	0.10
Cutoff		0.125	0.875	1/32	0.906	1.750	350	0.015	0.08
							Subtotal for machining:		4.93

Selection of Handling and Other Machine Cycle Data

Element	Time (min)
Start and stop machine	0.09
Advance turret and position raw stock	0.18
Place and remove oil guard	0.08
Speed change, assume 4 times, 4 × 0.04	0.16
Advance, index, and return turret, 6 times, 6 × 0.08	0.48
Inspect part with micrometer, irregular element, 0.30 × 1/5	0.06
Air-clean part	0.06
t_h = Subtotal handling and other machine data:	1.11
TOTAL CYCLE:	6.04

Setup Development

Element	Time (h)
Punch in and out, study drawing	0.2
First tool	1.3
Five additional tools	1.5
Collet fixture	0.2
TOTAL SETUP:	3.2

Operations Sheet Entry for Operation 10, Figure 25.5

Setup = 3.2
Hours per 100 units = 10.067

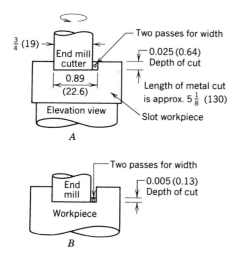

Figure 25.7
Vertical end mill operation on stainless-steel part. A, Rough pass. B, Finish pass.

as setup = 3.2 h and hours per 100 units = 10.067 are transferred to the op sheet as operation 10 in Figure 25.5.

Figures 25.7, 25.8, and 25.9 are intended for operations 20, 30, and 40 as required for the stainless-steel part. Tables 25.6, 25.7, and 25.8 correspond to these figures. (The setup hours and the cycle minutes are calculated and are shown on the tables.) The setup and hours per 100 units as determined on the tables are entered on the operations planning sheet are given by Figure 25.5.

This operations sheet can be altered to consider single parts, simple assemblies, or complicated products, but the approach remains the same. The preparation of the operations sheet is important for the finding of part operational costs. Notice that the part cost is the sum of the operational costs, and this fact allows us to concentrate on the important steps that are necessary for estimating operations. Some operations may not require setup. Flexible manufacturing systems, continuous production, or combined operations may not require setup time.

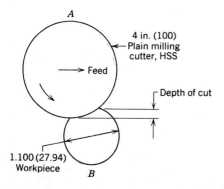

Figure 25.8
Horizontal mill operation on stainless-steel part. A, Cutter is 6 in. wide, 8 tooth, and one pass required. B, Workpiece has $^3/_8$ in. depth of cut. Length of metal cut = 1.05 in. Approach = $(^3/_8 (4 - ^3/_8))^{1/2} = 1.2$ in.

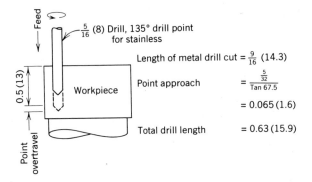

Figure 25.9
NC turret drill press operation on stainless-steel part.

MATERIAL ESTIMATING

Assumptions of a quantity, dimension of the material size and length, and specifications for the material are involved in estimating material costs. Various arrangements exist for delivery of material to the plant. In some cases protection is specified for special finished materials. Some materials are identified by chemical composition. The material specification and the quantity are the principal cost-affecting factors.

As an example of a material specification for a stainless-steel bar is as follows: "Bar, cold-rolled stainless steel; type 430F, 1 ³/₄ in. round OD × 12-ft lengths; unannealed, base packaging. FOB (free on board) service center for less than base quantity. FOB mill for base quantity."

Table 25.6 **Estimate for Operation 20**

	L	D_c	V	n_t	f_t	t_m (min)
A. Machining						
Rough	6.81	0.75	85	4	0.002	1.97
Finish	6.81	0.75	95	4	0.0015	2.35
					Subtotal:	4.31

Elements	t_n (min)
B. Handling	
Start and stop	0.03
Load and unload in collet	0.13
Engage machine	0.05
Change feeds and speeds, 2x	0.08
Blow chips and clean	0.06
Subtotal	0.40
TOTAL CYCLE:	4.71

C. Setup Time

                                                                                                          1.8

    0.2 + 1.6 =

D. Operation Sheet Entry for Operation 20, Figure 25.5
    Setup: 1.8 h
    Cycle: 7.850 h/100

Table 25.7 **Estimate for Operation 30**

	L	D_c	V	n_t	f_t	t_m (min)
A. Machining						
Finish	3.45	4	210	8	0.005	0.54
					Subtotal:	0.54

Elements						t_n (min)
B. Handling						
Start and stop						0.08
Pickup and aside						0.13
Clamp, unclamp vise						0.05
Engage machine feed						0.04
Blow chips and clean						0.06
					Subtotal:	0.36
					TOTAL CYCLE:	0.90

C. Setup Time
0.2 + 1.1 =   1.3

D. Operation Sheet Entry for Operation 30, Figure 25.5
Setup: 1.3 h
Cycle: 1.500 h/100

Table 25.8 **Estimate for Operation 40**

	L	D_c	f_{dt}	Lf_{dt}	Number of holes	t_m (min)
A. Machining						
Drilling	0.63	5/16	0.61	0.38	4	1.54
Tapping	0.5	3/8–16	0.33	0.17	4	0.66
					Subtotal:	2.20

Elements						t_n (min)
B. Handling						
Load part						0.13
Start machine						0.08
Start tape						0.02
Change tools, 2×						0.12
Raise tool, position to new location, advance × 8 × 0.06						0.48
Index turret, 2 × 0.03						0.06
Blow off chips						0.06
					Subtotal handling:	0.95
					TOTAL CYCLE:	3.15

C. Setup Time
0.2 + 0.14 + 0.25 + 0.07 =   0.66

D. Operation Sheet Entry for Operation 40, Figure 25.5
Setup: 0.66 h
Cycle: 5.245 h/100

Following this specification, the range of weights is listed and costs that correspond to these weights. The accompanying tabulation is an example.

Weight lb	Cost: ($/100 ℓb)
97.0 (= 1 bar)	321.00
2000	170.00
6000	166.00
10,000	162.00

Estimating materials is done for *direct materials;* that is materials appearing in the product. Indirect materials (e.g., cutting oils, supplies), which do not appear in the product, are included in the overhead cost. Customarily the weight, volume, length, or surface area is determined from drawings. This theoretically computed quantity is then increased by losses such as **waste, scrap,** and *shrinkage*. A general formula for material estimating is

$$C_{dm} = W(1 + L_1 + L_2 + L_3)C_m$$

where

C_{dm} = Cost of direct material for a part, $/unit
W = Theoretical finished weight in compatible dimensions, lb
L_1 = Percentage loss resulting from scrap, which is a consequence of errors in manufacturing or engineering
L_2 = Percentage loss resulting from waste, which is caused by manufacturing process such as chips, cutoff, overburden, skeleton, dross, etc.
L_3 = Percentage loss resulting from shrinkage (or theft and physical deterioration)

These three losses (L_1, L_2, L_3) are determined by measurement or historical information. Once these losses are computed and the decimal added to 1, a material estimate is possible using the aforementioned formula. The value for C_m would be taken from data tables such as those given in the tables.

Consider finding the material cost for the stainless-steel type 430 F pinion shown in Figure 25.4. Bar length for one unit = $1/32$ in. for safety stock + 0.015 in. for facing stock + 18.75 in. for design length + $3/16$ in. for cutoff tool width, or L = 18.984 in. Raw material is supplied as 12-ft bars, and thus seven pieces can be made per bar. Each bar weighs 97.0 lb, and 143 bars are required for the order of 1000 units. The weight of the bars to be purchased is 13,871 lb and thus qualifies for the minimum price per 100 lb as a base quantity. Total material cost = 13.871 × 1.62 or $22,471, and each bar will cost $22.47. This value is entered on the operations chart (Figure 25.5).

The amount of material is important to the estimate, because the planner must know whether one lot run, one order, one year's requirements, or a model life is selected as the quantity upon which to base an estimate. In estimating material unit cost, cutting or shearing charges, delivery unit cost, cutting or shearing charges, delivery of material in a single or consolidated shipment, and terms of the purchase agreement influence the cost. The *base quantity* is that weight at which price is no longer sensitive to additional amounts of purchase order.

ADVANCES IN OPERATIONS PLANNING

Traditional operations planning is done manually—that is, using pad, pencil, and tables. This results in differences between various planners, even for near-identical parts that have been processed over a period of time. Individuals have their own opinions about an optimal routing. These manual methods often have similar operations; for example, shear strips, shear blanks, blank, and form are operations for sheet metal starting with sheets of steel. Methods are available that can ultimately automate these processes.

Various programs in recent years have attempted to capture the logic, experience, and opinion that is so necessary for a manual preparation system. A computer-aided process planning system (*CAPP*) offers potential in reducing the clerical work, and can offer consistent routing that might be optimal. Computer systems are referred to as generative or retrieval.

Retrieval systems use part classification systems, or group classification, discussed in Chapter 17, as a major consideration. In this way parts are classified into families of a similar part, and they are identified according to their manufacturing characteristics. For each part family a standard routing is determined. This standard operations sheet is then stored in the computer for this parts family. Efficient retrieval is necessary and the parts classification is used as the call number. Variation from the standard plan is then thought through by the planner. If there is an extract match between the new part code number and an existing inventoried code, the retrieved op sheet is used.

Generative process plans use algorithms to create process plans from scratch. It is desirable that these plans do so without human assistance. Input to the system could include a complete set of specifications, perhaps even drawings. The analysis of the geometry, material, and reductions in the physical shape of the part lead to the selection of the sequence, machine, and time estimates.

Computerized cost estimating can be done on a terminal, where the planner interacts with the production requirements. One system, called the AM Cost Estimator, is a large data base involving data for a wide listing of manufacturing equipment. The planner calls for the program, sees the major equipment, follows a tree logic, selects the operation elements, and the computer does the work. Costs for the machine hour can be changed along with the content of the operations.

QUESTIONS AND PROBLEMS

1. Give an explanation of the following terms:

    Cost estimate   Perishable tooling   Handling cost   Operations sheet
    Production analysis   Operation unit cost   Bottleneck situations   Setup

    Cycle   Waste
    Fair work   Scrap
    Profit   CAPP

2. List the objectives for the operations sheet.

3. What are the principal elements of the metal cutting analysis?

4. (a) An operator earns $18/h, and time for handling and other metal cutting elements totals 1.35 min. What is the cost for the element?

   (b) The length of a machining element is 20 in. and the part diameter is 4 in. OD. Velocity and feed for this material are 275 ft/min and 0.020 ipr. What is the time to machine?

   (c) The Taylor tool life equation is $VT^{0.1} = 172$. What is the expected average tool life for $V = 275$ ft/min?

   (d) Tool changing time is 4 min, tool life equation is $VT^{0.1} = 172$, $V = 275$ ft/min, $C_o = \$0.25$/min, $L = 20$ in., $D = 4$ in., and $f = 0.020$ ipr. Determine the tool changing cost.

   (e) Using the information in part (d) and $C_t = \$5$, find the tool cost per operation. Would you recommend these machining conditions?

5. (a) Operator and machine expenses are $50/h and handling time is 1.65 min. Find the handling cost for this element.

   (b) The length and diameter of a gray-iron casting are 8.5 in. by 8.6 in. Velocity is 300 ft/min and feed is 0.020 ipr. Find the machining time.

   (c) The Taylor tool life equation is $VT^{0.15} = 500$. Find the expected tool life for 300 ft/min.

   (d) The time to remove an insert and index to another new corner is 2 min, the tool life equation is $VT^{0.15} = 500$, $V = 300$ ft/min, $C_o = \$1$/min, $L = 8.6$ in., $D = 8$ in., and $f = 0.020$ ipr. What is the cost to change tools?

   (e) Using the information in part (d), and that an eight-corner insert costs $24, find the tool cost per operation.

6. A simple job is estimated to require 2 h for setup, and three cycle elements are 0.15, 0.11, and 0.53 min. The lot quantity will be 83. Find the pieces per hour, hours per 100 units, and unit estimated time, including a fair share of the setup. The labor cost = $17.50/h, and the machine hour rate is 185%. Find the full cost if the material = $1.15 per unit. The cost plus markup is 10%. What is the price for the unit and the job?

7. The planner estimates that a work order will need several operations. One operation will need elements for the cycle consisting of 0.17, 0.28, 0.81, and 0.17 min. The setup time is 2 h and the quantity is 182 units. The cost for the machine operator is $14.50/h. The machine rate is 75% exclusive of the labor cost. The labor and overhead rate is 175%. Find the pieces per hour, hours per 100 units, the unit estimate, cost of labor for one unit, and the machine cost. What is the lot cost for this operation?

8. If labor and material costs $25/unit and the markup rate is 12.5%, find the price for the lot of 715 units.

9. A competitor produces 81 pieces per hour for an operation. What are the hours per 100 units and the unit time in minutes for this operation?

10. What factors affect the cost for materials?

11. A bar is 430F stainless steel and 1.750 in. OD. The part length is 7.25 in., and a cutoff width and facing stock allowance are $1/4$ in. and $1/32$ in., respectively. If the safety stock allowance is $1/16$ in., the bar is 12 ft long, and 15 parts are required, find the unit and lot cost for this material.

12. The part length = 22.44 in., safety = $1/32$ in., approach = 1.4 in., overtravel = 1.4 in., and the material is 430F steel. Find the cost for the following conditions: (a) $N = 75$ units and 12-ft bar length; (b) $N = 750$ units

and 12-ft bar length. Is it cheaper if the bar length is 16 ft long? (c) $N = 7$ units and 12-ft bar length.

13. Stainless-steel material is to be rough and finish turned. Diameter and length are 4 in. × 30 in. Recommended rough and finish cutting velocity and feed for tungsten carbide tool material are (350 ft/min, 0.015 ipr) and (350 ft/min, 0.007 ipr). Determine rough and finish cutting time.

14. Medium-carbon steel is to be rough and finish turned using high-speed steel tool material. The part diameter and cutting length are 4 in. and 20 in. Using Table 20.1, determine the total time to machine.

15. Gray cast iron is to be rough and finish turned with tungsten carbide tooling. Part diameter and cutting length are 8.5 in. and 8.6 in. What is the part revolutions per minute for the rough and finish? Using Table 20.1, find the turning time.

16. A stainless-steel part is to be drilled $5/16$ in. for a depth of 1 in. It is followed by a $3/16$-16 tap for a depth of $7/8$ in. Find the drilling length and the drilling and tapping time. Use Table 20.1. Repeat for steel.

17. Steel is tap drilled $3/8$ in. for a depth of 1 in. It is followed with a tapping element of $5/16$-10 hole for $7/8$ in. Find the drilling length and the drilling and tapping time. Repeat for stainless steel. Use Table 20.1.

18. Find the cutting time for a hard copper shaft 2 by 20 in. long. A surface velocity of 350 ft/min is suggested with a feed of 0.008 ipr. Convert to metric units and repeat.

19. An end facing cut is required for a 10-in. diameter workpiece. The revolutions per minute of the lathe are controlled to maintain 400 surface ft/min from the center out to the surface. Feed is 0.009 ipr. Find the time for the cut. Convert to SI and repeat.

20. If the tool is K-3H carbide, the material is AISI 4140 steel, depth of cut is 0.050 in., and the feed is 0.010 ipr, what is the turning speed in surface feet per minute for a 4-in. bar with a 6-min life if the tool life equation is $VT^{0.3723} = 1022$? Also, revolutions per minute.

21. Consider the Taylor tool life model, $VT^n = K$ for the following tool and work materials:

Tool	Work	n	K
High-speed steel	Cast iron	0.14	75
High-speed steel	Steel	0.125	47
Cemented carbide	Steel	0.20	150
Cemented carbide	Cast iron	0.25	130

For a tool life of 10 min for each of these combinations, what is the cutting velocity?

22. A rough-turning operation is to be performed on a medium-carbon steel. Tool material is high-speed steel. The part diameter is 4 in. OD and the cutting length will be 20 in. The tool point will cost $5. Time to reset the tool is 4 min. Part handling will be 2 min and the operator wage is $20/h. The Taylor tool life equation $VT^{0.1} = 172$. Feed of the turning operation is 0.020 ipr for a depth of cut = 0.25 in. (a) Plot the four items of cost to find total-cost curve and select optimum velocity. (b) Determine the minimum velocity analytically.

23. Let the cost of $C_t = 0$ for problem 22 and plot the elements of time to find optimum time. Compute V_{max} and Tl_{max} and T_u.

24. Construct individual cost curves similar to Figure 25.3 for the following machining work. A gray-iron casting having a diameter of 8.5 in. is rough turned to 8.020/8.025 in. for a length of 8.6 in. A renewable square carbide insert is used. The insert has eight corners suitable for turning work and costs $24. Operator and variable expenses less tooling costs are $60/h. The feed for this turning operation is 0.020 ipr. Taylor's tool life equation for part and tool material is $VT^{0.15} = 500$. The time for the operator to remove the insert, install another new corner, and qualify the tool ready to cut is 2 min. Part handling time is 1.65 min for a

casting mounted in a fixture. (a) If the y axis is unit cost and x axis if feet per minute, plot the curves and locate optimum velocity. (b) Determine optimum velocity analytically.

25. Assume that the cost of $C_t = 0$ for problem 24 and graphically find optimum time, V_{max}, and T_{max}.

26. Find the material cost and setup cycle time for the operations of a stainless-steel pinion similar in all respects to Figure 25.4 except for the following changes: Work each part separately. Complete the operations process sheet and find the unit cost as demonstrated by Figure 25.5. (a) Let the 18.750-in. dimension be 8.750 in.; the 1.100 in. dimension be 1.125 in., and no holes. (b) Let the 4.75-in. flat dimension be 8.00 in., and the raw material be 2 in OD instead of $1^3/_4$ in OD. (c) Let the 4-in. plain milling cutter be 3 in. in width, and operation 10 will use high-speed tool material instead of tungsten carbide. (d) Let the material be medium-carbon steel instead of stainless steel. Medium-carbon steel costs $0.72/lb.

27. A bar is purchased 12 ft long and 2 in. OD. The design calls for dimension 16 in. long. The manufacturing engineer suggests a facing dimension of $^1/_{16}$ in. and a cutoff width of $^3/_8$ in. Last part gripping requires 6 in. for this machine. Scrap and shrinkage is historically 1% and 0.5%, respectively. Material cost per pound is $0.55 and density is 0.29 lb/in.^{3} The job order requests 260 units. The labor standard is 1.825 h/100 units and the machine operator is paid $15.50/h. Analyze the job for direct costs of material and labor.

CASE STUDY
SUPER SNAP RING

Oil Head, Inc., a large transnational company specializing in manufacturing and installing long-distance pipelines, has developed a new scheme of attaching pipe ends together. The notion is that a snap ring, except much larger than the small-scaled snap rings, can be used to attach the bell ends of seamless steel pipe together. It is followed by tack-and-seam welding around the bell ends, and the snap ring is an important part of the field assembly. The welding is independent of the snap ring. Welding occurs in the field and often is in harsh and remote regions of the world. The ring has two drilled 0.403/0.405 in. holes, and a crosshole 0.251/0.250 in. A simplified sketch is given in Figure 25.10. The material is AISI 8640 alloy, and is received annealed as 10-in. bar stock × 12 ft long. Upon conclusion of the metal removal operations, a hardening operation is necessary.

Your job is to develop an operations process sheet along the lines of Figure 25.5. Especially develop the sequences, machine, process, or bench identifications and material removal statements. Inasmuch as not all dimensions are shown, you may scale roughly for missing dimensions. Find the approximate material yield. Find the time for each operation. Consider technical differences that might be necessary for an annual order of 180 versus 18,000 units.

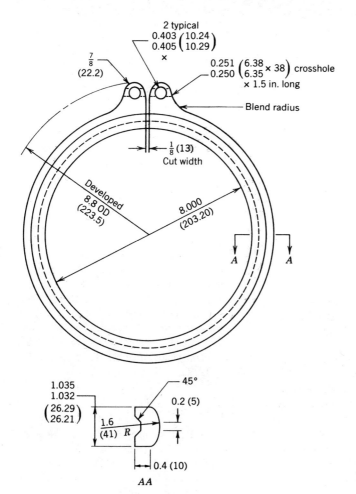

Figure 25.10
Case study.

BIBLIOGRAPHY

CHAPTER 1

Althing, L. *Manufacturing Engineering Processes*. New York: Dekker, 1982.

Bolz, R. W. *Production Processes*. 5th ed. New York: Industrial Press, 1981.

Datsko, J. *Material Properties and Manufacturing Processes*. New York: Wiley, 1966.

De Garmo, E. P. *Materials and Processes in Manufacturing*. 5th ed. New York: Macmillan, 1979.

Doyle, L. E., et al. *Manufacturing Processes and Materials for Engineers*. 2nd ed. Englewood Cliffs, N.J.: Prentice-Hall, 1961.

Edgar, C. *Fundamentals of Manufacturing Processes and Materials*. Reading, Mass.: Addison-Wesley, 1965.

Kalpakjian, S. *Manufacturing Processes for Engineering Materials*. Reading, Mass.: Addison-Wesley, 1984.

Lindberg, R. A. *Processes and Materials of Manufacturing*. 3rd ed. Boston: Allyn & Bacon, 1983.

Moore, H. D., and Kibbey, D. R. *Manufacturing: Materials and Processes*. Columbus, Ohio: Grid, 1975.

Patton, W. J. *Modern Manufacturing Processes and Engineering*. Englewood Cliffs, N.J.: Prentice-Hall, 1970.

Pollack, H. W. *Manufacturing and Machine Tool Operations*. 2nd ed. Englewood Cliffs, N.J.: Prentice-Hall, 1979.

Schey, J. A. *Introduction to Manufacturing Processes*. New York: McGraw-Hill, 1977.

Yankee, H. *Manufacturing Processes*. Englewood Cliffs, N.J.: Prentice-Hall, 1979.

CHAPTER 3

"AOD Finds a Home in a Foundry." *Mechanical Engineering*. February 1984, p. 25.

McGannin H. E. *The Making, Shaping, and Treating of Steel*. 9th ed. Pittsburgh, Pa.: United States Steel Corporation, 1971.

The Making of Steel. Washington, D.C. American Iron and Steel Institute, 1972.

Russell B., and Vaughn W. *Steel Production: Process, Products and Residuals*. Baltimore: Johns Hopkins University Press, 1976. pp. 85–102.

CHAPTER 4

AFS Transactions, Vol. 92. Des Plaines, ILL.: American Foundry Society, 1984.

Machine Design, 1984 Materials Reference Issue. Vol. 56, No. 8. Cleveland, Ohio: Penton/IPC, April 19, 1984.

Metals Handbook, "Properties and Selection of Metals," 8th ed. Vols. 1 and 7. Metals Park, Ohio: American Society for Metals, 1972.

CHAPTER 5

Beeley, P. R. *Foundry Technology,* New York: Wiley (Halsted Press), 1972.

Foundry Technology. Des Plaines, Ill.: American Society for Metals and American Foundryman's Society, 1982.

Hamilton, E. *Patternmaker's Guide*. Des Plaines, Ill.: American Foundrymen's Society, 1976.

"How to Make a Casting by the Full Mold Process." *Modern Castings*. October 1965, p. 56.

McGannin, H. E. *The Making, Shaping, and Treating of Steel,* 9th ed. Pittsburgh, Pa.: United States Steel Corporation, 1971.

Metals Handbook, 8th ed. Metals Park, Ohio: American Society for Metals, 1972.

Steel Castings Handbook, 5th ed. Des Plaines, Ill.: Steel Founders' Society of America, 1980.

Sylvia, B. J. *Cast Metals Technology*. Reading, Mass.: Addison-Wesley, 1972.

CHAPTER 6

Beeley, P. R. *Foundry Technology*. New York: Wiley (Halsted Press), 1972.
Foundry Technology. Des Plaines, Ill.: American Society for Metals and American Foundrymen's Society, 1982.
Kaye, A., Street, A. *Die Casting Metallurgy*. London: Butterworth, 1982.
McGannin, H. E. *The Making, Shaping, and Treating of Steel*, 9th ed. Pittsburgh, Pa.: United States Steel Corporation, 1971.
Metals Handbook, 8th ed. Metals Park, Ohio: American Society for Metals, 1972.
Schaum, J. H. "Electroslag Casting." *Modern Casting*. December 1974. pp. 44–45.

CHAPTER 7

Brown, D. *Metallurgy*. Albany, N.Y.: Delmar, 1981.
Carter, G. F. *Principles of Physical and Chemical Metallurgy*. Metals Park, Ohio: American Society for Metals, 1979.
Johnson, C. G., and Weeks, W. R. *Metallurgy*. 5th ed. New York: American Technical, 1977.
Leslie, W. C. *The Physical Metallurgy of Steels*. New York: McGraw-Hill, 1981.
McGannin, H. E. *The Making, Shaping, and Treating of Steel*, 9th ed. Pittsburgh, Pa.: United States Steel Corporation, 1971.
Neely, J. *Practical Metallurgy and Materials of Industry*. New York: Wiley, 1979.
Pollack, H. W. *Material Science and Metallurgy*, 3rd ed. Reston, Va.: Reston Publishing Company, 1980.

CHAPTER 8

Allen, K W. *Adhesion 5*. New York: Elsevier, 1981.
Cary, H. B. *Modern Welding Technology*. Englewood Cliffs, N.J.: Prentice-Hall, 1979.
Flax, R. F., Keith, R. E. and Randall, M. D. *Welding the HY Steels*. Columbus, Ohio: American Society for Testing and Materials, 1971.
Galyen, J. and Sear, G. *Welding: Fundamentals and Procedures*. New York: Wiley, 1984.
Juvinall, R. C. *Fundamentals of Machine Component Design*. New York: Wiley, 1983.
Lancaster, J. F. *The Metallurgy of Welding, Brazing and Soldering*. London: George Allen & Unwin, Ltd., 1970.
Little, R. L. *Welding and Welding Technology*. New York: McGraw-Hill, 1973.
Satas, D., ed. *Handbook of Pressure-Sensitive Adhesive Technology*. New York: Von Nostrand Reinhold, 1982.
Schneberger, G. L., ed. *Adhesives in Manufacturing*. New York: Dekker, 1983.
Schwartz, M. M. *Metals Joining Manual*. New York: McGraw-Hill, 1979.
Schwartz, M. M., consulting ed. *Source Book on Innovative Welding Processes*. Metals Park, Ohio: American Society for Metals, 1981.
Tweeddale, J. G. *Welding Fabrication*, Vols. 1 to 3. New York: Elsevier, 1969.
The Welding Encyclopedia, 18th ed. Lake Zurich, Ill.: Monticello Books, 1982.
Welding Handbook, 7th ed., Sec. 1–5. Miami, Fla.: American Welding Society, 1976.

CHAPTER 9

Hauser, H. H. *Forging of Powder Metallurgy Preforms*. Detroit, Mich.: American Powder Metallurgy Institute, 1973.
Hirschhorn, J. S. *Advanced Experimental Techniques in Powder Metallurgy*. New York: Plenum, 1970.
Lenel, E. V. *Powder Metallurgy*. Detroit, Mich.: Metal Powder Industries Division, 1980.
Source Book on Powder Metallurgy. Metals Park, Ohio: American Society for Metals, 1979.

CHAPTER 10

Levy, S. *Plastics Extrusion Technology Handbook.* New York: Industrial Press, 1981.
Robinson, J. S., ed. *Plastics Molding: Equipment, Processes and Materials.* Washington, NJ: Polymers and Plastics Technical Publishing House, 1981.
Schwartz, S. and Goodman, S. *Plastics Materials and Processes.* New York: Van Nostrand Reinhold, 1982.

CHAPTER 11

Grant, E. L., and Leavenworth, R. *Statistical Quality Control.* New York: McGraw-Hill, 1979.
Koben, S. and Rose, M. *Electronic Manufacturing Processes.* Reston, Va: Reston Publishing Co., 1982.
Sirohi, R. R., and Krishna, R. *Mechanical Measurements.* New York: Wiley, 1983.

CHAPTER 12

ASM Metals Reference Book. Des Plaines, Ill.: American Society for Metals, 1981.
Avitzur, B., and Van Tyne, C. J. eds. *Production to Near Net Shape.* Des Plaines, Ill.: American Society for Metals, 1983.
Caddell, M. J., and Hasford, W. F. *Metal Forming: Mechanics and Metallurgy,* Englewood Cliffs, N.J.: Prentice-Hall, 1983.
Metal Society Editors, *Hot Working and Forming Processes.* Brookfield, Vt.: Metal Society, 1980.
Meyers, M. A., and Chawla, K. K. *Mechanical Metallurgy Principles and Applications.* Englewood Cliffs, N.J.: Prentice-Hall, 1984.

CHAPTER 13

Crane, F. V. *Plastic Working of Metals and Non-metallic Materials in Presses,* 3rd ed. New York, Wiley, 1964.

Kalpakjiian, S. and Jain, S., eds. *Metalworking Lubrication,* New York: American Society of Mechanical Engineers, 1966.
Schey, J. A. ed. *Metal Deformation Processes: Friction and Lubrication.* New York, Marcel Dekker, 1970.
Source Book on Forming of Steel Sheet. Metals Park, Ohio, American Society for Metals, 1976.

CHAPTER 14

Dallas, D. B. *Progressive Dies, Design and Manufacture.* New York: McGraw-Hill, 1962.
Davis, R., and Austin, E. R. *Developments in High Speed Metal Forming.* New York: Industrial Press, 1962.
Eary, D. F., and Reed, E. A., *Techniques of Press-Working Sheet Metal.* Englewood Cliffs, N.J.: Prentice-Hall, 1958.
Sachs, G. *Principles and Methods of Sheet-Metal Fabricating.* 2nd ed. New York: Van Nostrand and Reinhold, 1966.
Schubert, P. B. *Die Methods, Design, Fabrication, Maintenance, and Application,* Books 1 and 2. New York: Industrial Press, 1966–1967.
Stasser, F. *Functional Design of Metal Stampings.* Dearborn, Mich.: Society of Manufacturing Engineers, 1971.

CHAPTER 15

Society of Manufacturing Engineers. *Tool and Manufacturing Engineers Handbook,* 3rd ed. New York: McGraw-Hill, 1976.
"What's New in Machining Centers?" *American Machinist,* Special Report 763, February 1984.

CHAPTER 16

Dyke, R. M. *Numerical Control.* Englewood Cliffs, N.J.: Prentice-Hall, 1967.
Howe, R. E., ed. *Introduction to Numerical Control in Manufacturing.* Dearborn,

Mich.: Society of Manufacturing Engineers, 1969.

Wilson, F. W., ed. *Numerical Control in Manufacturing*. Dearborn, Mich.: Society of Manufacturing Engineers, 1963.

CHAPTER 17

Boothroyd, G., Poli, C., Murch, L. E. *Automatic Assembly*. New York: Dekker, 1982.

Dorf, R. C. *Robotics and Automated Manufacturing*. Reston, Va.: Reston Publishing Co., 1983.

Groover, M. P., and Zimmer, E. W., Jr. *CAD/CAM: Computer-Aided Design and Manufacturing*. Englewood Cliffs, NJ: Prentice-Hall, 1984.

Koren, Y. *Computer Control of Manufacturing Systems*. New York: McGraw-Hill, 1976.

Tanner, W. F. *Industrial Robots*, Vol. 1. Dearborn, Mich.: Society of Manufacturing Engineers, 1979.

CHAPTER 18

Black, P. H. *Theory of Metal Cutting*. New York: McGraw-Hill, 1961.

Machining Data Handbook, 3rd ed. Cincinnati, Ohio: Machinability Data Center, 1972.

Schey, J. A. *Introduction to Manufacturing Processes*. New York: McGraw-Hill, 1977.

CHAPTER 19

Dallas, D. B., ed. *Tool and Manufacturing Engineers Handbook*. 4th ed. New York: McGraw-Hill, 1985.

Machining Data Handbook. 3rd ed. Cincinnati, Ohio: Metcut Research Associates, 1980.

CHAPTER 20

McCarthy, W. J., and Repp, V. E. *Machine Tool Technology*. Bloomington, Ill.: McKnight Publishing Co., 1978.

CHAPTER 21

Coes, L., Jr. *Abrasives*. New York: Springer-Verlag, 1971.

Contour Band Machining Handbook. Des Plaines, Ill.: DoAll Co., 1980.

Hine, C. *Machine Tools and Processes for Engineers*. New York: McGraw-Hill, 1971.

Jablonowski, J. "Fundamentals of Sawing." *American Machinist*. April 15, 1975. pp. 53–68.

Machining Data Handbook. 3rd ed., Vol. 1. Cincinnati, Ohio: Metcut Research Associates, 1980.

Psenka, J. A. "Cutting Tools/Broaches—Material Broachability." *Manufacturing Engineering*. November 1978. p. 41.

Tool and Manufacturing Engineers Handbook. 4th ed., Vol. 1. Dearborn, Mich.: Society of Manufacturing Engineers, 1983.

Vasilash, G. S. "Sawing Technology—Today and Tomorrow." *Manufacturing Engineering*. August 1980. pp. 118–120.

Worthington, B. *Specific Cutting Force Relationships in Broaching*. SME Technical Paper MR80-933, 1980.

CHAPTER 22

Juvinall, R. C. *Fundamentals of Machine Component Design*. New York: Wiley, 1983.

Merritt, E. *Gear Engineering*. New York: Wiley (Halsted Press), *Tool and Manufacturing Engineers Handbook*, 4th ed., Vol. 1. Dearborn, Mich.: Society of Manufacturing Engineers, 1983.

CHAPTER 23

McKee, R. L. *Machining with Abrasives*. New York: Van Nostrand Reinhold, 1983.

Tool and Manufacturing Engineers Handbook, 4th ed., Vol. 1. Dearborn, Mich.: Society of Manufacturing Engineers, 1983.

CHAPTER 24

Oldham, W. G. "The Fabrication of Microelectronic Circuits." *Scientific American*. September 1977, pp. 110–128.

Stern, L. *Fundamentals of Integrated Circuits*. Rochelle Park, N.J.: Hayden, 1968.

Thomas, H. E. *Handbook of Integrated Circuits*. Englewood Cliffs, N.J.: Prentice-Hall, 1971.

CHAPTER 25

Ostwald, P. F. *Cost Estimating*, 2nd ed. Englewood Cliffs, N.J.: Prentice-Hall, 1984.

Ostwald, P. F., "AM Cost Estimator, 1987–1983 Edition." *American Machinist*, New York: McGraw-Hill, 1987.

Ostwald, P. F., "AM Cost Estimator Software", two disks, American Machinists, New York: McGraw-Hill Book Co., 1985.

ACKNOWLEDGMENTS AND PHOTO CREDITS

Chapter 1. Fig. 1.1: Smithsonian Institution Photo No. 64173. Fig. 1.2: National Machinery Company. Fig. 1.6: Burgmaster Houdaille. Fig. 1.7: IBM Corporation.

Chapter 3. Fig. 3.2: Hojalata y Lamina. Fig. 3.12: A. M. Beyers Company. Fig. 3.14: Drop Forging Association. Fig. 3.15: Bethlehem Steel Corporation. Fig. 3.19B: The International Nickel Company, Inc. Fig. 5.16B: Davenport Machine and Foundry Company. Fig. 5.16C: Taccone Pneumatic Foundry Equipment Corporation. Fig. 5.16D: The Beardsley and Piper Company. Fig. 5.17: Wheelabrator Corporation.

Chapter 6. Fig. 6.1: Paramount Die Casting Division, Hayes-Albion Corporation. Fig. 6.2: Paramount Die Casting Division, Hayes-Albion Corporation. Fig. 6.4: Titan Metal Manufacturing, Division of Cerro Corporation. Fig. 6.5: HPM Division, Koehring Company. Fig. 6.7: Kaiser Aluminum and Chemical Corporation. Fig. 6.8: Easton Manufacturing Company. Fig. 6.9: American Cast Iron Pipe Company. Fig. 6.10: American Cast Iron Pipe Company. Fig. 6.11: American Cast Iron Pipe Company. Fig. 6.13: Casting Engineers. Fig. 6.14: Misco Precision Casting Company. Fig. 6.15: Universal Castings Corporation. Fig. 6.16: Universal Castings Corporation. Fig. 6.17: The Borden Company. Fig. 6.19: Dow Corning. Fig 6.20: American Smelting and Refining Company. Fig. 6.21: Aluminum Corporation of America.

Chapter 7. Fig. 7.1: Bureau of Standards, U.S. Department of Commerce. Fig. 7.11: J. L. Burns, T. L. Moore, R. S. Archer, "Quantitative Hardening," *Transactions ASM*, Vol. XXVI, 1938. Fig. 7.12: Bethlehem Steel Corporation. Fig. 7.16: The Ohio Crankshaft Company. Fig. 7.17: Linde Division, Union Carbide Corporation.

Chapter 8. Fig. 8.10: American Welding Society. Fig. 8.17: Linde Division, Union Carbide Corporation. Fig 8.18: Linde Division, Union Carbide Corporation. Fig. 8.20: Linde Division, Union Carbide Corporation. Fig 8.22: Skiaky Brothers. Fig. 8.23: Skiaky Brothers. Fig. 8.26: Metal and Thermit Corporation. Fig. 8.27: Koldweld Corporation. Fig. 8.28: Skiaky Brothers. Fig. 8.31: E. F. Industries, Inc. Fig. 8.35: E. I. du Pont de Numours and Company.

Chapter 9. Fig. 9.3: The United States Graphite Company, Division of the Wickes Corporation. Fig. 9.5: National Forge Company. Fig. 9.6: National Aeronautics and Space Administration, NASA SP-5060. Fig. 9.8: Japax Scientific Corporation. Fig. 9.10: Keystone Carbide Corporation. Fig. 9.11: Amplex Division, Chrysler Corporation. Fig. 9.12: Amplex Division, Chrysler Corporation.

Chapter 10. Fig. 10.1: HPM Division, Koehring Company. Fig. 10.2: General Electric Company. Fig. 10.3: HPM Division, Koehring Company. Fig. 10.5: HPM Division, Koehring Company. Fig. 10.8: Nalle Plastics. Fig. 10.9: F. J. Stokes Company, Division Pennsalt Chemicals Corporation. Fig. 10.11: Eastman Chemical Products. Fig. 10.12:

U.S. Industrial Chemicals Corporation. Fig. 10.13: U.S. Industrial Chemicals Company. Fig. 10.14: Phillips Petroleum Company. Fig. 10.17: Carbide Plastics Company. Fig. 10.18: Rohm and Haas Company. Fig. 10.19: Owens-Corning Fiberglas Corporation. Fig. 10.20: Owens-Corning Fiberglas Corporation. Fig. 10.21: Owens-Corning Fiberglas Corporation.

Chapter 11. Fig. 11.6: Browne and Sharpe Manufacturing Company. Fig. 11.7: Bausch and Lomb. Fig. 11.9: L.S. Starrett Company. Fig. 11.10: Pratt and Whitney, Division Miles-Bement-Pond Company. Fig. 11.11: Van Keuren Company. Fig. 11.12: American Society of Mechanical Engineers. Fig. 11.13: Bendix Automation and Measurement Division. Fig. 11.15: *Metalworking*, January 1964. Fig. 11.16: *Metalworking*, January 1964. Fig. 11.18: Perkin-Elmer Corporation. Fig. 11.21: Browne and Sharpe Manufacturing Company. Fig. 11.22: Nikon, Incorporated. Fig. 11.26: Bendix Automation and Measurement Division. Fig. 11.27: Bendix Automation and Measurement Division. Fig. 11.30: Bendix Automation and Measurement Division. Fig. 11.31: Bendix Automation and Measurement Division. Fig. 11.32: Webster Instrument, Incorporated.

Chapter 12. Fig. 12.3: Carnegie-Illinois Steel Corporation. Fig. 12.4: Chambersburg Engineering Company. Fig. 12.7: Chambersburg Engineering Company. Fig. 12.8: Wyman-Gordon Company. Fig. 12.10: The Ajax Manufacturing Company. Fig. 12.12: Edgewater Steel Company. Fig. 12.16: National Tube Division, U.S. Steel Corporation. Fig. 12.17: National Tube Division, U.S. Steel Corporation. Fig. 12.18: National Tube Division, U.S. Steel Division. Fig. 12.19: Hydropress. Fig. 12.20: National Tube Division, U.S. Steel Corporation. Fig. 12.21: The Holokrome Screw Corporation. Fig. 12.23: Reynolds Metal Corporation.

Chapter 13. Fig. 13.2: National Tube Division, U.S. Steel Corporation. Fig. 13.5: Bethlehem Steel Corporation. Fig. 13.7: Phoenix Products Company. Fig. 13.8: Cincinnati Milacron. Fig. 13.11: National Machinery. Fig. 13.13: National Machinery. Fig. 13.14: Universal Engineering Company. Fig. 13.19: The Yoder Company. Fig. 13.20: The Yoder Company. Fig. 13.24: Convair Division, General Dynamics. Fig. 13.25: General Atomic Division, General Dynamics. Fig. 13.26: General Atomic Division, General Dynamics. Fig. 13.28: USI-CLEARING, Division of U.S. Industries, Incorporated. Fig. 13.29: Wheelabrator Corporation.

Chapter 14. Fig. 14.1: Verson Allsteel Press Co. Fig. 14.3: E. W. Bliss Company. Fig. 14.4: E. W. Bliss Company. Fig. 14.5: Verson Allsteel Press Co. Fig. 14.6: Bethlehem Steel Corporation. Fig. 14.7: Wiedemann Division, The Warner and Swasey Company. Fig. 14.8: The Hydraulic Press Manufacturing Company. Fig. 14.9: E. W. Bliss Company. Fig. 14.10: American Machinist. Fig. 14.12: Niagara Machine and Tool Works. Fig. 14.16: Carbidex Corporation. Fig. 14.26: Verson Allsteel Press Co. Fig. 14.29: Cincinnati Milacron.

Chapter 15. Fig. 15.5: Giddings and Lewis, Incorporated. Fig. 15.6: Pope Machinery Corporation. Fig. 15.7A: Giddings and Lewis, Incorporated. Fig. 15.7B: Giddings and

Lewis, Incorporated. Fig. 15.8: Setco Industries. Fig. 15.9: Giddings and Lewis, Incorporated. Fig. 15.10: Master Machine Tools. Fig. 15.13: South Bend Lathe Works. Fig. 15.15: The Warner and Swasey Company. Fig. 15.17: The Warner and Swasey Company. Fig. 15.19: O. S. Walker Company, Incorporated. Fig. 15.20: Browne and Sharpe Manufacturing Company. Fig. 15.21: Bullard Machine Tool Co. Fig. 15.23: Burgmaster Houdaille.

Chapter 16. Fig. 16.1: Makino Machine Tool Company. Fig. 16.2: Jones & Lamson Machine Tool Company. Fig. 16.7: Electronic Industries Association. Fig. 16.9: General Electric Company. Fig. 16.12: *Modern Machine Shop 1974 NC Guide-book and Directory*, Modern Machine Shop, 1974, Cincinnati, OH.

Chapter 17. Fig. 17.9: Balance Engineering Division, General Motors Corporation. Fig. 17.13: *Automatic Assembly*, Marcel Dekker, New York. Fig. 17.14: Gillette Safety Razor Co. Fig. 17.15: Gillette Safety Razor Co. Fig. 17.16: Westinghouse Corporation. Fig. 17.21: *Industrial Robots—A Summary and Forecast*, Tech. Tran. Consulting, 1982. Fig. 17.22: Cincinnati Milacron. Fig. 17.32: Westinghouse Corporation.

Chapter 18. Fig. 18.5: Kestler Instrumente AG. Fig. 18.6: Kestler Instrumente AG. Fig. 18.11: Cincinnati Milacron. Fig. 18.22: The Warner and Swasey Company.

Chapter 19. Fig 19.1: LeBlond Lathe Company. Fig. 19.7: George Gorton Machine Company. Fig. 19.8: Greenlee Brothers and Company. Fig. 19.9: Giddings and Lewis, Incorporated. Fig. 19.11: South Bend Lathe Work. Fig. 19.12: Barnes Drill Company. Fig. 19.14: Giddings and Lewis, Incorporated. Fig. 19.16: Kearney and Trecker Company. Fig. 19.17: Bridgeport Machines, Incorporated. Fig. 19.19: Quality Machines, Incorporated.

Chapter 20. Fig. 20.6: National Twist Drill Co. Fig. 20.8A: Marvel Company. Fig. 20.12: National Twist Drill Co. Fig. 20.22: The Warner and Swasey Company. Fig. 20.23: The Warner and Swasey Company. Fig. 20.24: The Warner and Swasey Company. Fig. 20.28: Standard Tool Division, Lear Siegler. Fig. 20.34A: Cincinnati Milacron. Fig. 20.34B: Boston Digital Corporation. Fig. 20.35: Philip S. Ostwald, *Cost Estimating*, 2nd ed., Prentice-Hall, Inc., Englewood Cliffs, N.J., 1984. Fig. 20.37: Kennametal Company. Fig. 20.38: Kennametal Company.

Chapter 21. Fig. 21.6: Mesta Machine Company.

Chapter 22. Fig. 22.7: Gleason Works. Fig. 22.30: Gould and Eberhardt, Division of Norton Company.

Chapter 23. Fig. 23.10: Mattison Machine Works. Fig. 23.11: The Fosdick Machine Tool Company. Fig. 23.18: Queen Products Division, King-Seeley Thermos Company.

Chapter 24. Fig. 24.3: Cincinnati Milacron. Fig. 24.6: Hammund Machinery Builders. Fig. 24.10: Cincinnati Milacron. Fig. 24.14: Chemcut Corporation. Fig. 24.16: Chemcut Corporation. Fig. 24.17: The International Nickel Company, Inc. Fig. 24.18:

Bureau of Standards, U.S. Department of Commerce.   Fig. 24.19: The International Nickel Company, Inc.   Fig. 24.20: Metco, Incorporated.   Fig. 24.21: Metco, Incorporated.   Fig. 24.22: Metco, Incorporated.   Fig. 24.24: *Scientific American*, September 1977.   Fig. 24.25: Universal Instruments.

Chapter 25.   Fig. 25.1 to 25.8: Philip F. Ostwald *Cost Estimating*, 2nd ed., Prentice-Hall, Inc. Englewood Cliffs, N.J., 1984.

INDEX

Abrasion, 471
  wear, 468
Abrasive:
  belt grinding, 612
  disks, 558
  particles, 625
Abrasive-jet machining, 626
Absolute dimension, 396
Accuracy, 7, 242
Acetylene, 161
Acme threads, 574
Acoustic emission, 418
A–C introduction motor, 377
Adaptable programming assembly, 427
Adaptive control, 404, 443, 444
Addendum, 584
  circle, 584
Addition polymerization, 214
Additives, 21
Adhesion wear, 468
Adhesive bonding, 156, 191
Adjustable reamers, 513
Age hardening, 151
AGV, 417
Air logic robots, 434
Allotropic, 20
  changes, 132
Allowance(s), 83, 245, 675
  gage, 261
Alloy(s), 23, 60, 65
  aluminum, 67, 68
  cast nonferrous, 460
  composition, 287
  copper, 68
  designations, 48
  die casting, 70
  magnesium, 67, 69
  steel, 70, 131
  wrought, 68
Almandite, 619
Alnico magnet, 211
Alpha iron, 133

Alternating-current motors, 377
Alumina, 62
  brick, 38
Aluminum, 19, 108
  alloys, 67, 68
  base alloys, 71
  electrolytic, 62
  oxice, 615
  oxide, 609, 615, 617
  oxide particle, 626
  oxide powder, 461
  pellets, 307
  production, 62
  sheet, 306
Ammeter, 455
Amorphous, 19
Amplimeter, 256
Angle:
  back rake, 457
  cutting, 457
  side rake, 457
  side relief, 457
Angle-milling cutter, 518
Annealing, 22, 145
  full, 145
  isothermal, 145
  process, 146
Annular gear, 587
Anodized coatings, 60
Anodizing, 649
Antioch process, 121
AOD process, 42
APAS, 427
Approach length, 537
APT, 404
Arbor, 381
  cutters, 516
  threaded, 381
Arc:
  blow, 175
  welding, 174
Arch press, 340

INDEX

Areas tool interface, 465
Arm, 431
Artificial aging, 152
Artwork, 655
Asarco process, 125
ASCII standard, 398
Assembly, 428
    adaptable programming, 427
    lines, 428
    manual, 410
Assignable causes, 274
Aston process, 46
Atomic dislocation, 311
Atomization, 198
Atoms, 310
Attachments, 533
Attributes, 274
Austempering, 144
Austenite, 133, 137
    nonmagnetic, 130
    photomicrograph, 140
Automatic:
    arc welding, 179
    die heads, 577
    inspection, 270
    lathe, 486
    machine tools, 385
    programmed tools, 404
    screw machine, 424, 487
    transfer machine, 493
Automatically guided vehicles, 417
Automation, 8, 376, 410, 419
    full, 412
    partial, 412
    specific product machine, 426
Auxiliary punch, 306
Average, 275
    deviation, 256
    roughness, 256
    tool life, 668
    unit power, 540
Axis coordinates, 394

Back gears, 481
Backlash, 584
Back rake angle, 457
Bakelite, 617
Balanced core, 93
Band:
    friction cutting, 559
    polishing, 560
    sawing, 558
Band filing, 560
Bandsaw filing, 560

Bang-bang, 432
Bar, 496
    stock, 288, 379, 449
Barrel finishing, 612
Base, 371, 492
    circle, 583
    quantity, 683
Basic:
    elements, 371
    oxygen furnace, 37
BASIC, 404
Batch production, 412
Bauxite, 62, 616
Bayer process, 62
BCD, 398
Bed, 371
Belt:
    conveyor, 430
    overhead, 3
Bench lathe, 481
Bench molding, 77
Bench work, 410
Bending, 343, 353
    plate, 326
Bentonite, 87
Beryllium, 68
Bessemer converter, 37, 47
Bevel gear generators, 592
Bevel gear planer, 589
Bevel protractor, 252
Bilateral tolerance, 278
Billets, 288
Binary arithmetic, 399
Binary-coded decimal, 398
Binder, 76, 93
Black heart, 56
Black heart castings, 56
Blanking, 352
    chemical, 638
Blast furnace, 33, 34, 60, 65
Blind risers, 81
Blister copper, 65
Bloom, 47, 288
Board hammer, 292
BOF, 37
Bonding adhesive, 191
Borax, 161
Boring, 479
Boring machine, 370, 495
Boring tool, 479, 514, 515
Boron carbide, 615
Bottleneck situations, 670
Bottom board, 78
Bowl feeders, 424

INDEX 699

Box furnace, 153
Box tool, 525
Box-type jig, 532
Brake, 343
Brass, 20, 23, 46, 67, 108, 126
Brazed-tip tools, 506
Brazing, 156, 158, 175
Brazing joints, 159
Breakage:
  tool, 418
Brinell hardness, 29
Brinell hardness number, 48
Briquetted, 201
Broach:
  continuous, 564
  rotary, 564
  tunnel, 564
Broaching, 370, 546, 560, 589
  horizontal, 563
  keyway, 563
  pull, 561, 562
  push, 561, 562
  vertical, 561
Broaching classification, 561
Broaching tool, 449, 565
Broken tools, 418
Bronze, 67, 418
Brown and Sharpe taper, 521
Brushes:
  wire, 614
BUE, 462
Buffing, 615
Built-up edge, 462
Bulging, 316
Bullion, 65
Burnishing, 567, 595
Business objectives, 663
Buttress thread, 574
Butt welding, 171, 300

CAD/CAM, 411
Calcium carbide, 161
Caliper, 251
Calorizing, 649
CAM, 411, 440
Camera, 439
Cams, 348
Capacitance, 454
Capacity:
  press, 295
Capillary action, 465
Capital goods, 4
Capital tooling, 664
CAPP, 683

Carbide, 460
Carbide cutting tool, 460, 506
Carbon, 47, 56, 214
  effect of, 25
Carbon dioxide gas, 76
Carbon dioxide molds, 76
Carbon dioxide process, 122
Carbon electrode welding, 174
Carbonitriding, 148
Carbon monoxide, 36
Carburizing, 147
  flame, 163
  gas, 147
  liquid, 147
  pack, 147
Careers, 12
Carousel, 386
Carriage assembly, 481
Cartesian coordinates, 395
Cast iron, 25, 53, 134
  gray, 53
  mottled, 55
Cast metal, 33
Cast nonferrous alloy, 460
Casting:
  centrifugal, 112
  centrifuging, 115
  ceramic shell, 118
  cleaning, 99
  Corthias, 112
  electroslag, 112
  gravity, 111
  investment, 116
  lost wax, 117
  permanent-mold, 109
  plaster mold, 119
  precision, 116
  pressed, 112
  semicentrifugal, 115
  slush, 112
Casting alloy, 67
  aluminum, 68
Cavities mold, 322
Cell(s), 410
  machine, 416
  photoelectric, 398
  production, 413
Cementite, 55
  definition, 55, 134
Center, 479
Center gage, 523
Center rest, 378
Centrifugal casting,
  112

Centrifugal:
    compacting, 202
Centrifuging, 115
Ceramic, 461
Ceramic tools, 461
C-frame presses, 340
Chalcopyrite, 64
Chaplets, 93
Charpy test, 28
Chasers, 576
Chatter, 367, 520, 530
Cheek, 77
Chemical blanking, 638
Chemical coolants, 465
Chemical and electrolytic wear, 468
Chemical energy, 636
Chemical engraving, 641
Chemical milling, 636
Chills, 55, 82
Chip, 418, 449, 450, 625
    control, 464
    discontinuous, 461
    formation, 449
    segmental, 461
    shape, 461
    thickness ratio, 452
Chip breaker, 464
    step-type, 464
Chisel edge, 529
Chisel edge angle, 508
Chrome plating, 647
Chuck:
    combination, 379
    direct-current, 382
    drill, 379
    independent, 379
    magnetic, 382
    rotary, 382
    universal, 379
Chucking reamers, 513
Chucking tools, 483
Circular interpolation, 401
Circular pitch, 584
Circular sawing machines, 556
Cladding, 189
Clapper box, 547
Clay, 87
Clean room, 656
Clearance, 245, 584
Clearance diameter, 508
Clearance drill, 514
Climb milling, 530
Closed-impression dies, 295
Closed-loop control, 394

Cloth wheels, 614
Cluster drilling machines, 492
CNC, 403
Coalescence, 156
Coarsening temperature, 136
Coarse-grained steels, 136
Coated abrasives, 619
Coefficient of friction, 451, 453
Coining, 209, 323
Coke, 33, 41
Cold-chamber die casting, 108
Cold-drawing tubing, 312
Cold finished, 311
Cold heading, 319, 343
Cold working, 286, 310
Cold drawing, 312
Cold roll forming, 324
Cold sawing, 556
Cold-temperature machining, 636
Cold welding, 156, 187
Cold working, 68
Collapsible tap, 576
Collet(s), 380, 479
    stationary, 380
Column, 371, 492
Columnar grains, 50, 287
Combination chuck, 379
Combination die, 106
Combination stock stop and start drills, 525
Combined cuts, 484, 525
Components:
    robot, 431
Composition alloy, 287
Compound seam, 327
Compressed air, 465
Compressive strength, 27
Computer aided design, 391
Computer aided manufacturing, 391
Computer aided process planning system, 683
Computer assisted part programming, 392
Computerized cost estimating, 683
Computer numerical control, 403
Computers, 399
Computer storage disks, 390
Computer technology, 411
Condensation polymerization, 214
Conductive elastomer, 442
Constant errors, 242
Constant-speed motor, 481
Consumer demand, 4
Consumer goods, 5
Contact electrode, 177
Continuous broaching, 564
Continuous casting, 50, 124

Continuous chip type 2, 462
Continuous furnaces, 153
Continuous path, 400, 434
Continuous processing, 420
Continuous production, 677
Continuous-drawing processes, 313
Continuous-flow processes, 410
Continuous-path programming, 401
Continuous-slab, 51
Contouring operations, 401
Control:
    adaptive, 404, 443, 444
    closed-loop, 394
    open-loop, 393
    quality, 274
    real-time, 443
Control chart, 274
Controller, 432
Conventional machines, 370
Conversion rules, 11
Conveyor, 383, 429
Conveyor belt, 430
Coolant, 464, 465, 530
    chemical, 465
Coordinate motions, 435
Coordinate robot, 435
Coordinate systems, 435
Coordinates:
    cartesian, 395
    rectangular, 395
Cope, 77
Copper, 19
Copper alloys, 68
Copper base alloys, 71
Copper production, 64
Core, dry-sand, 80, 92
Core binders, 93
Core-blowing machines, 94
Core box, 86, 92
Core hardness, 87
Core making, 93
Core prints, 92
Core types, 92
Corrosion resistance, 60
Cortesian robot, 435
Corundum, 615
Cost estimate, 663
Cost estimator, 683
Cost minimization, 412
Cost plus, 677
Cost reduction, 664
Counterboring, 514
Countersinking, 514
CPP, 401

Crane, 383
Critical point, 131
Critical temperature, 47
Cross rail, 484
Cross-slide, 483
Crown gear, 587
Crucible, 42, 43
Crucible furnaces, 66
Cryolite, 62
C-shaped frame, 375
Cupola, 40, 47
    hot-blast, 42
Curling, 327
Curve, tool life, 470
Cut-and-strip masters, 655
Cutter:
    angle-milling, 518
    arbor, 516
    end mill, 518
    face, 516
    formed, 516
    form milling, 518
    gear generator, 590
    inserted-tooth, 518
    metal-slitting saw, 518
    milling, 479, 516
    multiple-point, 449
    multipoint, 505
    plain milling, 516
    profile, 516
    shank, 516
    side milling, 518
    single-point, 479
    steel, 516
    T-slot, 518
Cutting:
    oxyacetylene, 164
    three-dimension, 449
    transfer-arc, 181
    two-dimensional, 449
Cutting angle, 457
Cutting edge, 507, 512
Cutting fluids, 467
Cutting horsepower, 455
Cutting-off operations, 483
Cutting principles, 449
Cutting speed, 456, 458, 463, 467, 473, 535, 548
Cutting speed shaper, 548
Cutting time, 525
Cutting tool nose radius, 455
Cutting tools, 449, 505
    multiple-point, 449
    single-point, 367

702 INDEX

Cutting torch:
  Heliarc, 181
Cutting velocity, 473
Cut-weight, 308
Cyanide salt, 148
Cycle, 673
Cylindrical grinder, 370
Cylindrical turning, 520

Dead center, 378
Dead stops, 422
Debugging, 408
Deburr, 614
Dedendum, 584
Deep drawing, 360
Deep-hole drilling, 494, 510
Deep-hole drilling machine, 494
Deformation, plastic, 286
Deformation process, 286
Degassed, 45
Degrees of freedom, 435
Delays, 676
Dendrite, 21
Dendritic structure, 51
Deoxidizer, aluminum, 136
Deplating operation, 630
Depth of cut, 451, 456
Design, 4, 644
  computer aided, 391
  part, 422
  tooling, 408
Design attributes, 445
Design length, 537
Deviation, standard, 275
Devices, servo-controlled, 432
Diagram, free-body, 450
Dial:
  control, 404
  indicator, 262
  station feed, 351
Dimetral pitch, 584
Diamond, 461, 615
  band cutting, 560
  wheel grinding, 625
Diamond stylus, 255
Diaphram molding machine, 96
Die:
  combination, 106
  hob, 322
Die casting, 104
  cold chamber, 105, 108
  hot chamber, 105, 106
Die casting advantages, 105

Die casting alloys, 70
  aluminum, 71
  copper, 71
  lead, 72
  magnesium, 71
  tin, 72
  zinc, 70
Die casting dies, 105
Die casting limitations, 105
Dielectric fluid, 628
Dies, 70, 105, 351, 575
  multi-cavity, 106
  progressive, 358
  simple, 358
  steel rule, 361
  subpress, 359
Die sets, 358
Die stock, 576
Die wear, 309
Diffusion, 653
  wear, 468
  welding, 190
Dimension:
  absolute, 396
  incremental, 396
Dimensioning, 244
Dimension preset tools, 386
Direct arc, 38
Direct-chill process, 127
Direct computer information, 390
Direct-current chucks, 382
Direct-current motors, 377
Direct extrusion, 299
Direct iron, 33
Direct labor estimating, 672
Direct materials, 682
Direct numerical control, 403
Direct reduction, 35
Discontinuous chip, 461
Disposable clamped-on insert, 506
Disposable pattern, 75, 79
Distortion, 83
Divider, 251, 252
Dividing head, 533
DNC, 403, 413
Dog, 377
Dopants, 653
Double cut, 560
Double V-ways, 372
Double-acting presses, 341
Double-margin drills, 511
Double-seaming, 327
Dowel pins, 86

Down:
  milling, 530
Draft, 83
Drag, 77
Draw bar, 380
Draw-back collets, 380
Draw bench, 312
Draw-in chuck, 479
Drawing, 142, 303, 355
  block, 313
  deep, 360
Draw spike, 78
Drill(s), 508
  clearance, 514
  double-margin, 511
  gun, 508
  helix angle, 528
  sensitive, 490
  single-fluted, 508
  spade, 510
  special, 510
  step, 511
  twist, 508
  upright, 490
Drill chuck, 379
Drill cutting fluids, 530
Drill heads, 492
Drill hole size, 538
Drilling, 449, 479, 526
  deep-hole, 510
Drill jigs, 531
Drill point, 529
Drill point angle, 528
Drill press, 370, 490
Drive(s), 377
  friction disk, 490
  hydraulic, 349
  motor, 455
Drive mechanisms, 348
Drop forging, 291
Dross, 44, 65
Dry cyaniding, 148
Dry-sand molds, 76
Ductile iron, 56
Ductile materials, 311
Ductility, 27, 287, 466
Dump box, 121
Duplicating lathe, 486
Duplicating machines, 500
Durable goods, 408
Dynamic stiffness, 367
Dynamometer(s), 451, 467
  electronic load, 454

Eccentric drive, 348
Economics, system, 408
Eddy current testing, 273
Edge flanging, 327
EDM, 627
Efficiency in production, 6
Efficiency of spindle drive, 456
Ejector pins, 109
Ejector rod, 108
Elastic limit, 26, 287, 310
Elastomer, 442
Electrical component sequencer, 656
Electrical discharge machining, 627
Electric arc furnace, 35
Electric butt welding, 301
Electric furnace, 38
Electric gaging, 265
Electrochemical grinding, 631
Electrochemical machining, 630
Electrocoating, 620
Electrode(s), 175, 628
  graphite, 39
  tungsten, 177, 178
Electrode coatings, 176
Electrode materials, 628
Electroforming, 641
Electrohydraulic forming, 329
Electroless etching, 638
Electrolysis, 161
Electrolyte, 638
Electrolytic cells, 64
Electrolytic deposition, 198, 641
Electrolytic solution, 641
Electron beam machining, 633
Electron beam welding, 183
Electronic fabrication, 652
Electronic gaging, 263, 265
Electronic load dynamometers, 454
Electron microscope, 22
Electroplating, 647
Electroplating solutions, 642
Electroslag casting, 112
Electroslag welding, 182
Electrospark forming, 329
Elevated-temperature machining, 635
Eli Whitney, 368
Emboss, 343
Embossing, 323
  rotary, 324
Emery, 615
End clearance, 457
End effector, 431, 437, 438
End mill cutters, 518

End support column, 496
Endurance strength, 29
End-of-arm tooling, 437
Energy, 4
Engineer, 12
Engineering:
　industrial, 6
　manufacturing, 6
　systems, 408
　value, 7
Engineering economic methods, 413
Engineering materials, 7
Engineering properties, 25
Engine lathe, 480, 481
English units, 11, 242
Enterprise:
　manufacturing, 3
Environmental hot forming, 306
Equilibrium diagram, 23
Equipment:
　transfer, 422
Error, 22
Etch factor, 637
Etched, 22
Etching, 21
Etching reagent, 22
Eutectoid point, 133
Eutectoid steel, 133
Exothermics, 82
Expansion reamers, 513
Explosive compacting, 204
Explosive forming, 328
Explosive welding, 189
Extraction, 18
Extruded metal, 33
Extruding, 202
Extrusion, 298
　direct, 299
　impact, 299
　indirect, 299
　tube, 303

Fabricated parts, 505
Face cutters, 516
Face milling, 368
Face plate, 379, 479
Facing, 520
Factory, 408
Factory of the future, 442
Farrel-Sykes machines, 592
Fatigue life, 311
Fatigue strength, 29
Feed, 456, 474, 530
Feedback, 393, 432

Feed in inches per revolution, 372
Feed rate, 536, 668
Feed rod, 480
Feeler gage, 262
Fellows gear shaper, 590
Felting, 204
Ferrite, 51, 54, 133
Ferrite definition, 54
Ferromagnetic, 272
Ferromanganese, 57
Ferrous metals, 33
Fettling, 99
Fiber metal process, 204
Filler material, 214
Fillers, 214
Fillet, 84
Filters, 211
Finance and accounting, 5
Fine-grained steels, 136
Fine pearlite, 141
Finish, 83
Finish rolling, 286
Fixed capital, 4
Fixed zero, 395
Fixed-sequence robot, 432
Fixtures, 408, 418, 495, 505, 664
　lathe, 531
Flake graphite, 53
Flame carburizing, 163
Flame hardening, 150
Flame reducing, 163
Flank, 584
Flank wear, 469
Flash, 106
Flash welding, 172
Flat cable termination machines, 659
Flattening, 327
Flat-way, 374
Flexibility, 376
Flexible machine tool systems, 376
Flexible manufacturing systems, 416, 677
Flexible work stations, 410
Floating point, 395
Flooding, 465
Floor molding, 77
Flow line layout, 415
Flow welding, 187
Fluid drives, 377
Fluidized bed, 153
Fluorescent penetrants, 273
Fluorspar, 37, 41
Flutes, 575
Flux, 62, 157
　borax, 159

Fly cutter, 510
Flywheel, 339
FMS, 412, 416
Foam plaster mold, 120
FOB, 680
Foil, 288, 313
Force, 454
   frictional, 451
   horizontal cutting, 451
   normal, 450
   piezoelectric, 455
   tangential, 451
   tool, 455
   vertical, 451
Forces on chip, 450
Force triangles, 451
Force-feed lubrication, 374
Forging, 291
   drop, 291
   press, 295
   roll, 297
   smith, 291
   upset, 296
   warm, 305
Format, 401
Formed cutters, 516
Formed tooth process, 588
Forming, 353
   cold roll, 324
   electrospark, 329
   explosive, 328
   magnetic, 330
   roll, 324
   stretch, 318
Form milling cutter, 518
Foundry, 75
Foundry molding machines, miscellaneous, 98
Foundry sand, 86
Fourslide, 347
Fracture of tool, 471
Fragmentation, 311
Frame:
   C-shaped, 375
   machine, 371
Free-body diagram, 450
Free enterprise system, 4
Free on board, 680
Friction, 449
Frictional force, 451
Frictional resistance, 450
Friction disk drive, 490
Friction welding, 184
Fringe patterns, 254
Full automation, 412

Fullering, 309
Furan molds, 76
Furfuryl binders, 94
Furnace, 205
   acidic, 38
   air, 37
   AOD, 42
   basic, 38
   basic oxygen, 37
   bell top, 153
   blast, 34
   box, 153
   car bottom, 152
   continuous, 153
   converter, 37
   crucible, 37, 43, 66
   cupola, 37, 40
   electric, 37, 38, 42
   electric arc, 35, 36
   fluidized bed, 153
   heat treating, 152
   induction, 37, 42
   open hearth, 36, 37, 39
   puddling, 46
   regenerative, 39
   reverbatory, 37, 39
   rotary, 42
   salt bath, 153
   smelting, 60
   softening, 65
   vacuum, 44, 153

Gage, 260
   feeler, 262
   geometrical, 268
   inspection, 260
   manufacturing, 260
   plug, 262
   snap, 261
   strain, 267
   surface, 253
Gage allowance, 261
Gage blocks, 268
Gaging:
   electric, 265
   electronic, 263, 265
   machine, 268
   pneumatic, 264
Galvanizing, 648
Gamma iron, 133
Gang drilling machine, 491
Gangue, 60
Gannister, 38
Gap press, 340

706 INDEX

Garnet, 615
Gas-assisted laser, 634
Gas carburizing, 147
Gate, 78, 80, 88
Gate barrier, 385
Gathering:
   metal, 296
Gear, 571, 582
   annular, 587
   bevel, 586, 592
   crown, 587
   finishing, 595
   helical, 586
   herringbone, 586, 593
   hypoid, 587
   miter, 587
   pinion, 586
   rack, 587
   spur, 586
   types, 586
   worm, 587
   zerol, 587
Gear cutting, 368
Gear cutting machines, 424
Gear face, 584
Gear generating, 590
Gear grinding, 596
Gear hob cutter, 593
Gear hypoid, 593
Gear lapping, 596
Gear manufacture, 587
Gear nomenclature, 582, 583
Gear pitch, 584
Gear shaper, 590
Gear shaving, 596
Gear speed, 585
Gear spiral bevel, 593
General administration, 5
General purpose machinery, 8
Generative process plans, 683
Geometrical gage, 268
Geometry, 457
Glass substrate, 442
Go, 261
Gold, 65
Grain boundary, 21
Grain growth, 137
Grain size, 21, 136
Granular resins, 214
Granulation, 198
Graphite, 53
Graphite electrodes, 39
Graphitic carbon, 55
Gravity casting, 111

Gravity drop, 292
Gravity drop hammer, 293
Gravity feed, 351
Gravity segregation, 113
Gravity sintering, 203
Gray cast iron, 17, 53
Gray iron, 53
Green compact, 210
Green-sand molds, 75
Grinder:
   centerless, 602
   cylindrical, 602
   foundry, 100
   internal, 606
   jig, 609
   tool post, 604
Grinding, 599
   abrasive belt, 612
   bench, 608
   chip thickness, 601
   electrochemical, 631
   high energy, 612
   machines, 601
   off hand, 608
   pedestal, 608
   plunge cut, 604
   surface, 607
Grinding tools, 457
Grinding wheels, 616
   abrasives, 617
   bond, 618
   bond strength, 618
   grain spacing, 618
   marking system, 618
   shape, 617
   size abrasive, 618
Grinding wheel data, 617
Group classification, 683
Group technology, 444
Group technology layout, 414
Growing processes, 653
GT, 414, 444
Guerin process, 360
Guidance, 418
Gun drill, 492, 508
Gypsum, 120

Hacksaw blades, 555
Hacksaw power, 554
Hammer, 291
Hammer gravity drop, 293
Handling time, 665
Hand buttons, 386
Hand feed, 385

Hand forging, 291
Hand ladles, 99
Hand milling machine, 497
Hand-scraping, 374
Hard automation, 428
Hardenability, 138
Hardened steel constituents, 140
Hardened strain, 310
Hardening, 138
   age, 151
   flame, 150
   induction, 149
   nonferrous, 151
   precipitation, 151
   strain, 287
Hardness, 29, 466
   Moh's, 29
Hardness measurements, 271
Hardness of steel, maximum, 141
Hardness strength, 48
Hardness tester, 29
   Rockwell, 29
Hard surfacing, 650
Headstock, 371, 481
Hearth, 38
Heat treatment, 131, 467
Heat treatment definition, 131
Heavy stamping, 343
Heliarc, 181
Helical gears, 586
Helium-neon gas lasers, 258
Helix angle, 528
Hematite, 33
Hermaphrodite calipers, 251
Herringbone, 592
Herringbone gear, 586
Hexagon turret, 525
High-carbon steel, 459
High-carbon steel cutters, 516
High energy grinding, 612
High-energy forming, 305
High-energy rate forming, 327
Higher level languages, 404
High-frequency welding, 173
High-speed steel, 459, 516
High-speed steel cutter bits, 505
High-speed tools, 458
Honing, 609
Honing coolants, 610
Horizontal boring, 370
Horizontal boring machine, 496
Horizontal cutting force, 451
Horizontal milling machine, 370
Horizontal turrret lathes, 482

Horn press, 341
Horsepower, 455, 539
   cutting, 455
   idle, 455
Horsepower at motor, 456
Hot-blast cupola, 42
Hot box cores, 94
Hot-chamber die casting, 106
Hot finished, 288
Hot forming, environmental, 306
Hot-ladle cars, 36, 40
Hot pressing, 206
Hot spinning, 304
Hot working, 51, 68, 310
Hot-worked metal, 287
Hours per 100 units, 676
HSS, 459
Human resources, 5
Hunter process, 313
Hydraulic drives, 349
Hydraulic press, 345
Hydrogen, 36, 161, 177
Hydrogen arc welding, 177
Hypereutectoid steels, 134
Hypoid gears, 587

Idle horsepower, 455
Illite, 87
Impact extrusion, 299
Impacter forger hammer, 293
Impurities, 287
Inclined press, 339
Incremental dimension, 396
Independent chuck, 379
Index head, 533
Indexing, 670
Indexing table, 493
Indirect arc, 38
Indirect extrusion, 299
Induction furnaces, 42
Induction hardening, 149
Induction heating, 149
Induction welding, 173
Industrial engineering, 6
Industries, job lot, 409
Inert gas shielded arc welding, 178
Inference, 246
Infiltration, 207
In-line manufacturing cell, 416
Ingot, 33, 50, 127, 286
Inorganic materials, 8
Input, 4
Inserted-tooth cutter, 518
Insertion machines, 657

Inside calipers, 251
Inspection, radiographic, 272
Inspection gages, 260
Inspector, 275
Integrated circuit, 652
Integrated production systems, 505
Integrity, surface, 471
Intelligent robot, 432
Interchangeability, 533
Interchangeable manufacture, 274
Interchangeable parts, 7
Interchangeable production, 1, 242
Interference fringes, 254
Interferometer laser, 259
Interferometry, 253, 258
Intermetallic alloys, 21
Internal grinders, 606
Interpolation:
    circular, 401
    linear, 401
Interpolation routine, 401
Interstitial solid solution, 20
Intraforming, 321
Inventory, 376
Inverse rate curve, 131
Inverted drawing die, 356
Investment casting, 116
Involute, 583
Involute gears, 583
Iodine, 18
Iron, 8, 19
    alpha, 20
    basic, 35
    Bessemer, 35
    cast, 53
    ductile, 56
    foundry, 35
    gamma, 20
    malleable, 35, 55
    nodular, 56
    sponge, 36
    white cast, 55
Iron carbide, 133
Iron hands, 422
Iron-iron carbide diagram, 131
Iron pyrite, 33
Irregularly shaped grains, 310
ISO standards, 573
Isostatic molding, 203
Isothermal annealing, 146
Isothermal diagram, 137
Isotherms, 81

Jacksaw, 553
Jarno and Reed tapers, 521

Jaw chuck, 479
Jig, 408, 418, 479, 505, 532, 664
    box-type, 532
    drill, 531
    plate-drilling, 532
    table-type, 533
Jig boring machine, 495
Jig grinder, 609
Job lot industries, 409
Job lot production, 8, 408
Jointed spherical coordinate robot, 435
Jolt-molding machine, 95
Jolt-squeeze machine, 96
Jolt-squeeze rollover machine, 96
Jominy test, 138

Kaolinite, 87
Kerf, 181, 555
Kerosene lubricant, 465
Key seater, 550
Killed steel, 51
Kilogram, 11
Kinematic inversions, 367
Knuckle joint mechanism, 342, 349
Knuckle joint press, 342

Lance, 37
Lancing, 352
Land, 508, 566
Lap, 610
Lapping, 596, 610
Lap welding, 301
Lard oil compounds, 466
Laser, 258, 439, 632
    gas-assisted, 634
    interferometer, 259
    tooling, 259
Laser beam machining, 632
Laser interferometer, 259
Laser welding, 184
Lathe, 370, 479
    automatic, 486
    bench, 481
    duplicating, 486
    engine, 480, 481
    horizontal turret, 482
    multistation, 485
    ram-type turret, 482
    saddle-type, 482
    speed, 481
    step cone-driven, 481
    toolroom engine, 482
    tracer, 486

turret, 482
vertical turret, 484
Lathe fixtures, 531
Lathe operations, 520
Lathe size, 481
Lathe tools, 521
Lattice, 310
   body-centered, 20
   face-centered, 20
   hexagonal, 20
Lattice distortion, 311
Layout:
   group technology, 414
   machine center, 415
Le System International d'Unit, 11, 242
Lead, 19, 46, 106, 576
Lead-base alloys, 72
Lead production, 65
Lead screw, 480
Lead-through, 434
Lead-tin phase diagram, 158
Length of cut, 453, 536
Light section principle, 471
Lime, 37
Limestone, 33, 41
Limit:
   elastic, 287
   tolerance, 276
Limited-sequence robots, 432
Linear interpolation, 401
Linear variable differential transformer, 266
Lining, basic, 37
Lip relief angle, 508
Liquid carbonitriding, 148
Liquid carburizing, 147
Liquid nitriding, 148
Loading and unloading time, 524
Loam molds, 76
Longitudinal thrust, 456
Lost wax process, 117
Lower control limit, 277
Lubricants, 317
LVDT, 266

Machinability, 67, 466
Machinability rating, 467
Machinability tests, 466
Machine, 495
   automatic screw, 487
   automatic transfer, 493
   boring, 495
   cluster drilling, 492
   conventional, 370
   deep-hole drilling, 494

   duplicating, 500
   flat cable term, 659
   gang drilling, 491
   hand milling, 497
   horizontal boring, 496
   insertion, 657
   jig boring, 495
   metal removal, 367
   milling, 479, 497
   multispindle, 487, 497
   plain milling, 497
   planer-type mill, 499
   planetary milling, 500
   radial drilling, 491
   sawing, 553
   single-wire, 659
   transfer-type, 492
   tube-forming, 325
   turret, 492
   unmanned, 417
   upset forging, 297
   vertical milling, 497
   wave soldering, 658
Machine cell, 414, 416
Machine center layout, 415
Machine control unit, 393
Machine frame, 371
Machine gaging, 268
Machine handling time, 524
Machine hour cost, 676
Machine molding, 77
Machinery:
   general purpose, 8
   special purpose, 8
Machine tools, 367
   automatic, 385
   NC, 390, 410
Machine vision, 439
Machining:
   abrasive-jet, 626
   cold-temperature, 636
   EDM, 627
   electrochemical, 630
   electron beam, 633
   elevated-temperature, 635
   laser beam, 632
   ultrasonic, 625
   water jet, 627
Machining centers, 367, 376, 391, 499
Machining energy, 625
Machining systems, 414
Machining time, 667
Machinist, 11
Magazine feeds, 351

Magnaflux, 272
Magnesite, 38, 40, 42
Magnesium, 19, 69, 108
Magnesium alloys, 67, 69
Magnesium base alloys, 71
Magnesium production, 63
Magnetic chucks, 382
Magnetic forming, 330
Magnetic particle inspection, 271
Magnetic tape, 390, 398
Main frame, 431
Maintenance, 408
  prevention, 422
Malleable iron, 55, 467
Management, 4
Mandrel, 378
Manganese, 56
Manipulator, 431
  manual, 432
Manpower, 408
Manu factus, 2
Manual, 434
  assembly, 410
  manipulator, 432
Manufacture, 1
Manufacturing, 479
  computer aided, 391
  development, 4
  engineering, 6
  enterprise, 3
  gages, 260
  history, 1
  processes, 1
  systems, 408
Manuscript, 392
Marform process, 360
Margin, 508, 512
Marketing, 5
Markup, 677
Martempering, 144, 145
Martensite, 137, 141
  tempered, 142
Martensite photomicrograph, 141
Mass production, 8, 408, 422
Mass media abrasives, 620
Mass media finishing, 612
Master surface plate, 374
Match plates, 83
Material, 408
  allotropic, 20
  automobile, 18
  crystalline, 19
  direct, 682
  electrode, 628

engineering, 8
filler, 214
inorganic, 8, 17
metallic, 8, 16
nonmetallic, 8, 16
organic, 8, 17, 214
polymorphic, 20
properties, 16, 25
source, 18
synthetic organic, 214
tool, 458
unpolymerized, 654
Maudsley, Henry, 3
MCU, 393
Mean dimension, 276
Measurements, 274
  hardness, 271
Mechanical arms, 422
Mechanical loaders, 383
Mechanical molding equipment, 94
Mechanical work, 287
Mechanism, knuckle joint, 342, 349
Mechanization, 410, 418
Melting energy required, 45
Mercury, 119
Mesa, 442
Metal:
  alloys, 23
  cast, 33
  extruded, 33
  ferrous, 33
  lattice, 20
  monel, 23
  nonferrous, 60
  powdered, 33
  principal, 18
  pure, 23
  solidification, 23
  steel, 47
  structure, 19
  wrought, 33
  wrought iron, 46
Metal-cutting machines, 371
Metal cutting theory, 449, 473
Metal electrode welding, 175
Metal flow, 289
Metal gathering, 296
Metal hardness, 22
Metallic materials, 8
Metallurgy, 131
  definition, 131
Metal molds, 77
Metal powder spraying, 645
Metal properties, 20

INDEX 711

Metal removal machines, 367
Metal removal rate, 456
Metal-slitting saw cutter, 518
Metal spinning, 313
Metal spraying, 644
Metal surface finish, 22
Meter, 11, 246
Methane, 36
Metrology, 242
Microchips, 658
Microelectronic circuit, 653
Micrometer, 249
Midline-roughness datum, 473
MIG welding, 178
Mill:
   roll, 298
   vertical boring, 496
Milling, 449, 530
   chemical, 636
Milling cutter, 449, 479, 516, 519
Milling machine, 479, 497
Mineral oil, 466
Minerals, 214
Miter gears, 587
Model:
   orthogonal cutting, 449
   quantitative process, 444
Moderate production, 8, 408
Modular automation units, 421
Modulus of elasticity, 26
Moh's scale, 29
Moire fringe, 267
Mold, 75, 288
   carbon dioxide, 76, 122
   continuous, 126
   dry sand, 76
   foam, 120
   furan, 76
   graphite, 111
   green-sand, 75
   ingot, 50
   loam, 76
   metal, 77, 111
   metallic, 104
   palster, 120
   permanent, 104
   reciprocating, 124
   rubber, 123
   sand, 75
   Shaw, 123
   skin-dried, 76
   special, 77
   wood, 123
Molecule, plastic, 214

Monel, 23
Money, 408
Monitoring, on-line, 444
Monochromatic light, 254
Monomers, 214
Morse taper, 521
Motor:
   a-c induction, 377
   constant-speed, 481
   direct-current, 377
   drive, 455
   variable-speed, 481
Mottled cast iron, 55
Movable pallets, 416
Muffle furnace, 153
Mufflers, 90
Multiarmed assembly robot, 442
Multiple cuts, 483, 525
Multiple-point cutters, 449
Multiple-point cutting tools, 449
Multipoint cutters, 505
Multispindle automatic lathes, 424
Multispindle drilling machines, 492
Multispindle machines, 487

Natural aging, 152
NC, 390, 484
NC machine tools, 390, 410
NC program, 390
NC robots, 432
Negative image, 641
Net present worth, 413
Nicarbing, 148
Nickel, 19
Nitriding, 148
Nitriding liquid, 148
Nitriding iron, 56
No-bake cores, 94
No-go, 261
Nondestructive inspection, 29
Nondestructive testing, 271
Nonferrous material, 8
   aluminum, 17
   cast zinc, 17
   copper, 17
   lead, 17
   magnesium, 17
   properties, 60
   titanium, 17
Nonmagnetic, 50
Nonmagnetic steels, 50
Nonmetallic materials, 8
Nonrecurring, 413
Nonrobot costs, 413

712   INDEX

Nonservo, 432
Nonservomechanism robots, 432
Nonservo robots, 432
Nontraditional methods, 625
Nontransferred plasma, 181
Normal, force, 450
Normal curve, 277
Normalizing, 146
Nuclei, 21
Number of defectives, 279
Numerical control, 390
Numerical control machine, 9
Numerical data, 390

Oblique cutting tool, 455
Occupational Safety and Health Act, 383
On-line monitoring, 444
On-time delivery, 408
Op sheet, 663
Opaque printed pattern, 654
Open-back press, 339
Open hearth:
  acidic, 40
  basic, 40
Open-hearth furnace, 37, 39
  basic oxygen, 36
  electric arc, 36
Open-loop control, 393
Operation unit cost, 665
Operations:
  contouring, 401
  cutting-off, 483
  deplating, 630
  press, 351
  press-working, 338
Operations planning, 663
Operations sheet, 391, 672
Optical fiber based sensor, 441
Optical flat, 253
Optical instruments, 250
Optical laser, 633
Ore dressing, 60
Ores:
  argentite, 19
  bauxite, 19
  bornite, 19
  cassiterite, 19
  chalcocite, 19
  cryolite, 19
  dolomite, 19
  galena, 19
  hematite, 19
  limonite, 19
  magnesium, 19
  magnetite, 19
  pentlandite, 19
  seawater, 19
  siderite, 19
  sphalerite, 19
Organic materials, 8, 214
Organization, 5
Orthogonal cutting model, 449
Orthogonal tools, 449
Output, 4
Outside calipers, 251
Ovality, 269
Overtravel length, 537
Ownership, 5
Oxidation, 288
Oxidation wear, 468
Oxidizing flame, 163
Oxyacetylene cutting, 164
Oxyacetylene processes, 651
Oxyacetylene welding, 161
Oxygen, 161
Oxygen lance, 40
Oxyhydrogen welding, 164

Pack carburizing, 147
Pallet transfer, 417
Palletized transfer machine, 494
Palm buttons, 386
Paper tape, 390
  perforated, 393
Paraffin oil, 466
Parkerizing, 649
Part design, 422
Part families, 445
Partial automation, 412
Parting sand, 78
Part programming, computer assisted, 392
Parts, interchangeable, 7
Parts orientation, 423
Parts programmer, 391
Pattern, 82, 637
  disposable, 75, 79, 86
  polystyrene, 75
  removable, 75, 77, 82
  types, 82
Payback, 412
Payback robot, 432
P-Chart, 274
Pearlite, 51, 54
  definition, 54, 133
  fine, 141
Peening, shot, 334
Pellets, 306
Penthane, 86

Percentage elongagation, 28
Percussion presses, 349
Percussion welding, 173
Perforated paper tape, 393
Perforation, 358
Peripheral velocity, 535
Perishable tooling, 664
Permanent mold, 36, 104
Permanent-mold alloys, 68
Permanent-mold casting, 109
Personal requirements, 675
Pewter, 72
Phenol formaldehyde, 94, 121
Phenolic resin, 121
Phosphoric acid, 76
Phosphorus, 57
Photoelectric cells, 398
Photoelectric detector, 258
Photoelectric methods, 398
Photoelectric switch, 438
Photographic negatives, 640
Photographic resist process, 640
Photography, 656
Photolithography, 653
Photomask, 655
Photomicrographs, 134
Photoresist, 640, 654
Pick and place, 428
Pickling, 100
Pieces per hour, 676
Piercing, 301, 352
   progressive, 296
Piezoelectric force, 455
Pig iron, 33
   classification, 35
Pinion, 586
Pipe, 50
Pipe and tube manufacturing, 300
Pipe threads, 575
Pitch, 572
Pitch circle, 582
Pitch diameter, 579
Pit molding, 77
Pixels, 439
Plain milling cutter, 516
Plain milling machine, 497
Planer, 370, 499, 550
   bevel gear, 589
   double-housing, 551
   open-side, 551
   pit or edge, 552
   pit type, 552
Planer classification, 550
Planer drive, 550

Planers and shapers, differences, 552
Planer-type milling machine, 499
Planetary milling, 578
Planetary milling machine, 500
Planing, 449, 546
Planner, 671
Plant engineer, 6
Plant layout, 414
Plasma, 181
Plasma arc cutting, 635
Plasma flame spraying, 645
Plaster, 118
Plaster mold casting, 119
Plastic deformation, 286
Plasticity, 289
Plasticizers, 214
Plastic molecule, 214
Plastics, 214
Plastic state, 287
Plate, 288, 449
Plate bending, 326
Plate-drilling jig, 532
Plating, 209
Plug gages, 262
Plunge cut grinding, 581
Pneumatic gaging, 264
Pocketed chuck, 382
Point-of-operation guards, 385
Point to point, 434
Point-to-point programming, 400
Polar, 436
Polishing, 614
Polishing wheels, 614
Polymer, 214
Polymerization, 214
   addition, 214
   condensation, 214
Polymorphic, 20
Polystyrene, 75, 86
Polystyrene pattern, 79
Porosity, 287
Positioning method, 400
Pouring basin, 80, 81
Pouring temperature, 66
Powder, aluminum oxide, 461
Powdered metal, 33
Powdered metallurgy, 196
   advantages, 209
   apparent density, 197
   centrifugal, 202
   coining, 209
   compressibility, 197
   dies, 200

714 INDEX

Powder metallurgy (*Continued*)
  explosive, 204
  extruding, 202
  fiber metal, 204
  furnaces, 205
  gravity, 203
  green compact, 196, 201
  heat treatment, 209
  hot pressing, 206
  hydrostatic, 203
  impregnation, 207
  infiltration, 207
  isostatic, 203
  limitations, 209
  lubricants, 199
  plating, 209
  powder, 197
  pressing, 199
  rolling, 203
  sintering, 197, 205
  sizing, 209
  slip casting, 202
  uses, 211
Power:
  average unit, 540
  motive, 3
Power Age, 479
Power belts, 3
Power chucks, 379
Power and free material handling, 430
Precipitation hardening, 242
Precision, 11, 242
Precision casting, 116
Press:
  arch, 340
  C-frame, 340
  double-acting, 341
  drill, 490
  gap, 340
  horn, 341
  hydraulic, 345
  inclined, 339
  knuckle joint, 342
  open-backed, 339
  percussion, 349
  rack and gear, 348
  straight-side, 340
  transfer, 346
  turret, 345
  types, 338
Press brake, 343
Press capacity, 295
Pressed Corthias casting, 112

Press forging, 351
Press operations, 351
Pressure, 352
Press-working operations, 338
Preventive maintenance, 422
Price, 676
Primary operations, 411
Principle:
  cutting, 449
  light section, 471
Principles of cold working, 310
Printed circuit boards, 628, 639
Printed circuit board stuffing, 658
Process, 1
  deformation, 286
  manufacturing, 1
  marform, 286
  thermoelectric, 632
Process annealing, 146
Process efficiency, 411
Processing:
  continuous, 420
  rubber pad, 360
Processor language, 392
Production:
  batch, 412
  cell, 413
  interchangeable, 1, 242
  job lot, 8, 408
  mass, 8, 408, 422
  moderate, 408
Production of aluminum, 62
Production analysis, 664
Production of lead, 65
Production of magnesium, 63
Products, 449
Profile cutters, 516
Profit, 408
Profit maximization, 412
Program, 391
Programmer, parts, 391
Programming, 408
  continuous-path, 401
  point-to-point, 400
Progressive die, 358
Progressive piercing, 296
Projecting comparators, 263
Projection welding, 169
Properties:
  engineering, 25
  ferrous, 17
  malleable iron, 17
  nonferrous, 17

steel, 17
white cast iron, 17
wrought iron, 17
Proportion, 279
Proportional limit, 26
Prototype, 410
PTP, 400
Puddling process, 46
Pull-back devices, 386
Punch, 343, 351
auxiliary, 306
Punched tape, 396
Punching, 352
Purchasing, 5

Quality, 9
Quality assurance, 6
Quality control, 274
Quantitative process model, 444
Quartz element, 455
Quench, 138
Quenching, 22, 132
Quick-change gear box, 481
Quick-return mechanism, 547

Rack and gear presses, 348
Rack gears, 587
Rack tooth, 585
Radial, 456
Radial drilling machine, 491
Radian, 252
Radiofrequency, 635
Radiographic inspection, 272
Rake angle, 449, 451
Raker tooth, 556
Ram, 492
Ram-type turret lathe, 482
Random basis, 275
Random causes, 274
Rate of metal removal, 539
Rate of return, 413
Rating, machinability, 467
Readability, 244
Real time, 444
Real-time control, 443
Reamer, 511
adjustable, 513
chucking, 513
expansion, 513
rose, 513
shell, 513
taper pin, 513
Reciprocating mold process, 124

Recrystallization temperature, 286, 310
Rectangular coordinates, 395
Red hardness, 460
Reducing flame, 163
Reduction cells, 62
Refining, 18
Regenerative furnace, 39
Removable pattern, 75, 77, 82
Residual stresses, 311
Resin binders, 94
Resinoid process, 617
Resin, 214
granular, 214
Resistance, 454
frictional, 450
Resistance welding, 166
Resistors, 625
Resolution units, 403
Reverbatory furnace, 39, 63, 64
Reverse polarity, 174
Revolutions per minute, 372, 475
Rimmed steel, 51
Riser, 51, 79, 80, 81
Riveting, 323
Roasting oven, 62
Robot, 383, 408, 430
air logic, 434
cartesian, 435
coordinate, 435
fixed-sequence, 432
intelligent, 432
jointed spherical coordinate, 435
multiarmed assembly, 442
NC, 432
nonservo, 432
nonservomechanism, 432
servocontrolled, 433
spherical coordinate, 435
variable-sequence, 432
Rockwell hardness tester, 29
Rollability, 578
Roller conveyor, 429
Roll forging, 297
Roll forming, 301, 324
Rolling, 288
Rolling powdered metal, 203
Roll mill, 298
Root mean square, 256
Rose reamers, 513
Rotary broaching, 564
Rotary chucks, 382
Rotary embossing, 324
Rotary furnaces, 42
Rotary planer, 496

Rotary swaging, 319
Roughness, 269
  average, 256
  surface, 255, 471
Rounding, 11
Round-nose tools, 467
Route sheet, 392, 671
Rubber molds, 123
Rubber pad processing, 360
Rubber process, 617
Rules for conversion, 11
Runner, 80
Runways, 371

Saddle, 371
Saddle-type lathe, 482
Safety, 5, 385, 421, 537
Safety length, 537
Safety stock, 537
Sales, 5, 664
Salt bath, 153
Sample, 275
Sand:
  clay content, 89
  fineness, 88
  foundry, 86
  grain size, 87
  moisture, 89
  parting, 78
  permeability, 87, 89
  refractoriness, 87
Sandblasting, 100
Sand castings, 75
Sand conditioning, 90
Sand molds, 75
Sandpaper, 619
Sand reclamation, 90
Sandslinger, 98
Sandstone, 615
Sand strength, 87, 90
Sand testing, 87
Saw, 370
  feed, 554
Saw cutters, 520
Sawing, 360, 546
  abrasive, 558
  band, 558
  circular, 556
  cold, 556
  friction, 558
  hand, 553
  feed, 554
Sawing machines, 553
  classified, 554

Scale losses, 308
Scaling, 288
Scarfing, 160
Scrap, 682
Scraping, 374
Screw thread, 571
S-curves, 137
Seacoal, 122
Seam, 327, 343
Seamless tubing, 301
Seam welding, 170
Secondary assembly operations, 411
Segmental chip, 461
Selective metal-reducing, 638
Semicentrifugal casting, 115
Semiconductor, 653
Sensing device, 386
Sensitive drills, 490
Sensitive machines, 385
Sensitivity, 244
Sensor:
  optical fiber-based, 441
  robotic, 437
  silicon, 441
  tactile, 440
  vision, 438
  voice, 442
Series motor, 377
Servo controlled devices, 432
Servo controlled robots, 433
Servomotors, 394
Set, 556
Setup, 673
Setup cost, 376
Setup time, 524
Shake, 83
Shakeout, 99
Shank, 505
Shank cutters, 516
Shaper, 370, 546
  classification, 546
  cutting speed, 548
  draw cut, 549
  hydraulic, 548
  push cut, 547
  universal, 547
  vertical, 549
Shapers and planers, cutting tools, 553
Shaping, 546
Shareholders, 408
Sharp sand, 87
Shaving, 352, 595
Shaw process, 123
Shear, 450

Shear plane angle, 453
Shear spinning, 317
Shear strength, 27
Shearing, 351
Shear angle, 451, 463
Sheets, 288
　aluminum, 306
Shell molding, 121
Shell reamers, 513
Shellac, 86
Shellac process, 617
Shop efficiency, 421
Shore scleroscope, 29
Shot peening, 311, 334
Shotting, 47
Shrinkage, 50, 83, 682
Shrinkage cavity, 81
Shrink rule, 83
Shunt motor, 377
Shutdown, 421
Side milling cutter, 518
Side rake angle, 457
Side relief angle, 457
Silastic, 123
Silica sand, 86
Silicate process, 617
Silicon, 56, 652
Silicon carbide, 609, 615, 617
Silicon dioxide, 655
Silicon sensors, 441
Silicon wafer, 442, 652
Silver, 19, 65
Simple dies, 385
Sine bar, 252
Single crank, 348
Single-fluted drills, 508
Single-point cutter, 479, 505
Single-point cutting tools, 367
Single-point tool, 449, 464
Sintering, 196, 197, 205
Single-wire machines, 659
SI system, 242
Sizing, 209, 343
Skelp, 300
Skimming gates, 81
Skin-dried molds, 76
Slabs, 288
Slag, 18, 34, 46
Slag inducers, 62
Slag spout, 41
Slide rails, 371
Slip casting, 202
Slip planes, 310
Slitting, 352, 368

Slotters, 549
Slush casting, 112
Small-lot production, 242
Smear metal, 22, 596, 611
Smelting, 60
Smith forging, 291
Snagging wheel, 617
Snap gages, 261
Soaking pits, 288
Soda water, 265
Sodium silicate, 76, 122
Software, 408
Solder, 175
Soldering, 158, 175
Solidification, 80
Soluble oil, 465
Sonotrode, 188
Space lattice, 20
Spade drills, 510
Spare tools, 418
Spark sintering, 206
Spattering, 651
Special drills, 510
Special machines, 376
Special molds, 77
Special purpose machinery, 8
Specific heat, 45
Specific product machine automation, 426
Specimen, tensile, 25
Speed lathe, 481
Speeds, cutting, 535
Spencer, Christopher N., 487
Spherical coordinate robot, 435
Spheroidizing, 146
Spiegeleisen, 57
Spindle, 371
Spinning:
　hot, 304
　metal, 313
Spiral bevel gears, 593
Sponge iron, 36, 47
Spot facing, 514
Spot welder, 167
Spot welding, 168
Springback, 354
Sprue, 80
Sprue pin, 78
Spur gears, 586
Square insert, 506
Square threads, 574
Square-turret tools, 483
Squaring shears, 344
Squeeze molding machine, 96
Squeezer machine, 96

S-shaped, 414
Stack mold, 115
Stainless steels:
  austenitic, 50
  ferritic, 50
  martensitic, 50
Staking, 323
Stalk, 110
Standard, 242
Standard deviation, 275
Standard of measurement, 246
Standardization of tools, 418
Stapling, 323
Stationary collet, 380
Steadite, 54, 57
Steadite definition, 54
Steam engines, 3
Steam hammer, 291
Stearate, 197
Steel, 23, 25, 46, 47, 50
  carbon content, 36, 48
  high-carbon, 48, 459
  high-speed, 459
  killed, 51
  low carbon, 48
  medium carbon, 48
  microstructure, 51
  rephosphorized, 49
  resulfurized, 49
  rimmed, 51
Steel alloy, 131
Steel burnt, 137
Steel classification, 48
Steel friction drives, 558
Steel rule dies, 361
Step cone-driven lathe, 481
Step drills, 511
Stepping motor, 394
Step-type chip breaker, 464
Stiffness, dynamic, 367
Straightening, 343
Straight-fluted chucking reamers, 513
Straight polarity, 174
Straight-side press, 340
Straight tooth, 556
Strain, 25, 287
Strainer, 81
Strain gage, 267, 441
Strain hardened, 310
Strain hardening, 287
Strength:
  compressive, 26
  shear, 26
  torsional, 26
  yield, 287

Stresses, 444
  residual, 311
Stress-strain curve, 26
Stretch forming, 318
Strike rod, 77
Structural shapes, 288
Structures for cutting machines, 367
Stuccoing, 118
Stuffing, 658
Submerged-arc welding, 180
Subpress die, 359
Substitutional solid solution, 26
Substrate, 442, 652
Surface finish, 444, 466
Surface finishing, 609
Surface gage, 253
Surface hardening, 147
Surface integrity, 471
Surface plate, 253
Surface roughness(es), 255, 471, 472, 599
Swab, 78
Swaging, 319
Sykes machine, 592
Symbolic codes, 391
Synchronous transfer, 430
Synthetic material, 214
System:
  coordinate, 435
  flexible, 677
  flexible manufacturing, 416
  free enterprise, 4
  integrated prod, 505
  machining, 416
  manufacturing, 408
System economics, 408
System of measurement, 242
Systems engineering, 408
S-curves, 137
S-shaped, 414

Tab code, 401
Table, 370, 371
Table-type jig, 533
Tabulating cards, 440
Tactile sensors, 440
Tailstock, 480
Tampico brushes, 614
Tangential cutting, 456
Tangential force, 451
Tap and die holder, 526
Tape, punched, 396
Tape-controlled turret lathe, 484
Tap hole, 41
Tape preparation, 392

INDEX 719

Tape reader, 398
Taper pin reamers, 513
Taper pins, 521
Taper turning, 521, 522
Tapping machine, 577
Taps, 514, 575
Taylor, Frederick W., 1, 469
Taylor tool life equation, 470
Technician, 12
Technologist, 12
Technology:
    computer, 411
    group, 444
Teeming, 51
Temper carbon, 55
Temperature:
    forging, 295
    pouring, 60, 66
Tempering, 142
Template, 637
Template gear cutting, 589
Tensile strength, 25, 26, 30, 48
    hardness, 48
Testing:
    eddy current, 273
    endurance, 28
    impact, 28
    ultrasonic, 273
Tests, machinability, 466
Theory, metal cutting, 449, 473
Thermit welding, 185
Thermoelectric process, 632
Thermo-forging, 305
Thermospray, 645
Thickness of chip, 453
Thread, 571
    ACME, 574
    buttress, 574
    chasing, 576
    cutting, 575
    ISO, 573
    pipe, 575
    standard, 572
    square, 574
    types, 572
    UN, 572
    UNC, 573
    UNEF, 573
    UNF, 573
    unified, 572
    worm, 574
Three-dimensional cutting, 449
Three-high mill, 289
Throwaways, 507
TIG welding, 178

Time-temperature transformation, 137
Tin, 19, 20, 106
Tin base alloys, 72
Tin coating, 648
Titanium, 44
Toggle mechanism, 341, 349
Tolerance, 244, 391
    bilateral, 278
    unilateral, 245
Tolerance limits, 276
Tong hold, 308
Tonnage rating, 339
Tool, 370, 483, 505
    boring, 479, 514, 515
    brazed-tip, 506
    carbide, 506
    ceramic, 461
    chucking, 483
    cutting, 449, 505
    grinding, 457
    machine, 367
    oblique cutting, 455
    orthogonal, 449
    robot, 432
    round-nose, 467
    single-point, 449
    square-turret, 483
Tool angles, 458, 471, 507
Tool breakage, 418
Tool-changer magazine, 500
Tool changing, 386, 670
Tool changing cost, 669
Tool-chip interface, 535
Tool cost, 669
Tool designer, 531
Tool edge, 471
Tool face, 449
Tool forces, 455
Tool geometries, 472, 507
Tool hardness, 471
Tool interface areas, 465
Tool life, 481, 466–468, 668
Tool life curve, 470
Tool material, 458
Tool post grinder, 604
Tool signature, 507
Tool storage, 391
Tool storage rack, 386
Tool wear, 444, 467
Toolholder, 507
Tooling, 351, 421, 664
Tooling design, 408
Tooling end-of-arm, 437
Tooling laser, 259
Tooling lathe, 482

Torch, plasma, 181
Torque, 454
Torsional strength, 27
Tote box, 340
Towline, 429
Tracer, 256
Tracer lathes, 486
Traditional methods, 625
Transducer, 454, 625
Transducing device, 394
Transfer equipment, 422
Transfer press, 346
Transfer type, 493
Transferred-arc cutting, 181
Transferred plasma torch, 181
Transfer-type production drill machines, 492
Transformation diagram(s), 137, 144
Transverse variations, 269
Traveler, 392, 671
Traveling-wire electrical discharge machine, 629
Trepanning, 510
Trim, 343
Trimming, 352
Troikabot, 443
True centrifugal casting, 113
T-slot cutters, 518
T-slots, 380, 553
Tube extrusion, 303
Tube finishing, 312
Tube-forming machine, 325
Tube reducer, 321
Tubing, 302
   cold-drawing, 312
Tumbling, 612
Tumbling mill, 99
Tumbling nugget, 622
Tundish, 127
Tungsten carbide, 527
Tunnel broaching, 564
Turning, 449
Turret, 483, 492
Turret lathe, 482
   tape-controlled, 484
Turret lathe operation, 524
Turret machines, 492
Turret press, 345
Tuyeres, 41
Twining, 310
Twist drill, 508
Two-dimensional cutting, 449
Two-hand arm, 387
Two-high reversing mill, 289
Type 3 chip, 462
Types of presses, 338

U-die, 354
Ultimate strength, 26
Ultrasonic machining, 625
Ultrasonic testing, 273
Undercutting, 638
Underwater cutting, 165
Unified screw standards, 572
Unilateral tolerance, 245
Unimation, 427
Unit estimate, 676
Unit power, 457, 539
Units, 243
   machine control, 393
   modular automation, 421
Universal chuck, 379
Unmanned machines, 417
Unpolymerized material, 654
Untended night shift, 413
Up milling, 530
Upper control limit, 277
Upright drills, 490
Upset forging, 296
Upset forging machine, 297
Urea, 94
Urea formaldehyde, 121
U-shaped, 414

Vacuum furnace, 44
Vacuum melting, 120
Value engineering, 7
Vaporization, 465
Variable-sequence robot, 432
Variable-speed motors, 481
Variations, transverse, 269
V-channel, 354
Vector difference, 452
Vehicles, 418
Velocity of chip, 454
Velocity of shear, 454
Vernier caliper, 248
Vernier scale, 249
Vertical boring mill, 496
Verticle force, 451
Verticle milling machine, 497
Vertical multistation lathe, 485
Vertical turret lathe, 484
Vicker's hardness, 29
Vises, 381
Vision:
   machine, 439
   robot, 438
Vision sensors, 438
Visual feeding, 423
Vitrified process, 616

Voice sensor, 442
Volatilization, 651
V-way, 374

Wafer, silicon, 442
Wafer shaping, 653
Warm forging, 305
Waste, 286, 682
Water jet machining, 627
Water-soluble oil, 466
Water-soluble oil lubricant, 466
Watt-hour, 59
Wattmeter, 455
Wave set, 556
Wave soldering machines, 658
Wax welding, 118
Wear, 468
   abrasion, 468
   adhesion, 468
   chemical, 468
   diffusion, 468
   electrolytic, 468
   flank, 469
   oxidation, 468
Weld:
   finish seam, 171
   lap seam, 170
   mash seam, 170
   roll spot, 170
Weldability, 60, 161
Welded joint, 160
Welding, 156
   arc, 174
   arc spot, 179
   atomic, 177
   automatic, 179
   braze, 159
   butt, 171, 300
   cold, 156, 187
   diffusion, 190
   electric butt, 301
   electron beam, 183
   electroslag, 182
   explosive, 189
   flash, 172
   flow, 187
   forge, 160
   friction, 184
   gas, 161
   high-frequency, 173
   hydrogen, 177
   induction, 173
   inert gas, 178
   joints, 160
   lap, 301
   laser, 184
   metal electrode, 175
   oxyacetylene, 161
   oxyhydrogen, 164
   precussion, 173
   pressure gas, 164
   projection, 169
   resistance, 166
   seam, 170
   shielded arc, 178
   spot, 168
   stud, 180
   submerged arc, 180
   Thermit, 185
   types of, 156
   ultrasonic, 188
Welding bell, 300
Westinghouse, 427
White cast iron, 55, 467
White-heart castings, 55
Whitney, Eli, 1
Windbox, 41
Wire blade, 560
Wire brushing, 614
Wire guidance, 429
Work, mechanical, 287
Workholding devices, 553
Working capital, 4
Working, cold, 286
Workpiece, 375
Workstations, 376, 410
Worktable, 381, 492
Worm gearing, 587
Wrist, 431, 432
Wrought alloys, properties, 69
Wrought iron, 25
   advantages, 47
   photomicrograph, 47
Wrought metal, 33

X-bar chart, 274

Yield, material, 308
Yield strength, 26, 287
Young's modulus, 26

Zerol gears, 587
Zinc, 19, 65, 106
Zinc base alloys, 70
Zirconite wash, 86
Zirconium oxide, 615
Zones, 463

Inch-Millimeter Equivalents of Decimal and Common Fractions
From 1/64 to 1 in.

Inch	½'s	¼'s	8ths	16ths	32nds	64ths	Millimeter	Decimal Inch
						1	0.397	0.015 625
					1	2	0.794	0.031 25
						3	1.191	0.046 875
				1	2	4	1.588	0.062 5
						5	1.984	0.078 125
					3	6	2.381	0.093 75
						7	2.778	0.109 375
			1	2	4	8	3.175[a]	0.125 0
						9	3.572	0.140 625
					5	10	3.969	0.156 25
						11	4.366	0.171 875
				3	6	12	4.762	0.187 5
						13	5.159	0.203 125
					7	14	5.556	0.218 75
						15	5.953	0.234 375
		1	2	4	8	16	6.350[a]	0.250 0
						17	6.747	0.265 625
					9	18	7.144	0.281 25
						19	7.541	0.296 875
				5	10	20	7.938	0.312 5
						21	8.334	0.328 125
					11	22	8.731	0.343 75
						23	9.128	0.359 375
			3	6	12	24	9.525[a]	0.375 0
						25	9.922	0.390 625
					13	26	10.319	0.406 25
						27	10.716	0.421 875
				7	14	28	11.112	0.437 5
						29	11.509	0.453 125
					15	30	11.906	0.468 75
						31	12.303	0.484 375
	1	2	4	8	16	32	12.700[a]	0.500 0

Cont.